P9-AQP-887

Methods in Enzymology

Volume 383
NUMERICAL COMPUTER METHODS
Part D

METHODS IN ENZYMOLOGY

EDITORS-IN-CHIEF

John N. Abelson   Melvin I. Simon

DIVISION OF BIOLOGY
CALIFORNIA INSTITUTE OF TECHNOLOGY
PASADENA, CALIFORNIA

FOUNDING EDITORS

Sidney P. Colowick and Nathan O. Kaplan

QP601
M 49
V. 383

Methods in Enzymology

Volume 383

Numerical Computer Methods

Part D

EDITED BY

Ludwig Brand

JOHN HOPKINS UNIVERSITY
BALTIMORE, MARYLAND

Michael L. Johnson

UNIVERSITY OF VIRGINIA HEALTH SYSTEM
CHARLOTTESVILLE, VIRGINIA

ELSEVIER
ACADEMIC
PRESS

AMSTERDAM • BOSTON • HEIDELBERG • LONDON
NEW YORK • OXFORD • PARIS • SAN DIEGO
SAN FRANCISCO • SINGAPORE • SYDNEY • TOKYO
Academic Press is an imprint of Elsevier

NO LONGER THE PROPERTY
OF THE
UNIVERSITY OF R. I. LIBRARY

Elsevier Academic Press
525 B Street, Suite 1900, San Diego, California 92101-4495, USA
84 Theobald's Road, London WC1X 8RR, UK

This book is printed on acid-free paper.

Copyright © 2004, Elsevier Inc. All Rights Reserved.

No part of this publication may be reproduced or transmitted in any form or by any
means, electronic or mechanical, including photocopy, recording, or any information
storage and retrieval system, without permission in writing from the Publisher.

The appearance of the code at the bottom of the first page of a chapter in this book
indicates the Publisher's consent that copies of the chapter may be made for
personal or internal use of specific clients. This consent is given on the condition,
however, that the copier pay the stated per copy fee through the Copyright Clearance
Center, Inc. (www.copyright.com), for copying beyond that permitted by
Sections 107 or 108 of the U.S. Copyright Law. This consent does not extend to
other kinds of copying, such as copying for general distribution, for advertising
or promotional purposes, for creating new collective works, or for resale.
Copy fees for pre-2004 chapters are as shown on the title pages. If no fee code
appears on the title page, the copy fee is the same as for current chapters.
0076-6879/2004 $35.00

Permissions may be sought directly from Elsevier's Science & Technology Right
Department in Oxford, UK: phone: (+44) 1865 843830, fax: (+44) 1865 853333,
E-mail: permissions@elsevier.com.uk. You may also complete your request on-line
via the Elsevier homepage (http://elsevier.com), by selecting
"Customer Support" and then "Obtaining Permissions."

For all information on all Academic Press publications
visit our Web site at www.academicpress.com

ISBN: 0-12-182788-7

PRINTED IN THE UNITED STATES OF AMERICA
04 05 06 07 08 9 8 7 6 5 4 3 2 1

Table of Contents

Contributors to Volume 383

Article numbers are in parentheses and following the names of contributors.
Affiliations listed are current.

DAVID BAKER (4), *Department of Biochemistry and Howard Hughes Medical Institute, University of Washington, Seattle, Washington 98195*

NATHAN A. BAKER (5), *Department of Biochemistry and Molecular Biophysics, Washington University in St. Louis, St. Louis, Missouri 63110*

EMERY N. BROWN (16), *Department of Anesthesia and Critical Care, Division of Health Sciences and Technology, Harvard Medical School and Massachusetts Institute Technology, Boston, Massachusetts 02114*

JOSEPH C. CAPPELLERI (17), *Global Research and Development, Pfizer Inc., Groton, Connecticut 06340*

YONG CHOE (16), *Max Planck Institute of Neurobiology, Martinsried D-82152, Germany*

ADRIAN H. ELCOCK (8), *Department of Biochemistry, University of Iowa, Iowa City, Iowa 52242*

PATRICK J. FLEMING (3), *Jenkins Department of Biophysics, Johns Hopkins University, Baltimore, Maryland 21218*

ANGEL E. GARCÍA (6), *Theoretical Biology and Biophysics Group, Los Alamos National Laboratory, Los Alamos, New Mexico 87545*

S. GNANAKARAN (6), *Theoretical Biology and Biophysics Group, Los Alamos National Laboratory, Los Alamos, New Mexico 87545*

NORMA GREENFIELD (12), *Department of Neuroscience and Cell Biology, Robert Wood Johnson Medical School, University of Medicine and Dentistry of New Jersey, Piscataway, New Jersey 08854*

CRAIG HILL (15), *Department of Enzymology and High Throughput Screening, Celera Genomics, San Francisco, California 94080*

SVEN HOVMÖLLER (1), *Structural Chemistry, Stockholm University, Stockholm S-10691, Sweden*

JOHN H. IPSEN (9), *Physics Department, MEMPHYS Center for Biomembrane Research, University of Southern Denmark, Odense M DK-5230, Denmark*

ROGER E. ISON (1), *Manic Software, Loveland, Colorado 80537*

JAMES W. JANC (15), *Department of Enzymology and High Throughput Screening, Celera Genomics, San Francisco, California 94080*

ROBERT H. KRETSINGER (1), *Department of Biology, University of Virginia, Charlottesville, Virginia 22903*

PETR KUZMIČ (15), *Biokin Ltd., Pullman, Washington 99163*

JOSEPH LAU (17), *Institute for Clinical Research and Health Policy Studies, Tufts-New England Medical Center, Boston, Massachusetts 02111*

LING MIAO (9), *Physics Department, MEMPHYS Center for Biomembrane Research, University of Southern Denmark, Odense M DK-5230, Denmark*

KIRA M. S. MISURA (4), *Department of Biochemistry and Howard Hughes Medical Institute, University of Washington, Seattle, Washington 98195*

OLE MOURITSEN (9), *Physics Department, MEMPHYS Center for Biomembrane Research, University of Southern Denmark, Odense M DK-5230, Denmark*

JAY I. MYUNG (14), *Department of Psychology, Ohio State University, Columbus, Ohio 43210*

MORTEN NIELSEN (9), *Biocentrum-DTU, Technical University of Denmark, Lyngby DK-2800, Denmark*

HUGH NYMEYER (6), *Theoretical Biology and Biophysics Group, Los Alamos National Laboratory, Los Alamos, New Mexico 87545*

MARK A. PITT (14), *Department of Psychology, Ohio State University, Columbus, Ohio 43210*

DOUGLAS POLAND (18), *Department of Chemistry, The Johns Hopkins University, Baltimore, Maryland 21218*

JAMES POLSON (9), *Department of Physics, University of Prince Edward Island, Charllottetown, Prince Edward C1A 4P3, Canada*

CAROL ROHL (4), *Department of Biochemistry and Howard Hughes Medical Institute, University of Washington, Seattle, Washington 98195*

GEORGE D. ROSE (3), *Jenkins Department of Biophysics, Johns Hopkins University, Baltimore, Maryland 21218*

ROBERT SCHLEIF (2), *Biology Department, Johns Hopkins University, Baltimore, Maryland 21218*

CHRISTOPHER H. SCHMID (17), *Institute for Clinical Research and Health Policy Studies, Tufts-New England Medical Center, Boston, Massachusetts 02111*

VICTOR SOLO (16), *Department of Electrical Engineering and Computer Science, University of Michigan, Ann Arbor, Michigan 48101*

NARASIHMA SREERAMA (13), *Department of Biochemistry and Molecular Biology, Colorado State University, Fort Collins, Colorado 80523*

RAJGOPAL SRINIVASAN (3), *Jenkins Department of Biophysics, Johns Hopkins University, Baltimore, Maryland 21218*

MARTIN STRAUME (7), *Department of Internal Medicine, Division of Endocrinology and Metabolism, University of Virginia Health System, Charlottesville, Virginia 22904*

CHARLIE E. M. STRAUSS (4), *Biosciences Division, Los Alamos National Laboratory, Los Alamos, New Mexico 87545*

JOEL TELLINGHUISEN (11), *Department of Chemistry, Vanderbilt University, Nashville, Tennessee 37235*

JENIFER THEWALT (9), *Department of Physics, Simon Fraser University, Burnaby, British Columbia V5A 1S6, Canada*

ANTONIUS VANDONGEN (10), *Department of Pharmacology, Duke University, Durham, North Carolina 27710*

ILPO VATTULAINEN (9), *Laboratory of Computational Engineering, Helsinki University of Technology, Espoo 02150, Finland*

ROBERT W. WOODY (13), *Department of Biochemistry and Molecular Biology, Colorado State University, Fort Collins, Colorado 80523*

ZHENHUA ZHANG (16), *Human Computer Interaction Institute, Carnegie Melon University, Pittsburgh, Pennsylvania 15213*

HONG ZHU (9), *Department of Physics, McGill University, Montreal, Quebec H3A 2T8, Canada*

MARTIN ZUCKERMAN (9), *Department of Physics, Simon Fraser University, Burnaby, British Columbia V5A 1S6, Canada*

Preface

The speed of laboratory computers doubles every year or two. As a consequence, complex and time-consuming data analysis methods, that were prohibitively slow a few years ago, can now be routinely employed. Examples of such methods within this volume include wavelets, transfer functions, inverse convolutions, robust fitting, moment analysis, maximum-entropy, and singular value decomposition. There are also many new and exciting approaches for modeling and prediction of biologically relevant molecules such as proteins, lipid bilayers, and ion channels.

There is also an interesting trend in the educational background of new biomedical researchers over the last few years. For example, three of the authors in this volume are Ph.D. mathematicians who have faculty appointments in the School of Medicine at the University of Virginia.

The combination of faster computers and more quantitatively oriented biomedical researchers has yielded new and more precise methods for the analysis of biomedical data. These better analyses have enhanced the conclusions that can be drawn from biomedical data and they have changed the way that the experiments are designed and performed. This is our fourth "Numerical Computer Methods" volume for Methods in Enzymology. The aim of volumes 210, 240, 321, and the present volume is to inform biomedical researchers about some of these recent applications of modern data analysis and simulation methods as applied to biomedical research.

LUDWIG BRAND
MICHAEL L. JOHNSON

METHODS IN ENZYMOLOGY

VOLUME XXXV. Lipids (Part B)
Edited by JOHN M. LOWENSTEIN

VOLUME XXXVI. Hormone Action (Part A: Steroid Hormones)
Edited by BERT W. O'MALLEY AND JOEL G. HARDMAN

VOLUME XXXVII. Hormone Action (Part B: Peptide Hormones)
Edited by BERT W. O'MALLEY AND JOEL G. HARDMAN

VOLUME XXXVIII. Hormone Action (Part C: Cyclic Nucleotides)
Edited by JOEL G. HARDMAN AND BERT W. O'MALLEY

VOLUME XXXIX. Hormone Action (Part D: Isolated Cells, Tissues,
and Organ Systems)
Edited by JOEL G. HARDMAN AND BERT W. O'MALLEY

VOLUME XL. Hormone Action (Part E: Nuclear Structure and Function)
Edited by BERT W. O'MALLEY AND JOEL G. HARDMAN

VOLUME XLI. Carbohydrate Metabolism (Part B)
Edited by W. A. WOOD

VOLUME XLII. Carbohydrate Metabolism (Part C)
Edited by W. A. WOOD

VOLUME XLIII. Antibiotics
Edited by JOHN H. HASH

VOLUME XLIV. Immobilized Enzymes
Edited by KLAUS MOSBACH

VOLUME XLV. Proteolytic Enzymes (Part B)
Edited by LASZLO LORAND

VOLUME XLVI. Affinity Labeling
Edited by WILLIAM B. JAKOBY AND MEIR WILCHEK

VOLUME XLVII. Enzyme Structure (Part E)
Edited by C. H. W. HIRS AND SERGE N. TIMASHEFF

VOLUME XLVIII. Enzyme Structure (Part F)
Edited by C. H. W. HIRS AND SERGE N. TIMASHEFF

VOLUME XLIX. Enzyme Structure (Part G)
Edited by C. H. W. HIRS AND SERGE N. TIMASHEFF

VOLUME L. Complex Carbohydrates (Part C)
Edited by VICTOR GINSBURG

VOLUME LI. Purine and Pyrimidine Nucleotide Metabolism
Edited by PATRICIA A. HOFFEE AND MARY ELLEN JONES

VOLUME LII. Biomembranes (Part C: Biological Oxidations)
Edited by SIDNEY FLEISCHER AND LESTER PACKER

VOLUME LIII. Biomembranes (Part D: Biological Oxidations)
Edited by SIDNEY FLEISCHER AND LESTER PACKER

[1]   Prediction of Protein Structure

By ROBERT H. KRETSINGER, ROGER E. ISON, and SVEN HOVMÖLLER

Overview and Perspective

Extraction of Structural Information from Protein Sequence

The determination of crystal and of solution structures has been greatly rationalized over the past decade; however, it remains tedious, expensive work. In contrast, thousands of protein-encoding genes are sequenced each day. The deduced sequences of proteins provide invaluable insights into the functions of those proteins and the evolution of the organisms producing those proteins. However, much more information would be forthcoming if the structures of those proteins accompanied their sequences. This review is intended for the biologist who has no special expertise and who is not involved in the determination of protein structure. We have two goals:

1. To provide the generalist with enough background to understand the concepts, opportunities, and difficulties of protein structure prediction

2. To outline a general strategy that should allow the extraction and interpretion of structural information about a target sequence from publicly available databases, servers, and programs

The (nearly) complete DNA sequences of 84 bacteria, 16 archaea, and 15 eukaryotes, including *Anopheles gambiae, Arabidopsis thaliana, Caenorhabditis elegans, Drosophila melanogaster, Encephalitozoon cuniculi, Guillardia theta, Saccharomyces cerevisiae, Plasmodium falciparum,* and *Schizosaccharomyces pombe,* a total of 22 billion base pairs, were available as of January 2003 from the National Institutes of Health (NIH, Bethesda, MD) or are described in the Genome News Network (Table I[1–24]).

---

[1] J. Westbrook, Z. Feng, L. Chen, H. Yang, and H. M. Berman, *Nucleic Acids Res.* **31,** 489 (2003).

[2] J.-F. Gibrat, T. Madej, and S. H. Bryant, *Curr. Opin. Struct. Biol.* **6,** 377 (1996).

[3] A. G. Murzin, S. E. Brenner, T. Hubbard, and C. Chothia, *J. Mol. Biol.* **247,** 536 (1995).

[4] F. M. G. Pearl, D. Lee, J. E. Bray, I. Sillitoe, A. E. Todd, A. P. Harrison, J. M. Thornton, and C. A. Orengo, *Nucleic Acids Res.* **28,** 277 (2000).

[5] R. L. Tatusov, M. Y. Galperin, D. A. Natale, and E. V. Koonin, *Nucleic Acids Res.* **28,** 33 (2000).

[6] K. Mizuguchi, C. M. Deane, T. L. Blundell, and J. P. Overington, *Protein Sci.* **7,** 2469 (1998).

[7] R. Sowdhamini, D. F. Burke, J.-F. Huang, K. Mizuguchi, H. A. Nagarajaram, N. Srinivasan, R. E. Steward, and T. L. Blundell, *Structure* **6,** 1087 (1998).

[8] J. Moult, K. Fidelis, A. Zemla, and T. Hubbard, *Proteins* **5**(Suppl.), 2 (2001).

Copyright 2004, Elsevier Inc.
All rights reserved.
0076-6879/04 $35.00

TABLE I

DATABASES, SERVERS, AND PROGRAMS REFERRED TO IN TEXT

	Description/URL
	Databases
Entrez search	NCBI combined site to search PubMed, various sequence and structure databases
	http://www.ncbi.nlm.nih.gov/Entrez/
PubMed search	Literature searches; not a sequence or structure database per se
	http://www.ncbi.nlm.nih.gov/PubMed/
Genome sequences	Complete genome sequences and maps
	http://www.ncbi.nlm.nih.gov/PMGifs/Genomes/euk_g.html
Genome sequences	Quick guide to sequenced genomes
	http://gnn.tigr.org/sequenced_genomes/genome_guide_p1.shtml
PDB (Protein Data Bank)	Worldwide depository for biological macromolecular structures
	http://www.rcsb.org/pdb/
Nonredundant PDB	Selected highly dissimilar chains from the PDB, presumably nonhomologous or only distantly related
	http://www.ncbi.nlm.nih.gov/Structure/VAST/nrpdb.html
Select 25%	Representative selections from the PDB, about 1/15th of the full list, suitable for investigations when the full PDB with its many redundant entries is not required
	http://homepages.fh-giessen.de/~hg12640/pdbselect/recent.pdb_select25
VAST	Vector Alignment Search Tool, direct comparison of 3D structures, PDB entries classified by similarity
	http://www.ncbi.nlm.nih.gov/Structure/VAST/nrpdb.html
SCOP	Classification of tertiary structures
	http://scop.mrc-lmb.cam.ac.uk/scop/
CATH	Classification of tertiary structures
	http://www.biochem.ucl.ac.uk/bsm/cath/
COG	Clusters of orthologous groups of proteins
	http://www.ncbi.nlm.nih.gov/COG/
MembProtStruct	Membrane proteins of known structure, crystallization conditions, etc.
	http://www.mpibp-frankfurt.mpg.de/michel/public/memprotstruct.html
HOMSTRAD	Structure-based alignments of homologous families
	http://www-cryst.bioc.cam.ac.uk/homstrad/
CAMPASS	Sequence-based alignments of homologous families
	http://www-cryst.bioc.cam.ac.uk/campass/
Enzymes	Enzyme Classification System
	http://www.biochem.ucl.ac.uk/bsm/enzymes/index.html
Enzymes	Nomenclature
	http://www.chem.qmul.ac.uk/iubmb/enzyme/
Data search	Sanger Centre databases, BLAST servers
	http://www.sanger.ac.uk/DataSearch/
CASP4	Summary and prediction targets
	http://predictioncenter.llnl.gov/casp4/

(continued)

TABLE I *(continued)*

Description/URL

Servers and Programs

Homolog search and alignment

BLAST server	Sequence search and alignment
	http://www.ncbi.nlm.nih.gov/BLAST/
FASTA server	Sequence search and alignment
	http://www.ebi.ac.uk/fasta33/
DALI server	Find 3D from PDB similar to submitted 3D
	http://www2.ebi.ac.uk/dali/
Pfam server	Multiple sequence alignments
	http://pfam.wustl.edu/
	http://www.sanger.ac.uk/Software/Pfam/
BestFit server	Optimal alignments are found by inserting gaps to maximize the number of matches using the *local homology* algorithm of Smith and Waterman
	http://www.biology.wustl.edu/gcg/bestfit.html
	http://www.infobiogen.fr/doc/GCGdoc/Program_Manual/Comparison/bestfit.html

Secondary structure prediction

PSIPRED server	With links to MEMSAT and GenThreader
	http://bioinf.cs.ucl.ac.uk/psipred/
MEMSAT server	Transmembrane helices and strands; access via PSIPRED
	http://bioinf.cs.ucl.ac.uk/psipred/
BimCore	Useful links and tutorials
	http://www.bimcore.emory.edu/Tutorials/Online/
SAM-T02 server	Alignment to individual or family of proteins by HMM; protein database query; secondary structure prediction; homology-based structure prediction
	http://www.cse.ucsc.edu/research/compbio/HMM-apps/T02-query.html
PredictProtein server	Many tools available from this essential full-service site, including database search, alignment, sequence motif recognition, 2D prediction
	http://www.embl-heidelberg.de/predictprotein/predictprotein.html
PROF server	Secondary structure by profile-based neural network method
	http://www.aber.ac.uk/~phiwww/prof/
JPRED server	Secondary structure prediction by consensus method
	http://www.compbio.dundee.ac.uk/~www-jpred/
SSPRO server	Secondary structure by bidirectional recurrent neural networks
	http://promoter.ics.uci.edu/BRNN-PRED/

Domain definition

PSI-BLAST server	Search for conserved domains
	http://www.ncbi.nlm.nih.gov/BLAST/
ProDom server	Compare target sequence with database of domain families
	http://prodes.toulouse.inra.fr/prodom/2002.1/html/home.php

(continued)

TABLE I *(continued)*

Description/URL

Modeling of target to homolog in 3D
3D-Jigsaw server — Builds 3D models based on homologs
http://www.bmm.icnet.uk/servers/3djigsaw
MODELLER program — Program performs comparative structure modeling by satisfying spatial restraints; structure optimization, variety of other services
http://guitar.rockefeller.edu/modeller/modeller.html
CHARMM program — Program for energy optimization and molecular dynamics
http://yuri.harvard.edu/

Fold recognition
3D-PSSM server — Sequence or structure to fold
http://www.sbg.bio.ic.ac.uk/servers/3dpssm/
BIOINBGU server — Hybrid fold recognition combining sequence and evolutionary information
http://www.cs.bgu.ac.il/~bioinbgu/
FFAS server — Compare with PFAM and COG
http://bioinformatics.ljcrf.edu/FFAS/
PROSPECT program — Threading-based prediction when fold template has low similarity to target
http://compbio.ornl.gov/structure/prospect/

Threading of target to fold
GenTHREADER server — Fully automated; alignment, solvation potentials, neural network evaluation
http://bioinf.cs.ucl.ac.uk/psipred/
Links — FUGUE, CODA, JOY, COMPARER, RAMPAGE
http://www-cryst.bioc.cam.ac.uk/servers.html
FUGUE server — Profile library search alignments against HOMSTRAD
http://www-cryst.bioc.cam.ac.uk/~fugue/prfsearch.html
CODA server — Predicts loops. Runs two programs: FREAD, knowledge base of PDB fragments; and PETRA, *ab initio* from database of computer-generated conformers
http://www-cryst.bioc.cam.ac.uk/coda/
JOY server — Annotates protein sequence alignments by 3D
http://www-cryst.bioc.cam.ac.uk/cgi-bin/joy.cgi
COMPARER server — Alignment of sections by 3D, output in JOY format
http://www-cryst.bioc.cam.ac.uk/~robert/COMPARER/comparer.html

Adding side chains
SCWRL program — Side chain placement with a rotamer library
http://www.fccc.edu/research/labs/dunbrack/scwrl/

New/*ab initio*
PETRA server — Predicts conformation of fragments
http://www-cryst.bioc.cam.ac.uk//cgi-bin/coda/pet.cgi
FRAGFOLD server — *Ab initio* predictions by assembling fragments of supersecondary structure
http://www.bio.informatik.uni-muenchen.de/lehre/WS2002/Prakt_GOB/Gruppe_4/FRAGFOLD_Andrey_new-Dateien/v3_document.htm

(continued)

TABLE I *(continued)*

Description/URL

Evaluating models
  Geometry server,   Various macromolecular structural calculations
    programs   http://bioinfo.csb.yale.edu/geometry/
  CastP server   Calculates voids and pockets
                            http://cast.engr.uic.edu/cast/
  WHATIF   Versatile structure checker program
    program   http://www.cmbi.kun.nl/whatif/
  WHAT_CHECK   Many of the capabilities of WHATIF presented as stand-alone
    programs   programs
                            http://www.cmbi.kun.nl/gv/whatcheck/
  RAMPAGE   Ramachandran plot analysis of ϕ, ψ distribution
    server   http://raven.bioc.cam.ac.uk/rampage.php
  VERIFY3D   Visual analysis of the quality of a proposed structure in PDB format
    server   http://www.doe-mbi.ucla.edu/Services/Verify_3D/
  PROCHECK   Check stereochemical quality of structure, overall and by residue
    program (ftp)   http://www.biochem.ucl.ac.uk/~roman/procheck/procheck.html
  PROSA II   Check $C\alpha$, $C\beta$ potentials, backbone
    program (ftp)   http://www.lmcp.jussieu.fr/sincris-top/logiciel/prg-prosa.html
  KINEMAGE   500 PDB files with all hydrogen atoms added and optimized
    program   http://kinemage.biochem.duke.edu/
  MaxSub program   Assesses quality of predicted structures
                            http://www.cs.bgu.ac.il/~dfischer/MaxSub/MaxSub.html

Protein docking
  CAPRI   Results of critical assessment of predicted interactions
                            http://capri.ebi.ac.uk

Evaluating servers
  EVA LIVEBENCH   http://maple.bioc.columbia.edu/eva/
                  http://bioinfo.pl/LiveBench/

[8a] J. Moult, K. Fidelis, A. Zemla *et al.*, *Proteins* **53**(6): 334–339 (Suppl.), 6 (2003).
[9] L. Holm and C. Sander, *Science* **273**, 595 (1996).
[10] L. J. McGuffin, K. Bryson, and D. T. Jones, *Bioinformatics* **16**, 404 (2000).
[11] D. T. Jones, *FEBS Lett.* **423**, 281 (1998).
[12] P. A. Bates, L. A. Kelley, R. M. MacCallum, and M. J. E. Sternberg, *Proteins* **5**(Suppl.), 39 (2001).
[13] L. A. Kelley, R. M. MacCallum, and M. J. E. Sternberg, *J. Mol. Biol.* **299**, 501 (2000).
[14] D. Fischer, *Pac. Symp. Biocomp.* 119 (2000).
[15] L. Rychlewski, L. Jaroszewski, W. Li, and A. Godzik, *Protein Sci.* **9**, 232 (2000).
[16] D. Xu, O. H. Crawford, P. F. LoCascio, and Y. Xu, *Proteins* **5**(Suppl.), 140 (2001).
[17] D. T. Jones, *J. Mol. Biol.* **287**, 797 (1999).
[18] J. Shi, T. L. Blundell, and K. Mizuguchi, *J. Mol. Biol.* **310**, 243 (2001).
[19] R. L. Dunbrack, Jr., *Proteins* **3**(Suppl.), 81 (1999).
[20] Dean and Blundell (in press).
[21] R. Luthy, J. U. Bowie, and D. Eisenberg, *Nature* **356**, 83 (1992).
[22] J. M. Word, S. C. Lovell, J. S. Richardson, and D. C. Richardson, *J. Mol. Biol.* **285**, 1735 (1999).

Complete, high-resolution maps are also available for *Homo sapiens, Mus musculus, Rattus norvegicus, Danio rerio* (zebrafish), *Avena sativa* (oat), *Hordeum vulgare* (barley), *Oryza sativa* (rice), *Triticum aestivum* (wheat), and *Zea mays* (maize). From these and from more traditional sources, the amino acid sequences of about 1,300,000 proteins are available. About half of these protein sequences belong to 2000 recognized homolog families, of which about 1000 contain at least one determined tertiary structure. As Fischer *et al.*[25] noted, "What non-expert biologists need is to be able to apply automatic tools for their prediction needs, and on a large, genomic scale."

Second Half of Genetic Code

Since the pioneering experiments of Anfinsen[26] on the renaturation of RNase A, it has been shown for several proteins and inferred for many others that the amino acid sequence of a protein determines its tertiary structure. No other assembly mechanism is required, although chaperones may protect the protein from hydrolysis and/or rescue the folding protein from local minima. The genetically encoded amino acid sequence determines the tertiary structure of the protein. Understanding that folding process and the resultant structure is the second half of the genetic code.

Hierarchy of Structures

The linear sequence of amino acids, or residues, is referred to as the primary structure. Secondary structure refers to the spatial, or conformational, relationships of residues to those nearby in sequence; they fall predominantly into two regular patterns: α helix and β strand. The remaining residues are traditionally grouped together as coils or turns. These may be further classified into supersecondary structure elements; examples include α hairpin (consecutive α helices in a compact arrangement), α cornerI (consecutive α helices in a noncompact arrangement), β hairpin (hydrogen-bonded consecutive β strands), β corner (non-hydrogen-bonded consecutive β strands), β–α–β unit (parallel hydrogen-bonded β strands with intervening α helix), and split β–α–β unit (parallel non-hydrogen-bonded β strands with intervening α helix). Tertiary structure refers to the three-dimensional relationships of these elements of secondary structure within a single polypeptide chain, or monomer. Most proteins are parts of larger complexes that contain several constituent polypeptides chains; the relationship among these chains comprises the quaternary structure.

[23] Siew *et al.* (in press).

[24] J. Janin, *Proteins* **47,** 257 (2002).

[25] D. Fischer, A. Elofsson, L. Rychlewski, F. Pazos, A. Valencia, B. Rost, A. R. Ortiz, and R. L. Dunbrack, Jr., *Proteins* **5**(Suppl.), 171 (2001).

[26] C. Anfinsen, *Science* **181,** 223 (1973).

Parameterization of Protein Structure

All proteins consist of an unbranched polypeptide chain made from the 20 genetically encoded α-amino acids. The bond angles and bond lengths of the main-chain atoms are nearly the same, ± 0.01 Å, for all 20 and for their posttranslationally modified derivatives. The peptide bond is nearly planar and almost always trans; the few exceptions for proline are noted in the legend to Fig. 2. This means that models of proteins are built of units of standard dimensions. "Atoms" in this discussion means all atoms except hydrogen. In most cases the standard bond lengths and angles fix the position of the hydrogens. However, free rotation about the single bonds of methyl, amine, hydroxyl, and sulfhydryl groups leaves those hydrogens undefined; this is a concern for refinement of both crystal structures and of models. The posttranslational modifications of amino acids are not considered in most prediction schemes.

Dihedral Angles and Ramachandran (ϕ,ψ) *Plot*

Given common bond lengths and angles, the trace of the polypeptide chain and carbonyl oxygen, plus β-carbon for all residues excepting glycine, can be fully described by just two dihedral angles—ϕ,ψ—defined and illustrated in Fig. 1. As was predicted by Ramachandran and Sassiekharan[27] and subsequently observed in many crystal structures, the range of allowed ϕ,ψ values is restricted to a small area of the Ramachandran plot. Only more recently has it been fully appreciated, from examination of crystal structures refined with high-resolution (1.5 Å) X-ray diffraction data, that the area of the ϕ,ψ plot actually occupied by amino acids in proteins is even more restricted and significantly different from that originally predicted.[28] Further, these areas and their occupancies differ characteristically, almost like a fingerprint, for each residue (Fig. 2).

Just as evaluation of high-resolution crystal structures revealed discrete ϕ,ψ values, so too side chains assume discrete dihedral angles, χ, as summarized in rotamer libraries.[29] The number of χ angles available per amino acid are as follows: Gly, Ala, Pro = 0; Ser, Thr, Cys, Val = 1; Asp, Asn, Leu, Ile, His, Phe, Tyr, Trp = 2; Glu, Gln, Met = 3; Arg, Lys = 4. When normalized for average frequency of occurrence, there are ~ 1.75 degrees of freedom for the side chain of the average residue. Predictions of secondary structure do not consider χ angles. Structure predictions do not initially consider the dihedral values of side chains; however, these are important

[27] G. N. Ramachandran and V. Sassiekharan, *Adv. Protein Chem.* **28,** 283 (1968).
[28] S. Hovmöller, T. Zhou, and T. Ohlson, *Acta Crystallogr. D Biol. Crystallogr.* **58,** 768 (2002).
[29] S. C. Lovell, J. M. Word, J. S. Richardson, and D. C. Richardson, *Proteins* **40,** 389 (2000).

A

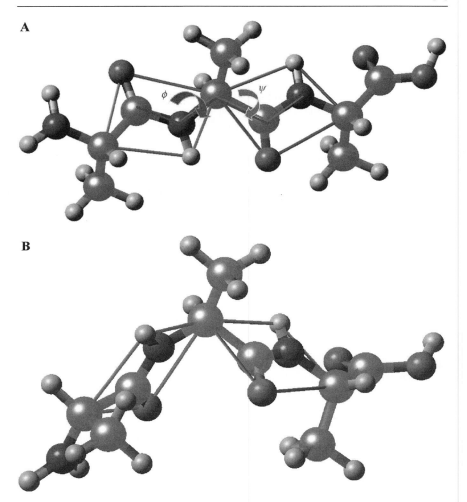

B

Fig. 1. Definition of ϕ and ψ with examples. (A) Trialanine. As drawn, $\phi_n = 180°$ and $\psi_n = 180°$. $C\alpha_n$, N_n, H_n, C_{n-1}, O_{n-1}, and $C\alpha_{n-1}$ are all contained in one plane; $C\alpha_n$, C_n, O_n, N_{n+1}, H_{n+1} and $C\alpha_{n+1}$ are contained in another. Positive rotation, increasing the value of the dihedral angle, is indicated by the arrows, proceeding along the peptide chain (N → C). Some find it easier to visualize ϕ by looking from $C\alpha$ toward N; in that case, positive rotation would be indicated by the arrow in the opposite sense; see (B). $\psi_{n-1} = 180°$. At the N terminus ϕ_n is not defined without the amide group being involved in a peptide bond. $\phi_{n+1} = 180°$. At the C terminus, ψ_{n+1} is not defined without the carboxylate group being involved in a peptide bond. (B) $\phi_n = -109°$ and $\psi_n = 121°$ as seen in the R, or polyproline II, conformation for alanine (Fig. 2A). (See color insert.)

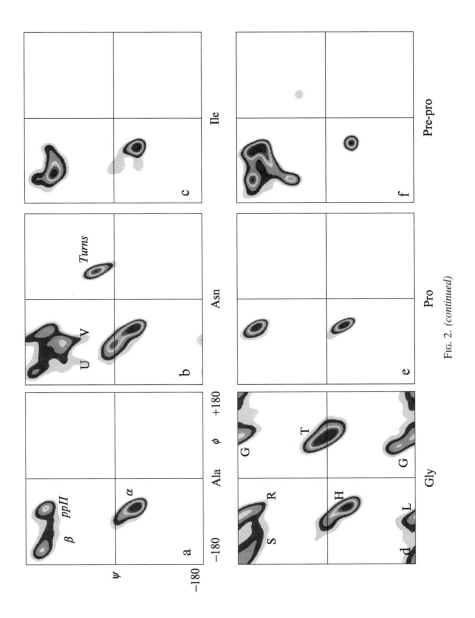

Fig. 2. (continued)

parameters when evaluating or refining a predicted structure. The side chains are the only difference among the amino acids; their order carries the hidden message that directs the polypeptide to fold into its tertiary structure.

It is generally assumed that linkers between domains (characterized in Section II.A, below) are more flexible than are the domains themselves and correspondingly that loops are more flexible than α helices and β strands. However, it is best to be alert to the exceptions. Dalal et al.[30] solved the "Paracelsus challenge" by changing a β-sheet protein to an α-helical protein by selectively changing less than 50% of the residues. Residues His40–Met67 of G-actin form an α helix in one crystal structure and a β turn in another[31]; see Fig. 3. The important point is that most proteins form an ensemble of structures; they must if they are to function like micromachines. It is misleading to speak of *the* structure of a protein. We assume that the sequences that enjoy alternate conformations in proteins are not strongly predicted to form one or another type of structure.

Protein Data Bank

The Protein Data Bank (PDB), as of January 2003 contained 18,482 protein, 1932 nucleic acid, and 18 carbohydrate structures. Many protein families have multiple entries, reflecting different complexes, different (site-directed) mutants, different isoforms, or space groups of

[30] S. Dalal, S. Balasubramanian, and L. Regan, *Nat. Struct. Biol.* **4,** 548 (1997).
[31] L. R. Otterbein, P. Graceffa, and R. Dominguez, *Science* **293,** 708 (2001).

Fig. 2. Ramachandran plot(s) for individual residues from structures determined at high resolution. (A) Alanine prefers to be in an α-helical conformation. Of the two regions in the β region, the left region (S, $\phi = -140°$ and $\psi = 140°$) is found in β sheets. (B) Asparagine has a complicated pattern of conformations with a large fraction in the turns, or left-handed α-helix, region (T at $\phi = 60°$ and $\psi = 35°$; see Fig. 1B) and in the two bridging areas U and V. It is rarely found in β strands. (C) Isoleucine prefers to be in β sheets (S, $\phi = -117°$ and $\psi = 128°$). (D) Glycine is the only amino acid without a Cβ and thus can have conformations that are sterically hindered in all other amino acids. Notice that the turn region, T, is the most occupied. The letters G, L, H, R, S, and T [together with U and V; see (B)] denote the eight discrete conformations assumed by amino acids in proteins. (E) Pro is the most restricted amino acid. Because its amine group is not available to form hydrogen bonds it is unusual in β sheets and is found in only the first three positions of α helices. (F) Although proline can have the conformation needed for an α helix, the amino acid just before proline nearly always has a β conformation, making proline a terminator of α helices (after Hovmöller et al.[28]). About 5% of prolines occur with the peptide bond *cis*. This is not explicitly stated in the PDB file. In some lower-resolution structures a *cis*-proline has been built *trans,* thereby distorting the local geometry. (See color insert.)

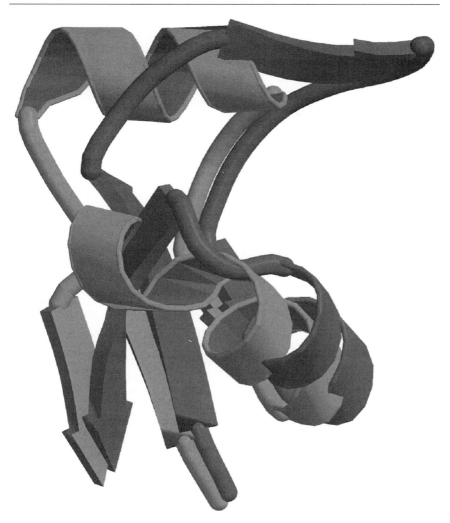

FIG. 3. Illustration of alternative shapes of regions of the same protein. Residues His40–Met67 of G-actin form an α helix in one crystal structure and a β turn in another. Redrawn after Otterbein *et al.*[31] (See color insert.)

crystallization. Conversely, there are classes of proteins that are under-represented, such as membrane proteins and large multidomain proteins. Distinct subfamilies are represented in the VAST nonredundant database with a 10^{-7} E (for expectation) value cutoff (http://www.ncbi.nlm.nih.gov/

Structure/VAST/vast.shtml) or in the pdbselect database (http://home-pages.fhgiessen.de/~hg12640/pdbselect/recent.pdb_select25) of 1949 chains with a 25% identity threshold used in the nonredundant PDB-select list. The Web sites of the PDB and of all other databases, servers, and programs are indicated in Table I and are not explicitly referenced in text.

The reliability and the errors associated with structures in the PDB vary. The atomic coordinates of atoms given in the PDB are, on average, accurate to within $\sim$0.2 Å for structures determined at 2.5-Å resolution and $R = 22\%$ and to within $\sim$0.15 Å for 1.5 Å resolution and $R = 19\%$. ($R = \text{i} = 1\text{n}$ [lobs $-$ lcalc]i, and i $= 1\text{n}$ lobs, where n $=$ number of reflections, lobs $=$ observed intensity, and lcalc $=$ calculated intensity). Temperature factor $B < 20$ Å^2 indicates good order; $B > 30$ Å^2 indicates high vibration and/or disorder of packing of the proteins within the crystal lattice. When modeling a target sequence against the known structure of a homolog, it is desirable to consider both the similarity of sequences of target and template and the reliability of the known structure, both overall and in specific regions of disorder, often in turns at the surface of the protein.

Structures of proteins in solution determined by nuclear magnetic resonance (NMR) provide invaluable information about the dynamics and mobilities of proteins. Usually those regions assigned high mobility by NMR accord with those having high B values in crystal structures. NMR also directly observes hydrogen atoms, unlike X-ray crystallography, and quantitatively measures the strength of proton interactions, as in hydrogen bonding. Unless explicitly stated, most prediction programs rely heavily or solely on crystal structures.

Physical versus Knowledge-Based Methods

The refinement of protein structures by energy minimization involves many empirical formulas that only approximate the exact functions. These empirical functions contain constants whose optimal values are not known. Further, it is especially difficult to estimate the entropic contribution to the free energy of the entire system. In spite of the theoretically attractive approach of predicting protein structures from physical principles, these methods cannot yet predict a protein model that yields a Ramachandran plot distribution similar to that empirically observed. Today the primary value of physical based methods in structure prediction is in identifying steric violations, internal voids, uncompensated charges, and unfulfilled hydrogen bond donors in either experimentally determined or in predicted structures. Thus, all of the structure prediction servers and programs—homolog comparison, fold recognition and threading, and new *(ab initio)*—are based on knowledge of the structures of other proteins, with little regard for ligands present or solvent employed in crystallization.

Classifications of Protein Structure

Classification by Sequence

Several sophisticated programs, for example, PSI-BLAST, can compare the sequence of a target(s) with another protein or with all the sequences in a database. If the E value is less than 10^{-2} of an alignment occurring by chance (this would roughly correspond to ~25% identity in an alignment of more than 200 residues), it is strongly inferred that the two proteins are homologs: they have evolved from a common precursor. Further, if the two proteins, for example, the target whose structure is to be predicted and the template of known structure are homologous, they certainly have similar structures. The greatest deviations occur in regions of insertion or deletion (indels). The root mean square deviation (RMSD) of aligned main-chain atoms increases monotonically with dissimilarity in sequence[32] within a protein (super) family. With few exceptions all members of one family are more similar in sequence and structure with all other members of that family than with any members of another protein family. Conversely, members of different families may or may not be homologous. Other lines of evidence (i.e., similar fold, function, cofactors) may indicate homology, but this cannot be demonstrated by statistical analysis of the sequences. This raises the fascinating question of the ultimate origin of protein families.

Many proteins consist of several distinct domains, each of which is characterized by a single hydrophobic core and/or by being a distinct unit as represented in an evolutionary dendrogram. The entire polypeptide chain originated by gene splicing. Each constituent domain has its own evolutionary history and is homologous to other members of its own family. Domains are frequently joined by flexible linkers and the relationships among these domains often change with the functional state of the protein. If success is encountered in predicting the structures of the domains of a multidomain monomer, there is still the problem of determining the spatial relationship between domains—a problem similar to prediction of quaternary structure.

Classifications of Structure

Proteins can be classified by structure, in addition to sequence. This is inherently much more difficult because sequences are discrete; structures are not. There is still controversy and a significant element of subjective judgment in defining the difference or similarity between two protein structures. It is reassuring that several protocols or algorithms provide

---

[32] T. C. Wood and W. R. Pearson, *J. Mol. Biol.* **291,** 977 (1999).

results similar to one another. Further, all proteins of one family, as defined by sequence analysis, fall into the same structure class. Sometimes two, or more, sequence families fall into a single structure class. Whether this reflects convergent evolution or loss, over eons, of detectable similarity in sequence of homologs remains a topic of spirited debate.

The three most widely used schemes for classification of structure are SCOP,[3] CATH,[4] and FSSP (in DALI).[9] SCOP considers classes (7), folds (701), superfamilies (1110), families (1940), and domains (44,327); numbers in parentheses indicate the number of entries in each group for the 1.61 release (September 1, 2002). All members of a family are strongly inferred to be homologous on the basis of both sequence alignment and (near) superposition of tertiary structures. Members of superfamilies share similar topologies of helices and strands and have "low sequence identity." They are probably related by divergent evolution from a common origin; however, convergent evolution cannot be precluded. SCOP has many links and servers, among them one that accepts a candidate sequence and assigns it to the most probable superfamily—the first half of structure prediction by fold (superfamily) recognition.

CATH classifies in terms of class (8), architecture (46), topology (1453), homologous superfamily (2066), and sequence family (4010). There is strong, but imperfect, correspondence between the 2066 homologous superfamilies in the 2.4 release (January 14, 2002) of CATH and the 1940 families of SCOP. CATH divides proteins into constituent domains by applying three programs—DETECTIVE, PUU, and DOMAK; if the three do not give the same result, the division (if any) is made by human judgment.

FSSP makes an all-against-all three-dimensional (3D) structure comparison of protein structures currently in the PDB, using C_α–C_α distance matrices. The resultant structure–structure alignment of proteins in the database is automatically maintained and continuously updated. The June 16, 2002 update compared 30,624 protein structures to generate 3242 families. Taylor[33] proposed a "periodic table" for protein structures on the basis of "... secondary links in the form of intra-chain hydrogen bonds and tertiary links formed by the packing of secondary structures." Unfortunately, there is not yet available an index indicating the correspondence, or lack thereof, among these various classifications; however, one is forthcoming.

---

[33] W. R. Taylor, *Nature* **416,** 657 (2002).

Concepts and Evaluations of Protein Predictions

Background: CASP

The fifth biennial Critical Assessment of Structure Prediction (CASP) meeting was held in December 2002.[8a] Our discussion is based on CASP4; the preliminary evaluation of CASP5 indicates modest progress with no major changes in concept or direction. Research groups that have determined the crystal or solution (NMR) structure of a protein can make its sequence available to the CASP organizers for several months, before the 3D structure is published. Any group in the world can try to predict the structures of these proteins. A panel then evaluates the blind predictions against the subsequently determined structures. The analyses are objective and common standards are applied to all predictions; the predictors exchange insights and hope to do better next time. However, as Moult *et al.*[8] wisely counseled, "A problem with the CASP format may be that it tends to reinforce conservative behavior—radical new methods are unlikely to be successful on the CASP cycle time scale...." The results of CASP IV and their analyses are summarized by a series of articles in Supplement V of *Proteins,* January 2001. We rely heavily on the CASP evaluations because most predictors participate and the evaluations are balanced and thorough.

The targets of prediction are considered in three classes: comparative (homolog) modeling, fold recognition and threading, and new fold *(ab initio)* methods. Further, within each class a distinction is made as to whether some human intervention is involved or whether the prediction is fully automatic: Critical Assessment of Fully Automated Structure Prediction (CAFASP). We summarize some of the approaches, concepts, and results. In Section IV we guide a potential user through a few examples from publicly available servers.

The evaluation of the results of CASP IV addressed seven issues (Moult *et al.*[8]):

1. Are the models produced similar to those of the corresponding experimental structure?
2. Is the mapping of the target sequence onto the proposed structure (i.e., the alignment) correct?
3. Have possible template structures been identified?
4. Are the details of the models correct?
5. Has there been progress from the earlier CASPs?
6. What methods are the most effective?
7. Where can future efforts be focused most productively?

These are appropriate concerns for both those who design prediction algorithms and for those who use them.

Comparative (or Homology) Modeling

Moult et al.[8] observed, "When the sequence of the target structure is clearly related to that of one or more structures, the structures will also be similar." The real question for the 21 predicted structures in this category of CASP is "... the extent to which there is improvement in the accuracy and detail of the models beyond simply copying the template ... the accuracy of alignments remains the key limitation on the quality and usefulness of the models, and has hardly improved since CASP2. Comparative models are still at best no more accurate than can be achieved by simply copying the appropriate regions of template structures." The optimal placement of indels is the limiting factor. Tramontano et al.[34] added, "Any sequence showing a significant expectation value [<0.02] with a protein of known structure [http://PredictionCenter.llnl.gov/casp4] after a PSI-BLAST run was considered a target for comparative modeling ... In most cases the structural similarity between model and target is worse, even much worse, than that between target and template. The factors primarily responsible for this effect are the selection of a non-optimal parent [template] structure and the errors in the alignment ... In general, methods using multiple parents perform better than those based on a single parent ... It is still rare, even for the best methods to achieve an accuracy above 80% for targets with sequence identity less than 50% ... We have once more to conclude that rarely is a model closer to the experimental structure than is the structural parent." That is, if the sequence identity of the target is >30% of that of the template(s), superimpose the backbone of the target on the template and then evaluate the indels and adjust the χ values of side chains to relieve bad contacts, fill voids, and form hydrogen bonds. Biologically important regions tend to be predicted better than other parts of the model; this probably reflects higher conservation of sequence in those regions. Several evaluators emphasized that it is better to have no model than a (grossly) wrong model. Unfortunately, models do not come color-coded for good and bad regions. Otherwise, efforts could be continued until the right answer was obtained. Modeling against several templates, using several different programs, increases confidence in the common regions and hardens suspicions regarding discrepancies. Servers (Table I) that monitor modeling servers are especially valuable. We emphasize the importance of evaluating models; if it is necessary to wait for the crystal structure to be certain, why bother modeling?

[34] A. Tramontano, R. Leplae, and V. Morea, *Proteins* 5(Suppl.), 22 (2001).

The boundaries between homolog, fold, and new modeling are in practice blurred.[12] Venclovas[35] emphasized that there are "... no inherent limits as to the application of comparative modeling. In practice, effective limits are imposed by the ability to detect relatedness between query protein sequence [target] and any of the proteins with known 3D structure [templates]... In the case of very distant homology, these issues are overshadowed by the necessity to correctly map the target sequence onto the conserved regions of the template(s) in the first place... Alignment remains the major source of error in all models based on less than about 30% sequence identity... At one end of the spectrum, targets with a high level of sequence identity to a known structure can be modeled with relatively small errors [typically <1 Å for C_α atoms at $>60\%$ sequence identity], while at the other, many new folds are still very hard to predict, and all models of these folds may be close to random."

Fold Recognition

Fold recognition relies heavily on the ever more sophisticated methods of sequence comparison. The assessments of success in CASP4 considered 52 target domains from the 40 target proteins in four subcategories: comparative modeling/fold recognition (CM/FR), borderline sequence recognizable folds/homologous folds (FR/H), analogous folds (FR/A), and analogous folds/new folds (FR/NF). As Moult *et al.*[8] summarized, "Increasingly, new structures deposited in the PDB turn out to have folds that have been seen before; even though there is no obvious sequence relationship between the related structures." This raises two questions: How successful are the different methods at identifying fold relationships? When successful, what is the quality of the models produced? This prompts a third question: What is the asymptote; how many folds are there in the biosphere? Moult *et al.*[8] said, "Techniques for fold recognition include advanced sequence comparison methods,... comparison of predicted secondary structure strings with those of known folds, tests of the compatibility of sequences with 3D folds (threading), and the use of human expert knowledge."

Koretke *et al.*[36] noted "... it has become progressively harder to identify instances where traditional fold recognition methods are superior to sensitive sequence database search tools" and added that on "... the basis of our experience in CASP3, we focused on sequence comparisons to make fold predictions in CASP4... Briefly, we first perform a PSI-Blast search with the target sequence... Proteins identified in this search are divided

[35] Č. Venclovas, *Proteins* **5**(Suppl.), 47 (2001).
[36] K. K. Koretke, R. B. Russell, and A. N. Lupas, *Proteins* **5**(Suppl.), 68 (2001).

into a significant sequence space, containing those sequences with an E-value $<10^{-3}$ and a "trailing end" of sequences above [but with $E < 10$]. We then perform transitive PSI-Blast searches to expand the significant sequence space by using proteins within the space that have $<25\%$ identity to the PSI-Blast. During the search, we continuously monitor the sequence space for proteins of known three-dimensional structure; the first one identified provides the template used to predict the fold." Murzin and Bateman[37] acknowledged that "The average quality of our fold predictions was far less than the quality of our distant homology recognition models...."

The prediction of secondary structure is of general interest. It is essential to that strategy for fold recognition that depends on alignment of predicted secondary structure with one or several candidate folds, followed by threading the target into the candidate. Lesk *et al.*[38] concluded, "In the case of secondary structure predictions, few groups performed consistently well, independently of the set of targets considered, and according to different evaluation schemes ... The popular Q3 score, for example, can vary between 57.3 and 95.9 for the same group over 40 targets [excluding targets belonging to the comparative modeling category] ... Q3 gives the percentage of residues correctly predicted in one of the three states: helix, strand, other."

Most integral membrane proteins consist of a distorted cylinder of 6 to 8 α helices or a warped barrel of 8 to 16 β strands, forming a channel, that spans the phospholipid membrane. Nearly all the residues of the α helices and of the outer, lipid-contacting surfaces of the channels are hydrophobic. The prediction of these spanning helices and strands is especially valuable in characterizing or even identifying integral membrane proteins (MEMSAT in PSIPRED). The few membrane structures available do not indicate significantly different guidelines for the prediction of membrane structures than those developed for proteins in aqueous media.

New (ab Initio) *Folds*

The traditional term, *"ab initio,"* has been replaced by "new" or "novel" fold because "... most of the methods ... do make extensive use of available structural information, both in devising scoring functions to distinguish between correct and incorrect predictions, and in choosing fragments to incorporate in the model."[8] That is, all these methods are knowledge based; even if an appropriate fold is not recognized, the predictor is making use of extensive information about the conformations of individual amino

[37] A. G. Murzin and A. Bateman, *Proteins* 5(Suppl.), 76 (2001).
[38] A. M. Lesk, L. Lo Conte, and T. J. P. Hubbard, *Proteins* 5(Suppl.), 98 (2001).

acids and short peptides. Lesk et al.[38] added that "The distinction [between *ab initio* and novel] ... is that the understanding achieved has been distilled into general principles that can be applied without looking up specific information in databases." It might be noted that the *ab initio* methods are using knowledge from many chemical studies of model compounds; they certainly are not making predictions on the basis of first principles of quantum mechanics. Twelve targets in the FR/NF category and only 5 truly new folds were considered. Again, it is clear that the distinction between analogous folds/new folds (AF/NF) and new folds is blurred.

Although prediction of new folds is exciting for the expert, its results hardly warrant use by the general user. Hence, we focus our discussion on comparative modeling. The prediction of protein–protein or protein–small molecule docking or of quaternary structure, and the related problem of the relationship between two distinct domains in a single polypeptide chain (CAPRI; Table I) are not covered in CASP, nor are they elaborated in this chapter.

Bonneau et al.[39] had some success in predicting novel folds using the Rosetta algorithm: "Large segments were correctly predicted [>50 residues superimposed within an RMSD of 6.5 Å] for 16 of the 21 domains under 300 residues for which models were submitted... These promising results suggest that Rosetta may soon be able to contribute to the interpretation of genome sequence information... The Rosetta method of *ab initio* structure prediction is based on the assumption that the distribution of conformations sampled by a local segment of the polypeptide chain is reasonably well approximated by the distribution of structures adopted by that sequence and closely related sequences in known protein structures. Fragment libraries for all possible three- and nine-residue segments of the chain are extracted from the protein structure database by using a sequence profile comparison method."

Jones[40] employed a similar strategy: "The method is based on the assembly of supersecondary structural fragments taken from highly resolved protein structures using a simulated annealing algorithm... In many cases, however, even when a new fold is discovered, it is observed that many new folds are still composed of common structural motifs at the supersecondary structural level... The PSIPRED predictions for all of the CASP4 targets were also submitted in the secondary structure prediction category and achieved an average Q3 score of 80.6% across the 40 submitted target domains with no obvious sequence similarity to structures present in

[39] R. Bonneau, J. Tsai, I. Ruczinski, D. Chivian, C. Rohl, C. E. M. Strauss, and D. Baker, *Proteins* **5**(Suppl.), 119 (2001).
[40] D. T. Jones, *Proteins* **5**(Suppl.), 127 (2001).

PDB . . . These predictions, along with the associated PSI-BLAST multiple sequence alignments, were used as inputs to FRAGFOLD. Note that apart from biasing the selection of fragments, secondary structure prediction information was not used elsewhere in FRAGFOLD . . . In addition to the sequence-specific fragment list, a general fragment list is also constructed from all tripeptide, tetrapeptide and pentapeptide fragments from the library of highly resolved structures. These smaller fragments are not preselected.''

Process of Extracting Information about Protein Structure from Sequence

Preliminary Evaluation

In Section III we summarized the concepts and achievements of predictors from the perspective of CASP. We now consider actual databases, servers, and programs in the context of comparative, fold, and new modeling. We chose for illustration a few of the servers that performed well in CASP, and/or are frequently used, and/or are especially easy to use in order to illustrate the process; not all are mentioned explicitly.

To begin, evaluate the target sequence to devise a strategy.

1. Identify and delete from the target regions of low complexity, that is, regions with highly biased use and/or repetition of patterns of residues; their inclusion distorts the statistics of sequence alignment.[41] These low-complexity regions usually occur before, sometime after, globular domains; if between, they may serve as linkers or spacers. Search the sequence databases with PSI-BLAST or FASTA to determine what homologous sequences are known. From them glean whatever information might be available about domain boundaries, conserved residues, variable regions, sites of posttranslational modification, preferred sites of proteolysis (perhaps indicating flexible loops), and so on. All this information can contribute to the process and, more important, to evaluation of the predictions.

2. Predict the distribution of α helices, β strands, and coils, including regions of low complexity, for the target and for several of its nearest homologs. If the homologs share >30% sequence identity with the target, the predicted secondary structures should align, with allowance for indels. Lesk et al.[38] admonished, "For easy homology models, it is better to take your secondary structure assignment directly from the three-dimensional

---

[41] J. C. Wootten, *Comput. Chem.* **18,** 269 (1994).

structure [of the template]... For all other cases, direct secondary structure prediction is more reliable [this sounds obvious], with several interesting exceptions coming from the tertiary structure prediction field...."

For example, we submitted the sequence of a protein of unknown function and structure to PSI-BLAST. It has many homologs among the $(\beta/\alpha)_8$ or TIM (triose isomerase phosphate-like) barrel proteins (Fig. 4).

3. Predict the domain boundaries within the target. The comparisons of predicted secondary structures should accord with the domain boundaries and choice of homologs. Sippl *et al.*[42] emphasized that "Target characterization starts with the definition of domains and domain boundaries. Then the extent of sequence similarity of each domain to proteins in PDB is evaluated by using PSI-BLAST and the structural similarity is determined by using the program, ProSup, which is based on rigid body superposition." ProDom and CAPRI were also valuable in good definition of cleavage. The first 94 residues of our target have no corresponding residues in the closest homolog of known structure, 1fws. A subsequent BLAST search with just these 94 residues found an alignment with the first 94 residues of AE016951_293 (phospho-2-dehydro-3-deoxyheptonate synthase, $E = 6.0 \times 10^{-15}$); this may reflect another domain.

4. Find what structures are available in the PDB for the nearer homologs of the target. If one or several homolog structures at >25% structural identity are known, proceed with homolog modeling confident that the backbone of the target, exclusive of indels, can be modeled to within a few angstroms, RMSD, of its true structure. Low-complexity regions are usually found before, between, or after the domains recognized in homolog searches. If no homologs are found in the PDB, try fold recognition, which may include advanced homology searches.

Among the homologs of known structure, 3-deoxy-D-*manno*-octulosonate 8-phosphate (KDOP) synthase (1fws in the PDB) is most similar to our unknown, with 34% sequence identity. Their alignment, observed secondary structure of the template, and predicted secondary structure of the target are shown in Fig. 4.

Comparative (or Homology) Modeling

In comparative modeling the first step is to select one or several templates. Usually, but not always, the sequence of a known structure most closely resembling the target is the best target as selected by PSI-BLAST. PSI-BLAST is used by most groups to determine the optimal alignment of

[42] M. J. Sippl, P. Lackner, F. S. Domingues, A. Prlic, R. Malik, A. Andreeva, and M. Wiederstein, *Proteins* 5(Suppl.), 55 (2001).

```
KDOPS (1fws, template) vs. DAHPS-thermatoga(target = TH)


  1 MIVVLKPGSTEEDIRKVVKLAESYNLKCHISKGQERTVIGIIGDDRYVVV 50  TH
    CEEEECCCCCHHHHHHHHHHHHHCCCCEEEEEECCCCEEEEEEECCCCEEEE    PROF
    CEEEECCCCCHHHHHHHHHHHHHCCCCEEEEEECCCCEEEEEEECCCCCCCH A PSIPRED

 51 DKFESLDCVESVVRVLKPYKLVSREFHPEDTVIDLGDVKIGNGY 95   TH
    EECCHCCCCEEEEECCCCCCCECCCCCCCCCCCCCEEEEECEECCCCC     PROF
    HHHHHCCCCCHHHHCHHHHHHHHHHHHCCCCCCEEECCCEEECCCC      PSIPRED

    RRSSSSSRHSVHSHHHHHHHHHHHHHHHHHHHHVHHSSRSRSSRHUVHURHSHH   D8
    CCEEEEEECSCCCHHHHHHHHHHHHHHCEEEEEEEECCCCCCCCCCCC
  4 FLVIAGPCAIESEELLLKVGEEIKRLSEKFKEVEFVFKSSFDKANRSSIH 53 FWS
    |  :||||| .:|  |:|:.    :    |   |        |  :    |  |.|  :
 96 FTIIAGPCSVEGREMLMETAHFLSELGVK......VLRGGAYKP.RTSPY 138 TH
    EEEEECCCCCCCHHHHHHHHHHHHHHHCH      HEEEEECCC CCCCC    PROF
    EEEEEECCCCCCHHHHHHHHHHHHHHHCCC     . EECCCCCC CCCCC   PSIPRED

    RRHSUGHHHHHHHHHHHHHHHHHHTSRSSSSSHSHHHHHHHHHHHHRHSSSSRHH   D8
    CCCCCHHHHHHHHHHHHHHHHHHHCEEEEEECCCCHHHHHHCCEEECCCCC
 54 SFRGHGLEYGVKALRKVKEEFGLKITTDIHESWQAEPVAEVADIIQIPAF 103 FWS
    ||.| |  |  |.. ||.  :.:|: :  |:        |||  ||||||  |
139 SFQGLG.EKGLEYLREAADKYGMYVVTEALGEDDLPKVAEYADIIQIGAR 187 TH
    HHCHHE EEEECCCCCCCCCCCCCCEEEEECCCCCHHHHHHHHHCCHH     PROF
    HCCCEE ECCCCCCCCCCCCCCCCCEEECCCCCCHHHHHHHHHCCEEECCCH  PSIPRED

    VHHVHHHHHHHHHHTRRSSSSSRRHHRRHHHHHHHHHHHHHTRHUSSSSS   D8
    CCCHHHHHHHHHHHHCCCEEEEEECCCCCCCCHHHHHHHHHHHHCCCEEEEEE
104 LCRQTDLLLAAAKTGRAVNVKKGQFLAPWDTKNVVEKLKFGGAKEIYLTE 153 FWS
     .    ||  |  | .|:|         ||   |   |  |  .|  |.| |
188 NAQNFRLLSKAGSYNKPVLLKRGFMNTIEEFLLSAEYIANSGNTKIILCE 237 TH
    HCCCHHHHHHHHHCCCEEEEECCCCCCHHHHHHHHHHHHCCCCEEEEEE    PROF
    HCCCHHHHHHHHHCCCEEEEECCCCCCHHHHHHHHHHHHCCCCCEEEEC    PSIPRED

    UG SRU GRTSSSSVHHHHHHHHHUS  RSSSVHHHHHRS     D8
    CC CCC CCCEEEECCCHHHHHHHCCC  EEEEECCCCCCcccccccccC
154 RG.TTF.GYNNLVVDFRSLPIMKQWA..KVIYDATHSVQLPGGLGDKSGG 199 FWS
    ||   ||      .|  ..||.:.  .    ::  |  .|         |||
238 RGIRTFEKATRNTLDISAVPIIRKESHLPILVDPSH...........SGG 276 TH
    CCCCCCCCCCCCCCEEEEHHHHHCCCCEEEEECC          CCC   PROF
    CCCCCCCCCCEECHHHHHHHHHCCCCEEEECCCCCC         CCH   PSIPRED

    HHHHHHHHHHHHHHHGSHSSSSSSSHUHHHHRHSTHHHRRRHHHHHHHHHH   D8
    CCCHHHHHHHHHHHHHHCCEEEEEEECCCCCCCCCCCCCEEECHHHHHHHH
200 MREFIFPLIRAAVAVGCDGVFMETHPEPEKALSDASTQLPLSQLEGIIEA 249 FWS
    |:  :  ||  |||:|||    |:  .|  ||||||||||      |    :::
277 RRDLVIPLSRAAIAVGAHGIIVEVHPEPEKALSDGKQSLDFELFKELVQE 326 TH
    CCCCCHHHHHHHHHHCEEEEEEECCCCCCCCCCCCCCCCCCCHHHHHHHH    PROF
    CCHHHHHHHHHHCCCCEEEEEEECCCCCCCCCCCCCCCCCCCHHHHHHHH    PSIPRED

    HHHHHHHHHHHHRRR    D8
    HHHHHHHHHHHHCCC
250 ILEIREVASKYYETIPVK 267 FWS
    . .: :
327 MKKLADALGVKVN 339     TH
    HHHHHHHHHCCCC          PROF
    HHHHHHHHHCCCCC         PSIPRED
```

FIG. 4. Summary of results of modeling an unknown structure. The sequence dahps-th, of target structure TH, is aligned with its closest homolog of known structure, 1fws (FWS), using BestFit, part of GCG. The two are 34% identical, 46% similar, over 238 residues of the target; hence, there is confidence in the general conformation of the predicted structure (see Fig. 5). The secondary structure of the target, dahps-th, was predicted by two different servers, PROF and PSIPRED; the two predictions agree in 257 of 339 positions, $Q_3 = 76\%$. More important,

the sequence of the target with those of the closer homologs. Several groups have emphasized that misalignment is the greatest source of error in homolog modeling. Although, in general, structural similarity correlates with sequence similarity, there are numerous exceptions. To address this problem, Venclovas[35] "... used multiple templates whenever they were available. That was one of the reasons to choose MODELLER, a program that is capable of automatically taking into account structural data from multiple templates." 3D-JIGSAW also functioned well in CASP4 to build the target to the chosen templates. DALI is also useful for the super-position of templates and subsequent manual adjustment in a graphics program such as O.

The least reliable parts of the resulting models are the loops, often the sites of indels and often poorly defined in the crystal structures of the templates. The program CHARMM, which can be licensed, is often used to build and refine such loops.

Our target has been modeled to the known structure of KDOP synthase (1fws) using BestFit in the CGG package. Modeller gives a similar result. The structures of the automatically generated structure and of the manually adjusted structure of the unknown are shown in Fig. 5. The remaining residues of our target, save the last five, are aligned with residues 4–257 of 1fws. If it is accepted as reasonable that the unknown will contain the 8 β strands and the 8 following α helices of the template, then the secondary prediction schemes identified, in approximately the correct position, all 16 elements except α-helix 2. All seven indels between the two alignments occur between these strands and helices. The single large deviation, and inferred error, in the automatically generated structure involves β-strand 2 (Fig. 5A). The (inferred) correct structure is predicted if the first deletion of six residues, the second of one, and the third of one are condensed into a single deletion of eight and VLRGG*** (residues 125–129) of the target are forced to align with FKEVE*** (residues 39–43) of the template (Fig. 5B). As Tolstoy observed, "Every unhappy family is unhappy in its own way";

---

both predictions of secondary structure have similar patterns of helix (blue), extended (β strand; red), and coil as assigned by a PDB routine to the template FWS, using criteria both of general ϕ, ψ value and of hydrogen-bonding pattern. There is added faith in the general pattern of PROF and PSIPRED predictions because they map well onto the pattern actually observed in the template, a $(\beta/\alpha)_8$ or TIM barrel protein. There is an insertion of 11 residues, $_{186}$SVQLPGGLGDK$_{196}$, in the template, FWS, relative to the target, TH. This corresponds to a loop, residues 192–198, not visible in the crystal structure (see Fig. 5). The program that assigns secondary structure in the PDB assigns coil *(c)*, often with justification, to such nonvisible regions. The top line, D8, assigns each residue to one of eight discrete conformations (see Fig. 2), as determined solely by ϕ, ψ. A BLAST search with the first 95 residues of the target indicates their optimal alignment with phospho-2-dehydro-3-deoxyheptonate synthase. (See color insert.)

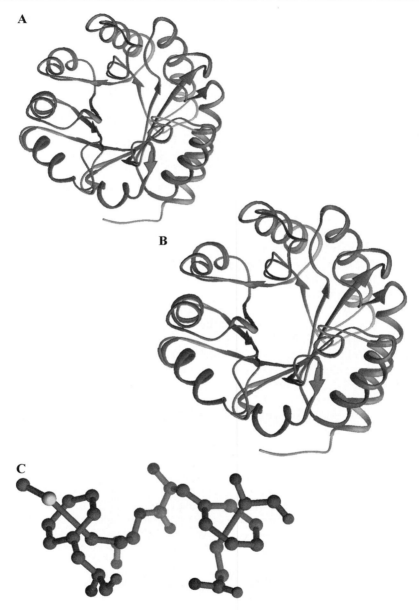

FIG. 5. Backbone drawing of the known template, kdops-1fw, with the automatically predicted model dahps-th, superimposed. The proteins are viewed approximately down the axis of the $(\beta/\alpha)_8$ barrel; the C termini of the β strands are near the viewer. (A) One element, residues ~40 to ~55 of the target, is obviously misplaced. Note that the loop not resolved in the

so every unhappy protein has its own special grievance. Yet there are common themes, such as infidelity of alignment. Manual intervention by a structure-based sequence alignment of the target and template structures after a DALI alignment can often avoid this sort of pitfall.

The "correct" backbone of the unknown (Fig. 5B) takes no account of the orientation of side chains. For instance, as seen in Fig. 5C, many side chains of the model are placed in obviously unacceptable positions, for instance, Gln190 is superimposed on the benzene ring of Phe211. In retaliation, Met212 is thrust through Phe218. These and many other steric violations, potential hydrogen bond donors without recipients, and interior voids are identified by Procheck.

Fold Recognition, Prediction, and Threading

Fold recognition was evaluated in CASP in four categories: (1) comparative modeling/fold recognition (CM/FR), (2) borderline sequence recognizable folds/homologous folds (FR/H), (3) analogous folds (FR/A), and (4) analogous folds/new folds (FR/NF). This gradation is reflected in three sorts of approaches: (1) advanced sequence comparison methods, (2) comparison of predicted secondary structure strings with those of known folds, and (3) tests of the compatibility of sequences with 3D folds (threading). Massive parallel application of all three methods might be employed in evaluations of genomes. The category (CM/FR . . . FR/NF) is known only after finding the answer; start by assuming CM/FR and work toward FR/NF only if the easier procedures do not produce reasonable results. Critical thinking is permitted at all stages.

3D-PSSM, BIOINBGU, FFAS, and PROSPECT employ various algorithms to match a target sequence to a known fold. Given one or several candidate folds, the challenge is then to align or thread the target sequence to the fold(s). GenTHREADER, and the suite of servers—FUGUE, CODA, JOY, and COMPARER—tackle threading. SCWRL can add, change, and/or optimize the orientations of side chains.

---

electron density (residues 192–198) is represented by a green line connecting residue 191 to residue 199. The corresponding region of the target (see Fig. 4), is also guessed. (B) The misplaced element of the target has been manually corrected by shifting the alignment as described in text. (C) Many side chains of the model of (B) are placed in obviously unacceptable positions; for instance, the side chain of Gln190 is superimposed on the benzene ring of Phe211. Met212 is thrust through Phe218. (See color insert.)

New (ab Initio) Folds

Lasciate ogni speranza, voi ch'entrate (Abandon all hope, ye who enter—Dante[43]). If secondary structure is predicted on the basis of the three crude categories—helix, strand, and coil (H, E, and C in Fig. 4), the best results approach $Q_3 = 80\%$. The averaged protein has about 20% β strand, 35% α helix, and 45% other structures all lumped together as coil. The random assignment of 20% H, 35% E, and 45% will yield on average $Q_3 = 37\%$; assigning 100% C yields $Q_3 = 45\%$. If those residues, such as Ala, Leu, Glu, and so on, with higher propensities for α helix, are always assigned to helices; Val is always assigned to β strands; and Gly, Pro, His, Ser, Thr, Cys, Asn, and Asp are always assigned to coils, Q_3 increases to 48%. If the propensities of pairs and triplets of residues are considered, Q_3 increases to 62%. That is, about half the forces determining secondary conformation come from local sequence. The other half comes from interactions of side chain and main chain atoms more distant in sequence, such as those involved in the lateral interactions of β strands to form β sheets. It has taken 20 years to get from 64 to 80%. We need close to 100% to predict tertiary structure *ab initio*. The problem is further complicated by the existence of real alternative structures (Fig. 3) in these nanomachines. Perhaps we are asking the wrong question; we may realize $Q_3 > 95\%$ only from better 3D predictions made independent of initial assignments of H, E, and C.

ROSETTA and FRAGFOLD have enjoyed some success in *ab initio* prediction, especially for those targets that turn out to have some relationship to known folds (FR/NF category). It is valuable to predict correctly half the target with a new fold; however, it is necessary to know which half.

Refinement

All loops are considered for replacement in homolog and fold modeling using database fragment searches. CHARMM is used to adjust the conformations of loops to match ends of helices and strands. Manual adjustments address (1) maintenance of a hydrophobic core, (2) equivalencing of known core residues, (3) preservation of continuity of secondary structure elements, (4) maintenance of spatial arrangements of residues of the active site, (5) alignment of known motifs, and (6) maintenance of distance between cysteine residues. Although refinement and evaluation are treated as different concepts, in fact the refinement is part of the evaluation. We caution that an overrefined model can easily become enshrined in the literature as gospel.

---

[43] "L'Inferno," Canto III, line 9.

Evaluation

The CASP evaluator knows the correct answer. The generalist making the prediction wants to use the predicted structure *now*. Be assured that the prediction will deviate from the subsequently determined crystal structure. Atomic resolution in a target is almost never achieved with the exception of high sequence similarity with a template that itself was determined at high resolution. However, modeling a newly discovered gene by homology and confirming that a reasonable hydrophobic core is formed and that the active site residues are in general agreement can provide powerful evidence toward biologically relevant conclusions.

Although some servers score better than others, all sows find a few acorns. Use several different servers and compare results. Consistency does not ensure a good prediction; however, inconsistency is serious cause for concern.

A few predictions will be totally wrong; a few, nearly perfect. However, most will have regions that resemble the true structure(s) and regions that do not. Several servers and programs—CastP, RAMPAGE, MaxSub, WHAT_CHECK, PROCHECK, VERIFY3D, and PROSA II—check for various biases and errors in the predicted structures. They may also be applied to experimentally determined structures. Incorporate the chemical insights from the initial evaluation to help establish the reasonableness of the various regions of the several predictions.

Address the specific problem: Why was the prediction made? What sort of information is wanted? How reliable is this information? Is it worth the investment to purify, crystallize, and determine the structure of the unknown target in order to be certain? For instance, the arrangement of ligands about an active site might be correct with the overall shape of the predicted structure quite distorted, or just the inverse.

EVA and LIVEBENCH do not predict or evaluate structures. They perform the valuable service of monitoring the performance of various servers—well worth checking.

Future Directions

Some reviewers have expressed disappointment about the rate of improvement in CASP predictions. Others suggest that what progress has been observed from CASP3 to CASP5 reflects not better algorithms but simply richer databases essential to all these knowledge-based approaches. In either case three predictions can be safely made. First, the PDB of known structures will continue to double every 3 years. Second, the number of sequences in want of a predicted structure will continue to double every few months—Moore's law of genomics. Finally, ever more predicted structures will appear in the literature: *caveat emptor*.

[2] Modeling and Studying Proteins with Molecular Dynamics

By ROBERT SCHLEIF

Introduction

Structure determines function, but it is not easy to deduce many of the activities or properties of proteins by mere visual inspection of their structures. We do not even know which questions to ask about structure so as to gain the most insight into the behavior of a protein. It does seem likely, though, that answering questions involving geometry, energetics, and atomic motions will be important in understanding protein function. Typical questions that might be asked are as follows: Which residues interact with each other, and how strongly? How exposed to solvent is a particular residue? Is a part of the protein held rigidly or loosely in position with respect to the rest of the protein? How much change in the structure will be generated by a particular mutation? If a mechanism has been postulated for the function of a protein, how would a particular mutation alter that function?

Quite a number of graphical display computer programs are capable of answering geometric questions about proteins, using as input the atomic coordinates as determined by X-ray diffraction, nuclear magnetic resonance (NMR), or model building. Some of these programs can, in addition, perform energetics calculations, using the atomic coordinates and the identities of the atoms. Finally, still more powerful programs can perform geometric and energetics calculations as well as allow movement of atoms in response to interatomic forces. In this third group of programs, atoms may be allowed to move in the course of energy minimization, or all the atoms in a system may be given velocities characteristic of a chosen temperature, and the trajectory of each atom in response to all the forces acting on it may be calculated for a large number of short time steps. This latter technique simulates what we think are the atomic motions in a system, although we call it molecular dynamics. The value of molecular dynamics simulations in understanding the activities of proteins remains to be determined, but model building and calculations concerning proteins of known structure or proteins not too distantly related to proteins of known structure that involve geometry, energetics, and that allow movement of atoms, seem almost certain to be indispensable. Therefore, it seems likely that growing numbers of biochemists, molecular biologists, and biophysicists

Copyright 2004, Elsevier Inc.
All rights reserved.
0076-6879/04 $35.00

will choose to include in their arsenal of research tools the capability of using a molecular dynamics program. With one program, they will be able to handle a wide variety of calculations.

Several commercial programs for model building, structure analysis, and molecular dynamics that run on workstations are available.* These programs smoothly combine excellent graphical display with model-building facilities, powerful protein structure analysis tools, and computational facilities. The core of much of the computational work done by both programs is the molecular dynamics program CHARMM (Chemistry at Harvard, Molecular Modeling).[1,2] Some scientists in need of these facilities may, however, prefer to do molecular dynamics calculations on a personal computer and without the front end that is provided by the commercial programs. Using a molecular dynamics program in this way does require the writing of scripts, which are primitive programs, but in return the total costs are low (just the personal computer and the nominal the cost of the CHARMM program) and computational versatility is high, and considerable safety is gained because the program is directly controlled rather than being controlled through an interface that freely allows the performance of meaningless calculations. By the time the user can make CHARMM itself do a calculation, the user, more likely than not, knows enough to direct it to perform sensible calculations.

The molecular dynamics programs Amber,[3] CHARMM,[1,2] Gromos,[4] and NAMD[5] all appear suitable for the types of calculations outlined above, although the strengths of each program are slightly different. Because the four programs share much in philosophy, functions, features, and some of their file structures, much of what is true of one program is likely to be true of the others. The remainder of this chapter focuses on CHARMM. This program has been continuously upgraded and expanded by scores of experts over the years, and its availability, accuracy, and

* Quanta and Insight are available from Accelrys (San Diego, CA).
[1] B. R. Brooks, R. E. Bruccoleri, B. D. Olafson, D. J. States, S. Swaminathan, and M. Karplus, *J. Comput. Chem.* **4,** 187 (1983).
[2] A. D. MacKerell, Jr., B. Brooks, C. L. Brooks III, L. Nilsson, B. Roux, Y. Won, and M. Karplus, *in* "The Encyclopedia of Computational Chemistry" (P. V. R. Schleyer, N. L. Allinger, T. Clark, J. Gasteiger, P. A. Kollman, H. F. Schaefer III, and P. R. Schreiner, eds.), Vol. 1, p. 271. John Wiley & Sons, Chichester, 1998.
[3] D. A. Pearlman, D. A. Case, J. W. Caldwell, W. S. Ross, T. E. Cheatham III, S. DeBolt, D. Ferguson, G. Seibel, and P. Kollman, *Comput. Phys. Commun.* **91,** 1 (1995).
[4] W. R. P. Scott, P. H. Hünenberger, I. G. Tironi, A. E. Mark, S. R. Billeter, J. Fennen, A. E. Torda, T. Huber, P. Krüger, and W. F. van Gunsteren, *J. Phys. Chem. A* **103,** 3596 (1999).
[5] L. Kalé, R. Skeel, M. Bhandarkar, R. Brunner, A. Gursoy, N. Krawetz, J. Phillips, A. Shinozaki, K. Varadarajan, and K. Schulten, *J. Comput. Phys.* **151,** 283 (1999).

reputation are all well known. Because it runs on a wide variety of Unix computers, as well as on inexpensive personal computers running Linux, it is widely used.

Most of the authors of chapters in this volume are producers rather than consumers. That is, they originated and wrote the computer programs they write about. The author of this chapter, however, is solely a consumer. For many years research in the author's laboratory has been centered on regulation of the L-arabinose operon in *Escherichia coli,* and lately it has been directed at understanding the atomic details in its regulatory protein, AraC. It became apparent that computational abilities were needed to help analyze and guide the biochemical and molecular genetics approaches chosen. This chapter describes the author's experience in using CHARMM for this purpose.

Sampling of CHARMM Capabilities

Below are listed a few examples of protein structure–function questions that can be addressed with a program such as CHARMM.

Simple Structural Questions

- Simple geometric quantities, such as distance between two atoms, the angle between three atoms, the dihedral angle between four atoms, locations of holes in a protein, or the surface area of a protein, can be determined.
- The interaction energies between any atoms or residues and any other atoms or residues can be calculated.
- Complex selections can be included in quantities to be calculated. For example, it is possible to calculate the strength of the electrostatic interactions between the backbone atoms of a portion of a protein and the solvent.
- The data for residue–residue contact maps and residue–residue interaction energy maps can easily be calculated.
- The solvent exposure of each residue with and without ligand present or the relative solvent accessibility of each residue in the protein can easily be determined.

Model-Building Operations Possible

- Simple molecules such as a carbohydrate or complex molecules such as a protein can be "constructed" *de novo* and studied computationally in the same way as molecules whose representation in the computer is based on a real structure.

- It is possible to "mutate" one or more of the residues in a protein and to determine the subsequent structural accommodation of the rest of the protein to the change.
- The rotameric state of the side chain of a residue can be adjusted.
- A peptide or protein or ligand can be moved around in space.
- Any set of atoms can be fixed in space and another part of the peptide or protein can be pulled to any desired position, all the while allowing all the unfixed atoms to move subject to forces exerted by existing bonds and nonbonding interactions.
- Two peptides or proteins can be fused.
- The homologous regions of two different proteins can be overlaid through RMS fitting.

Operations That Involve Molecular Dynamics

- Dynamics of a protein can be run in a shell of water or in an infinite array with periodic boundary conditions. Any temperature can be simulated. When periodic boundary conditions are used, simulations can be done at constant pressure, constant temperature, or constant volume.
- From a molecular dynamics trajectory, it is possible to determine average structure and determine RMS fluctuations in atom positions. This can locate flexible and rigid portions of a protein. The fluctuations can be assigned to the *B* value or the crystallographic temperature factor, written to an output file in the Protein Data Bank[6] (PDB) format, and the flexibility of the protein can be visualized with a molecular display program that colors the backbone according to the *B* value.
- It is possible to see how the structure and interactions in a protein respond to a mutational change in a residue or multiple residues at physiologically meaningful temperatures in the presence of water.
- The energy of the system or any interaction energy can be determined for any or all of the frames of a dynamics trajectory.
- During a molecular dynamics run, it is possible to gently pull on a part of the protein so as to generate a desired conformational change. The work done in generating the conformational change can then be calculated.
- The CHARMM program can also carry out more advanced operations such as Monte Carlo studies, Poisson–Boltzman

---

[6] H. M. Berman, J. Westbrook, Z. Feng, G. Gilliland, T. N. Bhat, H. Weissig, I. N. Shindyalov, and P. E. Bourne, *Nucleic Acids Res.* **28,** 235 (2000).

calculations, and normal mode analysis; determine free energy differences by thermodynamic integration or perturbation techniques; and examine issues of protein folding and denaturation.

Program Operation Basics

Molecular Dynamics: Theory

We deal with atoms that have reasonably high mass compared with electrons, at reasonably high temperatures, and we consider reactions that do not include the making or breaking of chemical bonds. Therefore, reasonably accurate results can be obtained by classic mechanics rather than quantum mechanics and by considering the locations of the atomic nuclei rather than the densities of electron clouds. The basis of the calculations can then be Newton's equation of motion as shown in Eq. (1):

$$F_i = m_i a_i \tag{1}$$

where F_i is the sum of forces acting on the ith particle, m_i is the mass of the ith particle, and a_i is the acceleration of the ith particle. A molecular dynamics program is said to integrate these equations, one for each particle. The word *integrate* as used here does not mean to find the mathematical expressions for particle motions that solve the differential equations; rather, it means to determine the numerical values of the coordinates of each particle at successive, closely spaced, time points, all the while obeying Newton's equation of motion. In practice, these time points are about 10^{-15} s apart.

The principle of numerical integration is that given the position $x_i(t_0)$ and velocity $v_i(t_0)$ of the ith particle at the starting time t_0, the position and velocity of the particle a short time later, at t_1, are given by

$$x_i(t_1) = x_i(t_0) + v_i(t_0) \times (t_1 - t_0) \tag{2}$$

$$v_i(t_1) = v_i(t_0) + \left. \frac{dv_i(t_0)}{dt} \right|_{t_0} \times (t_1 - t_0) \tag{3}$$

(3) becomes the following from Newton's equation of motion,

$$v_i(t_1) = v_i(t_0) + \frac{F_i(t_0)}{m_i} \times (t_1 - t_0) \tag{4}$$

Hence, the positions and velocities of all the particles at time t_1 can be calculated from their values at time t_0. The total force acting on each particle, F_i, depends on the positions of all the other particles in the system and, in addition, may include externally imposed forces. In actual practice, the equations used by molecular dynamics programs are slightly different but are still based on the principles outlined above.

A crucial component in the trajectory calculations is the force that is felt by each atom. A precise representation of the force may yield simulations that closely approximate nature, but they take impossibly long to compute. Alternatively, representations of the forces that lead to rapid computation may be poor approximations of the real forces. Different molecular dynamics programs choose different compromises to this difficult tradeoff. The potential function used by CHARMM, from which energies and the forces needed for trajectory calculations are derived, approximates the forces on an atom as the sum of its pairwise interactions with the other atoms in the system.[7] The energy of an atom is approximated as the sum of bond stretching, bond bending, bond twisting, improper torsions (which act to maintain planar bonds), and van der Waals, and electrostatic interactions. The parameters describing these terms are tabulated and read in when CHARMM starts. Much of the work of developing a molecular dynamics program goes into determining the best possible values for the many terms in a parameter table. These potential functions used by CHARMM are approximated as follows.[7]

Bond-stretching energy is approximated as

$$V_{\text{bond}} = K_{\text{b}}(b - b_0)^2 \tag{5}$$

where K_{b} is a constant that depends on the two atoms sharing the bond, b is the length of the bond, and b_0 is the unstrained bond length. The energy in the bond or the force exerted by the bond is approximated as depending on the coordinates of only the two atoms sharing the bond and on the values of the constant K_{b}, which depend on the types of atoms that are involved. These and other constants used to describe the other atomic interactions are obtained by a combination of *ab initio* calculations and hand fitting of results obtained by simulation of small model compounds to physical measurements.[7]

Bond-bending energy is approximated as

$$V_{\text{angle}} = K_\theta(\theta - \theta_0)^2 \tag{6}$$

where K_θ is a constant that depends on the three atoms defining the angle, θ is the angle, and θ_0 is the unstrained angle.

[7] A. D. MacKerell, Jr., D. Bashford, M. Bellott, R. L. Dunbrack, Jr., J. D. Evanseck, M. J. Field, S. Fischer, J. Gao, H. Guo, S. Ha, D. Joseph-McCarthy, L. Kuchnir, K. Kuczera, F. T. K. Lau, C. Mattos, S. Michnick, T. Ngo, D. T. Nguyen, B. Prodhom, W. E. Reiher III, B. Roux, M. Schlenkrich, J. K. C. Smith, R. Stote, J. Straub, M. Watanabe, J. Wiórkiewicz-Kuczera, D. Yin, and M. Karplus, *J. Phys. Chem. B* **102**, 3586 (1998).

Bond twisting requires four atoms, A, B, C, and D, to define the rotation or torsion angle of the B–C bond. This term is also known as the dihedral energy, and it is approximated as

$$V_{\text{dihedral}} = K_\chi[1 + \cos(n\chi - \delta)] \tag{7}$$

where K_χ and δ are constants that depend on the adjacent atoms, n is an integer (2, 3, 4, or 6) that depends on the four atoms, and χ is the value of the dihedral angle between the planes defined by atoms A, B, and C, and by atoms B, C, and D. Thermal energies are usually sufficient to shift some of these bonds from one energy minimum to another. The term provides a preference for the torsion angle to place the atoms bonded to atoms B and C in noneclipsed positions, and describes the different rotameric states of the side chains of amino acids.

A group consisting of one central atom, A, that is bonded to three other atoms, B, C, and D, can be held in a particular configuration by what is called an improper energy term,

$$V_{\text{improper}} = K_\psi(\psi - \psi_0)^2 \tag{8}$$

where K_ψ and ψ_0 are constants and ψ is the angle between the plane containing atoms A, B, and C and the plane containing atoms B, C, and D. Most often an improper dihedral term is used to hold a set of four atoms in a plane. In contrast to the regular dihedrals, the energy constants for improper dihedrals are large enough to prevent substantial deviation of ψ from ψ_0.

The van der Waals interaction between two atoms is approximated with a Lennard–Jones potential as

$$V_{\text{Lennard–Jones}} = \varepsilon_{i,j}\left[\left(\frac{R_{\text{mi},i,j}}{r}\right)^{12} - 2\left(\frac{R_{\text{mi},i,j}}{r}\right)^{6}\right] \tag{9}$$

where

$$\varepsilon_{i,j} = \sqrt{\varepsilon_i \times \varepsilon_j} \tag{10}$$

and where ε_i and ε_j are constants characteristic of the two atoms,

$$R_{\text{mi},i,j} = \frac{R_{\text{mi},i}}{2} + \frac{R_{\text{mi},j}}{2} \tag{11}$$

where $R_{\text{mi},i}$ and $R_{\text{mi},j}$ are also constants characteristic of the two atoms, and r is the distance between the centers of the two atoms.

Finally, the electrostatic interaction between two atoms is

$$V_{\text{electrostatic}} = \frac{q_i q_j}{4\pi\varepsilon r} \tag{12}$$

where q_i and q_j are the charges of the two atoms, r is their separation, and ε is the dielectric constant. Ordinarily, energy calculations and simulations are performed with the protein immersed in water molecules. The parameter tables in CHARMM have been developed for this condition and with ε set equal to 1.

Program Operation: Practice

Because a molecular dynamics program is capable of generating a path of motion for every atom of a protein and a surrounding bath of water molecules, it needs much input information. Some of this information comes from a file of atomic coordinates that have been determined by X-ray crystallography, NMR, or model building. Such files provide the IUPAC names of the atoms in the residues and provide the positions of the atoms in space. Of course, coordinates that are determined by X-ray crystallography lack positions for the hydrogen atoms, and the program must place such atoms on its own.

In addition to positional information, a program needs to know every chemical bond in a structure, the bending, stretching, and twisting strengths of these bonds, the strengths of the improper restoration forces, the radius of each type of atom, and the partial electric charge on each atom. These latter classes of information are made available to the CHARMM program in the form of two types of table. One table is called the residue topology file. This contains a list of the amino acid residues and/or nucleotide bases and commonly encountered small molecules. It lists the bonding pattern for every atom in a residue or base, the partial charge of every atom, the IUPAC name of each atom, and the CHARMM atom type for the same atom. The topology file also lists the geometric position of each atom with respect to other atoms to which it is bonded. This local coordinate system is called the internal coordinate system. It allows deduction of the likely positions of atoms whose coordinates may be missing from the input coordinate files and greatly simplifies some coordinate manipulation operations. The second table of information used for input of information to CHARMM is the parameter table. This contains the masses of the atoms, the stretching, bending, and twisting strengths of the chemical bonds, the strength of improper forces, the parameters recommended for determining how the interaction force between two atoms will be calculated, and the strengths of van der Waals forces and radii for all atom types.

From the input coordinates, the residue topology file, and the parameter file, the program can construct two files specific to a protein. These are used as the program performs further operations on the protein such as energy determination, energy minimization, alteration of residues, structure modification, or the running of dynamics simulations. One of the protein-specific files is a list of the coordinates of every atom in the system. The second is known as the principal structure file. It lists every residue in the protein, lists each atom, provides the type of each atom, and gives the mass and partial charge of each atom. It also lists every pair of atoms connected by a covalent bond, each atom triplet that defines an angle energy term, each set of four atoms defining a dihedral energy term, each set of four atoms that define an improper angle, the hydrogen bond donors, and the hydrogen bond acceptors. Splitting the files that describe a protein like this is logical, as the information in the principal structure file of a protein almost never changes, whereas that in the coordinate file changes with many operations and during a dynamics simulation.

In addition to the forces that are generated by the bonds between atoms, electrostatic and van der Waals forces also act on atoms. In principle, every atom in a system interacts with every other atom through these two types of forces. Because calculating every one of these forces would be far too time consuming and because these forces become negligible at longer separation distances, atom pairs are considered for calculation of these forces only if their separation is less than a user-set cutoff distance. Molecular dynamics programs maintain and constantly update this nonbonded list and then calculate the interactions between only those pairs on the list.

CHARMM is run by commands that are entered at a command prompt. For a reasonably complex calculation, the number of commands required is too great for practical keyboard entry. Instead, commands are written into command files called scripts. These are directed to the program by the Linux or Unix redirect command <. Similarly, CHARMM produces voluminous output, and usually this is directed into another file that is examined after the run with a text editor. Hence, CHARMM is typically invoked at the command prompt as charmm <script.inp>script.out. Data are often extracted from the output files with the Linux and Unix commands grep and awk. The resulting output can then be imported into spreadsheet programs and analyzed. Files containing coordinates and coordinate trajectories that are produced by CHARMM often are also used as input to molecular display programs.

Example Analysis

To communicate some of the flavor of using CHARMM, the steps necessary to run a dynamics simulation of AraC protein, a transcriptional regulator of the L-arabinose operon in *Escherichia coli,* are provided and described. Suppose we have just obtained the structure of its arabinose-binding domain in the presence of arabinose and this is available in a file in PDB format. Examination of the protein with a molecular display program such as Rasmol or VMD shows that arabinose is bound inside the protein and that the N-terminal 15 amino acids of the protein close over the arabinose. It seems of interest to learn the strength of interaction of each residue with arabinose to compare with those of mutated sites leaving the protein unable to respond to arabinose, but still capable of folding. Because protein structures fluctuate and the energetics of a static structure *in vacuo* might be highly deceiving, it seems best to determine the average strength of interaction of each residue with the arabinose at 300 K while the protein is surrounded with water. The following sections show the steps required to perform such a calculation.

Preparation of Input Coordinate Files

CHARMM usually cannot directly use PDB files containing atomic coordinates, and several small modifications are necessary. These can be made with an editor, or with a word processor, and saved in ASCII format. When many files must be modified, it is convenient to write a file that directs the actions of the Linux line-editing program called awk to make the modifications. A word of warning concerning the transfer of files between Windows and Linux–Unix: on occasion a problem is caused by the fact that Windows files end lines with two characters, a carriage return, and a new line character, whereas Linux and Unix files end lines with just the new line character. Sometimes a file that originated on a Windows system and has been transferred to a Linux system looks perfectly good, but the nonprinting and nonapparent carriage return characters present at the ends of lines cause problems for CHARMM. These characters can be removed from a file named *dirty* and written to a file named *clean* with the Linux command translate, tr -d '\r' <dirty> clean. The option -d instructs translate to delete, and '\r' is the trouble-making return character that is to be deleted.

We begin with the PDB file for AraC, named 2ARC.[8] This contains the coordinates of both subunits of the dimeric protein, two molecules of arabinose, and tightly bound water molecules. Although coordinates are given for residues 7 through 167 of one subunit and through residue 170 of the other subunit, this example uses only residues 7 through 120, as the remainder of the residues form a subdomain that is well separated from arabinose. It proves simplest to split the input PDB file into separate files,

---

[8] S. M. Soisson, B. MacDougall-Shackleton, R. Schleif, and C. Wolberger, *Science* **276**, 421 (1997).

one for each segment of structure. Thus, in this case, one file contains the coordinates of residues 7 through 120, the second contains the coordinates of the first arabinose molecule, and the third contains the coordinates of the oxygen atoms of the crystallographic water molecules. Although much useful information is contained in the header lines, and they should be read, they need not be included with the separated files.

The header lines of 2ARC mention that alternative conformations are present for a number of the residues. The alternatives must be removed as CHARMM requires only a single coordinate value for each atom. Thus lines such as

```
ATOM  149 N AGLU A 27    14.382 47.461 27.991 0.63 14.88      N
ATOM  150 N BGLU A 27    14.433 47.417 27.971 0.39 14.03      N
```

must be replaced with single lines:

```
ATOM  149 N GLU   A 27    14.382 47.461 27.991 1.00 14.88      N
```

All residue names HIS must be changed to HSD. Even though the charge state of some residues may in reality flicker, CHARMM approximates each residue with a fixed charge. Thus, the program needs to be told which charge state the user is specifying. In the case of histidine, using the residue name HSD instead of HIS indicates that the uncharged state of histidine is to be used, and in this case all the histidine residues will be simulated as being uncharged. Perhaps for historical reasons, CHARMM does not work with the IUPAC base designation of ILE atom CD1, and these entries need to be changed to CD before reading in the coordinates. Thus, the line

```
ATOM  324 CD1 ILE A 46    10.698 35.637 40.610 1.00 12.62      C
```

becomes

```
ATOM  324 CD  ILE A 46    10.698 35.637 40.610 1.00 12.62      C
```

The C-terminal oxygen atoms need to be renamed OT1 and OT2 rather than O and OXT, as they may be in PDB files. In our case, the oxygen atom of PHE 120 becomes one of the C-terminal oxygen atoms, and will automatically be replaced with two C-terminal oxygen atoms that will be named OT1 and OT2.

In the PDB file, the segment of protein ends with the line

```
TER     1346    ILE A 167
```

and at the end of the file is a line containing the word *end*. Thus, the end of the file should be changed to the following:

```
TER     1346    PHE A 120
END
```

The atom numbers in column 2 need not be adjusted as they are ignored when CHARMM reads the input files. Two additional changes must be made to most of the lines. In column 23 of the input file is the letter A for the A subunit. This must be replaced with a blank. In this PDB file, column 73 contains symbols for the atom that is described on that line. Other PDB files contain data bank identification labels such as 2ARC in columns 73–76. CHARMM requires one label in this position, a label that must be provided later in scripts. In this example the label PROT is being used.

```
ATOM  324 CD ILE  46    10.698 35.637 40.610 1.00 12.62    PROT
```

Which of the two arabinose molecules is the one that is bound to the A subunit can be determined by close examination of the coordinate files or by the use of a molecular display program. In this file, the term HETATM must be replaced with ATOM, and the segment identifier in columns 73–76 must be provided. This example uses LIGA. Thus, the final lines of the PDB file for arabinose become the following:

```
ATOM  2705 O5 ARA 100   8.802 37.405 36.359 1.00 13.87    LIGA
END
```

Changes also are made in the file containing the bound water molecules. HETATM must be changed to ATOM, columns 14–16 are changed to OH2, columns 18–21 are changed from HOH to TIP3, and SOLV must be written in columns 73–76. TIP3 specifies the highly refined model for the water molecule that is used in CHARMM simulations.[7] The topology and parameter tables contain special entries for it.

Handling Structure of Arabinose

Programs that calculate interaction energies need to know the identity and properties of every atom in a structures. As mentioned earlier, in CHARMM information about the identities of atoms contained in a residue, their partial charges, the bonding pattern of the atoms, and their spatial positions with respect to one another are all obtained from the residue topology file. CHARMM will be unable to read the PDB file of arabinose described above because its residue topology file contains no mention of arabinose. Therefore it is necessary to add an arabinose entry to the residue topology file. The *de novo* construction of a complete topology file entry for a small molecule can require substantial effort and is not described here. In the present case, however, the construction of a workable topology file entry is not difficult. Examination of the PDB file for arabinose shows that it contains coordinates for all the heavy atoms. Hence, CHARMM will not have to construct the locations of any missing atoms through the use of the internal coordinates that may be provided in a top-ology file entry. Therefore, the internal coordinates can be omitted. Exam-ination of the arabinose molecule with a graphics display program shows that the sugar in the protein is in a α-pyranose ring. Glucose also forms a pyranose ring and, therefore, much of a topology file entry for glucose can be used in writing a new entry for arabinose (Fig. 1). In practice, the entries for many carbohydrates will differ from one another only in internal coordinates of one or more oxygen and hydrogen atoms. Their bonding patterns, atom types, partial charges, and hydrogen bond donors and acceptors will all be the same.

It proves easiest to model the new arabinose topology file entry after that given for glucose given in a separate topology file, top_all22_sugar.inp, and then to place this in the top_all27_prot_na.rtf file because this file, and its partner parameter file, par_all27_prot_na_prm, can be used for calcula-tions and simulations involving proteins, nucleic acids, and combinations of the two. Examination of the sugar file and Fig. 1 shows that the structure given for glucose becomes that of arabinose by removing all references to H61, H62, O6, and HO6 changing C6 to H51 and H5 to H52, and adjusting the stereochemistry at C4. The CHARMM type and partial charge of each atom in arabinose must also be specified. The types and partial charges are developed together, along with the parameter table. As the arabinose entry is to be added to the top_all27_prot_na.rtf file, this file should be used for examples of atom types and partial charges. The file contains entries for ribose in ribonucleotides, and thus the types and partial charges for the atoms in arabinose can be determined by example.

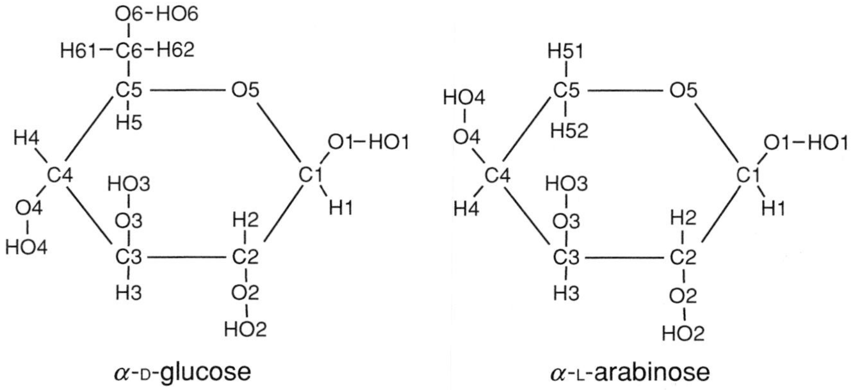

FIG. 1. Structures and Protein Data Bank identification labels of the atoms of glucose and arabinose.

Energy Determination Script for CHARMM

The script for determination of the residue–arabinose interaction energies is described below. Scripts and output files begin with header lines that can provide useful archival information about the functioning of the script. These title lines begin with an asterisk and the line following the last title line contains only an asterisk. Thus, the title lines for the energy determination script would be

```
*Example script for energy calculation
*
```

The first step in the determination of the interaction energies is the reading of the modified topology and parameter files. A file that is to be opened is named, and a label in the form of a unit number is provided so that the program can later refer to this file. The specification of card in the open statement indicates that the data are in ASCII format. This is probably a holdover from the days of punched cards. In the script, the topology and parameter files are opened and read with the following commands:

```
open read card unit 20 name top_all27_prot_na.rtf
read rtf card unit 20
close unit 20
open read card unit 22 name par_all27_prot_na.prm
read parameter card unit 22
close unit 22
```

Next, the protein to be analyzed is read in. This is done in two steps. First, the sequence of the protein is read. From the amino acid sequences, and by reference to the residue topology file, CHARMM generates arrays with a slot for each atom in the protein. These arrays include the x, y, and z coordinates of each atom; a duplicate coordinate array; an array for the internal coordinates; also arrays for the x, y, and z components of the forces on each atom, frictional forces that may be assigned to each atom, whether or not an atom is to be constrained; as well as additional arrays that can be used in computation. The generate command instructs CHARMM to set up these arrays, and the setup option instructs the program to place in the internal coordinate array of the protein the appropriate internal coordinate entries from the various residues in the residue topology file. The generate command must include a segment identifier for the molecule being read. In this case, the identifier is PROT. The relevant commands are

```
open read unit 23 card name protein.pdb
read sequence pdb unit 23
generate prot setup
rewind unit 23
```

Because the N-terminal and C-terminal amino acids of a protein possess structures different from those of the internal amino acids, special treatment is automatically given to the first and last residues of a structure. Water, of course, does not require such treatment as is reflected in the first none and last none options in the generate statement. Also, the TIP3 water molecules, which are specially devised approximations to water, are not allowed to bend or twist, hence the noangle and nodihedral options. Sequences of the arabinose and crystallographic water molecules are read in and the necessary arrays are generated as follows:

```
open read unit 24 card name arabinose.pdb
read sequence pdb unit 24
generate liga setup
rewind unit 24
open read unit 25 card name xwater.pdb
read sequence pdb unit 25
generate solv setup first none last none noangle nodihedral
rewind unit 25
```

Having generated the necessary slots, the coordinates can be read in:

```
read coordinate pdb unit 23
close unit 23
read coordinate pdb append unit 24
close unit 24
read coordinate pdb append unit 25
close unit 25
```

The side chains of some hydrophilic residues on the surface of a protein are likely to extend into the solvent and not to have fixed positions. As a result, such side chains often are not seen in structures determined by X-ray diffraction, and their coordinates will therefore be missing from the PDB files. Provision is provided in CHARMM for the generation of such missing coordinates. This same capability is also used when the program is used to construct structures *de novo*. The first of the operations necessary in the construction of the coordinates of missing atoms is to issue the ic fill preserve command. This instructs CHARMM to generate values for internal coordinates of all atoms for which Cartesian coordinates were present in the PDB file and that are now in the coordinate array. All such new internal coordinate values are to replace internal coordinate values already in the table while preserving the remainder of the internal coordinate values. At this point, then, the internal coordinate table consists mostly of entries that have been calculated from the actual coordinates of the protein and a few entries for missing atoms whose entries derive from the residue topology file. In some cases the internal coordinate entries in the residue topology file may not have been fully specified and, therefore, at this point the internal coordinate table may still be incomplete. Therefore the ic parameter command instructs CHARMM to use the parameter table as a source of last resort and to use it for rather generic values for the internal coordinate table. Finally, when the internal coordinate table has been filled, the ic build command instructs CHARMM to reconstruct the Cartesian coordinates of all the atoms. If coordinates for an atom exist in the coordinate arrays, these values are retained, but the coordinates of all other atoms are calculated from the internal coordinate values. The final step in constructing the entire protein is the addition of any hydrogen atoms that have not been placed already. The commands used in the reconstruction process are as follows:

```
ic fill preserve
ic parameters
ic build
hbuild
```

In our example, only a part of one of the two subunits of AraC is being simulated, but crystallographic water molecules were associated with both subunits. Therefore those water molecules that were associated with the parts of the protein that are not being simulated must be deleted. This is accomplished with the delete command, by which entire water molecules are removed if their oxygen atoms are not within 8 Å of the segment we are studying. The hyphen is a line continuation symbol and is used for long lines.

```
delete atom select ( .byres. ( (.not. (segid prot .around. 8 ) )
  . and. - ( segid solv .and. type OH2 ) ) ) end
```

After input of the protein, arabinose, and crystallographic water molecules and removal of water molecules not associated with the protein fragment, the complex is to be centered at the origin. This is accomplished with the coordinate orient command:

```
coordinate orient select segid prot .or. segid liga .or.
  segid solv end
```

To immerse the protein in a shell of water, a box of water molecules centered at the origin is read and followed by deletion of water molecules that either overlap the complex or that lie more than 8 Å from the protein segment. The water box was prepared separately and was stored in the CHARMM coordinate format called crd. Either the Protein Data Bank format, PDB, or the CHARMM format, CRD, can be used for most purposes, although the specification of specific atoms for use in calculations is sometimes easier if the crd format of the coordinates is available. The box is read in, the deletion operations are performed, and the designations of the two types of water, solv for the water molecules seen in the X-ray studies, and boxx for the water molecules of the box, are also coalesced for convenience as follows:

```
open read unit 26 form name box.crd
read sequence unit 26
generate boxx setup first none last none noangle nodihedral
rewind unit 26
read coordinate card append unit 26
close unit 26
delete atom select ( .byres. ( ( ( segid prot .around. 2.5 ) )
  .and. - (segid boxx .and. type OH2 ) ) ) end
delete atom select ( .byres. ( ( .not. (segid prot .around.
  8 ) ) .and. - (segid boxx .and. type OH2 ) ) ) end
join solv boxx renumber
```

At this point the psf file can be written out for later use. The commands for doing so are as follows:

```
open write unit 41 card name protwat.psf
write psf unit 41 card
*psf of protein, arabinose, crystallographic waters, and
  water box
*
```

The above-described manipulations likely have placed some atoms very near one another or abnormally stretched some bonds. Dynamics simulations cannot be done with such a system because on starting a simulation some atoms will rapidly acquire impossibly high velocities. Therefore, an energy minimization is required to reduce the net force on each atom to nearly zero. In a minimization, the coordinates of all atoms in the system will be allowed to be move. Here 1000 steps of minimization are used and the results are printed out every 100 steps. For a system containing 10,000 atoms, this operation takes about 1 h on a 1-GHz machine.

```
minimize abnr nstep 1000 nprint 100
```

The molecular dynamics can now be run. Because they are so light, the hydrogen atoms vibrate at high frequencies. As a result, small time steps are required. Longer time steps can be used if the shake command is issued first. It alters the way hydrogen atoms are considered. A restart file is also opened. This contains the atom positions and velocities and allows a dynamics run to be restarted and continued from the time point at which the file was written. A file for recording the dynamics trajectory is also opened. In this phase of the simulation, the system will be heated from near by 0 K to about 300 K. The heating is specified to take place in 6000 time steps of 0.001×10^{-12} s, coordinates are to be saved every 100 steps, the status of the system is to be recorded every 100 steps, and atom velocities are to be increased each 100 steps so as to increase the temperature of the system by 5 K. A 1-GHz machine can process about 4000 time steps per hour of a 10,000-atom system.

```
shake bonh tolr 1e-09 mxit 500 param
open write unit 31 card name heat.rst
open write unit 32 file name heat.dcd
dynamics leap verlet start nstep 6000 timestep 0.001 nprint
  100 -
  nsavc 100 nsavv 0 inbfrq -1 iprfrq 100 ihtfrq 100 -
  iunread -1 iunwri 31 iuncrd 32 iunvel -1 kunit -1 -
  wmin 1.5 first t 0.00000 finalt 300 teminc 5 -
  iasors 1 iasvel 1 iscvel 0 ichecw 0
```

After heating, the dynamics simulation is restarted from the heating re-start file. The system is allowed to equilibrate for 10,000 steps. Periodically during this time, velocities will be adjusted so as to bring the system temperature back to 300 K if it has drifted. The necessary commands for doing this are as follows:

```
open read unit 30 card name heat.rst
open write unit 31 card name equil.rst
open write unit 32 file name equil.dcd
dynamics leap verlet restart -
  nstep 6000 time .001 nprint 100 nsavc 100 -
  nsavv 0 inbfrq -1 iprfrl 100 ihtfrq 0 -
  ieqfrq 100 ntrfrq 100 -
  iunrea 30 iunwri 31 iuncrd 32 iunvel -1 kunit -1 -
  wmin 15 finalt 300.0 -
  iasors 1 iasvel 1 iscvel 0 ichecw 0
```

Once the system has been equilibrated, the simulation is restarted from the equilibration restart file and allowed to run free. The following commands specify this as well as the storing of coordinates after the run.

```
open read unit 30 card name equil.rst
open write unit 31 card name dyn.rst
open write unit 32 file name dyn.dcd
dynamics leap verlet restart -
```

```
    nstep 100000 time 0.001 nprint 1000 nsavc 1000 -
    nsavv 0 inbfrq -1 iprfrq 1000 ihtfrq 0 -
    ieqfrq 0 ntrfrq 0 -
    iunread 30 iunwri 31 iuncrd 32 iunvel 01 kunit -1 -
    wmin 1.5 final 300 iasors 1 iasvel 1 iscve10 ichecw 0
 open write unit 41 card name s.crd
 write coordinate unit 41 card
 *Coordinates after simulation dynamics
 *
```

Ordinarily the dcd file from the simulation run is analyzed later. In this example, however, the same script both generates the trajectory file and then analyzes it. The analysis portion begins by opening a file for the output of the energy data. Then the trajectory file is opened for reading. The program must be told at which time step to begin reading data and the number of time steps between data points to be read. This is specified as follows:

```
    open write unit 51 card name energy.tst
    open read unit 52 file name dyn.dcd
    trajectory iread 52 begin 13000 skip 1000
```

Two loops will be used to calculate the energies. The outer loop controls the reading of the trajectory frames, uses the loop variable i, and the label *outer*. The frames will be read one by one to the end of the dcd file. The inner loop calculates the interaction energy between successive residues of the protein and the arabinose and uses the loop variable j and the label *inner*. Successive energy calculations will write data to the energy.tst file. Each time data are to be written, the data will be preceded by a title line that includes the current values of j and of i. This structure allows simple extraction of the energy for any particular residue. The interact command specifies the calculation of an interaction energy, and the select portion specifies that the calculation be of the interaction between residue j and the ligand.

```
 set i 1
 label outer
 trajectory read
 set j 1
 label inner
 write title unit 51
 *Interaction energy of residue @j with arabinose frame @i
```

```
*
interact select segid prot .and. resid @j end select —
  segid liga end unit 51
increment j by 1
if j lt 113 goto inner
increment i by 1
if i lt 101 goto outer
stop
```

A small portion of the output file, energy.tst is as follows:

```
INTERACTION ENERGY OF RESIDUE 1 WITH ARABINOSE FRAME 1
INTE ENR: Eval#    ENERgy   Delta-E    GRMS
INTE EXTERN:       VDWaals      ELEC  HBONds     ASP     USER
------------- --------- -------- ------- ------- -------
INTE>        1  -0.68143  0.00000 0.02584
INTE
 EXTERN>        -0.07480 -0.60663 0.00000 0.00000 0.00000
------------- --------- -------- ------- ------- -------
INTERACTION ENERGY OF RESIDUE 2 WITH ARABINOSE FRAME 1
INTE ENR: Eval# ENERgy Delta-E GRMS
INTE EXTERN:       VDWaals      ELEC  HBONds     ASP     USER
------------- --------- -------- ------- ------- -------
INTE>        2  -3.98527  3.30384 0.10683
INTE
 EXTERN>        -0.32970 -3.65557 0.00000 0.00000 0.00000
------------- --------- -------- ------- ------- -------
```

The following will extract the interaction energies between residue 1 and arabinose and list them in a file it creates called energy1.out. The grep command finds lines in energy.tst containing "RESIDUE 1 WITH" and pipes the line that matches and the next four lines to the awk command that finds the lines containing INTE> and outputs the third column to the file energy1.out.

```
grep —A 4 ''RESIDUE 1 WITH'' energy.tst | awk '/INTE>/ {print
 $3}' >energy1.out
```

The energy1.out file may then easily be imported into a spreadsheet program for analysis or plotting.

Acknowledgments

The author thanks Lenny Brand, Ken Johnson, George Rose, and Michael Rodgers for comments, guidance, and suggestions on the preparation of this chapter, and acknowledges NIH Grant GM18277 for support.

[3]   *Ab Initio* Protein Folding Using LINUS

By Rajgopal Srinivasan, Patrick J. Fleming, and George D. Rose

Introduction

Prediction of the native conformation of a globular protein from its amino acid sequence is one of the more important problems in biology.[1] Structure prediction based on comparison with sequences of known structure has enjoyed much success,[2] but *ab initio* methods based solely on physical principles have not fared as well.

LINUS[3–5] is an *ab initio* method for simulating the folding of a protein on the basis of simple physical principles. The approach emphasizes the organizing role of steric exclusion and conformational entropy in guiding folding. Although many have recognized the importance of steric interactions,[6–8] their full impact on chain organization has not been realized until more recently.[9,10] It is now apparent that unfolded proteins are not random coils[9,11,12] and may, in fact, contain a substantial polyproline II population.[10,13–15]

In addition to steric interactions, which are repulsive, LINUS also includes attractive forces resulting from hydrogen bonding and hydrophobic burial. Although LINUS does utilize a few familiar structural motifs—α helix, β strand, β turn—the procedure does not rely on quantitative surface/interior tendencies, and so on.

A primary goal in creating LINUS was to devise a facile system for conducting "thought experiments" about protein folding, using simple,

[1] C. B. Anfinsen, *Science* **181,** 223 (1973).

[2] J. Moult, T. Hubbard, K. Fidelis, and J. T. Pedersen, *Proteins* **3**(Suppl.), 2 (1999).

[3] R. Srinivasan and G. D. Rose, *Proteins Struct. Funct. Genet.* **22,** 81 (1995).

[4] R. Srinivasan and G. D. Rose, *Proc. Natl. Acad. Sci. USA* **96,** 14258 (1999).

[5] R. Srinivasan and G. D. Rose, *Proteins Struct. Funct. Genet.* **47,** 489 (2002).

[6] S. J. Leach, G. Nemethy, and H. A. Scheraga, *Biopolymers* **4,** 369 (1966).

[7] G. N. Ramachandran and V. Sasisekharan, *Adv. Protein Chem.* **23,** 283 (1968).

[8] F. M. Richards, *Annu. Rev. Biophys. Bioeng.* **6,** 151 (1977).

[9] R. V. Pappu, R. Srinivasan, and G. D. Rose, *Proc. Natl. Acad. Sci. USA* **97,** 12565 (2000).

[10] R. Pappu and G. D. Rose, *Protein Sci.* **11,** 2437 (2002).

[11] D. Shortle and M. Ackerman, *Science* **293,** 487 (2001).

[12] T. R. Sosnick, personal communication (2003).

[13] M. L. Tiffany and S. Krimm, *Biopolymers* **6,** 1379 (1968).

[14] T. P. Creamer and M. N. Campbell, *Adv. Protein Chem.* **62,** 263 (2002).

[15] Z. Shi, C. A. Olson, G. D. Rose, R. L. Baldwin, and N. R. Kallenbach, *Proc. Nat. Acad. Sci. USA* **99,** 9190 (2002).

Copyright 2004, Elsevier Inc.
All rights reserved.
0076-6879/04 $35.00

qualitative simulations. The method seeks to capture conformational biases rather than detailed energetics. LINUS is implemented in Python (http://www.python.org/) and runs in all common operating systems.

Anatomy of LINUS Simulation

LINUS is a Monte Carlo[16] procedure for exploring the conformational space of a polypeptide chain and extracting both conformational biases and structure. A typical LINUS simulation begins with a sequence of interest modeled as an extended polypeptide chain ($\phi = -120°$, $\psi = 120°$), with all nonhydrogen atoms included. New conformations are generated by choosing a three-residue segment at random and modifying its torsion angles, both backbone and side chain. Every new conformation is evaluated according to the Metropolis criterion,[17] using a simple scoring function to calculate the energy. A *cycle* is defined as N Monte Carlo steps, where N is the total number of residues in the sequence. Some secondary structure becomes apparent within a few thousand cycles, and additional structure emerges as the procedure progresses. These components are now described in detail.

Metropolis Criterion

In LINUS, each new chain conformation is evaluated according to the Metropolis criterion[17] as shown in Scheme 1. Specifically, a new conformation is generated with a "smart" move set (described below). This trial conformation is rejected if there are steric violations (infinite energy). If not, the energy of the new conformation, E_{trial}, is compared with the energy of the initial conformation, $E_{initial}$, and the new conformation is accepted if $E_{trial} < E_{initial}$. Otherwise, the new conformation is accepted if the Boltzmann-weighted difference in energies, $e^{-\beta\Delta E}$, is greater than a uniformly distributed random number between 0 and 1. Whenever the new conformation is not accepted, the original conformation is retained. β is usually set to 2 but may be changed to explore temperature effects.

The smart move set in LINUS generates new conformations by perturbing three consecutive residues at a time. With a three-residue move, a turn of helix or a β turn can be nucleated in a single Monte Carlo step. The move set is composed of five commonly occurring structural motifs: α helix, β strand, β turn, polyproline II (P_{II}), and sterically allowed coil.

[16] K. Binder and D. W. Heerman, *in* "Monte Carlo Simulation in Statistical Physics: An Introduction." Springer Series in Solid-State Sciences, Springer-Verlag, New York, 1997.
[17] N. Metropolis, A. W. Rosenbluth, M. N. Rosenbluth, A. H. Teller, and E. Teller, *J. Chem. Phys.* **21**, 1087 (1953).

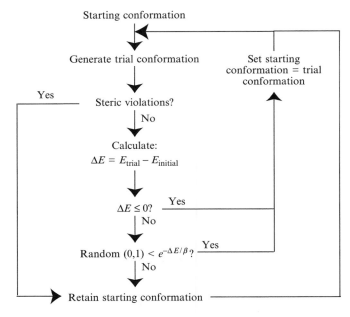

Starting conformation

Generate trial conformation

Set starting
conformation = trial
conformation

Yes

Steric violations?

No

Calculate:
$\Delta E = E_{\text{trial}} - E_{\text{initial}}$

$\Delta E \leq 0$?   Yes

No

Random $(0,1) < e^{-\Delta E / \beta}$?   Yes

No

Retain starting conformation

SCHEME 1

Initially, all five move types are equiprobable ($p = 0.2$), but after a short, exploratory round of simulation, LINUS-based conformational biases are computed for each residue. These biases are reckoned as the fraction of the ensemble that residues populate in each structure: helix, strand, turn, P_{II}, or coil. This fraction is given by the persistence ratios for each move type at every residue position. For example, if residues $(i - 1)$, i, and $(i + 1)$ are helical in 60% of the ensemble during this exploratory round, then the probability of choosing a helical move for residue i is modified from 0.2 to 0.6 in subsequent rounds. Using this strategy, segmental structural propensities emerge as the simulation progresses. A significant finding of LINUS is that large fractions of a protein chain exhibit a bias toward their native conformations, purely on the basis of sequentially local interactions.[4]

Chain Representation

The protein chain is represented solely by heavy atoms—no hydrogen atoms (either polar or nonpolar) are included. In addition, a pseudo-atom, situated at the centroid of each six-membered ring, is appended to the aromatic residues phenylalanine, tyrosine, and tryptophan. This pseudo-atom is used when calculating the contact energy but ignored when determining steric clashes. Bond lengths and bond angles are held fixed at ideal values,[18] with the exception of the τ angle ($N-C_\alpha-C$), which is allowed to vary in the range of 108 to 114°. Atom radii are as follows: carbon with double bond character, 1.50 Å; all other carbons, 1.65 Å; nitrogen, 1.35 Å; oxygen, 1.35 Å; cysteinyl sulfur, 1.85 Å; methionyl sulfur, 1.80 Å; and aromatic pseudo-atom, 0.0 Å.

Move Set

LINUS perturbs three consecutive residues at a time to generate a new conformation. Five move types constitute the move set: α helix, β strand, β turn, P_{II}, and coil. Specifically, the five move types are as follows.

1. α Helix: All three residues are assigned backbone torsion angles within a ϕ, ψ bin centered on $\phi = -64°$, $\psi = -43°$ (labeled H in Fig. 1).

2. β Strand: All three residues are assigned backbone torsion angles within a ϕ, ψ bin centered on $\phi = -135°$, $\psi = 145°$ (labeled S in Fig. 1).

3. β Turn: Either the first two residues or the last two residues of the three-residue segment are assigned to one of six β-turn conformations, and the remaining residue is assigned a conformation from the sterically allowed region of ϕ, ψ space (i.e., coil). The backbone torsion angles for the two positions in turn conformations are chosen from within the following ϕ, ψ bin combinations, where the different bins are shown in Fig. 1:

Type I turn = choice {TO, TP}
Type I′ turn = choice {To, Tp; Tp, Tp}
Type II turn = choice {TR, Tp; TR, To}
Type II′ turn = choice {Tr, TP; Tr, TJ}

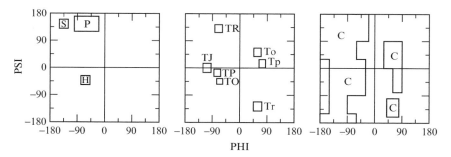

FIG. 1. LINUS move types. *Left:* ϕ, ψ bins for helix (H), strand (S), and P_{II} (P) moves. *Middle:* ϕ, ψ bins used to make the different types of turn moves. *Right:* Allowed ϕ, ψ areas (C) for nonproline, nonglycine residues.

[18] R. A. Engh and R. Huber, *Acta Crystallogr.* **47**, 392 (1991).

Type III turn = TP, TP

Type III' turn = To, To

4. P_{II}: All three residues are assigned backbone torsion values from within a ϕ, ψ bin centered on $\phi = -60°$, $\psi = 145°$ (labeled P in Fig. 1).

5. Coil: For nonglycine, nonproline residues each of the three residues is assigned backbone torsion angles with values from the allowed region of ϕ, ψ space (labeled C in Fig. 1) chosen at random.[19]

The allowed region of ϕ, ψ space for glycine is approriately expanded and for proline it is limited to two regions centered on helical or extended chain conformation. In addition, the coil move for proline samples *cis*-peptide conformations. For each move, the peptide bond angle ω is set to $180 \pm 5°$ and side-chain χ angles (if any) are chosen at random from a list of sterically acceptable combinations.

LINUS Scoring Function

The LINUS scoring function includes three components: hydrogen bonding, hydrophobic contacts, and a backbone torsion. The hydrogen-bonding and hydrophobic contact scores are calculated between pairs of atoms from residues with a sequence separation between two and N residues, where N is the total number of residues in the polypeptide chain. Sequentially local interactions are considered to be those within a six-residue interval.[20] A typical LINUS simulation proceeds hierarchically, starting with local interactions and progressing incrementally to 18-residue and N-residue intervals, as illustrated in the following examples.

1. Hydrogen-bonding score. In LINUS, hydrogen bond donors are backbone nitrogens, and acceptors are carbonyl oxygens, both backbone and side chain. When identifying hydrogen bonds,[21] both distance and orientation criteria are applied. As diagrammed in Fig. 2, a nitrogen and an oxygen are considered to be hydrogen bonded when $3.5 \text{ Å} \leq$ N–O distance

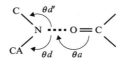

Fig. 2. Hydrogen bond criteria. Both the N–O distance and angles shown must satisfy the criteria discussed in text.

[19] S. C. Lovell, I. W. Davis, W. B. I. Arendall, P. I. W. Bakker, J. M. Word, M. G. Prisant, J. S. Richardson, and D. C. Richardson, *Proteins Struct. Funct. Genet.* **50,** 437 (2003).

[20] T. P. Creamer, R. Srinivasan, and G. D. Rose, *Biochemistry* **36,** 2832 (1997).

[21] D. F. Stickle, L. G. Presta, K. A. Dill, and G. D. Rose, *J. Mol. Biol.* **226,** 1143 (1992).

≤ 5.0 Å; the angle θa is $180 \pm 40°$; and the angles $\theta d'$ and θd are greater than $90°$. The score scales linearly with distance, reaching a maximal value at 3.5 Å and decreasing monotonically to zero at 5.0 Å. At the user's option, the hydrogen bond score may be limited to backbone–backbone interactions, backbone–side-chain interactions, or both. A maximum score of 0.5 is assigned to backbone–backbone interactions, and a maximum score of 1.0 is assigned to backbone–side-chain interactions. The lower score for the former compensates for the fact that a three-residue move favors hydrogen bond-stabilized α-helical conformers.

2. Hydrophobic contact score. Hydrophobic contacts are rewarded when specific atoms in the side chains of hydrophobic residues are juxtaposed. For nonaromatic residues, these atoms are Ala–C_β; Cys–S_γ; Ile–$C_{\gamma 2}$, $C_{\delta 1}$; Leu–$C_{\delta 1}$, $C_{\delta 2}$; Met–C_ε; each is assigned an effective contact radius of 2 Å. For the aromatic rings in phenylalanine, tryptophan, and tyrosine, the appended pseudo-atom at the ring centroid serves as the contact atom. For two proximate atoms with radii r_1 and r_2, a maximum score of 1.0 is assigned when the atoms are in contact (i.e., distance between the atom centers $\leq r_1 + r_2$), and it decreases monotonically to zero at a separation distance equal to the diameter of a water molecule (i.e., distance between the atom centers $= r_1 + r_2 + 2.8$ Å).

3. Backbone torsion score. Proteins favor conformers on the left side of a Ramachandran plot ($\phi < 0$); α helix, β strand, P_{II}, and most turn moves have negative ϕ values. The most notable exceptions to this trend involve glycine and asparagine residues. Glycines are distributed approximately equally on both sides of the Ramachandran plot, whereas asparagine residues with $\phi > 0$ are found in about 11% of the structural data set.[22] Accordingly, LINUS penalizes moves in which a residue adopts a ϕ value > 0 by assigning a backbone torsion score of -1.0, except in the cases of glycine, which is rewarded with a positive score of 1.0, and asparagine, which is neither rewarded nor penalized.

The total energy of a conformation used in the Metropolis criterion is given by the negative sum of the three preceding scores: hydrogen bonding, hydrophobic contact, and backbone torsion.

Secondary Structure Assignment

Secondary structure is assigned to every accepted conformation in order to maintain a structural propensity list, which is needed to determine each residue's conformational bias (described in Metropolis Criterion, above). Secondary structure assignments are based solely on backbone

[22] S. Hovmoller, T. Zhou, and T. Ohlson, *Acta Crystallogr. D Biol. Crystallogr.* **58**, 768 (2002).

TABLE I
PARTITIONING OF ϕ, ψ SPACE INTO 36 BINS

ψ	ϕ						
	$-180°$	$-120°$	$-60°$	$0°$	$60°$	$120°$	$180°$
$180°$	A	G	M	S	m	g	A
$120°$	F	L	R	X	n	h	F
$60°$	E	K	Q	W	o	i	E
$0°$	D	J	P	V	p	j	D
$-60°$	C	I	O	U	q	k	C
$-120°$	B	H	N	T	r	l	B
$-180°$	A	G	M	S	m	g	A

torsion angles. For this purpose, ϕ, ψ space is partitioned into 36 bins, each represented by a one-letter code as shown in Table I.

We refer to these bins as *mesostates* and to the one-letter codes as *mesostate codes*. To assign secondary structure, ϕ, ψ values for each residue are mapped onto the corresponding mesostate code. In turn, the mesostate codes are mapped into a secondary structure class. The default value for each residue is initialized to coil, with mesostate code $\rightarrow$ secondary structure mappings in precedence order as follows: P_{II} is assigned to any residue in M or R; a turn is assigned to any consecutive pair from the set {OO, OP, OJ, PO, PP, PJ, JO, JP, JJ, Mo, Mp, Mj, Ro, Rp, Rj, oo, op, oj, po, pp, pj, jo, jp, jj, mO, mP, mJ, rO, rP, rJ}; helix is assigned to five or more repetitions of O or P; and strand is assigned to three or more repetitions of L or G.

Implementation

LINUS is implemented in the Python (http://www.python.org/) programming language, with computationally intensive parts also written as an extension module in C. LINUS does not require the C module, but performance is enhanced if it is available. Python is a widely used, object-oriented programming language with a shallow learning curve, and it is portable across all common operating systems. A LINUS simulation is specified by a simple Python language command script. LINUS can be downloaded from http://www.roselab.jhu.edu/. The distribution includes documentation and example scripts. Source code for the extension module in C is also included and may be compiled with a standard C compiler such as gcc (http://gcc.gnu.org/). Several utility programs written in Python are included as well; these are useful for creating initial chain conformations and for viewing and analyzing saved conformations.

A distinct advantage of the current modular implementation of LINUS in Python is that all of the parameters described above can be changed easily in the Python source code. This affords great flexibility when developing code for hypothesis testing.

Simulation Examples

In this section we illustrate the use of LINUS in simulations of peptides and proteins. The minimal input needed to run LINUS is a Python command file and a protein structure in PDB format. Optionally, LINUS input may also include a secondary structure propensity (biases) list or a set of conformational constraints in the form of a mesostate list. The initial protein structure is usually an extended conformation, but any unknotted conformation is valid. A companion utility program (RIBOSOME), included with the LINUS distribution, can be used to generate this initial coordinate file in PDB format, using a residue sequence that is either extracted from PDB header records or ATOM records or from a FASTA sequence file. LINUS command files for each example described below are listed in Appendix I. Many additional examples are included in the distribution.

Simulation of Short Homopolymers

To appreciate the relative importance of the energy terms, consider the following examples that compare the α-helix, β-strand, β-turn, and P_{II} propensities of four nonapeptides—polyalanine, polyvaline, polyleucine, and polylysine. We discuss four simulations.

1. A pure hard sphere simulation
2. A simulation in which only the hydrogen bond energy term is enabled
3. A simulation in which only the hydrophobic contact energy term is enabled
4. A simulation in which both hydrogen bond and hydrophobic contact energy terms are enabled

For the case of polyalanine and polylysine, where by construction no hydrophobic contacts are possible, there are only two simulations (1 and 2) to be performed. The rationale for this choice of four peptides is as follows.

- Polyalanine—examine the behavior of the peptide backbone
- Polylysine—examine the behavior of a peptide with a nonbranched side chain
- Polyvaline—examine the behavior of a β-branched hydrophobic side chain

- Polyleucine—examine the behavior of a non-β-branched hydrophobic side chain

The relevant parts of Python command scripts for the simulations are shown in Appendix I (examples 1a–1d). In all cases, the energy terms are evaluated only for local interactions, that is, atoms in residue i can "feel" the effect only of other atoms situated within a sequential interval of five residues in either chain direction. We refer to this interval as $\Delta = 6$, but note that the window size is actually 11 residues, $i \pm 5$. Residues $i - 1$ and $i + 1$ are excluded from evaluation.

$$\longleftarrow \quad\quad \Delta = 6 \quad\quad \longrightarrow$$
$$..., i - 5, i - 4, i - 3, i - 2, i - 1, \boldsymbol{i}, i + 1, i + 2, i + 3, i + 4, i + 5, ...$$
$$\longleftarrow \quad\quad \text{11 residues} \quad\quad \longrightarrow$$

An interaction interval of $\Delta = 6$ is the LINUS default, but it can be incremented as discussed below. A *cycle* is defined as N Monte Carlo moves, where N is the number of residues in the polypeptide. For every three-residue segment, the procedure seeks to generate a sterically allowed conformation; a maximum of 1000 tries is attempted. Typically, 1000 to 50,000 cycles are performed during a simulation. Conformations are saved periodically for later inspection if desired; a LINUS utility can be used to view the ordered set of saved conformations in RASMOL.[23]

The secondary structure propensities resulting from these simulations are displayed graphically in Figs. 3–6. In each graph, the x axis corresponds to the residue number, and the y axis plots the fractional distribution of residues in each of the five secondary structural states: α helix (red), β strand (green), β turn (blue), P_{II} (black), and coil (cyan).

1. Hard Sphere Simulation. In a hard sphere simulation, the helix content (red curves) is negligible in all cases (Fig. 3). Extended chain conformations (green, black) are preferred. Strand propensity (green) is highest for polyvaline and is approximately equal to P_{II} propensity (black) for polyleucine.

2. Hydrogen Bond Term. On introduction of a hydrogen bond term, the helix content increases dramatically for polyalanine, but not for polyvaline (Fig. 4). Polylysine has an increased helix content as well, but with some remaining strand propensity. Polyleucine exhibits no specific backbone structural propensity.

[23] R. A. Sayles and E. J. Milner-White, *Trends Biochem. Sci.* **220,** 374 (1995).

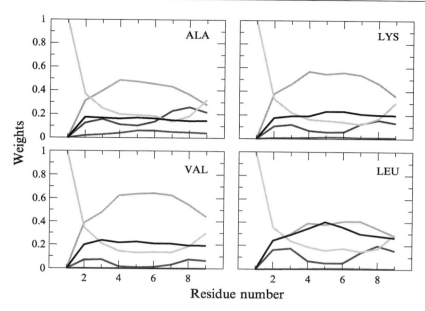

FIG. 3. Plots of fractional distribution (weights) of residues in helix (red), strand (green), turn (blue), P$_{II}$ (black), and coil (cyan) states for polyalanine, polylysine, polyvaline, and polyleucine in a hard sphere simulation. (See color insert.)

3. Contact Energy Term. With only a contact energy term, the helix content of both polyvaline and polyleucine is negligible (Fig. 5). Qualitatively, these plots resemble the hard sphere simulations in Fig. 3.

4. Hydrogen Bond and Contact Energy Terms. In polyvaline, strand still dominates, but in polyleucine, helix content exceeds strand, and turn content also increases significantly (Fig. 6) in comparison with simulations using contact energy alone.

For this set of simulations, extended conformations dominate in the absence of attractive interactions (Fig. 3). The long γ-branched side chain in leucine is restricted by steric clash more often in strand than in P$_{II}$, which is why P$_{II}$ competes well with strand for occupancy. On inclusion of hydrogen bond terms (Fig. 4), a significant helix population is observed for polyalanine and polylysine, with a much smaller population for polyleucine. But, polyvaline still exhibits negligible helix propensity. On inclusion of contact energy terms (Fig. 5), no significant helical population is observed, suggesting that helix formation is not induced by hydrophobic interactions alone. However, when both hydrogen bond and hydrophobic contacts are included (Fig. 6), the helix content of polyleucine increases beyond the level seen with hydrogen bonds alone.

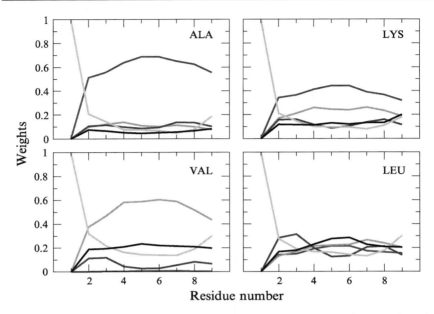

FIG. 4. Plots of fractional distribution (weights) of residues in helix (red), strand (green), turn (blue), P_{II} (black), and coil (cyan) states for polyalanine, polylysine, polyvaline, and polyleucine with hydrogen bond energy calculation enabled. (See color insert.)

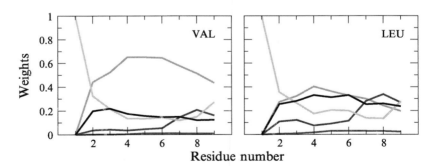

FIG. 5. Plots of fractional distribution (weights) of residues in helix (red), strand (green), turn (blue), P_{II} (black), and coil (cyan) states for polyvaline and polyleucine with hydrophobic contact energy calculation enabled. (See color insert.)

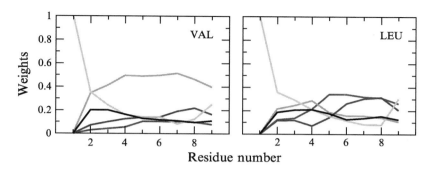

Fig. 6. Plots of fractional distribution of residues in helix (red), strand (green), turn (blue), P_{II} (black), and coil (cyan) states for polyvaline and polyleucine with hydrogen bond energy and hydrophobic contact energy calculations enabled. (See color insert.)

One overall conclusion suggested by these experiments is that isolated hydrophobic interactions are insufficient to induce helix formation, but once a helix is formed, they can stabilize it. Another notable conclusion is that polyvaline has no significant tendency to form helical structures, even when all attractive interactions are included. This latter result underscores the dominant steric effect that β-branched side chains have on the backbone conformation.[24]

An important application of computational tools such as LINUS is to generate and test hypotheses. Our simple simulations illustrate the use of LINUS to perform such computer experiments. An obvious advantage of this approach is that one can selectively toggle interactions on and off at will, an easy task in the computer but an impossibility in the test tube. Furthermore, these peptide simulations take only minutes to perform on a desktop workstation.

Simulation of Small Proteins

We turn now to illustrative simulations of two small proteins, immunoglobulin-binding protein G (PDB code 1 pgb)[25] and the villin headpiece subdomain (PDB code 1 vii). Each has been studied extensively in simulations by others.[26–28] For both proteins, LINUS reveals protein segments with native-structure propensities arising solely from sequentially local

[24] T. P. Creamer and G. D. Rose, *Proc. Natl. Acad. Sci. USA* **89,** 5937 (1992).

[25] H. M. Berman, J. Westbrook, Z. Feng, G. Gilliland, T. N. Bhat, H. Weissig, I. N. Shindyalov, and P. E. Bourne, *Nucleic Acids Res.* **28,** 235 (2000).

[26] J. Shimada and E. I. Shakhnovich, *Proc. Natl. Acad. Sci. USA* **99,** 11175 (2002).

[27] Y. Duan and P. Kollman, *Science* **282,** 740 (1998).

[28] B. Zagrovic, C. D. Snow, M. R. Shirts, and V. S. Pande, *J. Mol. Biol.* **323,** 927 (2002).

interactions. The simulation protocol for each protein starts with local interactions ($\Delta = 6$) only, to detect the locally determined secondary structure propensities for each residue. These propensities, or "weights," are then used to establish biased moves in a second round of simulation. In this second round, the interval of interaction, Δ, is increased in two hierarchic steps. The first step is performed at $\Delta = 18$, an interval that is intended to capture supersecondary structure interactions. The second step is performed at $\Delta = N$, where all possible interactions between residues are included. In this latter step, 5000-cycle simulations are performed 50 times. In each, the last accepted conformation is retained, with initial move weights taken from the simulation at $\Delta = 6$. Conformations are saved every 500 cycles for later inspection.

Protein G. Protein G is a 56-residue α/β structure with two β hairpins separated by a central helix. The hairpins form a four-stranded β sheet that packs against the helix. Figure 7 is a plot of the secondary structure biases for each residue calculated from five simulations at $\Delta = 6$. The secondary structure propensities are plotted beneath a color-coded bar that shows the

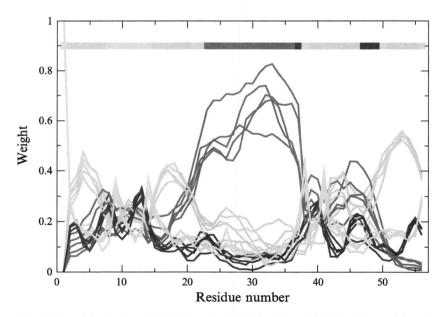

FIG. 7. Plot of the fractional distribution (weights) of residues in helix (red), strand (green), turn (blue), and coil (cyan) states in protein G from five separate simulations with only local interactions enabled. The corresponding secondary structure segments for 1 pgb are indicated by the color-coded bar. (See color insert.)

experimentally determined secondary structure, calculated from ϕ, ψ angles and the mesostate mapping described above. In these simulations, the sequence from residues 23–36 has a clear helical propensity (red), in good agreement with the observed crystal structure. Strand propensities (green) also correlate with the crystal structure for strands 1, 2, and 4, but the bias in strand 3 is less pronounced. Chain turns (blue) are clearly evident between the first two strand segments (residues 8–14), between the helix and strand 3 (residues 37 and 38), and between the last two strands (residues 47–50). In 1 pgb, the turns between the second strand and the helix (residues 21 and 22) and between the last two strands (residues 47–50) have backbone torsion angles that fall within the α-helical ϕ, ψ region; LINUS sometimes finds a segment of five such ϕ, ψ angles, and in these cases the residues are assigned as helices instead of turns.

The population of β-hairpin supersecondary structure is approximately 1% in the saved conformations. Figure 8 shows one such β-hairpin conformer, together with the central α helix that is also present in most conformations. Although the β hairpins do emerge in this simulation, they fail to persist because the torsion angles in subsequent turn moves function like lever arms, resulting in large displacements of one strand relative to the other.

Villin. The villin headpiece subdomain is a small 36-residue protein containing three short helices and a hydrophobic core. The first and second helices are joined by a six-residue loop, and the second and third helices are joined by a three-residue segment, two residues of which are in the P_{II}

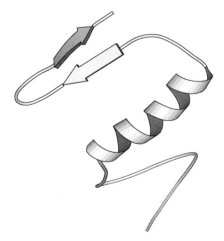

FIG. 8. Example of a β-hairpin supersecondary structure and central α helix occurring in a LINUS simulation of protein G.

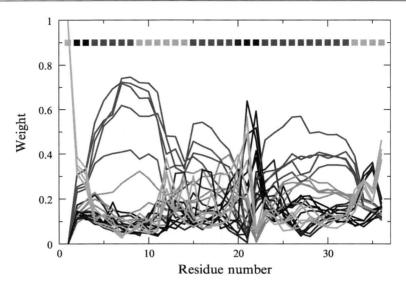

FIG. 9. Plot of the fractional distribution (weights) of residues in helix (red), strand (green), turn (blue), P_{II} (black), and coil (cyan) states in villin headpiece subdomain from five separate simulations with only local interactions enabled. The corresponding secondary structure segments for 1 vii are indicated by the color-coded bar. (See color insert.)

backbone conformation. Figure 9 plots the secondary structure propensities for each residue in this subdomain from simulations confined to local interactions only. The three helices are readily apparent as well as the P_{II} segment in residues 21 and 22. The loop region between the first two helices (residues 8–13) is not well predicted by LINUS, although a significant bias toward coil conformations is found in this segment.

Hydrophobic "collapse" to the overall villin headpiece structure is seen occasionally in the ensemble of saved conformations, but it does not persist for many cycles. Figure 10 shows one such example.

Conclusion

Almost half a century ago, Fermi, Pasta, and Ulam[29] conducted the first dynamics simulation on a computer, a new kind of thought experiment. Unlike an actual experiment, simulations can either include or exclude interactions of interest, and the ensuing result can be assessed. For

[29] E. Fermi, J. Pasta, and S. Ulam, in "Studies of Nonlinear Problems." Los Alamos Report LA-1940, 1955. Reproduced in "Nonlinear Wave Motion" (A. C. Newell, ed.). American Mathematical Society, Providence, RI, 1974.

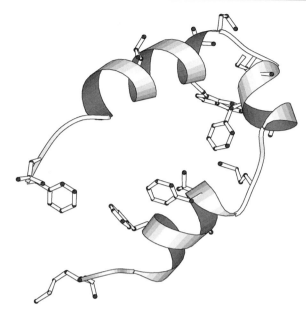

FIG. 10. Example of villin subdomain conformation with hydrophobic core from a LINUS simulation. Hydrophobic side chains are shown extending from the backbone schematic trace.

example, to evaluate the effect of steries and hydrogen bonding on chain organization, a LINUS simulation can be performed with hydrophobic interactions "turned off." The preceding section on simulation of short homopolymers illustrates the power of this approach.

Many investigators are now exploring the kinetics, dynamics, and energetics of small proteins in long simulations, some extending to microsecond time scales.[27, 28, 30–34] In contrast, LINUS was devised to conduct fast simulations that capture qualitative effects, such as segmental biases toward secondary and supersecondary structure. As such, LINUS does not require a detailed forcefield, with its many adjustable parameters, but relies instead on a simple, robust scoring function. Despite this simplicity, the method has proven effective in predicting the overall conformation of large, supersecondary-sized protein fragments.[5]

[30] A. R. Dinner, T. Lazaridis, and M. Karplus, *Proc. Natl. Acad. Sci. USA* **96**, 9068 (1999).
[31] A. E. Garcia and K. Y. Sanbonmatsu, *Proteins* **42**, 345 (2001).
[32] W. F. van Gunsteren, R. Burgi, C. Peter, and X. Daura, *Angw. Chem. Int. Ed.* **40**, 353 (2001).
[33] A. Cavalli, P. Ferrara, and A. Caflisch, *Proteins* **47**, 305 (2002).
[34] C. Clementi, A. E. Garcia, and J. N. Onuchic, *J. Mol. Biol.* **326**, 933 (2003).

In closing, we emphasize that LINUS is not intended to be used as a "black box" for protein prediction. The examples in Appendix I demonstrate the flexibility of LINUS, but the price of flexibility is the need to "learn the system" in order to use it to greatest advantage. We urge users not to apply the protocol in example 2a to their favorite protein blindly. Instead, use LINUS to probe the sequence for conformational biases and likely substructures that provide a basis for thinking about the protein and designing meaningful experiments.

Appendix I

Example 1a Hard sphere simulation of a polypeptide

```
# set the simulation parameters
sim.set_simulation_parameters(numsave=100,numcycle=5000)

# disable hydrogen bonding energy calculation
sim.set_hbond_parameters(use_hbond=0,
                         use_sidechain_hbond=0)

# disable contact energy calculation
sim.set_contact_parameters(use_contact=0)

# enable phi torsion penalty calculation
sim.set_torsion_parameters(use_torsion=1)
```

Example 1b Simulation with hydrogen bond energy calculation enabled

```
# set the simulation parameters
sim.set_simulation_parameters(numsave=100,numcycle=5000)

# enable hydrogen bonding energy calculation
sim.set_hbond_parameters(use_hbond=1,
                         use_sidechain_hbond=1)

# disable contact energy calculation
sim.set_contact_parameters(use_contact=0)

# enable phi torsion penalty calculation
sim.set_torsion_parameters(use_torsion=1)
```

Example 1c Simulation with contact energy calculation enabled

```
# set the simulation parameters
sim.set_simulation_parameters(numsave=100,
                              numcycle=5000)

# disable hydrogen bonding energy calculation
sim.set_hbond_parameters(use_hbond=0,
                         use_sidechain_hbond=0)

# enable contact energy calculation
sim.set_contact_parameters(use_contact=1)

# enable phi torsion penalty calculation
sim.set_torsion_parameters(use_torsion=1)
```

Example 1d Simulation with both hydrogen bond energy and contact energy calculations enabled

```
# set the simulation parameters
sim.set_simulation_parameters (numsave=100,
                                numcycle=5000)

# enable hydrogen bonding energy calculation
sim.set_hbond_parameters (use_hbond=1,
                          use_sidechain_hbond=1)

# enable contact energy calculation
sim.set_contact_parameters (use_contact=1)

# enable phi torsion penalty calculation
sim.set_torsion_parameters (use_torsion=1)
```

Example 2a Simulation of a protein with local interactions only (the LINUS default) to determine secondary structure propensities

```
# set the simulation parameters
sim.set_simulation_parameters (numsave=5000,
                                numcycle=5000)

# enable hydrogen bonding energy calculation
sim.set_hbond_parameters (use_hbond=1,
                          use_sidechain_hbond=1)

# enable contact energy calculation
sim.set_contact_parameters (use_contact=1)

# enable phi torsion penalty calculation
sim.set_torsion_parameters (use_torsion=1)
```

Example 2b Simulation of a protein with increasing interaction Δ, multiple passes, and output of protein conformations

```
# set the initial simulation parameters
sim.set_simulation_parameters (numsave=5000,
                                numcycle=5000)

# enable hydrogen bonding
sim.set_hbond_parameters (use_hbond=1,
                          use_sidechain_hbond=1)

# enable contact scoring
sim.set_contact_parameters (use_contact=1)

# enable phi torsion penalty calculation
sim.set_torsion_parameters (use_torsion=1)

# read the move set propensities from initial simulation
sim.read_weights_from_file ('weights')

# run simulation at interaction delta of 18
sim.set_hbond_parameters (hbond_winmax=18)
sim.set_contact_parameters (contact_winmax=18)
sim.run ()

# run 50 simulations at interaction delta of N, and save
# conformations every 500 cycles
```

```
sim.set_simulation_parameters (numsave=500, trials=50)
NUMRES = sim.protein.num_residues
sim.set_hbond_parameters (hbond_winmax=NUMRES)
sim.set_contact_parameters (contact_winmax=NUMRES)
sim.run ()
```

Acknowledgments

We thank Nicholas Fitzkee, Haipeng Gong, Nicholas Panasik, and Timothy Street for critical reading of the manuscript. Support from the NIH and Mathers Foundation is gratefully acknowledged.

[4] Protein Structure Prediction Using Rosetta

By CAROL A. ROHL, CHARLIE E. M. STRAUSS, KIRA M. S. MISURA, and DAVID BAKER

Introduction

Double-blind assessments of protein structure prediction methods, held biannually in the community-wide critical assessment of structure prediction (CASP) experiments, have documented significant progress in the field of protein structure prediction and have indicated that the Rosetta algorithm is perhaps the most successful current method for *de novo* protein structure prediction.[1-3] In the Rosetta method, short fragments of known proteins are assembled by a Monte Carlo strategy to yield native-like protein conformations. Using only sequence information, successful Rosetta predictions yield models with typical accuracies of 3–6 Å Cα root mean square deviation (RMSD) from the experimentally determined structures for contiguous segments of 60 or more residues. In such low- to moderate-accuracy models of protein structure, the global topology is correctly predicted, the architecture of secondary structure elements is generally correct, and functional residues are frequently clustered to an active site region. Models obtained by *de novo* prediction methods have been demonstrated to have utility for obtaining biological insight, either through

[1] R. Bonneau, J. Tsai, I. Ruczinski, D. Chivian, C. Rohl, C. E. M. Strauss, and D. Baker, *Proteins Struct. Funct. Genet.* **45**(Suppl. 5), 119 (2001).

[2] P. Bradley, D. Chivian, J. Meiler, K. M. S. Misura, C. A. Rohl, W. R. Schief, W. J. Wedemeyer, O. Schueler-Furman, P. Murphy, J. Schonbrun, C. E. M. Strauss, and D. Baker, *Proteins Struct. Funct. Genet.* **53**(Suppl. 6), 457 (2003).

[3] A. M. Lesk, L. Lo Conte, and T. J. Hubbard, *Proteins Struct. Funct. Genet.* **S5**, 98 (2001).

Copyright 2004, Elsevier Inc.
All rights reserved.
0076-6879/04 $35.00

functional site recognition or functional annotation by fold identification.[4,5] The Rosetta method is sufficiently fast to make genome-scale analysis possible: a recent study predicted structures for ≈500 PfamA families with no link to known structure.[6] On the basis of previous performance, one of the five models reported for each Pfam family is expected to be a reasonable match to the true structure for about 50–60% of the families, and many of these predictions suggest a homology unapparent in their sequences

Because of its success in *de novo* structure prediction, the Rosetta method has also been successfully extended to other protein-modeling problems including structure determination using limited experimental constraints,[7,8] *de novo* protein design,[9,10] protein–protein docking,[11] and loop modeling.[12] Structure determination by using Rosetta in combination with limited experimental constraints generally yields structures of higher overall accuracy, often with an RMSD of 2–3 Å over the entire protein. Loop modeling is carried out in the context of a homology-based template that is also frequently only ≈2 Å from the true structure. For design of novel protein structures, sequence selection algorithms require backbone structures of accuracy equivalent to experimentally determined X-ray crystal structures. To address these problems, as well as to refine *de novo* models, improvements to the Rosetta method have focused on increased detail in the potential functions and finer control of chain motion in the search algorithm.

Although *de novo* structure prediction with the Rosetta algorithm has been previously described, here we summarize the current method in its entirety. The benefits and limitations of the fragment assembly strategy utilized by Rosetta are discussed, and we describe adaptations of the Rosetta method for structural modeling with finer resolution. Enhancements to the fragment assembly strategy that allow more local modifications of protein conformation are described, and the effectiveness of

[4] R. Bonneau, J. Tsai, I. Ruczinski, and D. Baker, *J. Struct. Biol.* **13,** 186 (2001).

[5] J. A. Di Gennaro, N. Siew, B. T. Hoffman, L. Zhang, J. Skolnick, L. I. Neilson, and J. S. Fetrow, *J. Struct. Biol.* **134,** 232 (2001).

[6] R. Bonneau, C. E. M. Strauss, C. A. Rohl, D. Chivian, P. Bradley, L. Malmström, T. Robertson, and D. Baker, *J. Mol. Biol.* **322,** 65 (2002).

[7] P. M. Bowers, C. E. M. Strauss, and D. Baker, *J. Biomol. NMR* **18,** 311 (2000).

[8] C. A. Rohl and D. Baker, *J. Am. Chem. Soc.* **124,** 2723 (2002).

[9] B. Kuhlman, J. W. O'Neill, D. E. Kim, K. Y. Zhang, and D. Baker, *J. Mol. Biol.* **315,** 471 (2002).

[10] B. Kuhlman, G. Dantas, G. Ireton, G. Varani, B. Stoddard, and D. Baker, *Science* **302,** 1364 (2003).

[11] J. J. Gray, S. Moughon, C. Wang, O. Schueler-Furman, B. Kuhlman, C. A. Rohl, and D. Baker, *J. Mol. Biol.* **331,** 281 (2003).

[12] C. A. Rohl, C. E. M. Strauss, D. Chivian, and D. Baker, *Proteins Struct. Funct. Genet.* in press (2004).

these operators for energy function minimization is illustrated. In addition, in Appendix I we derive a new, efficient approach to screening local moves; that is, finding short sets of torsional angle changes that permit local changes in a protein chain while collectively minimizing global changes. Our formulation is computationally fast while offering better correlation to global distance changes appropriate to the atomic interaction potentials than previous popular methods (e.g., Gunn[12a]). The method is applicable to both the problem of screening discrete moves as well as allowing gradient descent of continuous multiangle moves.

Rosetta Strategy

A guiding principle of the Rosetta algorithm is to attempt to mimic the interplay of local and global interactions in determining protein structure. The method is based on the experimental observation that local sequence preferences bias but do not uniquely define the local structure of a protein. The final native conformation is obtained when these fluctuating local structures come together to yield a compact conformation with favorable nonlocal interactions, such as buried hydrophobic residues, paired β strands, and specific side-chain interactions. In the Rosetta algorithm, the structures sampled by local sequences are approximated by the distribution of structures seen for those short sequences and related sequences in known protein structures: a library of fragments that represent the range of accessible local structures for all short segments of the protein chain are selected from a database of known protein structures. Compact structures are then assembled by randomly combining these fragments, using a Monte Carlo simulated annealing search. The fitness of individual conformations with respect to nonlocal interactions is evaluated on the basis of a scoring function derived from conformational statistics of known protein structures.

Rosetta utilizes a torsion space representation in which the protein backbone conformation is specified as a list of backbone ϕ, ψ, and ω torsion angles. Conformation modification occurs in torsion space, although for purposes of evaluating the energy of the conformation the corresponding Cartesian space protein representation is generated with atomic coordinates for all heavy atoms in the protein backbone, assuming ideal bond lengths and angles for individual residues.[13] Two alternate representations of side chains are utilized depending on the requirements of the energy function in use (see below). For residue-based potential terms, a reduced

[12a] J. R. Gunn, *J. Chem. Phys.* **106**, 4270 (1997).
[13] R. Engh and R. Huber, *Acta Crystallogr. A* **47**, 392 (1991).

description is used in which each side chain is represented by a centroid located at the side-chain center of mass (Cβ and beyond). For glycine, the centroid is coincident with the Cα atom. Centroid positions for each residue type are determined by averaging over observed side-chain conformations in known protein structures. For increased detail, atomic coordinates for all side-chain atoms, including hydrogens, are utilized. Side chains are restricted to discrete conformations as described by a backbone-dependent rotamer library.[14] Side-chain conformations are added to the backbone structure by means of a Monte Carlo simulated annealing search.[15]

Derivation of the Rosetta scoring function or potential energy surface (PES) is based on a Bayesian separation of the total energy into components that describe the likelihood of a particular structure, independent of sequence, and those that describe the fitness of the sequence given a particular structure.[16,17] The terms in this scoring function in their current form are summarized in Table I. The original Rosetta scoring function uses a fairly coarse-grained or low-resolution description of structure: terms corresponding to solvation and electrostatic effects are based on observed residue distributions in protein structures. Hydrogen bonding is not described explicitly, but probabilistic descriptions of β-strand pairing geometry and β-sheet patterns are included. Steric overlap of backbone atoms and side-chain centroids is penalized, but favorable van der Waals interactions are modeled only by rewarding globally compact structures. The scoring function does not explicitly evaluate local interactions because these interactions are implicitly included in the fragment library (see below).

For applications requiring finer resolution, more detailed descriptions of the determinants of protein structure are needed and have motivated the development of a more physically realistic, atomic-level potential function that attempts to model the primary contributions to stability and structural specificity (Table II). van der Waals interactions are modeled with a 6–12 Lennard–Jones potential, attenuated to a linear function in the repulsive regime to compensate for the discrete rotamer representation of side chains. Solvation effects are included, using the model of Lazaridis and Karplus,[18] and hydrogen bonding is explicitly included, using a secondary structure- and orientation-dependent potential derived from analysis of hydrogen bond geometries in high-resolution protein structures.[19–21]

[14] R. L. Dunbrack and F. E. Cohen, *Protein Sci.* **6,** 1661 (1997).

[15] B. Kuhlman and D. Baker, *Proc. Natl. Acad. Sci. USA* **97,** 10383 (2000).

[16] K. T. Simons, C. Kooperberg, E. Huang, and D. Baker, *J. Mol. Biol.* **268,** 209 (1997).

[17] K. T. Simons, I. Ruczinski, C. Kooperberg, B. Fox, C. Bystroff, and D. Baker, *Proteins Struct. Funct. Genet.* **35,** 82 (1999).

[18] T. Lazaridis and M. Karplus, *Proteins Struct. Funct. Genet.* **35,** 133 (1999).

[19] T. Kortemme, A. V. Morozov, and D. Baker, *J. Mol. Biol.* **326,** 1239 (2003).

TABLE I
COMPONENTS OF ROSETTA ENERGY FUNCTION[a]

Name	Description (putative physical origin)	Functional form	Parameters (values)
env[b]	Residue environment (solvation)	$\sum_i -\ln[P(aa_i|nb_i)]$	i = residue index aa = amino acid type nb = number of neighboring residues[c] (0, 1, 2 ... 30, >30)
pair[b]	Residue pair interactions (electrostatics, disulfides)	$\sum_i \sum_{j>i} -\ln\left[\dfrac{P(aa_i, aa_j|s_{ij}d_{ij})}{P(aa_i|s_{ij}d_{ij})\,P(aa_j|s_{ij}d_{ij})}\right]$	i, j = residue indices aa = amino acid type d = centroid–centroid distance (10–12, 7.5–10, 5–7.5, <5 Å) s = sequence separation (>8 residues)
SS[d]	Strand pairing (hydrogen bonding)	Scheme A : $SS_{\phi,\theta} + SS_{hb} + SS_d$ Scheme B : $SS_{\phi,\theta} + SS_{hb} + SS_{d\sigma}$ where $SS_{\phi,\theta} = \sum_m \sum_{n>m} -\ln[P(\phi_{mn}, \theta_{mn}|d_{mn}, \mathrm{sp}_{mn}, s_{mn})]$ $SS_{hb} = \sum_m \sum_{n>m} -\ln[P(\mathrm{hb}_{mn}|d_{mn}, s_{mn})]$ $SS_d = \sum_m \sum_{n>m} -\ln[P(d_{mn}|s_{mn})]$ $SS_{d\sigma} = \sum_m \sum_{n>m} -\ln[\mathrm{P}(d_{mn}\sigma_{mn}|\rho_m, \rho_n)]$	m, n = strand dimer indices; dimer is two consecutive strand residues V = vector between first N atom and last C atom of dimer m = unit vector between $\hat{V}_m$ and $\hat{V}_n$ midpoints x = unit vector along carbon–oxygen bond of first dimer residue y = unit vector along oxygen–carbon bond of second dimer residue ϕ, θ = polar angles between $\hat{V}_m$ and $\hat{V}_n$ (36° bins) hb = dimer twist, $\sum_{k=m,n}^{k=m,n} 0.5(|\hat{m}\cdot\hat{x}_{kl}| + |\hat{m}\cdot\hat{y}_{kl}|)$ (< 0.33, 0.33–0.66, 0.66–1.0, 1.0–1.33, 1.33–1.6, 1.6–1.8, 1.8–2.0) d = distance between $\hat{V}_m$ and $\hat{V}_n$ midpoints (< 6.5 Å) σ = angle between $\hat{V}_m$ and $\hat{M}$ (18° bins) sp = sequence separation between dimer-containing strands (< 2, 2–10, > 10 residues) s = sequence separation between dimers (>5 or >10) ρ = mean angle between vectors $\hat{m}$, $\hat{x}$ and $\hat{m}$, $\hat{y}$ (180° bins)

sheet[e]	Strand arrangement into sheets	$-\ln\left[P(n_{\text{sheets}}n_{\text{lonestrands}}\mid n_{\text{strands}})\right]$	n_{sheets} = number of sheets $n_{\text{lone strands}}$ = number of unpaired strands n_{strands} = total number of strands
HS	Helix–strand packing	$\sum_m \sum_n -\ln\left[P(\phi_{mn}, \psi_{mn}\mid sp_{mn}d_{mn})\right]$	m = strand dimer index; dimer is two consecutive strand residues n = helix dimer index; dimer is central two residues of four consecutive helical residues $\hat{V}$ = vector between first N atom and last C atom of dimer ϕ, θ = polar angles between $\hat{V}_m$ and $\hat{V}_n$ (36° bins) sp = sequence separation between dimer-containing helix and strand (binned < 2, 2–10, >10 residues) d = distance between $\hat{V}_m$ and $\hat{V}_n$ midpoints (< 12 Å)
rg	Radius of gyration (vdw attraction; solvation)	$\sqrt{\langle d_{ij}^2\rangle}$	i, j = residue indices d = distance between residue centroids
cbeta	Cβ density (solvation; correction for excluded volume effect introduced by simulation)	$\sum_i \sum_{sh} -\ln\left[\dfrac{P_{\text{compact}}(\text{nb}_{i,sh})}{P_{\text{random}}(\text{nb}_{i,sh})}\right]$	i = residue index sh = shell radius (6, 12 Å) nb = number of neighboring residues within shell[f] P_{compact} = probability in compact structures assembled from fragments P_{random} = probability in structures assembled randomly from fragments
vdw[g]	Steric repulsion	$\sum_i \sum_{j>i} \dfrac{\left(r_{ij}^2 - d_{ij}^2\right)^2}{r_{ij}}; \ d_{ij} < r_{ij}$	i, j = residue (or centroid) indices d = interatomic distance r = summed van der Waals radii[h]

[a] All terms originally described in Refs. 16 and 17.
[b] Binned function values are linearly interpolated, yielding analytic derivatives.

(continued)

TABLE I (continued)

[c] Neighbors within a 10-Å radius. Residue position defined by $C\beta$ coordinates ($C\alpha$ for glycine).

[d] Interactions between dimers within the same strand are neglected. Favorable interactions are limited to preserve pairwise strand interactions, that is, dimer m can interact favorably with dimers from at most one strand on each side, with the most favorable dimer interaction ($SS_{\phi,\theta} + SS_{hb} + SS_d$) determining the identity of the interacting strand. $SS_{d\sigma}$ is exempt from the requirement of pairwise strand interactions. SS_{hb} is evaluated only for m, n pairs for which $SS_{\phi,\theta}$ is favorable. $SS_{d\sigma}$ is evaluated only for m, n pairs for which $SS_{\phi,\theta}$ and $S\tilde{S}_{hb}$ are favorable. A bonus is awarded for each favorable dimer interaction for which $|m - n| > 11$ and strand separation is more than eight residues.

[e] A sheet is composed of all strands with dimer pairs <5.5 Å apart, allowing each strand having at most one neighboring strand on each side. Discrimination between alternate strand pairings is determined according the most favorable dimer interaction. Probability distributions fitted to $c(n_{strands}) - 0.9n_{sheets} - 2.7n_{lone\ strands}$ where $c(n_{strands}) = (0.07, 0.41, 0.43, 0.60, 0.61, 0.85, 0.86, 1.12)$.

[f] Residue position defined by $C\beta$ coordinates ($C\alpha$ for glycine).

[g] Not evaluated for atom (centroid) pairs whose interatomic distance depends on the torsion angles of a single residue.

[h] Radii determined from (1) 25th closest distance seen for atom pair in pdbselect25 structures, (2) the fifth closest distance observed in X-ray structures with better than 1.3-Å resolution and $<40\%$ sequence identity, or (3) X-ray structures of <2-Å resolution, excluding $i, i + 1$ contacts (centroid radii only).

TABLE II

COMPONENTS OF ROSETTA ALL-ATOM ENERGY FUNCTION[a]

Name	Description (physical origin)	Functional form	Parameters	Ref.
rama	Ramachandran torsion preferences	$$\sum_i -\ln[P(\phi_i, \psi_i | aa_i, ss_i)]$$	i = residue index; ϕ, ψ = backbone torsion angles (36° bins); aa = amino acid type; ss = secondary structure type[b]	7, 12
LJ[c]	Lennard–Jones interactions	$$\sum_i \sum_{j>i} \left\{ \begin{array}{l} \left[\left(\frac{r_{ij}}{d_{ij}} \right)^{12} - 2\left(\frac{r_{ij}}{d_{ij}}\right)^6 \right] e_{ij}, \quad \text{if } \frac{d_{ij}}{r_{ij}} > 0.6 \\[2mm] \left[-8759.2\left(\frac{d_{ij}}{r_{ij}}\right) + 5672.0 \right] e_{ij}, \quad \text{else} \end{array} \right.$$	i, j = residue indices; d = interatomic distance; e = geometric mean of atom well depths[d]; r = summed van der Waals radii[e]	15
hb[f]	Hydrogen bonding	$$\sum_i \sum_j \left(-\ln[P(d_{ij}|h_j ss_{ij})] \right.$$ $$-\ln[P(\cos\theta_{ij}|d_{ij}h_j ss_{ij})]$$ $$\left. -\ln[P(\cos\psi_{ij}|d_{ij}h_j ss_{ij})] \right)$$	i = donor residue index; j = acceptor residue index; d = acceptor–proton interatomic distance; h = hybridization (sp², sp³); ss = secondary structure type[g]; θ = proton–acceptor–acceptor base bond angle; ψ = donor–proton–acceptor bond angle	19–21
solv	Solvation	$$\sum_i \left[\Delta G_i^{ref} - \sum_j \left(\frac{2\Delta G_i^{free}}{4\pi^{3/2}\lambda_i r_{ij}^2} e^{-d_{ij}^2} V_j + \frac{2\Delta G_j^{free}}{4\pi^{3/2}\lambda_j r_{ij}^2} e^{-d_{ij}^2} V_i \right) \right]$$	i,j = atom indices; d = distance between atoms; r = summed van der Waal radii[e]; λ = correlation length[h]; V = atomic volume[h]; $\Delta G^{ref}, \Delta G^{free}$ = energy of a fully solvated atom[h]	15, 18

(continued)

TABLE II (continued)

pair	Residue pair interactions (electrostatics, disulfides)	$$\sum_i \sum_{j>i} -\ln\left[\frac{P(\text{aa}_i,\ \text{aa}_j | d_{ij})}{P(\text{aa}_i | d_{ij})\, P(\text{aa}_j | d_{ij})}\right]$$	i, j = residue indices aa = amino acid type d = distance between residues[i]	15
dun	Rotamer self-energy	$$\sum_i -\ln\left[\frac{P(\text{rot}_i | \phi_i, \psi_i)\, P(\text{aa}_i | \phi_i, \psi_i)}{P(\text{aa}_i)}\right]$$	i, j = residue indices rot = Dunbrack backbone-dependent rotamer aa = amino acid type ϕ, ψ = backbone torsion angles	14, 15
ref	Unfolded state reference energy	$$\sum_{\text{aa}} n_{\text{aa}}$$	aa = amino acid type n = number of residues	15

[a] All binned function values are linearly interpolated, yielding analytic derivatives, except as noted.

[b] Three-state secondary structure type as assigned by DSSP.[22]

[c] Not evaluated for atom pairs whose interatomic distance depends on the torsion angles of a single residue.

[d] Well depths taken from CHARMm19 parameter set.[23]

[e] Radii determined from fitting atom distances in protein X-ray structures to the 6-12 Lennard-Jones potential using CHARMm19 well depths.

[f] Evaluated only for donor acceptor pairs for which $1.4 \leq d \leq 3.0$ and $90° \leq \psi$, $\theta \leq 180°$. Side-chain hydrogen bonds in involving atoms forming main-chain hydrogen bonds are not evaluated. Individual probability distributions are fitted to eighth-order polynomials and analytically differentiated.

[g] Secondary structure types for hydrogen bonds are assigned as helical ($j - i = 4$, main chain); strand ($|j - i| > 4$, main chain), or other.

[h] Values taken from Lazaridis and Karplus.[18]

[i] Residue position defined by Cβ coordinates (Cα of glycine).

Electrostatics are modeled using a residue-based pair potential, similar to that utilized at low resolution. In addition, because finer modifications of conformation permit introduction of angles not part of the original discrete fragment set (see below), energetic description of local interactions is included in the total energy, including an amino acid- and secondary structure-dependent torsion potential for the backbone dihedral angles.

De Novo Structure Prediction with Rosetta

Fragment Selection

The basic conformation modification operation employed by Rosetta is termed a "fragment insertion." For each fragment insertion, a consecutive window of three or nine residues is selected, and the torsion angles of these residues are replaced with the torsion angles obtained from a fragment of a protein of known structure. For each query sequence to be predicted, a customized library of fragments defining the conformational space to be searched is selected by comparison of short windows of the query sequence with known protein structures. All three- and nine-residue windows in the query are scored against all windows in a nonredundant database of proteins of known structure composed of X-ray structures of 2.5 Å resolution or better and <50% sequence identity. All bond lengths and bond angles in these structures have been set to ideal values and the backbone torsion angles fine tuned using small perturbations to maintain agreement with the X-ray-determined atomic coordinates, minimize steric overlap (Table I; vdw), and maintain favorable values of backbone torsion angles as evaluated by the torsion potential (Table II; rama).

Sequence profiles for the query sequence and each sequence in the structure database are constructed by two rounds of PSIBLAST[24] with a cutoff of 9×10^{-4}. Over each sequence window, a profile–profile similarity score is calculated as the sum of the absolute value of the differences of the probabilities of each amino acid at each position (L1 norm, widely known as the city block or taxi cab distance). In addition, the predicted secondary structure of the query sequence is compared with the DSSP[22]-assigned secondary structure of the known structure in each sequence window. Currently, three secondary structure predictions are utilized: Psipred,[25]

[20] W. Wedemeyer and D. Baker, *Proteins Struct. Funct. Genet.* **53**, 262 (2003).

[21] J. Schonbrun, C. A. Rohl, and D. Baker, unpublished results (2003).

[22] W. Kabsch and C. Sander, *Biopolymers* **22**, 2577 (1983).

[23] E. Neria, S. Fischer, and M. Karplus, *J. Chem. Phys.* **105**, 1902 (1996).

[24] S. F. Altschul, T. L. Madden, A. A. Schaffer, J. Zhang, Z. Zhang, W. Miller, and D. J. Lipman, *Nucleic Acids Res.* **25**, 3389 (1997).

SAM-T99,[26] and JUFO.[27] The similarity score for each secondary structure prediction is calculated as the negative sum of the three-state confidence for the correct secondary structure type at each position in the sequence window. For each secondary structure prediction, the overall similarity score is the sum of the sequence similarity score and half the secondary structure similarity score. Fragments containing backbone torsion angles inconsistent with the torsional preferences of the residue types specified by the query sequence are also discarded (e.g., *cis* peptide bonds are allowed only at proline residues in the query sequence).

A ranked list of the top fragments in each sequence window is assembled iteratively, adding the top scoring fragment according to each secondary structure prediction to the combined ranked list and eliminating redundancies. Fragments selected according to Psipred secondary structure prediction are incorporated into the list with a threefold greater frequency than fragments selected on the basis of other secondary structure predictions. As this round robin assembly of the fragment list proceeds, the proportion of helix, strand, and other secondary structure types at each residue is balanced to be consistent with the average three-state prediction of all secondary structure predictions utilized, supplementing the final list as needed with fragments with the desired secondary structure type at a particular position, ranked according to their agreement with the average secondary structure prediction and sequence profile. The final fragment list for a query sequence is composed of 200 nine-residue and 200 three-residue fragments for every overlapping insertion window in the query.

Fragment Assembly

The assembly of fragments into protein-like structures occurs by a Monte Carlo search. The search is arbitrarily started with the protein in a fully extended conformation. A 9-residue fragment insertion window is randomly selected and a fragment for this window is randomly selected from the top 25 fragments in the ranked list for this position. After replacing the torsion angles in the protein chain with the torsion angles from the selected fragment, the energy of the resulting conformation is evaluated. Moves that decrease the energy are retained; those that increase the energy are retained according to the Metropolis criterion. If no moves are accepted in 150 attempted insertions, the probability of accepting a move of increased energy is incrementally increased. After an accepted

---

[25] D. T. Jones, *J. Mol. Biol.* **292,** 195 (1999).

[26] K. Karplus, R. Karchin, C. Barrett, S. Tu, M. Cline, M. Diekhans, L. Grate, J. Casper, and R. Hughey, *Proteins Struct. Funct. Genet.* **S5,** 86 (2001).

[27] J. Meiler, M. Müller, A. Zeidler, and F. Schmäschke, *J. Mol. Model* **7,** 360 (2001).

move, the acceptance probability is returned to its initial value. Each simulation begins from a different random seed and attempts 28,000 nine-residue fragment insertions.

The complete scoring function used for *de novo* prediction is given in Table I. During the course of the simulated annealing protocol, terms are progressively added to the total potential. Initially, only the steric overlap term (Table I; vdw) is evaluated, and this stage continues until all initial torsion angles have been replaced. Over the next 2000 fragment insertion attempts, secondary structure is accumulated in the chain, and all terms except those rewarding compactness (Table I; cbeta, rg) are evaluated. The strand pairing score is evaluated according to Scheme A (Table I; SS) at 0.3 of its final weight. For the next 20,000 attempted fragment insertions, the SS score is increased to its full weight, and the cbeta term is added at half its final weight. During this stage of the simulation, the SS score alternates every 2000 cycles between encouraging local strand pairing and relaxing it. This relaxation in strand pairing requirements is accomplished by evaluating interactions only between residues separated by more than 10 residues in sequence. For the last 6000 attempted moves, strand pairing is encouraged, using the standard sequence separation cutoff of 5 residues (see Table I). For the final 4000 attempted moves, the complete scoring function as described in Table I is utilized with all terms at their full weight. The SS score is evaluated according to Scheme B, using the 5-residue sequence separation cutoff. After the assembly of decoy structures from 9-residue fragments, each decoy is subjected to a short refinement of 8000 attempted 3-residue fragment insertions of the "gunn" type (see below), using the complete scoring function.

For each structure prediction, many short simulations starting from different random seeds are carried out to generate an ensemble of "decoy" structures that have both favorable local interactions and protein-like global properties. This set is then clustered by structural similarity to identify the broadest free energy minima; the structure predictions for a sequence are generally the centers of the largest clusters.[6] Examples of successful predictions made by means of this strategy at CASP 5 are shown in Fig. 1.

Structure Prediction by Fragment Assembly

The fragment assembly approach has multiple benefits for *de novo* protein structure prediction. First, and foremost, the fragment library approximates Gibbs sampling of the populated regions of the local potential energy surface of the backbone. The Rosetta philosophy is that during the folding process of real proteins, the local structure fluctuates between

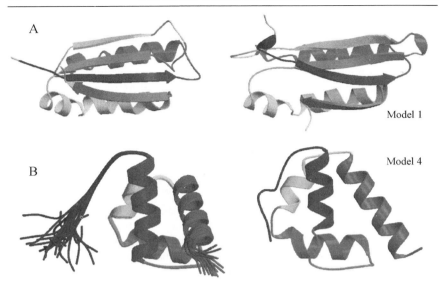

FIG. 1. Rosetta-predicted protein structures for CASP 5 targets. *Right:* Models predicted using the *de novo* prediction protocol. *Left:* Experimental structure of each protein. Protein chains are colored in a blue-to-red gradient along the length of the chain to highlight correctly predicted secondary structure elements. (A) T0135. The predicted model has 54 residues (of 106 total) predicted at a Cα RMSD of 4 Å to the experimental structure. (B) T0171. The predicted model has 60 residues (of 69 total) predicted at a Cα RMSD of 4 Å to the experimental structure. The global Cα RMSD between the prediction and the experimental structure is 4.2 Å. (See color insert.)

alternative local conformations and each fragment is a likely conformation of the local sequence. The use of a preset library of low-energy local structures means the local interaction energy need not be explicitly calculated with each move. This simplification is both efficient and crucial; computing the interaction energy assumes that an accurate potential energy surface is known, which may not be possible. Fragments, on the other hand, allow an accurate, but implicit, representation of the potential energy surface for local interactions. In the Rosetta fragment move set, a single-fragment substitution moving the protein from one topological isomer to another is like instantly transporting from one local energy minimum on the local PES to another; something a more continuous molecular dynamics or gradient search algorithm would be hard pressed to mimic.

Having dispensed with the need for an accurate local PES, the remaining global PES in Rosetta can be coarse grained in distance, and discrete in the combinatorics of strand pairing (Table I). Such a potential is well suited to the large search space inherent in the folding problem, and a second major advantage of the fragment insertion strategy is that it trades precise atomic positioning in favor of rapidly and coarsely sampling the large conformational space. Because fragment insertion modifies a consecutive set of backbone torsion angles, the effect of a move is not localized as it would be with a Cartesian move: the orientation and displacement of atoms on either side of an insertion position can change dramatically as a result of a single backbone torsion angle rotation. Angular changes made in the insertion window are not continuous, but rather are selected from a discrete library. Multiple angles are changed simultaneously, and the original values of the torsion angles being changed are not considered when selecting the angles that will be used for the new move. Consequently, the angular changes from move to move can be quite dramatic, allowing the conformation to evolve rapidly and escape local minima.

The single-fragment insertion approach makes many global conformers dynamically inaccessible on the search trajectory. In effect, the space of accessible conformations is cut off and thereby dramatically reduced. Whether this effect is beneficial depends only on whether the set of included conformations contains a close neighbor of the native protein conformer. The Rosetta philosophy has been shaped by empirical observations that the folding dynamics of protein domains (see e.g., Plaxco et al.[28]) are consistent with a process dominated by a quick quench: proteins may repeatedly quickly collapse from an extended chain to a compact structure and either unfold if the structure is distant from the native conformation or, in the rare case that the chain collapses to the native free energy basin, stay folded rather than sampling many energy basins while in a compact form. The "single move at a time" philosophy is not consistent with the physical fact that all the torsion angles are free to move simultaneously, but empirically and intuitively it does bias the final set of accessible structures toward the set achievable by a rapid, unorchestrated collapse.

Enhancements of Fragment Insertion Strategy

For *de novo* fold prediction, the benefits of fragment insertion allow rapid convergence on collapsed structures of plausible topology. Once this initial collapse has occurred, however, the fragment insertion strategy hinders efficient model refinement. Within a compact structure, any randomly selected, rigid body transformation of part of the chain is likely to create a clash with neighboring atoms or break favorable contacts. In

[28] K. W. Plaxco, I. S. Millet, D. J. Segel, S. Doniach, and D. Baker, *Nat. Struct. Biol.* **6,** 54 (1999).

addition, once the structure is coarsely established, the scale of conformation modification must be appropriate for optimizing more fine-grained potentials. It is thus desirable to define conformation modification operators for which the scale of global perturbation can be adjusted. Single-fragment insertion also lacks the notion that two or more consecutive moves taken together might offset the harmful effect of each other, yielding a net improvement. Each single insertion is rejected or accepted on the basis of the new energy of the entire conformation, and two moves can be coupled only indirectly via the Metropolis acceptance criteria: a move that increases the energy of the structure is on occasion accepted, setting the stage for a subsequent compensating move. Finally, the Monte Carlo fragment insertion strategy lacks the concept of using the gradient of the potential function to bias conformation modifications toward those that are more likely to be accepted.

For modeling scenarios requiring finer sampling about compact structures such as loop modeling, model refinement, and protein design, we have supplemented the original fragment insertion move set of Rosetta with additional conformation modification operators. Five basic concepts are combined to generate these novel operators: (1) random torsion angle perturbation, (2) selection of globally nonperturbing fragments, (3) rapid torsion angle optimization to offset global backbone perturbations, (4) optimization of the scoring function by gradient descent after a backbone modification (Monte Carlo plus minimization), and (5) rapid optimization of side-chain rotamers. In some of these operations, detailed below, deviations from backbone torsion angles of the fragment library are permitted, but such deviations are small so that the assumption that the fragment structures approximate low-energy local interactions is not violated.

Random Angle Perturbation

The simplest approach to local sampling about a compact structure is to perturb the torsion angles from their current values. We employ either small perturbations of randomly selected (ϕ, ψ) pairs ("small") or perturbation of a randomly selected ϕ angle coupled with a compensating rotation of equal magnitude but opposite direction of the preceding ψ angle. The latter case effects a "shear" motion in which the intervening peptide plane is rotated with minimal perturbation to the rest of the chain. Modification of residues in α helices is not allowed, and random perturbations have an upper limit of 2° for residues in β strands and 3° for all other residues. In addition, perturbations that increase the Ramachandran score (Table II; rama) are discriminated against, using a Metropolis criterion.

Selection of Globally Nonperturbing Fragments: A Local Move

The second approach retains reliance on fragment insertion, but biases fragment selection toward those that are most similar to the existing fragment in the model. Two different methods are used to estimate the similarity of two fragments. In the "chuck" strategy, the relative displacement of all the atoms in the protein on fragment replacement is measured, whereas in the "gunn" method, the net rotation and translation effected by different fragments are directly compared.

In the chuck method of fragment selection, the rigid body displacement of the downstream chain resulting from a fragment replacement is computed. The smaller the mean square deviation (MSD) of the atoms in the rigid body, the more the move is regarded as local. The MSD computation can be done extremely efficiently for a library of candidate fragments by relying on the fact that the differential rotation and translation need not be applied to every atom in order to compute the MSD, but only to the inertial ellipsoid of the rigid body (Appendix I). Further advantage is obtained by pretabulating the rotation and translation implied by every fragment in the discrete library. The library is winnowed to those fragments with a total downstream MSD change below a specified threshold, and a fragment in this set is then chosen at random for insertion.

In the gunn strategy, the rotation and translation effected by a fragment on the downstream portion of the chain are summarized by six degrees of freedom that are defined such that they are independent of the absolute origin and orientation of the coordinate system. Consequently, two fragments with similar net rotations and translations will have six-parameter descriptions that are almost numerically equal. The parameterization originally described by Gunn[12a] is used, but the arbitrary cost function is chosen such that large parameter differences are attenuated more than smaller ones: in other words, a fragment with five closely matching parameters and one poor match is preferred over a fragment with six mediocre matches (Appendix II). The gunn cost is computed for each library fragment at the selected insertion window, and a random selection is made among fragments with costs lower than the specified threshold.

Torsion Angle Variation to Offset Global Perturbation

In the "wobble" operation, the global perturbation of a fragment insertion (or any initial conformation modification) is offset by continuous variation of backbone (ϕ, ψ) angles within or adjacent to the insertion window. This operation is similar to the chuck strategy in that the MSD of downstream atoms is the measure of global perturbation. The wobble gradient descent is accelerated by analytic derivatives and, conveniently, our choice of the downstream MSD is differentiable and efficiently computed (Appendix I). In addition to the MSD, the cost function for optimization includes the Ramachandran potential (Table II; rama) that is derived from a smoothed, highly flattened version of the residue- and secondary structure-specific frequency with which a given (ϕ, ψ) pair occurs. The flattening deliberately weakens discrimination in allowed (ϕ, ψ) regions so that the MSD term dominates the result. The smoothing erases local minima and creates gradients leading from low-frequency areas to allowed regions. The look-up table is linearly interpolated between adjacent bins, making the derivative analytic.

The wobble operation is typically combined with fragment insertion in order to generate a complete conformation modification operator. In the standard combination, a fragment is selected by the chuck method, using an MSD cutoff of 60 Å^2. Subsequently, the torsion angles of a residue at the edge of the insertion window are modified by a wobble operation (wobble move in Fig. 3). Multiple fragment insertions and wobble operations can also be combined to generate more complex modification operators. The "crank" move in Fig. 2 is one such example. The move is initiated by making a chuck insertion at a selected insertion window. Torsion angles of an adjacent residue are then perturbed, using the wobble operation. Finally, at a second site not adjacent to the insertion window, torsion angles of two additional residues are perturbed by a second wobble operation. This type of move also attempts to select insertion windows where perturbations are more likely to be tolerated, biasing selection to residues not part of regular secondary structure elements. We note that the chuck methodology can be easily extended to double-fragment insertions, although in practice we find such operations inefficient and prefer to combine chuck insertions with wobble operations.

Monte Carlo Plus Minimization

Finally, any of the above described operations can be combined with direct optimization of the Rosetta potential energy function, replacing the Monte Carlo search strategy with the Monte Carlo-plus-minimization strategy described by Li and Scheraga.[29] After application of the initial modification, we attempt to rescue conformations with slightly increased energy by gradient descent to a local minimum. Either a single line minimization is carried out along the initial gradient ("lin"; Fig. 3) or an iterative descent to the local minimum is employed, using the variable metric method of Davidon, Fletcher, and Powell ("dfp"; Fig. 3).[30] In either case,

[29] Z. Li and H. A. Scheraga, *Proc. Natl. Acad. Sci. USA* **84,** 6611 (1987).

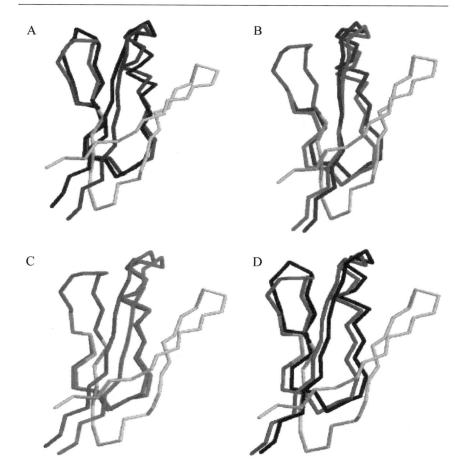

FIG. 2. Modified "crank" fragment insertion into 1 dan. (A) Superposition of the protein conformations preceding (black) and following (blue) insertion of a nine-residue fragment. The fragment insertion window is shown in red. The portion of the chain unperturbed by insertion is shown in gray. (B) Superposition of the protein conformations preceding (blue) and following (green) optimization of angles at a wobble site (cyan) adjacent to the insertion window. (C) Superposition of the protein conformations preceding (green) and following (magenta) optimization of angles at a second wobble site (orange) nonadjacent to the insertion window. (D) Superposition of the original (black) and final (magenta) conformations. (See color insert.)

[30] W. H. Press, S. A. Teukolski, W. T. Vetterling, and B. P. Flannery, "Numerical Recipes in Fortran 77: The Art of Scientific Computing," 2nd Ed. Cambridge University Press, New York, 2001.

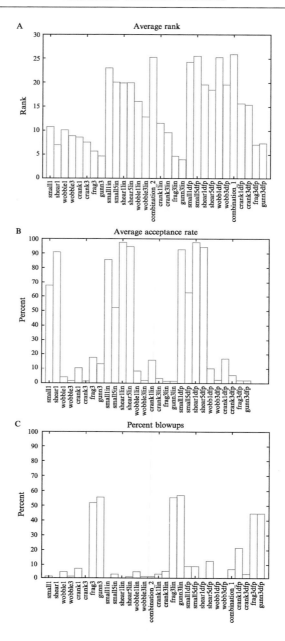

(ϕ, ψ) torsion angles within five residues of the site(s) of initial perturbation are varied. Torsion angles that are part of helical segments are held fixed. After minimization of the potential function, moves are accepted according to the usual energy criteria.

Rapid Side-Chain Optimization

The methods described above implement modifications of the backbone torsion angles and are combined in a variety of ways to generate backbone conformation modifications. In addition to modifying backbone conformation, operators must also allow for changes in side-chain conformation. Using a simulated annealing protocol, side-chain rotamers can be completely reoptimized.[15] In combination with operators that modify the backbone conformation, however, side-chain rotamers are rapidly optimized by cycling through each side-chain position in random order and replacing the current rotamer with the lowest energy rotamer available at that position. Combining both backbone and side-chain modification, complete conformation modification operators follow this general progression of steps: (1) initial backbone modification, either by random angle perturbation or selection of a globally nonperturbing fragment, (2) wobble of selected torsion angles to offset the global perturbation caused by the

---

FIG. 3. Comparison of move types in optimizing the all-atom energy function. Moves are named according to the type of perturbation made and the number of residues in the original perturbation (see text for details): small, random perturbation of one or more nonconsecutive (ϕ, ψ) pairs; shear, random compensating changes in a ϕ angle and the preceding ψ angle; wobble, insertion of a chuck fragment followed by a wobble of one residue; crank, insertion of a chuck fragment followed by a wobble of one residue adjacent to the insertion window and then by a wobble of two residues nonadjacent to the insertion window (illustrated in Fig. 2); frag, unmodified fragment insertion; gunn, insertion of a fragment selected using the gunn strategy. Addition of lin to the move name indicates the move is followed by a single-line minimization along the gradient of the potential function before evaluation of the Metropolis criterion. Addition of dfp indicates the move is followed by variable metric optimization of the potential function before evaluation of the Metropolis criterion. For combination 1, the attempted moves were cycled between small1dfp, small5dfp, shear5dfp, and wobble3dfp. For combination 2, the attempted moves were cycled between small1lin, shear5lin, wobble1lin, and wobble3lin. (A) Average rank of moves. For each starting decoy in the test set, the energies of the lowest energy decoy obtained from application of each move were sorted from highest energy (1) to lowest (30). The histogram reports the average overall decoys for each move type. (B) Percentage of moves accepted. Acceptance rates are reported for each move type, averaged over all decoys. The percentage was scaled on the basis of the percentage of independent simulations that resulted in an expanded structure, in order to account for the dramatic increase in acceptance rate into expanded models relative to compact models. (C) Frequency of simulations resulting in expanded structures. (See color insert.)

initial modification, (3) rapid optimization of side-chain rotamers, and (4) optimization of the scoring function by gradient descent.

Effectiveness of Conformation Modification Operators for Energy Function Optimization

Modified fragment insertions of the gunn type have been incorporated into the *de novo* prediction protocol, as described above, and permit significant optimization of the scoring function that is often accompanied by improvements in decoy accuracy and/or discrimination of near-native decoys.[31] When the Rosetta strategy is combined with structural constraints, experimentally determined by nuclear magnetic resonance (NMR), the incorporation of the modified moves described here is essential for refining initial decoy conformations generated by fragment assembly.[7,8] Experimental constraints provide an unusually effective scoring function for identifying accurate structures, and optimization of the agreement between the experimental data and the model structure nearly always results in models of increased accuracy.

To more generally evaluate the effectiveness of the modified move sets described here, we applied each move in isolation to a small test set of model structures and evaluated the frequency with which moves were accepted and the overall ability of each move type to optimize the value of the all-atom energy function (Table II). Most of the modified move sets evaluated here involve perturbation of backbone torsion angles to values not described by the original fragment set, necessitating the inclusion of the torsion potential term (Table II; rama) to describe local interactions. In addition, all terms in the all-atom function are differentiable as binned terms are linearly interpolated.

The test set consisted of eight models generated by Rosetta for each of seven small proteins. These models were generated by the standard *de novo* protocol (see above). The backbones were subjected to a short relaxation protocol to remove steric clashes (using the high-res radii set for the vdw term; see Table I), and then side chains were added to the decoys by means of a simulated annealing repacking algorithm.[15] For each decoy, a total of 200 times the number of residues in the sequence moves was attempted for each move type or combination of move types. During the course of the attempted backbone moves, side chains were completely reoptimized, using the simulated annealing protocol. Five independent simulations were performed for each move type for each decoy. Simulations in which the final model differed from the starting structure by more than 4 Å C_α RMSD

---

[31] K. T. Simons, C. Strauss, and D. Baker, *J. Mol. Biol.* **306,** 1191 (2001).

were discarded, and the simulation resulting in the lowest energy model was selected for analysis.

Figure 3 illustrates the relative effectiveness of each move type for optimizing the all-atom energy function. For each starting decoy in the test set, the energies of the lowest energy decoy obtained from application of each move type were sorted from highest energy (1) to lowest (30). The average rank over all decoys for each move type is reported in Fig. 3A. The fraction of attempted moves accepted for each move type is reported in Fig. 3B, and Fig. 3C reports the frequency of simulations that resulted in expanded structures. As expected, moves that cause smaller global perturbations (small, shear) are generally accepted with higher frequency than more globally perturbing moves. The critical conclusion is that most of the modified fragment insertion move types, although not accepted with high frequency, are significantly more effective than standard fragment insertions in optimizing the cost function, and are less likely to result in expanded protein conformations. Addition of the Monte Carlo-plus-minimization strategy increases the efficacy of the moves as well. Furthermore, application of the most effective moves in combination further increases both the acceptance rate and the extent of optimization of the cost function (see combination 1 and combination 2 in Fig. 3).

Conclusions

Although any protein-modeling strategy must attempt to find an optimal tradeoff between cost of computation of each move and the effectiveness of modifications in optimizing a cost function, the optimal tradeoff is specific to the particular problem of interest. The random selection of fragment insertions without consideration of gradient information or likelihood of the modification being accepted allows fragment insertion to be an extremely rapid operation and is well suited for *de novo* structure prediction, where coarse topological information is of interest. Conversely, careful selection of a move with a higher probability of acceptance or an operation that modifies many degrees of freedom simultaneously can be expensive to compute, yet may be a significantly more effective modification. Many structural modeling problems, including model refinement and loop modeling, require finer searches of a localized region of conformational space. For such problems, backbone modifications designed to effectively search local conformations are likely worth the extra computational expense.

Our goal in developing Rosetta is to assemble a unified platform for structural modeling that provides scoring functions and move sets applicable to a wide range of resolutions. The scoring function is modular, and individual terms of the energy functions can be combined with weights optimized for specific applications. In addition, the move sets described here provide options for both coarse-grained and local conformational searches. Different combinations of these move sets and scoring functions provide Rosetta modules for *de novo* structure prediction, model refinement, loop modeling, rigid body protein–protein docking, and protein design. Combinations of modules allow complex modeling problems to be approached. For example, combination of the *de novo* prediction, design, and model refinement modules has been used to design a protein with a novel topology.[10] Combining the docking and design modules allows alternative rigid body orientations to be explored for interface design, whereas combining the docking module with model refinement is applicable to docking with backbone flexibility.

Supplemental Materials

Licensing information for Rosetta may be obtained by e-mail (rosettaNMR@rosetta.bakerlab.org, rosettaABINITIO@rosetta.bakerlab.org, rosettaFRAGMENTS@rosetta.bakerlab.org). In addition, automated Rosetta predictions can be obtained from the Rosetta server[32] at http://robetta.bakerlab.org. The Rosetta server uses a combination of *de novo* prediction and homology modeling to produce complete three-dimensional models for proteins. Rosetta fragment libraries can be obtained from the automated server at http://rosetta.bakerlab.org/fragmentssubmit.jsp.

Appendix I

Efficient Computation of MSD Induced by Rigid Body Transformation

After fragment insertion, the relative positions of the protein chains before and after the insertion point undergo a rigid body motion. A convenient and differentiable measure of this global change is to fix one of the chains in space and compute the mean squared deviation (MSD) of the atoms in the other, "downstream" chain. Commonly, we want to compute this quantity quickly for all members of the fragment library at a given insertion position and then screen out the ones with a large MSD. This appendix discusses a general method for accelerating this screen and the computation of the derivative.

---

[32] D. Chivian, D. E. Kim, L. Malmström, P. Bradley, T. Robertson, P. Murphy, C. E. M. Strauss, R. Bonneau, C. A. Rohl, and D. Baker, *Proteins Struct. Funct. Genet.* **53**(Suppl. 6), 524 (2003).

To compute the MSD, we apply the fragment replacement transform to each downstream atom, find its squared deviation, and compute the mean. Specifically, if $\{\hat{V}_k\}$ are the initial coordinates, (V_x, V_y, V_z), of the N_{atoms} atoms downstream from the end of the insertion window; $\hat{A}_1$ and $\hat{U}_1$ are, respectively, the rotation and translation caused by the existing torsion angles in the insertion window; and $\hat{A}_2$ and $\hat{U}_2$ are, respectively, the rotation and translation induced by the new torsion angles in the insertion window, then after the fragment replacement the new coordinates of the N_{atoms} atoms are $\hat{V}_k' \equiv {}^T\hat{A}_2\left[\hat{A}_1\left(\hat{V}_k - \hat{U}_1\right)\right] + \hat{U}_2$ and the MSD is given by

$$\text{MSD} = \frac{1}{N_{\text{atoms}}} \sum_{k=1}^{N_{\text{atoms}}} \left\| \hat{V}_k - {}^T\hat{A}_2\left[\hat{A}_1\left(\hat{V}_k - \hat{U}_1\right)\right] - \hat{U}_2 \right\|^2 \qquad (A1)$$

As written, this calculation is slow for large values of N_{atoms}, but it can be significantly accelerated by the insight that all relevant properties of the atom positions, (V_x, V_y, V_z), can be summarized by the inertial tensor, $\hat{E}$, and the center of mass position, $\hat{V}_{\text{ave}}$.

Expanding all of the multiplications implied by the square and rearranging the sum yields

$$\text{MSD} = \left\| {}^T\hat{A}_2\hat{A}_1\left(\hat{V}_{\text{ave}} - \hat{U}_1\right) - \left(\hat{V}_{\text{ave}} - \hat{U}_2\right) \right\|^2$$
$$+ \frac{2}{N_{\text{atoms}}} \sum_{k=1}^{N_{\text{atoms}}} {}^T\left(\hat{V}_k - \hat{V}_{\text{ave}}\right)\left(\hat{1} - {}^T\hat{A}_2\hat{A}_1\right)\left(\hat{V}_k - \hat{V}_{\text{ave}}\right) \qquad (A2)$$

where the first term is simply $\|\hat{V}_{\text{ave}} - \hat{V}'_{\text{ave}}\|^2$, the MSD change of the center of mass, and the last term is the contribution to the MSD from the rotation about the center of mass. We can pull the rotational dependence out of the summation as follows. Let $\hat{R} \equiv {}^T\hat{A}_2\hat{A}_1$ be the net rotation, and let $\hat{E}$ be the second moments of $\{\hat{V}_k\}$. For example,

$$E_{xy} \equiv \sum_{k=1}^{N_{\text{atoms}}} (V_{x_k} - V_{x_{\text{ave}}})(V_{y_k} - V_{y_{\text{ave}}})$$

The summation of the second term in Eq. (A2) may then be written as the dot product of the $\hat{1} - \hat{R}$ and $\hat{E}$ matrices viewed as vectors:

$$E_{xx} + E_{yy} + E_{zz} - [R_{xx}E_{xx} + R_{xy}E_{xy} + R_{xz}E_{xz} + R_{yx}E_{yx}$$
$$+ R_{yy}E_{yy} + R_{yz}E_{yz} + R_{zx}E_{xx} + R_{zy}E_{zx} + R_{zz}E_{zz}]$$

After precomputation of $\hat{E}$, the cost of computing the MSD for each candidate insertion fragment is independent of N_{atoms}, resulting in more than an order of magnitude speed improvement [over Eq. (A1)] in evaluating the MSD for typical downstream chain lengths.

This MSD is asymmetric because the chain on one side of the fragment insertion is treated as fixed while the chain on the "downstream" side undergoes the rigid body motion. To avoid a directional preference, we fix whichever chain is the larger of the two, on average halving the precomputation of the inertial tensor. The transformation factors, $\hat{A}$, $\hat{U}_1$, $\hat{A}_2$, and $\hat{U}_2$, are dependent on the arbitrary initial protein orientation, but this dependence can be eliminated by moving the fragment insertion position to a standard origin and orientation: this transformation only requires rotating the inertial moments and also allows $\hat{A}_2$ and $\hat{U}_2$ to be precomputed for the entire fragment library. As a final refinement, $\hat{A}_1$ is preapplied to $\hat{E}$, reducing $\hat{R}$ trivially to ${}^T\hat{A}_2$.

The decomposition of the MSD expression into translational and rotational motion is not only illustrative, but also allows fragments to be screened for motions that are both small in MSD and also primarily shearing (or nonshearing) motions. For example, one might seek a motion that twists a helix in place or, conversely, a pure shearing motion that frame shifts a strand pairing without twisting it. Screening for fragments with these effects is possible in this framework by unevenly weighting the relative contribution of individual terms of Eq. (A2).

Efficient Computation of MSD Partial Derivatives

Following a fragment insertion that transforms the downstream chain from $\{\hat{V}_k\}$ to $\{\hat{V}'_k\}$, this transformation can be counteracted by continuously varying a selected set of torsion angles to minimize the MSD by gradient descent (wobble). Here $\{\hat{V}'_k\}$ are the coordinates of the downstream chain before minimization, $\{\hat{V}_k\}$ are the target coordinates of the downstream chain (i.e., the coordinates before the initial fragment insertion), and we must compute the partial derivatives of the MSD with respect to each torsion angle. The differential motion $\partial\hat{V}'$ of a point in space when rotated by $\partial\theta$ about a bond axis $\hat{B}$ passing through an atom $\hat{V}_{\text{atom}}$ is

$$\partial\hat{V}' = \hat{B} \times \left(\hat{V}' - \hat{V}_{\text{atom}}\right)\partial\theta$$

The change in the distance between any two points V' and V caused by this infinitesimal motion is its projection along the line joining them:

$$\partial\|\hat{V} - \hat{V}'\| = \partial\hat{V}' \cdot \left(\hat{V} - \hat{V}'\right)/\|\hat{V} - \hat{V}'\|$$

Thus the partial derivative of the MSD with respect to any torsion angle is

$$\frac{\partial \text{MSD}}{\partial \theta} = \frac{1}{N_{\text{atoms}}} \sum_{k=1}^{N_{\text{atoms}}} \frac{\partial}{\partial \theta} \|\hat{V}_k - \hat{V}'_k\|^2 = \frac{2}{N_{\text{atoms}}} \sum_{k=1}^{N_{\text{atoms}}} \|\hat{V}_k - \hat{V}'_k\| \frac{\partial \|\hat{V}_k - \hat{V}'_k\|}{\partial \theta}$$

$$= \frac{2}{N_{\text{atoms}}} \sum_{k=1}^{N_{\text{atoms}}} \hat{B} \times \left(\hat{V}_k - \hat{V}_B\right) \cdot \left(\hat{V}_k - \hat{V}'_k\right) = \frac{-2}{N_{\text{atoms}}} B \cdot \sum_{k=1}^{N_{\text{atoms}}} \hat{V}_B$$

$$\times \left(\hat{V}_k - \hat{V}'_k\right) + \hat{V}_k \times \hat{V}'_k \tag{A3}$$

Applying the cross-product associativity rule,

$$\frac{\partial \text{MSD}}{\partial \theta} = \frac{2}{N_{\text{atoms}}} \sum_{k=1}^{N_{\text{atoms}}} -\hat{B} \cdot \hat{V}_{\text{atom}} \times \left(\hat{V}_k - \hat{V}'_k\right) - \hat{B} \cdot \hat{V}_k \times \hat{V}'_k$$

Finally, the center of mass motion and the rotation about the center of mass can be factored into two separate components:

$$\frac{\partial \text{MSD}}{\partial \theta} = -2\hat{B} \cdot \left(\hat{V}_{\text{ave}} - \hat{V}_{\text{atom}}\right) \times \left(\hat{V}'_{\text{ave}} - \hat{V}_{\text{atom}}\right)$$

$$- \frac{2}{N_{\text{atoms}}} \hat{B} \cdot \sum_{k=1}^{N_{\text{atoms}}} \left(\hat{V}'_k - \hat{V}_{\text{ave}}\right) \times \left(\hat{V}'_k - \hat{V}'_{\text{ave}}\right) \tag{A4}$$

As before, the last sum can be collapsed to a simple vector depending only on R and E. For example, the x coordinate of this sum is

$$R_{zx}E_{xy} + R_{zy}E_{yy} + R_{zz}E_{zy} - R_{yy}E_{xz} - R_{vv}E_{yz} - R_{yz}E_{zz}$$

In computing the gradient, one evaluates the partial derivative for each torsion axis. This calculation can be done efficiently by using the form in Eq. (A4): $\hat{V}_{\text{atom}}$ and $\hat{B}$ differ for each torsion angle, but $\hat{V}_{\text{ave}}$, $\hat{V}'_{\text{ave}}$, R, and E are determined by the configuration and are the same for every torsion angle.

Applications of Efficient MSD Evaluation

Although the MSD is discussed above in the context of single-fragment replacement and subsequent continuous perturbation (wobble) of selected torsion angles, numerous other applications of these methods exist. In loop modeling, for example, the starting and ending points of a loop segment are known and the goal is to locate a fragment that will join these end points. This problem is isomorphic to computing the MSD in fragment replacement; although there is no existing fragment being replaced, the targets $\hat{A}_1$ and $\hat{U}_1$ are defined by the fixed segments. In addition, in the case of incomplete loop closure, the downstream chain position implied by the torsion angles of the terminal loop residue differs from the fixed template coordinates of the downstream chain. Gradient descent minimization of this MSD can be used to perturb torsion angles of loop residues to close this chain break. These methods are in fact incorporated into the loop-modeling strategy used in Rosetta (Rohl et al.[12]). In general, the methods described here are applicable to computation and minimization of MSD between any two coordinate sets, $\{\hat{V}_k\}$ and $\{\hat{V}_k'\}$.

Appendix II

Gunn Estimation of Global Perturbation Induced by Fragment Replacement

The net translation and rotation of the chain resulting from a fragment is described by six degrees of freedom that can be chosen such that if two fragments have nearly the same values for these six parameters, they produce nearly the same rotation and translation. Let $\hat{x}_1$ be the unit vector along the N-to-C_α bond at the N terminus of the fragment, let $\hat{x}_2$ be the analogous unit vector along the C-to-C_α bond at the C terminus of the fragment, and let $\hat{R}$ be the vector between the C_α atoms at the fragment N and C termini. Let $\hat{y}_1$ and $\hat{y}_2$ be the normals to the planes defined by N, C_α, and C of the N- and C-terminal (respectively) residue of the fragment. The six parameters to describe the net rotation and translation are chosen such that they are independent of the absolute position and orientation. q_1 and q_2, that is,

$$q_1 = \hat{x}_1 \cdot \hat{R}$$
$$q_2 = \hat{x}_2 \cdot \hat{R}$$

are proportional to the cosine of the polar angle of final and initial bond vectors with respect to the R-axis;

$$|q_3| = \arccos \left[\frac{\hat{x}_1 \cdot \hat{x}_2 - (\hat{x}_1 \cdot \hat{R})(\hat{x}_2 \cdot \hat{R})}{\sqrt{(1 - q_1^2)(1 - q_2^2)}} \right]$$

is the dihedral angle between $\hat{x}_1$ and $\hat{x}_2$ along the $\hat{R}$ axis;

$$|q_4| = \arccos \left(\frac{\hat{y}_1 \cdot \hat{R}}{\sqrt{(1 - q_1^2)}} \right)$$

and

$$|q_5| = \arccos\left(\frac{\hat{y}_2 \cdot \hat{R}}{\sqrt{\left(1 - q_2^2\right)}}\right)$$

are the angles between $\hat{y}_1$, $\hat{y}_2$ and the $\hat{R}$ axis; and

$$q_6 = \|\hat{R}\|$$

is the fragment length. These definitions match those of Gunn,[12a] other than the specific atoms used to define the coordinate system.

These parameters are independent of the orientation and origin of the coordinate system, allowing two fragments to be compared term by term:

$$\text{cost} \equiv c_1 \ln\left(1 + |\Delta q_1| + |\Delta q_2|\right) + c_2 \ln\left(1 + |\Delta q_3|\right)$$
$$+ c_3 \ln\left(1 + |\Delta q_4| + |\Delta q_5|\right) + c_4 \ln\left(1 + |\Delta q_6|\right)$$

where Δq is the difference between the respective q values of the original and replacement fragment and the absolute value also implies modulo-π for the angular Δq values. The $\{c\}$ coefficients were determined by regression on a test set of single-domain proteins to make the cost function a good discriminator between small and large RMS deviations in the downstream chain induced the fragment swap. In Rosetta, $\{c\} = \{5.72, 2.035, 3.84, 0.346\}$, and typical lower and upper cost thresholds are 0.03 and 4.08, respectively. In ballpark terms, these limits correspond to pure angular changes between ~ 0.5 and $\sim 90°$, or pure displacements of less than ~ 2 Å. When all the terms contribute equally to the cost, the upper limit on the deviation of any one angular degree of freedom falls to about $10°$.

In the above description, the effect of modifying ψ (and ω) of the C-terminal residue of the insertion window is not included in the six q parameters. These angles are inserted into the chain, however, to avoid potential violations of allowed Ramachandran space that may occur if the C-terminal residue (ϕ, ψ) angles are effectively drawn from different library fragments. Although this inaccuracy limits the discriminatory power of the cost function and could be corrected by modification of the q parameter definitions, the cost function works well in practice, presumably because when the rotation and translation of two fragments are similar, the terminal torsion angles are likely to be similar as well.

[5] Poisson–Boltzmann Methods for Biomolecular Electrostatics

By Nathan A. Baker

Introduction

The understanding of electrostatic properties is a basic aspect of the investigation of biomolecular processes. Structures of proteins and other biopolymers are being determined at an increasing rate through structural genomics and other efforts; furthermore, the specific roles of these biopolymers in cellular pathways and supramolecular assemblages are being detected by genetic and other experimental efforts. The integration of this information into physical models for drug discovery or other applications requires the ability to evaluate intra- and interbiomolecular energetics. Among the various components of molecular interactions, electrostatic energetics and forces are of special importance because of their long range and the substantial charges of amino and nucleic acids.

Because of the ubiquitous nature of electrostatic interactions in biomolecular systems, a variety of computational methods have been developed for elucidating these interactions (see Refs. 1–5 and references therein). Popular computational electrostatics methods for biomolecular systems can be loosely grouped into two categories: "explicit solvent" methods, which treat the solvent in full molecular detail, and "implicit solvent" methods, which include solvent influences in averaged or continuum fashion. Although explicit solvent approaches offer more detailed insight into solvent-mediated biomolecular interactions, the necessity to integrate over the numerous solvent degrees of freedom often limits the ability of these methods to calculate thermodynamic quantities for large biomolecular systems.

[1] N. A. Baker and J. A. McCammon, in "Structural Bioinformatics" (P. Bourne and H. Weissig, eds.), John Wiley & Sons, New York, 2002.

[2] M. Gilson, in "Biophysics Textbook Online" (D. A. Beard, ed.), Biophysical Society Bethesda; T. A. Darden, in "Computational Biochemistry and Biophysics" (O. M. Becker, A. D. J. MacKerell, B. Roux, and M. Watanabe, eds.), pp. 91–114. Marcel Dekker, New York, 2001.

[3] B. Roux, in "Computational Biochemistry and Biophysics" (O. M. Becker, A. D. J. MacKerell, B. Roux, and M. Watanabe, eds.), pp. 133–152. Marcel Dekker, New York, 2001.

[4] M. E. Davis and J. A. McCammon, Chem. Rev. 94, 7684 (1990).

[5] B. Honig and A. Nicholls, Science 268, 1144 (1995).

Copyright 2004, Elsevier Inc.
All rights reserved.
0076-6879/04 $35.00

Brief Overview of Implicit Solvent Methods

Because of the sampling issues associated with explicit solvent treatments, implicit solvent methods have gained increasing popularity for elucidating the electrostatic properties of biomolecules in solution (see Refs. 1–5). As their name implies, these methods implicitly average over the configuration space of solvent and counterion species surrounding the biomolecule. The result is a polarizable continuum representation for the solvent and a mean field charge "cloud" for the counterion distribution. Despite some artifacts arising from this continuum treatment (see Roux[3] and references therein), implicit solvent methods offer a significant advantage over traditional explicit solvent approaches and have become standard techniques for investigating the energetics and dynamics of biomolecular systems.

The importance of electrostatic interactions in protein behavior was recognized decades ago in work by Linderstrom-Lang[6] in developing protein titration models; Tanford and Kirkwood[7] in investigating the effects of pH and ionic strength on enzyme activity; and Flanagan et al.[8] in studying the energetics of dimer–tetramer assembly in hemoglobin. However, electrostatic models for protein systems were improved dramatically in 1982 by Warwicker and Watson.[9] Drawing on the increased knowledge of the three-dimensional structure of proteins increased computer power; they introduced a grid-based, finite difference approach for calculating the electrostatic potential of a nonspherical protein by solving the Poisson–Boltzmann (PB) equation, a nonlinear partial differential equation that incorporates detailed information about the biomolecular shape and charge distribution. Since this pioneering work, the PB equation has become a standard method for the detailed investigation of biomolecular electrostatics.[1,4,5,10]

In addition to PB methods, simpler approximate models have also been constructed for continuum electrostatics, including distance-dependent dielectric functions,[11] analytic continuum methods,[12] and generalized Born

[6] K. Linderstrom-Lang, *C. R. Travl. Lab. Carlsberg* **15** (1924).

[7] C. Tanford and J. G. Kirkwood, *J. Am. Chem. Soc.* **79**, 5333 (1957).

[8] M. A. Flanagan, G. K. Ackers, J. B. Matthew, G. I. H. Hanania, and F. R. N. Gurd, *Biochemistry* **20**, 7439 (1981).

[9] J. Warwicker and H. C. Watson, *J. Mol. Biol.* **157**, 671 (1982).

[10] C. Holm, P. Kekicheff, and R. Podgornik, eds. *in* "Electrostatic Effects in Soft Matter and Biophysics," Vol. 46. NATO Science Series. Kluwer Academic, Boston, 2001.

[11] A. D. J. MacKerell and L. Nilsson, *in* "Computational Biochemistry and Biophysics" (O. M. Becker, A. D. J. MacKerell, B. Roux, and M. Watanabe, eds.), Marcel Dekker, New York, 2001; A. R. Leach, *in* "Molecular Modelling: Principles and Applications," 2nd Ed., pp. xxiv and 744, and plate 16. Prentice Hall, New York, 2001.

models.[13–15] Among the most popular of these simpler methods is the generalized Born model, introduced by Still et al. in 1990[13] and refined by several other researchers.[14,15] This method is based on the Born ion, a canonical electrostatics model problem describing the electrostatic potential and solvation energy of a spherical ion.[16] The generalized Born method uses an analytical expression based on the Born ion model to approximate the electrostatic potential and solvation energy of small molecules. Although it fails to capture all the details of molecular structure and ion distributions provided by more rigorous models,[15,17,18] such as the Poisson–Boltzmann equation, it has gained popularity as a rapid method for evaluating approximate forces and energies for solvated molecules and continues to be vigorously developed.

Poisson–Boltzmann Methods for Biomolecular Electrostatics

As mentioned above, the Poisson–Boltzmann equation is derived from a continuum model of the solvent and counterion environment surrounding a biomolecule.[1,4,5,10] Although there are numerous derivations of the PB equation based on statistical mechanics (see Holm et al.[10] for review), the simplest begins with Poisson's equation[19]:

$$-\nabla \cdot \varepsilon(x)\nabla\phi(x) = \rho(x) \quad \text{for} \quad x \in \Omega \qquad \text{where} \quad \phi(x) = g(x) \quad \text{for} \quad x \in \partial\Omega$$

$$(1)$$

the canonical equation for describing the dimensionless electrostatic potential $\phi(x)$ generated by a charge distribution $\rho(x)$ in a polarizable continuum with dielectric constant $\varepsilon(x)$. Equation (1) is generally solved in some finite domain Ω with a fixed potential (Dirichlet boundary condition) $g(x)$ on the boundary $\partial\Omega$. In a biomolecular system, it is useful to consider two types of charge distributions. First, the partial atomic charges are typically lumped into a "fixed" charge distribution:

[12] M. Schaefer and M. Karplus, J. Phys. Chem. **100**, 1578 (1996).
[13] W. C. Still, A. Tempczyk, R. C. Hawley, and T. Hendrickson, J. Am. Chem. Soc. **112**, 6127 (1990).
[14] B. N. Dominy and C. L. Brooks, J. Phys. Chem. B **103**, 3765 (1999); D. Bashford and D. A. Case, Annu. Rev. Phys. Chem. **51**, 129 (2000); K. Osapay, W. S. Young, D. Bashford, C. L. Brooks, and D. A. Case, J. Phys. Chem. **100**, 2698 (1996).
[15] A. Onufriev, D. A. Case, and D. Bashford, J. Comput. Chem. **23**, 1297 (2002).
[16] M. Born, Z. Phys. **1**, 45 (1920).
[17] L. David, R. Luo, and M. K. Gilson, J. Comput. Chem. **21**, 295 (2000).
[18] R. Luo, M. S. Head, J. Moult, and M. K. Gilson, J. Am. Chem. Soc. **120**, 6138 (1998); J. A. Given and M. K. Gilson, Proteins Struct. Funct. Genet. **33**, 475 (1998).
[19] J. D. Jackson, in "Classical Electrodynamics," 2nd Ed. John Wiley & Sons, New York, 1975.

$$\rho_{\mathrm{f}}(x) = \frac{4\pi e_{\mathrm{c}}^2}{kT} \sum_{i=1}^{M} Q_i \delta(x - x_i) \tag{2}$$

which models the M atomic partial charges of the biomolecule as delta functions $\delta(x - x_i)$ located at the atom centers $\{x_i\}$ with magnitudes $\{Q_i\}$. The scaling coefficients ensure the dimensionless form of the potential and include e_{c}, the charge of an electron, and kT, the thermal energy of the system. Second, the contributions of counterions are modeled in a continuous (or "mean field") fashion by a Boltzmann distribution, giving rise to the "mobile" charge distribution

$$\rho_{\mathrm{m}}(x) = \frac{4\pi e_{\mathrm{c}}^2}{kT} \sum_{j}^{m} c_j q_j \exp\left[-q_j \phi(x) - V_j(x) \right] \tag{3}$$

for m counterion species with charges $\{q_j\}$, bulk concentrations $\{c_j\}$, and steric potentials $\{V_j\}$ (i.e., potentials that prevent biomolecule–counterion overlap). In the case of a one-to-one electrolyte such as NaCl, Eq. (3) reduces to

$$\rho_{\mathrm{m}}(x) = \bar{\kappa}^2(x) \sinh \phi(x) \tag{4}$$

where the coefficient $\bar{\kappa}^2(x)$ describes both ion accessibility (indirectly via $\exp[-V(x)]$) and bulk ionic strength. Combining the expressions for the fixed [Eq. (2)] and mobile [Eq. (4)] counterion distributions with Poisson's equation [Eq. (1)] gives the Poisson–Boltzmann equation for a one-to-one electrolyte:

$$-\nabla \cdot \varepsilon(x) \nabla \phi(x) + \bar{\kappa}^2(x) \sinh \phi(x) = \frac{4\pi e_{\mathrm{c}}^2}{kT} \sum_i q_i \delta(x - x_i)$$

$$\text{for} \quad x \in \Omega; \quad \text{where} \quad \phi(x) = g(x) \quad \text{for} \quad x \in \partial\Omega \tag{5}$$

As described above, details of the biomolecular structure enter into the coefficients of the PB equation (see Fig. 1). Most obviously, the atom locations appear in the delta functions as the set of points $\{x_i\}$; the atomic positions and radii also enter into the definitions of coefficients $\varepsilon(x)$ and $\bar{\kappa}^2(x)$. The dielectric function $\varepsilon(x)$ has been represented by a variety of models, including discontinuous transitions at the molecular surface,[20,21] smooth spline-based definitions,[22,23] and Gaussian-based descriptions.[24]

[20] B. Lee and F. M. Richards, *J. Mol. Biol.* **55,** 379 (1971).
[21] M. L. Connolly, *J. Mol. Graphics* **11,** 139 (1993).
[22] W. Im, D. Beglov, and B. Roux, *Comput. Phys. Commun.* **111,** 59 (1998).
[23] C. L. Bajaj, V. Pasucci, R. J. Holt, and A. N. Netravali, *in* "Fourth Issue of the Special Series of Discrete Applied Mathematics on Computational Biology," 2001.

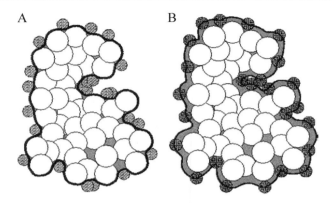

FIG. 1. Surface definitions used to evaluate coefficients of PB equation. (A) Representation of the molecular surface, a popular definition for the dielectric coefficient $\varepsilon(x)$. (B) Representation of the inflated van der Waals surface, the common definition for the ion accessibility coefficient $\bar{\kappa}^2(x)$.

Among these various dielectric coefficient models, the most popular have been definitions based on the molecular[21] and Lee–Richards surfaces.[20] In these models, $\varepsilon(x)$ is discontinuous along the biomolecular surface and assumes solute dielectric values inside and bulk solvent values outside the surface. The ion accessibility function $\bar{\kappa}^2(x)$ has typically been characterized in terms of the inflated van der Waals surface: the boundary of the union of spheres centered at the atomic positions with radii equal to the atomic van der Waals radii plus the counterion species radius.

The "full" or nonlinear form of the problem given in Eq. (5) is often simplified to the linearized PB equation by replacing the $\sinh\phi(x)$ term with its first-order approximation, $\sinh\phi(x) \approx \phi(x)$, to give

$$-\nabla \cdot \varepsilon(x)\nabla\phi(x) + \bar{\kappa}^2(x)\varphi(x) = \frac{4\pi e_c^2}{kT}\sum_i q_i\delta(x - x_i) \quad \text{for}$$

$$x \in \Omega \quad \text{where} \quad \phi(x) = g(x) \quad \text{for} \quad x \in \partial\Omega \tag{6}$$

However, although the linearized version of the problem can be somewhat simpler to solve, it is only as useful as the underlying approximation $\sinh\phi(x) \approx \phi(x)$. Specifically, this linearization is appropriate only at small potential values where the nonlinear contributions to $\sinh\phi(x)$ are negligible.

It is worthwhile to note that the PB equation (nonlinear or linearized) is an approximate theory and therefore cannot be applied blindly to all

---

[24] J. A. Grant, B. T. Pickup, and A. Nicholls, *J. Comput. Chem.* **22,** 608 (2001).

biomolecular systems. Specifically, the PB equation is derived from mean field or saddle point treatments of the electrolyte system and therefore neglects counterion correlations and fluctuations that can affect the energetics of highly charged biomolecular systems such as DNA, RNA, and some protein systems. A good review of the scope and impact of these deviations from PB theory can be found in Holm *et al.*[10] In short, Poisson–Boltzmann theory gives reasonable quantitative results for biomolecules with low linear charge density in monovalent symmetric salt solutions; however, PB theory can be qualitatively incorrect for highly charged biomolecules or more concentrated multivalent solutions. Therefore, application of PB theory and software requires some discretion on the part of the user.

Since the PB equation was first applied to biomolecular systems, methods for the solution of the PB equation have been repeatedly revisited and refined to further improve the efficiency of electrostatics calculations, including improved finite difference, finite element, and boundary element methods; as well as a host of new and/or improved algorithms for using PB data in energy and force evaluations, pK_a calculations, and biomolecular simulations. The following sections describe the various aspects of these methods for obtaining and using solutions to the PB equation.

Numerical Solution of Poisson–Boltzmann Equation

Few analytical solutions of the PB equation exist for realistic biomolecular geometries and charge distributions. Therefore, this equation is usually solved numerically by a variety of computational methods. These methods typically rely on a discretization to project the continuous solution down onto a finite dimensional set of basis functions. In the case of the linearized PB equation [see Eq. (6)], the resulting equations take the usual linear matrix vector form, which can be solved directly. However, the nonlinear equations obtained from the full PB equation require more specialized techniques, such as Newton methods, to determine the solution to the discretized algebraic equation.[25] Specifically, Newton methods start with an initial solution guess and iteratively improve this guess by solving related linear equations for corrections to the current solution. The remainder of this section describes the most common discretization and solver methods used for both linearized and nonlinear PB models.

[25] M. J. Holst and F. Saied, *J. Comput. Chem.* **16,** 337 (1995).

Finite Difference Discretization

Some of the most popular discretization techniques employ Cartesian meshes to subdivide the domain in which the PB equation is to be solved. Of these, the finite difference (FD) method has been at the forefront of PB equation solvers (see Refs. 5, 26–28, and references therein). In its most general form, the finite difference method solves the PB equation on a non-uniform Cartesian mesh, as shown in Fig. 2A for a two-dimensional domain. In this general setting, the Laplacian or Poisson differential operator is transformed into a sparse difference matrix by means of a Taylor expansion. The resulting matrix equations are then solved (either directly or over the course of a Newton iteration) by a variety of linear algebra techniques. Although FD grids offer relatively simple problem setup, they provide little control over how unknowns are placed in the solution domain. Specifically, as shown by Fig. 2A, the Cartesian or tensor-product nature of the mesh makes it impossible to locally increase the accuracy of

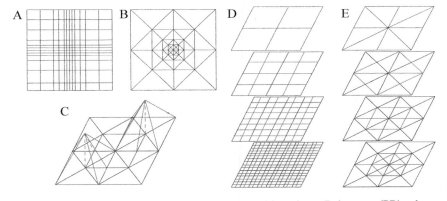

FIG. 2. Discretization schemes and hierarchies used in Poisson–Boltzmann (PB) solvers. (A) Cartesian mesh suitable for finite difference (FD) calculations. (B) Finite element (FE) mesh exhibiting adaptive refinement. (C) Examples of typical piecewise linear basis functions used in FE methods. (D) The multilevel hierarchy used to solve the PB equation for an FD discretization; red lines denote the additional unknowns added at each level of the hierarchy. (E) The multilevel hierarchy used to solve the PB equation for FE discretizations; red lines denote the simplex subdivisions used to introduce additional unknowns at each level of the hierarchy. (See color insert.)

[26] M. E. Davis, J. D. Madura, B. A. Luty, and J. A. McCammon, *Comput. Phys. Commun.* **62,** 187 (1991).

[27] W. Rocchia, S. Sridharan, A. Nicholls, E. Alexov, A. Chiabrera, and B. Honig, *J. Comput. Chem.* **23,** 128 (2002).

[28] M. E. Davis and J. A. McCammon, *J. Comput. Chem.* **12,** 909 (1991).

the solution in a specific region without increasing the number of unknowns across the entire grid.

Adaptive Finite Element Discretization

Unlike finite difference methods, adaptive finite element (FE) discretizations[29] offer the ability to place computational effort in specific regions of the problem domain. Finite element meshes (see Fig. 2B) are composed of simplices (e.g., triangles or tetrahedra) that are joined at edges and vertices. The solution is constructed from piecewise polynomial basis functions (see Fig. 2C), which are associated with mesh vertices and typically are nonzero only over a small set of neighboring simplices. Solution accuracy can be increased in specific areas by locally increasing the number of vertices through simplex refinement. As shown in Fig. 2E, the number of unknowns (vertices) is generally increased only in the immediate vicinity of the simplex refinement and not throughout the entire problem domain, as in FD methods. This ability to locally increase the solution resolution is called "adaptivity" and is the major strength of finite element methods applied to the PB equation.[30–33] As with the FD method, this discretization scheme leads to sparse symmetric matrices with a small number of nonzero entries in each row.[29]

Multilevel Solvers

Multilevel solvers,[34] in conjunction with the Newton methods described above, have been shown to provide the most efficient solution of the algebraic equations obtained by discretization of the PB equation with either finite difference or finite element techniques.[1,25,30,32,35,36] Most sizable

[29] O. Axelsson and V. A. Barker, *in* "Finite Element Solution of Boundary Value Problems: Theory and Computation." Academic Press, San Diego, CA, 1984; D. Braess, "Finite Elements: Theory, Fast Solvers, and Applications in Solid Mechanics. Cambridge University Press, Cambridge, 1997; S. C. Brenner, and L. R. Scott, "The Mathematical Theory of Finite Element Methods," 2nd Ed., pp. xv and 361. Springer-Verlag, New York, 2002.

[30] M. Holst, N. Baker, and F. Wang, *J. Comput. Chem.* **21,** 1319 (2000).

[31] N. Baker, M. Holst, and F. Wang, *J. Comput. Chem.* **21,** 1343 (2000).

[32] N. A. Baker, D. Sept, M. J. Holst, and J. A. McCammon, *IBM J. Res. Dev.* **45,** 427 (2001).

[33] C. M. Cortis and R. A. Friesner, *J. Comput. Chem.* **18,** 1591 (1997); C. M. Cortis and R. A. Friesner, *J. Comput. Chem.* **18,** 1570 (1997).

[34] W. Hackbusch, *in* "Multigrid Methods and Applications." Springer-Verlag, Berlin, 1985; W. L. Briggs, *in* "A Multigrid Tutorial," pp. ix and 88. Society for Industrial and Applied Mathematics, Philadelphia, PA, 1987.

[35] M. Holst, *Adv. Comput. Math.* **15,** 139 (2001).

[36] N. A. Baker, D. Sept, S. Joseph, M. J. Holst, and J. A. McCammon, *Proc. Natl. Acad. Sci. USA* **98,** 10037 (2001).

algebraic equations are solved by iterative methods, which start with an initial guess and repeatedly apply a set of operations to improve this guess until a solution of the desired accuracy is reached. However, the speed of traditional iterative methods has been limited by their inability to quickly reduce low-frequency (long-range) error in the solution.[34] This problem can be overcome by projecting the discretized system onto meshes (or grids) at multiple resolutions (see Fig. 2D and E). The advantage of this multiscale representation is that the slowly converging low-frequency components of the solution on the finest mesh are quickly resolved on coarser levels of the system. This coupling of scales gives rise to a "multilevel" solver algorithm, where the algebraic system is solved directly on the coarsest level and then used to accelerate solutions on finer levels of the mesh.

As shown in Fig. 2D and E, assembly of the multilevel hierarchy depends on the method used to discretize the PB equation. For FD types of methods, so-called multigrid methods are used: the nature of the FD grid lends itself to the assembly of a hierarchy with little additional work.[25,34] In the case of adaptive finite element discretizations, "algebraic multigrid" methods[32,35] are employed. For FE meshes, the most natural multiscale representation is constructed by refinement of an initial mesh, which typically constitutes the coarsest level of the hierarchy.

The advantage of multiscale algorithms in solving numeric problems cannot be overstated as they enable the solution of large-scale numerical systems in an optimal fashion; that is, the time required to solve an $N \times N$ linear algebraic system scales as $O(N)$.[25,30,34] Figure 3 illustrates the theoretical scaling behavior of a variety of numerical methods (both traditional and multilevel) implemented in popular PB software (see Table I). This plot was calibrated on the basis of timing data from real-world multigrid calculations. For comparison, the PB equation for a typical small protein (i.e., lysozyme) could be solved to reasonable accuracy (0.5–1.0 Å) using a finite difference discretization on a relatively small ($65 \times 65 \times 65$) grid by any of the methods shown in Fig. 3; the expected calculation time would be roughly 1 s. However, to study systems only five times larger (i.e., ribosomes, microtubules, polymerases, etc.) could require up to six orders of magnitude more time with a standard iterative method (e.g., successive overrelaxation) than with multigrid. On the basis of this scaling behavior, it should be clear that multilevel methods are essential to the efficient solution of the PB equation in an era in which the scale of experimentally resolved structures is continuously increasing.

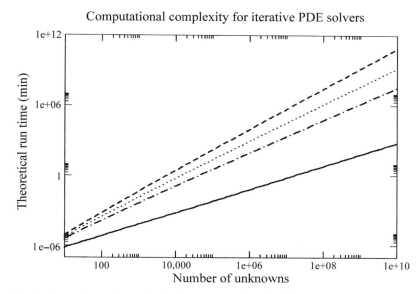

FIG. 3. Computational complexity for common iterative PD equation solvers: multigrid (—), successive overrelaxation (- • -), conjugate gradient (•••), and Gauss–Seidel (- - -). Theoretical run times were calculated from asymptotic scaling estimates for these methods[25,37] and calibrated by the observed multigrid run time with APBS software[36,38] for 65 × 65 × 65 unknowns.

Parallel Methods

Regardless of the scalability of the numerical algorithm used to solve the PB equation, there are some systems that are simply too large to be solved sequentially (i.e., on one processor). For example, although small-to medium-sized protein systems (100–1000 residues) are amenable to sequential calculations, there are an increasing number of structures of macromolecular assemblages with tens to hundreds of thousands of residues (e.g., microtubules, entire viral capsids, ribosomes, polymerases, etc.). Studies of these large systems are not feasible on most sequential platforms; instead, they require multiprocessor computing platforms to solve the PB equation in a parallel fashion.

Traditional parallel algorithms for solving the PB equation introduced parallelism at the "fine-grain" level[16,39]; for example, during matrix–vector

[37] G. H. Golub and C. F. van Loan, *in* "Matrix Computations," 3rd Ed. Johns Hopkins University Press; M. D. Baltimore, 1996; W. H. Press, S. A. Teukolsky, W. T. Vetterling, and B. P. Flannery, *in* "Numerical Recipes in C." Cambridge University Press, New York, 1992.
[38] N. A. Baker, *in* "Adaptive Poisson–Boltzmann Solver (APBS)." Washington University in St. Louis, St. Louis, MO, 1999–2003. http://agave.wustl.edu/apbs/

TABLE I

POISSON–BOLTZMANN SOFTWARE FOR BIOMOLECULAR SYSTEMS[a]

Software package	Description	URL	Availability
APBS[36,38]	Solves PBE in parallel with FD MG and FE AMG solvers	http://agave.wustl.edu/apbs/	Windows, all Unix; free, open source
DelPhi[27,40]	Solves PBE sequentially with highly optimized FD GS solver	http://trantor.bioc.columbia.edu/delphi/	SGI, Linux, AIX; $250 academic
GRASP[41]	Visualization program with emphasis on graphics; offers sequential calculation of qualitative PB potentials	http://trantor.bioc.columbia.edu/grasp/	SGI; $500 academic
MEAD[42]	Solves PBE sequentially with FD SOR solver	http://www.scripps.edu/bashford	Windows, all Unix; free, open source
UHBD[26,43]	Multipurpose program with emphasis on SD; offers sequential FD SOR PBE solver	http://mccammon.ucsd.edu/uhbd.html	All Unix; $300 academic
MacroDox	Multipurpose program with emphasis on SD; offers sequential FD SOR PBE solver	http://pirn.chem.tntech.edu/macrodox.html	SGI; free, open source
Jaguar[33,44]	Multipurpose program with emphasis on QM; offers sequential FE MG, SOR, and CG solvers	http://www.schrodinger.com/Products/jaguar.html	Most Unix; commercial
CHARMM[45]	Multipurpose program with emphasis on MD; offers sequential FD MG solver and can be linked with APBS	http://yuri.harvard.edu	All Unix; $600 academic

[a] Notation: PBE, Poisson–Boltzmann equation; MG, multigrid; AMG, algebraic multigrid; FD, finite difference; FE, finite element; GS, Gauss–Seidel; CG, conjugate gradient; SOR, successive overrelaxation; SD, stochastic dynamics; QM, quantum mechanics; MM, molecular mechanics; MD, molecular dynamics.

multiplication and other linear algebra operations. Although such methods are among the most common form of parallelization, they require substantial communication. An optimal parallel algorithm using P processors performs P times more work than the same algorithm using one processor; that is, the work performed by an optimal algorithm scales linearly with P. However, the substantial communication required by traditional fine-grain parallel PB solver algorithms destroys the parallel performance and offers only sublinear (suboptimal) parallel scaling. In other words, there is a diminishing return in the amount of work achieved by increasing the number of processors. Therefore, suboptimal parallel algorithms are not suitable for scaling PB calculations to larger biomolecular systems. However, two new methods have been developed for the coarse-grained parallel solution of the PB equation.[32,36] These highly efficient parallel algorithms were designed for use with both the FD and FE discretizations explained above and are suitable for the study of large biomolecular systems consisting of millions of atoms. The two methods are described in detail in the following sections.

Parallel Finite Element Methods. In 2000, a new algorithm (Bank–Holst) was described for the parallel adaptive finite element solution of elliptic partial differential equations with negligible interprocess communication.[46] As mentioned above, adaptive finite element techniques generate accurate solutions of these equations by locally enriching the basis set in regions of high error through refinement of the domain discretization. The Bank–Holst algorithm exploits this local refinement and constructs a parallel algorithm by confining local mesh refinement to regions of the problem domain assigned to particular processors.

---

[39] A. Ilin, B. Bagheri, L. R. Scott, J. M. Briggs, and J. A. McCammon, *in* "Parallelization of Poisson–Boltzmann and Brownian Dynamics Calculations." American Chemical Society Symposium Series, Vol. 592, pp. 170–185. American Chemical Society, Washington, DC, 1995.

[40] A. Nicholls and B. Honig, *J. Comput. Chem.* **12,** 435 (1991).

[41] A. Nicholls, K. A. Sharp, and B. Honig, *Proteins* **11,** 281 (1991).

[42] D. Bashford, *in* "Scientific Computing in Object-Oriented Parallel Environments" (Y. Ishikawa, R. R. Oldehoeft, J. V. W. Reynders, and M. Tholburn, eds.). Springer-Verlag, Berlin, 1997.

[43] J. D. Madura, J. M. Briggs, R. C. Wade, M. E. Davis, B. A. Luty, A. Ilin, J. Antosiewicz, M. K. Gilson, B. Bagheri, L. R. Scott, and J. A. McCammon, *Comput. Phys. Commun.* **91,** 57 (1995).

[44] G. Vacek, J. K. Perry, and J.-M. Langlois, *Chem. Phys. Lett.* **310,** 189 (1999).

[45] B. R. Brooks, R. E. Bruccoleri, B. D. Olafson, D. J. States, S. Swaminathan, and M. Karplus, *J. Comput. Chem.* **4,** 187 (1983); A. D. J. MacKerell, B. Brooks, C. L. I. Brooks, L. Nilsson, B. Roux, Y. Won, and M. Karplus, *in* "The Encyclopedia of Computational Chemistry" (P. V. R. Schleyer, ed.), pp. 271–277. John Wiley & Sons, Chichester, 1998.

[46] R. E. Bank and M. Holst, *Siam J. Sci. Comput.* **22,** 1411 (2000).

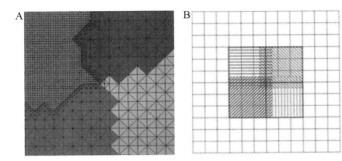

Fig. 4. Domain decomposition for parallel methods; colors denote mesh partitions belonging to individual processors of a parallel computer. (A) The Bank–Holst parallel FE method; the mesh shown has been refined from a coarser mesh by the green processor. (B) The parallel focusing method; each processor focuses from the larger coarse mesh to its particular smaller colored region. (See color insert.)

The algorithm (illustrated in Fig. 4A) proceeds as follows: the first step in the Bank–Holst parallel finite element algorithm is the solution of the equation over the entire problem domain, using a coarse resolution basis set by each processor. This solution is then used, in conjunction with an error estimator, to partition the problem domain into P subdomains that are assigned to the P processors of a parallel computer. The algorithm is load balanced by equidistributing the error estimates across the subdomains. Each processor then solves the partial differential equation over the global mesh, but confines its adaptive refinement to the local subdomain and a small surrounding overlap region. This procedure results in an accurate representation of the solution over the local subdomain. After all processors have completed their local adaptive solution of the equation, a global solution is constructed by the piecewise assembly of the solutions in each subdomain.

It can be rigorously shown that the piecewise-assembled solution is as accurate as the solution to the problem on a single global mesh.[35] Because each processor performs all of its computational work independently, the Bank–Holst algorithm requires little interprocess communication and exhibits excellent parallel scaling.

Parallel Focusing. Unfortunately, although the Bank–Holst algorithm works well for adaptive techniques, such as finite elements, it is not directly applicable to the fixed-resolution finite difference methods that are the standard methods used by the biological community. However, by combining the Bank–Holst method with other techniques, a new parallel focusing method was developed by Baker and co-workers[36] to study large biomolecular systems in parallel, using traditional finite difference discretizations of the PB equation.

Electrostatic "focusing" is a popular technique in finite difference methods for generating accurate solutions to the PB equation in subsets of the problem domain, such as a binding or titratable sites within a protein.[4,5,47] The first step in electrostatic focusing is the calculation of a low-accuracy solution on a coarse finite difference mesh spanning the entire problem domain. This coarse solution is then used to define the boundary conditions for a much more accurate calculation on a finer discretization of the desired subdomain. As noted previously,[46] this focusing technique is superficially related to the Bank–Holst algorithm, where the local enrichment of a coarse, global solution has been replaced by the solution of a fine, local multigrid problem using the solution from a coarse, global problem for boundary conditions.

Parallel focusing is simply a combination of standard focusing techniques with the Bank–Holst algorithm into a new method for the parallel solution of the PB equation with finite difference discretization. This algorithm begins with each processor independently solving a coarse global problem. The per-processor subdomains are chosen in a heuristic fashion as outlined in Fig. 4. Like standard focusing calculations, only a subset of the global mesh surrounding the area of interest is used for the parallel calculations. This subset is partitioned into P approximately equal subdomains that are distributed among the processors. Each processor then performs a fine-scale finite difference calculation over this subdomain and an overlap region that usually spans about 5–10% of the neighboring subdomains. The overlap regions are included to compensate for inaccuracies in the boundary conditions derived from the global coarse solution; however, these regions are not used to assemble the fine-scale global solution and do not contribute to calculations of observables such as forces.

Unlike previous parallel algorithms for solving the PB equation, this method has excellent parallel complexity, permitting the treatment of large biomolecular systems on massively parallel computational platforms. Furthermore, the finite difference discretization on a regular mesh allows for fast solution by highly efficient multigrid solvers. Because of the efficiency of the parallel and underlying multigrid algorithms, implementation of the parallel focusing algorithm in APBS software (see Table I) has enabled the largest PB calculations to date.[36] Specifically, Baker and co-workers used APBS on the National Partnership for Advanced Computational Infrastructure (NPACI). Blue Horizon supercomputer to solve the PB equation for the 88,000-atom small ribosomal subunit,[48] the 95,000-atom

[47] M. K. Gilson and B. H. Honig, *Nature* **330,** 84 (1987).
[48] A. P. Carter, W. M. Clemons, D. E. Brodersen, R. J. Morgan-Warren, B. T. Wimberly, and V. Ramakrishnan, *Nature* **407,** 340 (2000).

large ribosomal subunit,[49] and a 1.2-million atom microtubule structure derived from coordinates by Nogales and co-workers.[50] Figure 5 illustrates some of the specific results obtained from the microtubule calculations and the optimal parallel scaling of the algorithm. The use of parallel focusing in routine APBS calculations is straightforward; users simply specify

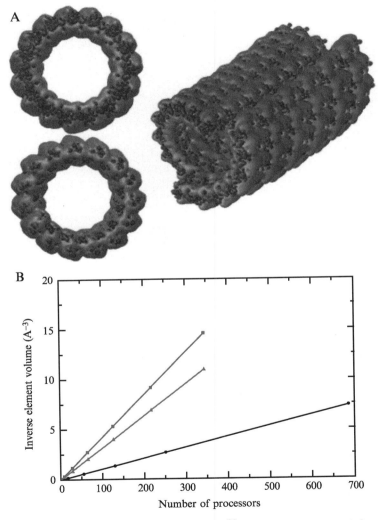

FIG. 5. Results from parallel focusing calculations.[36] (A) Views of electrostatic isocontours for a 1.2-million atom microtubule structure (blue, $+1$ kT/e; red, -1 kT/e). (B) Graph illustrating the optimal efficiency of parallel focusing algorithm. (See color insert.)

the desired number of processors and an overlap parameter and the remaining aspects of the run are calculated for them.

Software for Computational Electrostatics

Table I presents a list of the major software currently used to solve the Poisson–Boltzmann equation for biomolecular systems. There are a variety of such programs, ranging from multipurpose computational biology packages (e.g., CHARMM, Jaguar, UHBD, and MacroDox) to specialized PB solvers (e.g., APBS, MEAD, and DelPhi).

In addition to the traditional "stand-alone" software packages listed in Table I, Web-based services are also becoming available for solving the PB equation. Such services often have the advantage of removing the troublesome details of software installation from the user. Quite often, Web services also simplify and/or automate the process of setting up calculations. The first of these Web-based packages is the APBS Web Portal (https://gridport.npaci.edu/apbs/), a service for preparing, submitting, and organizing electrostatics calculations on supercomputing platforms. This Web portal is aimed at quantitative electrostatics calculations and currently does not offer substantial analysis or visualization options. On the other hand, the GRASS (http://trantor.bioc.columbia.edu/) Web service supports *qualitative* electrostatics calculations with extensive visualization capabilities. GRASS allows users to easily calculate and visualize electrostatic potentials and other properties using uploaded structures or PDB entries.

Applications to Biomedical Sciences

There have been far too many contributions of electrostatics calculations to the biological community to list them all in detail; see Refs. 1, 4, and 5, and the numerous references therein. However, we provide a brief overview of some of the applications of computational electrostatics with a particular focus on the practical application of PB methods to common biomolecular electrostatics problems.

Qualitative and Quantitative Potential Analysis

Perhaps the most readily recognizable aspects of computational electrostatics are the images produced by coloring biomolecular surfaces according to electrostatic potential (see Fig. 6A) or by plotting electrostatic isosurfaces surrounding a molecule (see Fig. 6B). By far, the most popular

[49] N. Ban, P. Nissen, J. Hansen, P. B. Moore, and T. A. Steitz, *Science* **289,** 905 (2000).
[50] E. Nogales, M. Whittaker, R. A. Milligan, and K. H. Downing, *Cell* **96,** 79 (1999).

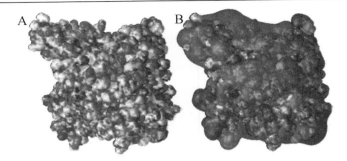

FIG. 6. Electrostatic potential of acetylcholinesterase (PDB ID 1MAH[51]). Potential calculation by solution of the PB equation with APBS. (A) Electrostatic potential mapped onto molecular surface; regions with negative potential are shaded red, regions with positive potential are shaded blue. (B) Isocontours of electrostatic potential; blue contour at $+1kT/e$, red contour at $-1kT/e$. (See color insert.)

software package for generating such packages is GRASP,[41] developed by the Honig laboratory. Images of the electrostatic potential such as in Fig. 6 provide insight into the electrostatic properties of the protein—both at the molecular surface and surrounding the biomolecule. Such insight cannot simply be obtained by mapping atomic partial charges onto the surface of the biomolecule; the complex shapes of proteins, together with their low dielectric interior, have been shown to perturb the overall electrostatic potential, often focusing it into functionally important regions.[4,5,52] In addition to visual presentation, the electrostatic potentials calculated by the PB and other methods have also been analyzed to identify active and ligand-binding sites,[53–55] to predict protein–protein[56,57] and protein–membrane interfaces,[58–60] and to categorize biomolecules on the basis of the potential

[51] Y. Bourne, P. Taylor, and P. Marchot, *Cell* **83,** 503 (1995).

[52] I. Klapper, R. Hagstrom, R. Fine, K. Sharp, and B. Honig, *Proteins* **1,** 47 (1986).

[53] A. H. Elcock, *J. Mol. Biol.* **312,** 885 (2001); M. J. Ondrechen, J. G. Clifton, and D. Ringe, *Proc. Natl. Acad. Sci. USA* **98,** 12473 (2001).

[54] Z. Y. Zhu and S. Karlin, *Proc. Natl. Acad. Sci. USA* **93,** 8350 (1996).

[55] A. M. Richard, *J. Comput. Chem.* **12,** 959 (1991).

[56] R. Norel, F. Sheinerman, D. Petrey, and B. Honig, *Protein Sci.* **10,** 2147 (2001); S. M. de Freitas, L. V. de Mello, M. C. da Silva, G. Vriend, G. Neshich, and M. M. Ventura, *FEBS Lett.* **409,** 121 (1997); L. Lo Conte, C. Chothia, and J. Janin, *J. Mol. Biol.* **285,** 2177 (1999); J. Janin and C. Chothia, *J. Biol. Chem.* **265,** 16027 (1990); V. A. Roberts, H. C. Freeman, A. J. Olson, J. A. Tainer, and E. D. Getzoff, *J. Biol. Chem.* **266,** 13431 (1991).

[57] R. C. Wade, R. R. Gabdoulline, and F. De Rienzo, *Int. J. Quantum Chem.* **83,** 122 (2001); J. Novotny and K. Sharp, *Prog. Biophys. Mol. Biol.* **58,** 203 (1992); A. J. McCoy, V. Chandana Epa, and P. M. Colman, *J. Mol. Biol.* **268,** 570 (1997).

[58] A. Arbuzova, L. B. Wang, J. Y. Wang, G. Hangyas-Mihalyne, D. Murray, B. Honig, and S. McLaughlin, *Biochemistry* **39,** 10330 (2000).

at and surrounding their surfaces.[55,57,60,61] It is expected that such methods for automatic processing of structural will become increasingly important as the number of experimentally resolved structures increases dramatically as the result of structural genomics efforts.[1,53,62]

Biomolecular Energetics

Evaluation of biomolecular energetics is another common application of PB and other computational electrostatics methods. PB methods have been used for numerous medicinal chemical and ligand/ion-binding calculations. In particular, the MM/PBSA method (a hybrid molecular mechanics–PB technique), developed by Massova and Kollman, is popular for the calculation of binding free energies.[63] In addition, Poisson–Boltzmann solvers have also aided in the investigation and simulation of several supramolecular systems, including protein–protein,[64,65] protein–nucleic acid,[66] and protein–membrane interactions.[59,60] Poisson–Boltzmann and continuum electrostatics energetics have been central to protein titration and pK_a calculations.[7,67] Such pK_a calculations play a role in several aspects of computational biology, including the setup of biomolecular simulations, the investigation of mechanisms for ligand binding and catalysis, and the identification of enzyme active sites. Finally, PB methods have also been used to investigate DNA and RNA systems, examining

[59] J.-H. Lin, N. A. Baker, and J. A. McCammon, *Biophys. J.* **83**, 1374 (2002); C. Fleck, R. R. Netz, and H. H. von Grunberg, *Biophys. J.* **82**, 76 (2002).

[60] D. Murray and B. Honig, *Mol. Cell* **9**, 145 (2002).

[61] S. A. Botti, C. E. Felder, J. L. Sussman, and I. Silman, *Protein Eng.* **11**, 415 (1998); E. Demchuk, T. Mueller, H. Oschkinat, W. Sebald, and R. C. Wade, *Protein Sci.* **3**, 920 (1994); L. T. Chong, S. E. Dempster, Z. S. Hendsch, L. P. Lee, and B. Tidor, *Protein Sci.* **7**, 206 (1998); N. Blomberg, R. R. Gabdoulline, M. Nilges, and R. C. Wade, *Proteins Struct. Funct. Genet.* **37**, 379 (1999); L. P. Lee and B. Tidor, *Protein Sci.* **10**, 362 (2001); E. Kangas and B. Tidor, *Phys. Rev. E* **59**, 5958 (1999).

[62] H. M. Berman, T. N. Bhat, P. E. Bourne, Z. Feng, G. Gilliland, H. Weissig, and J. Westbrook, *Nat. Struct. Biol.* **7**(Suppl.), 95 (2000).

[63] I. Massova and P. A. Kollman, *Perspect. Drug Discov. Design* **18**, 113 (2000); L. T. Chong, Y. Duan, L. Wang, I. Massova, and P. A. Kollman, *Proc. Natl. Acad. Sci. USA* **96**, 14330 (1999).

[64] A. H. Elcock, D. Sept, and J. A. McCammon, *J. Phys. Chem. B* **105**, 1504 (2001).

[65] J. A. McCammon, *Curr. Opin. Struct. Biol.* **8**, 245 (1998).

[66] S. W. W. Chen and B. Honig, *J. Phys. Chem. B* **101**, 9113 (1997); F. Fogolari, A. H. Elcock, G. Esposito, P. Viglino, J. M. Briggs, and J. A. McCammon, *J. Mol. Biol.* **267**, 368 (1997); V. K. Misra, J. L. Hecht, A. S. Yang, and B. Honig, *Biophys. J.* **75**, 2262 (1998); K. A. Sharp, R. A. Friedman, V. Misra, J. Hecht, and B. Honig, *Biopolymers* **36**, 245 (1995).

[67] J. E. Nielsen and G. Vriend, *Protein Struct. Funct. Genet.* **43**, 403 (2001); J. Antosiewicz, J. A. McCammon, and M. K. Gilson, *Biochemistry* **35**, 7819 (1996); D. Bashford and M. Karplus, *Biochemistry* **29**, 10219 (1990).

issues of nucleic acid–ion binding,[68] nucleic acid structure minimization, and conformational analysis.[69]

The remainder of this section describes how electrostatic energies are obtained from solution of the PB equation and used to study biological systems. The total electrostatic energy $G[\phi]$ can be obtained by integrating the solution $\phi(x)$ to the PB equation over the solution domain Ω[70]:

$$G[\phi] = \int_\Omega \left[\rho_f \phi - \frac{\varepsilon}{2}(\nabla\phi)^2 - \bar{\kappa}^2(\cosh\phi - 1)\right] dx \qquad (7)$$

The first term of Eq. (7) is the energy of inserting the protein charges into the electrostatic potential and can be interpreted as the interaction energy between charges. However, unlike analytic representations of charge–charge interactions, this energy also includes large "self-energy" terms associated with the interaction of a particular charge with itself. These self-energy terms are highly dependent on the discretization of the problem; as the mesh spacing increases, these terms become larger. In general, self-energies are treated as artifacts of the calculation and are removed by a reference calculation using the same discretization (see later section). The second term of Eq. (7) can be interpreted as the energy of polarization for the dielectric medium. Finally, the third term is the energy of the mobile counterion distribution. Both the dielectric and mobile ion energies do not include self-energy terms and therefore do not need to be corrected by reference calculations. As with the PB equation, the energy can be linearized by noting $\cosh\phi(x) - 1 \approx \phi^2(x)/2$ to give the simplified expression

$$G[\phi] = \int_\Omega \left[\rho_f \phi - \frac{\varepsilon}{2}(\nabla\phi)^2 - \frac{\bar{\kappa}^2}{2}\phi^2\right] dx = \frac{1}{2}\int_\Omega \rho_f \phi\, dx \qquad (8)$$

where the second equality is derived from substitution of the linearized PB equation [Eq. (6)] into the integrand. The following sections provide a few examples showing the application of Eq. (7) and Eq. (8) to biomolecular systems.

[68] V. A. Bloomfield and I. Rouzina, *Methods Enzymol.* **295,** 364 (1998); G. Lamm, L. Wong, and G. R. Pack, *Biopolymers* **34,** 227 (1994); V. K. Misra and D. E. Draper, *J. Mol. Biol.* **294,** 1135 (1999).

[69] R. A. Friedman, and B. Honig, *Biopolymers* **32,** 145 (1992); J. L. Hecht, B. Honig, Y. K. Shin, and W. L. Hubbell, *J. Phys. Chem.* **99,** 7782 (1995); V. K. Misra, K. A. Sharp, R. A. Friedman and B. Honig, *J. Mol. Biol.* **238,** 245 (1994).

[70] K. A. Sharp and B. Honig, *J. Phys. Chem.* **94,** 7684 (1990); M. Gilson, K. Sharp, and B. Honig, *Proteins* **4,** 7 (1988); A. M. Micu, B. Bagheri, A. V. Ilin, L. R. Scott, and B. M. Pettitt, *J. Comput. Phys.* **136,** 263 (1997).

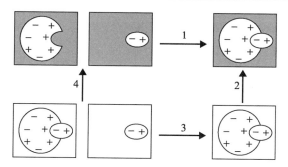

FIG. 7. Typical free energy calculations for a biomolecular system. The white background indicates an environment identical to the biomolecular interior with a low dielectric constant $\varepsilon(x) = \varepsilon_p$ and zero ionic strength $\bar{\kappa}^2(x) = 0$. The gray background indicates an environment identical to the biomolecular exterior with (generally) higher dielectric coefficient $\varepsilon(x) = \varepsilon_s$ and nonzero ionic strength $\bar{\kappa}^2(x) > 0$. Arrow 1 indicates the binding of two biomolecules in their physical state (with different internal and external environments). Arrow 2 indicates the solvation energy calculation for the biomolecular complex; the reference state has the same internal and external environment. Arrow 3 indicates the binding of the biomolecules in a constant environment identical to the biomolecular interior. Arrow 4 denotes the solvation energy calculation for the isolated biomolecular components.

Calculating Solvation Energy. The most basic component of an electrostatic energy calculation is the solvation energy, shown as processes 1 and 4 of the thermodynamic cycle depicted in Fig. 7. Specifically the solvation energy is computed as

$$\Delta G_{\text{solv}} = G_{\text{sys}} - G_{ref} \qquad (9)$$

Here G_{sys} is the electrostatic free energy of the system of interest with differing dielectric constants inside (ε_p) and outside (ε_s) the biomolecule, a fixed charge distribution corresponding to the atom locations and charges, and a varying ion accessibility coefficient ($\bar{\kappa}^2$) that is zero inside the biomolecule and equal to the bulk ionic strength outside. The reference free energy G_{ref} uses the same fixed charge distribution but has a constant dielectric coefficient equal to the value in biomolecular interior* $\varepsilon(x) = \varepsilon_p$ and a constant zero ion accessibility coefficient. Therefore a solvation energy calculation consists of the following steps:

---

* Note that ε_p may not necessarily be equal to 1 (vacuum) and therefore the computed quantity may not be directly comparable to an experimentally observed solvation free energy. See later sections for information on how the desired solvation energy can be calculated when $\varepsilon_p \neq 1$.

1. Calculate ϕ_{sys} by solving the PB equation for the system of interest, using a particular FD or FE discretization.

2. Calculate ϕ_{ref} by solving the PB equation for the reference, using the same FD or FE discretization.

3. Calculate the solvation free energy from these two solutions:

$$\Delta G_{solv} = \int_{\Omega} \left[\rho_f \left(\phi_{sys} - \frac{\phi_{ref}}{2} \right) - \frac{\varepsilon}{2} (\nabla \phi_{sys})^2 - \bar{\kappa}^2 (\cosh \phi_{sys} - 1) \right] dx \quad (10)$$

An important aspect of this calculation is the use of the same discretizations (i.e., same FD grid spacing or FE refinement) in steps 1 and 2. Both the G_{sys} and G_{ref} energies contain large self-energy terms that are dependent on the discretization; however, by choosing the same meshes or grids for the both calculations, these self-energies cancel in computing ΔG_{solv} via Eq. (10).

The electrostatic solvation energy is often supplemented with an estimate of the nonpolar contribution to biomolecular energetics. There are several approximate methods for evaluating the nonpolar term (see Ref. 3), but the most popular is simply a linear function of the solvent-accessible surface area:

$$\Delta G_{solv} = \int_{\Omega} \left[\rho_f \left(\phi_{sys} - \frac{\phi_{ref}}{2} \right) - \frac{\varepsilon}{2} (\nabla \phi_{sys})^2 - \bar{\kappa}^2 (\cosh \phi_{sys} - 1) \right] dx + \gamma A \quad (11)$$

where A is the solvent-accessible surface area and γ is the energetic coefficient or surface tension.

Given the basic role of solvation energy calculations in PB methods, most PB software packages offer some sort of automatic way to calculate Eq. (11) without requiring users to be too involved in the setup of the reference and system calculations. However, knowledge of the details of these steps is important when designing and troubleshooting more complex energy calculations.

Calculating Binding Energies. The concept of removing self-energies through solvation energy reference calculations is a common theme in electrostatic energetics. Figure 7 illustrates this procedure in the context of a binding energy calculation thermodynamic cycle. Following the usual conventions, the binding free energy is calculated as

$$\Delta G_{bind} = \Delta G_1 = \Delta G_2 + \Delta G_3 - \Delta G_4 \quad (12)$$

where ΔG_2 is the solvation energy of the complex, ΔG_4 is the total solvation energy of the isolated components, and ΔG_3 is the energy of the component charge distributions interacting in a uniform dielectric continuum. In practice, the solvation energies ΔG_2 and ΔG_4 are obtained by numerical solution of the PB equation and Eq. (11). Although ΔG_3 can also be evaluated numerically, it is typically desirable to use the analytical expression for this energy (i.e., Coulomb's law) to achieve more accurate results.

In general, both binding energy calculations and other types of energy calculations (see below) are not automatically calculated by most PB software. Therefore, users need to be aware of the details (especially the importance of removing self-energies) when setting up these computations and basing their calculations on free energy cycles as in Fig. 7.

Other Energy Calculations. As mentioned above, there are several types of energetic calculations that can be performed with electrostatic methods. However, most of these use the same general formula as outlined for binding in later sections. Consider a general transition of a biomolecular system from state A to state B (e.g., ligand binding, conformational change, protonation, etc.). To calculate the free energy for the transition $G_{sys}^{A \rightarrow B}$, one must calculate

- ΔG_{solv}^{A}: The solvation energy of state A. This is determined from a numerical solution to the PB equation and Eq. (11).
- ΔG_{solv}^{B}: The solvation energy of state B. This is determined from a numerical solution to the PB equation and Eq. (11) (with the same discretization as for state [A]).
- $\Delta G_{ref}^{A \rightarrow B}$: The energy of going from state A to B in the reference environment; homogeneous dielectric $\varepsilon(x) = \varepsilon_p$ and zero ionic strength $\bar{\kappa}^2(x) = 0$. This is determined either from a numerical solution of the PB equation or an analytical expression such as Coulomb's law.

The desired energy for the transition is then calculated from these three quantities:

$$\Delta G_{sys}^{A \rightarrow B} = \Delta G_{ref}^{A \rightarrow B} + \Delta G_{solv}^{B} - \Delta G_{solv}^{A} \qquad (13)$$

Biomolecular Dynamics

Not surprisingly, PB calculations have also found numerous applications in the study of biomolecular dynamics. In general, solutions to the PB equation are used to derive forces[22,71,72] for incorporation in an implicit solvent stochastic dynamics simulation. Among the most popular dynamics

---

[71] M. K. Gilson, M. E. Davis, B. A. Luty, and J. A. McCammon, *J. Phys. Chem.* **97**, 3591 (1993).

[72] B. Z. Lu, W. Z. Chen, C. X. Wang, and X. J. Xu, *Proteins* **48**, 497 (2002); M. E. Davis and J. A. McCammon, *J. Comput. Chem.* **11**, 40 (1990); M. Friedrichs, R. H. Zhou, S. R. Edinger, and R. A. Friesner, *J. Phys. Chem. B* **103**, 3057 (1999).

applications of PB theory is the Brownian dynamics (BD) simulation[73–75] of diffusional encounter. There have been numerous applications of BD simulations to biomolecular systems, but the most common use of these methods is the calculation of diffusive reaction rates for biomolecule–biomolecule or biomolecule–ligand binding for a variety of protein, membrane, and nucleic acid systems.[65,73,76] In general, such diffusional encounter simulations offer little or no atomistic flexibility and generally consider the rigid body motion of the biomolecules. However, PB electrostatics have also been used to supply forces for atom-level stochastic dynamics simulations using either Brownian or Langevin[75,77] equations of motion. The size of the molecules studied by stochastic dynamics has generally been limited by the calculation of electrostatic interactions—the computational bottleneck for these algorithms. Therefore previous atom-level simulations were constrained to much smaller molecules and generally used for the purposes of conformational sampling.[17,78] However, several improvements to the solution of the PB equation and calculation of forces have been made to accelerate such calculations and facilitate the stochastic dynamics simulations of much larger molecules. These improvements include new coarse-grain parallelization methods described later, effective charge methods for approximating electrostatic potentials,[79] and optimization of accuracy and update frequencies for PB force calculation methods.[80]

PB forces can be derived from functional differentiation of the free energy integral (see Refs. 22 and 71):

$$\mathbf{F}_i = -\int_\Omega \left[\phi\left(\frac{\partial \rho_f}{\partial \mathbf{y}_i}\right) - \frac{(\nabla\phi)^2}{2}\left(\frac{\partial \varepsilon}{\partial \mathbf{y}_i}\right) - (\cosh\phi - 1)\left(\frac{\partial \overline{\kappa}^2}{\partial \mathbf{y}_i}\right) \right] dx \qquad (14)$$

where $\mathbf{F}_i$ is the force on atom i, and $\partial/\partial \mathbf{y}_i$ denotes the derivative with respect to displacements of atom i. The first term of Eq. (14) represents

[73] D. L. Ermak and J. A. McCammon, *J. Chem. Phys.* **69,** 1352 (1978).

[74] T. J. Murphy and J. L. Aguirre, *J. Chem. Phys.* **57,** 2098 (1972); A. C. Branka, and D. M. Heyes, *Phys. Rev. E* **58,** 2611 (1998); W. F. van Gunsteren and H. J. C. Berendsen, *Mol. Phys.* **45,** 637 (1982).

[75] C. W. Gardiner, in "Handbook of Stochastic Methods for Physics, Chemistry, and the Natural Sciences," 2nd Ed., pp. xix and 442. Springer-Verlag, New York, 1985.

[76] S. H. Northrup, S. A. Allison, and J. A. McCammon, *J. Chem. Phys.* **80,** 1517 (1984); R. R. Gabdoulline and R. C. Wade, *Methods* **14,** 329 (1998).

[77] W. F. van Gunsteren and H. J. C. Berendsen, *Mol. Simul.* **1,** 173 (1988).

[78] J. L. Smart, T. J. Marrone, and J. A. McCammon, *J. Comput. Chem.* **18,** 1750 (1997); T. Y. Shen, C. F. Wong, and J. A. McCammon, *J. Am. Chem. Soc.* **105,** 8028 (2001); M. K. Gilson, *J. Comput. Chem.* **16,** 1081 (1995).

[79] R. R. Gabdoulline and R. C. Wade, *J. Phys. Chem.* **100,** 3868 (1996).

[80] R. Luo, L. David, and M. K. Gilson, *J. Comput. Chem.* **23,** 1244 (2002).

the force density for displacements of atom i in the potential; it can also be rewritten in the classic form $q_i \nabla \phi$ for a charged particle in an electrostatic field. The second term is the dielectric boundary pressure; that is, the force exerted on the biomolecule by the high dielectric solvent surrounding the low-dielectric interior. Finally, the third term is equivalent to the osmotic pressure or the force exerted on the biomolecule by the surrounding counterions.

Not surprisingly, like their energetic counterparts, force evaluations must also be performed in the context of reference calculation due to the presence of self-interactions in the charge distribution. Specifically, the first term of Eq. (14) contains large artificial "self-force" contributions from the charge distribution. Therefore, to calculate the PB-based forces on a biomolecule one must calculate the following three quantities:

- $\mathbf{F}_i^{sys}$: The total electrostatic force on atom i for the system of interest due to all atoms. This is calculated from the numerical solution of the PB equation and Eq. (14); the dielectric and ion accessibility coefficients are inhomogeneous.
- $\mathbf{F}_i^{ref}$: The total electrostatic force on atom i for the reference system due to all atoms. This is calculated from the numerical solution of the PB equation and Eq. (14); the dielectric and ion accessibility coefficients are homogeneous; namely, $\varepsilon(x) = \varepsilon_p$ and $\bar{\kappa}^2(x) = 0$. As with the computation of free energies, this calculation must be performed using the same discretization as $\mathbf{F}_i^{sys}$.
- $\mathbf{f}_i^{ref}$: The total electrostatic force on atom i due to all atoms except atom i (i.e., self-interactions are removed) for the reference system. Because the reference system is a set of point charges in a homogeneous dielectric medium, this force is obtained analytically from Coulomb's law.

Given these quantities, the electrostatic force on atom i with self-interactions removed is

$$\mathbf{f}_i^{sys} = \mathbf{f}_i^{ref} + \mathbf{F}_i^{sys} - \mathbf{F}_i^{ref} \tag{15}$$

Like PB energy evaluations, continuum electrostatics force calculations are often supplemented with approximations for nonpolar influences on the biomolecule. Using the solvent-accessible nonpolar term described in Section III.B2, we have an alternate form for the force that is directly analogous to the energy expression Eq. (11):

$$\mathbf{f}_i^{sys} = \mathbf{f}_i^{ref} + \mathbf{F}_i^{sys} - \mathbf{F}_i^{ref} - \int_\Omega \gamma \left(\frac{\partial A}{\partial \mathbf{y}_i}\right) dx \tag{16}$$

This extra apolar term plays an important role in electrostatic force calculations. The solvation forces obtained from PB calculations work to maximize the solvent–solute boundary surface area, thereby providing the maximum solvation for the biomolecule. Apolar forces, on the other hand, tend to drive the system to a conformation with minimum surface area. Therefore, solvation and apolar forces typically act in a delicate balance by exerting opposing forces on molecular structure.

PB force evaluation is currently available only in a small number of software packages: in particular APBS, CHARMM, and UHBD. Of these, CHARMM currently provides the most straightforward integration of PB forces into dynamics and other molecular mechanics applications.

Conclusions

The importance of electrostatics to molecular biophysics is well established; electrostatics have been shown to influence various aspects of nearly all biochemical reactions. Evaluation of the electrostatic properties of proteins and nucleic acids has a long history in the study of biopolymers and continues to be a standard practice for the investigation of biomolecular structure and function. Foremost among the models used for biomolecular electrostatics is the Poisson–Boltzmann equation. Several numerical methods have been developed to solve the PB equation; of these, finite difference is the most popular. However, new finite element techniques offer potentially better efficiency through adaptivity and may gain popularity over finite difference methods in future.

Despite over two decades of use, PB methods are still actively developed. The increasing size and number of biomolecular structures have necessitated the development of new parallel finite difference and finite element methods for solving the PB equation. These parallel techniques allow users to leverage supercomputing resources to determine the electrostatic properties of single-structure calculations on large biological systems consisting of millions of atoms. In addition, advances in improving the efficiency of PB force calculations have paved the way for a new era of biomolecular dynamics simulation methods using detailed implicit solvent models. It is anticipated that PB methods will continue to play an important role in computational biology as the study of biological systems grows from macromolecular to cellular scales.

Acknowledgment

Support for this work was provided by the National Partnership for Advanced Computational Infrastructure and Washington University in St. Louis, School of Medicine.

[6] Atomic Simulations of Protein Folding, Using the Replica Exchange Algorithm

By Hugh Nymeyer, S. Gnanakaran, and Angel E. García

Introduction

Molecular dynamics simulations of biomolecules are limited by inadequate sampling and possible inaccuracies of the semiempirical force fields. These two limitations cannot be overcome independently. The best way to optimize force fields is to validate against experiments such that simulation results are consistent with experimental data in a broad range of system's; however, validation cannot be done in the absence of adequate sampling. The energy landscape of biomolecular systems is rough. It contains many energetic barriers that are much larger than thermal energies, and these large barriers can trap biomolecular systems in local minima.[1,2] Overcoming this sampling problem is a crucial step needed to advance the field of biomolecular simulations. New techniques have been developed to enhance sampling and speed the determination of kinetic and equilibrium properties in these types of rough landscapes.[3] These techniques fall into several classes, some designed to study kinetic properties and some designed to study equilibrium properties. Many of these techniques are well suited for use on the parallel clusters common today and with distributed computing schemes.[4–6] In this chapter we provide a tutorial on the use of the replica exchange molecular dynamics (REMD) method in the context of biomolecular simulation. To place this method in context, we describe briefly some of the different methods that are being applied to sample protein conformational space.

One of the oldest methods to enhance the calculation of static properties is umbrella sampling.[7] In the umbrella sampling method a biasing potential is added to the system to enhance sampling in regions of high energy or away from equilibrium, and the sampled configurations are corrected to

[1] J. N. Onuchic, Z. Luthey-Schulten, and P. G. Wolynes, *Annu. Rev. Phys. Chem.* **48,** 545 (1997).

[2] J. N. Onuchic, H. Nymeyer, and A. E. García, *Adv. Protein Chem.* **53,** 87 (2000).

[3] S. Gnanakaran, H. Nymeyer, J. J. Portman, K. Y. Sanbonmatsu, and A. E. García, *Curr. Opin. Struct. Biol.* **13,** 168 (2003).

[4] M. R. Shirts and V. S. Pande, *Phys. Rev. Lett.* **86,** 4983 (2001).

[5] D. K. Klimov, D. Newfield, and D. Thirumalai, *Proc. Natl. Acad. Sci. USA* **99,** 8019 (2002).

[6] M. R. Shirts and V. S. Pande, *Science* **290,** 1903 (2000).

[7] G. Torrie and J. Valleau, *J. Comp. Phys.* **23,** 187 (1977).

Copyright 2004, Elsevier Inc.
All rights reserved.
0076-6879/04 $35.00

account for this bias. Usually several separate simulations are carried out, each with a different biasing potential. This method can be combined with the weighted histogram analysis method[8,9] to describe the equilibrium free energy of a system in terms of order parameters describing the system's structural properties, such as radius of gyration and number of native contacts. In the context of protein folding, this method has been most extensively utilized by Brooks and co-workers.[10-13] The umbrella sampling method parallelizes trivially because multiple simulations can be performed independently, and each calculation can be biased to selectively sample a different region of the free energy map.

A less obvious method of umbrella sampling is to use a biasing potential that is solely a function of the potential energy, the multicanonical potential function. Determining this bias self-consistently so that all potential energies are equally sampled allows the system to make a random walk in potential energy space and surmount large enthalpic barriers. This method, the multicanonical sampling method, was invented by Berg and Neuhaus[14] to enhance sampling across discontinuous phase transitions. It was first applied to peptides by Hansmann and Okamoto.[15] Although widely used, the determination of the biasing function is difficult, especially for systems with explicit solvent. The self-consistent determination of the biasing function limits the application of this method, although iterative procedures have been proposed to speed this process.[16]

Highly parallel methods have also been used to enhance the determination of protein kinetics. The simplest parallel sampling method is to run many uncoupled copies of the same system under different initial conditions.[17] The massive parallelism inherent in this method has been useful in projects such as Folding@Home,[6] which uses the excess compute cycles of weakly coupled private computers. This simple parallel simulation method is most successful in systems with implicit solvent, in which the slowest relaxation rates are increased.[18,19]

[8] A. M. Ferrenberg and R. H. Swendsen, *Phys. Rev. Lett.* **63**, 1195 (1989).

[9] S. Kumar, D. Bouzida, R. H. Swendsen, P. A. Kollman, and J. H. Rosenberg, *J. Comp. Chem.* **13**, 1011 (1992).

[10] E. M. Boczko and C. L. Brooks III, *Science* **269**, 393 (1995).

[11] C. L. Brooks III, *Acc. Chem. Res.* **35**, 447 (2002).

[12] J. E. Shea and C. L. Brooks III, *Annu. Rev. Phys. Chem.* **52**, 499 (2001).

[13] J. E. Shea, J. N. Onuchic, and C. L. Brooks III, *Proc. Natl. Acad. Sci. USA* **99**, 16064 (2002).

[14] B. Berg and T. Neuhaus, *Phys. Lett. B* **14**, 249 (1991).

[15] U. Hansmann and Y. Okamoto, *J. Comp. Chem.* **14**, 1333 (1993).

[16] T. Terada, Y. Matsuo, and A. Kidera, *J. Chem. Phys.* **118**, 4306 (2003).

[17] I. C. Yeh and G. Hummer, *J. Am. Chem. Soc.* **124**, 6563 (2002).

[18] P. Ferrara, J. Apostolakis, and A. Caflisch, *Proteins* **46**, 24 (2002).

[19] M. Shen and K. Freed, *Biophys. J.* **82**, 1791 (2002).

A more sophisticated method to enhance the time-domain sampling problem is the parallel replica method.[20] In this method independent simulations are started from the same conformational basin. When one simulation exits the basin, all the other simulations are restarted with random velocities from the new basin position. In the ideal case that barrier crossing is fast and waiting times are exponential, this method yields a linear speed-up in the rates of conformational transitions.[4] A rigorous application of this method to biomolecular systems has not been carried out because of the difficulty of defining transition states and assessing when they have been crossed.

The list of methods we have briefly described is not exhaustive. We have merely highlighted some of the methods that have been more widely applied to biomolecular systems.

In this chapter we focus on the description of a particular technique that has been effective for enhancing sampling in biomolecular simulations. This method, commonly referred to as the simulated tempering or replica exchange (RE) method, was invented independently on several occasions.[21–24] In this method, several copies or replicas of a system are simulated in parallel, only occasionally exchanging temperatures through a Monte Carlo (MC) move that maintains detailed balance. This algorithm is ideal for a large cluster of poorly communicating processors because temperature exchanges can be relatively infrequent and require little data transfer. A parallel version of this algorithm was first proposed in 1996.[24] The first application of this algorithm to a biological system was a study of Met-enkephalin.[25] It was adapted for use with molecular dynamics and named the replica exchange molecular dynamics (REMD) method.[26] This REMD method has numerous advantages. It is easy to implement and requires no expensive fitting procedures; it produces information over a range of temperature; and it works well on systems with explicit solvent as well as implicit solvent. A comparison of this algorithm with constant temperature molecular dynamics applied to peptides at room temperature showed that this algorithm decreased the sampling time by factors of 20 or more.[27] REMD has been used to study the equilibrium of protein folding and binding in models using explicit[27–30] and implicit[31–34] solvent. REMD

[20] A. Voter, *Phys. Rev. B* **57**, 13985 (1998).
[21] R. Swendsen and J. Wang, *Phys. Rev. Lett.* **57**, 2607 (1986).
[22] C. Geyer and E. Thompson, *J. Am. Stat. Assoc.* **90**, 909 (1995).
[23] E. Marinari and G. Parisi, *Europhys. Lett.* **19**, 451 (1992).
[24] K. Hukushima and K. Nemoto, *J. Phys. Soc. Jpn.* **65**, 1604 (1996).
[25] U. Hansmann, *Chem. Phys. Lett.* **281**, 140 (1997).
[26] Y. Sugita and Y. Okamoto, *Chem. Phys. Lett.* **314**, 141 (1999).
[27] K. Y. Sanbonmatsu and A. E. García, *Proteins* **46**, 225 (2002).

has been extended to include constant pressure,[35] extended to sample in the grand canonical ensemble,[31] and combined with other enhanced sampling methods.[36–39]

In the RE and REMD methods, all copies of the peptide system are identical except for temperature. Temperature exchanges are attempted at specified intervals of time. These exchanges allow individual replicas to bypass enthalpic barriers by moving to high temperature. Although the RE method enhances sampling, there currently exists no rigorous method for extracting information about folding kinetics from its application. In most instances we must resort to the energy landscape theory to compute kinetics.[40] These calculations require a measure of the configurational diffusion coefficient, which depends on the order parameters used to describe the energy surface. Lattice[41] and off-lattice[42] folding simulations on minimalist models have shown that the energy landscape theory describes the kinetics for folding and unfolding to about an order of magnitude, which is remarkable considering the many orders of magnitude that such rates can span.

We briefly outline the REMD method and discuss details necessary for applying REMD effectively to biological systems.

Replica Exchange Molecular Dynamics

We motivate the replica exchange method as a special type of Monte Carlo move for parallel umbrella sampling.

Umbrella sampling usually begins with the simulation of N copies or replicas of the system. Each replica has a unique potential chosen to enhance the sampling of some region of phase space. These simulations can be subsequently combined to "fill in" regions of phase space that would

[28] S. Gnanakaran and A. E. García, *Biophys. J.* **84,** 1548 (2003).
[29] A. E. García and K. Sanbonmatsu, *Proteins Struct. Funct. Genet.* **42,** 345 (2001).
[30] R. Zhou, B. J. Berne, and R. Germain, *Proc. Natl. Acad. Sci. USA* **98,** 14931 (2001).
[31] M. Fenwick and F. Escobedo, *Biopolymers* **68,** 160 (2003).
[32] R. Zhou and B. Berne, *Proc. Natl. Acad. Sci. USA* **99,** 12777 (2002).
[33] A. Mitsutake and Y. Okamoto, *Chem. Phys. Lett.* **332,** 131 (2000).
[34] A. Mitsutake and Y. Okamoto, *J. Chem. Phys.* **112,** 10638 (2000).
[35] T. Okabe, Y. Okamoto, M. Kawata, and M. Mikami, *Chem. Phys. Lett.* **335,** 435 (2001).
[36] Y. Sugita, A. Kitao, and Y. Okamoto, *J. Chem. Phys.* **113,** 6042 (2000).
[37] Y. Sugita and Y. Okamoto, *Chem. Phys. Lett.* **329,** 261 (2000).
[38] A. Mitsutake, Y. Sugita, and Y. Okamoto, *J. Phys. Chem.* **118,** 6676 (2003).
[39] A. Mitsutake, Y. Sugita, and Y. Okamoto, *J. Phys. Chem.* **118,** 6664 (2003).
[40] J. Bryngelson and P. G. Wolynes, *J. Phys. Chem.* **93,** 6902 (1989).
[41] N. D. Socci, J. N. Onuchic, and P. G. Wolynes, *J. Chem. Phys.* **104,** 5860 (1996).
[42] N. Hillson, J. N. Onuchic, and A. E. García, *Proc. Natl. Acad. Sci. USA* **96,** 14848 (1999).

otherwise be poorly sampled. In general, replica $i \in [1, N]$ has a unique potential $U_i(\mathbf{x}_i)$, which is a function of its coordinates $\mathbf{x}_i$. In many instances, $U_i(\mathbf{x}_i)$ is the sum of a generic molecular dynamics potential, which is the same for all replicas, and a restraint potential, which is unique to each replica.

More general potentials can be used for replicas other than this. For example, it is often useful to simulate one system at many different temperatures and to combine the results. In these simulations each replica shares the same potential $U(\mathbf{x}_i)$, but has its own temperature T_i. Because the potential energy and temperature always enter in the ratio $U(\mathbf{x}_i)/RT_i$ in MC, this is equivalent (in coordinate space) to simulating the N replicas at the same unit temperature but at different scaled potentials $E_i(\mathbf{x}_i) = U(\mathbf{x}_i)/RT_i$. In this sense, this is a type of umbrella sampling in temperature space.

The umbrella sampling method is ideally suited for parallel computers, because it involves the simulation of multiple noninteracting (and hence noncommunicating) replicas. This is conceptually equivalent to simulating one large system with many noninteracting subsystems. This supersystem is described by coordinates $\mathbf{x} = \mathbf{x}_1 \otimes \cdots \otimes \mathbf{x}_N$ and potential $U(\mathbf{x}) = U_1(\mathbf{x}_1) + \ldots + U_N(\mathbf{x}_N)$.

In umbrella sampling the N component replicas of the supersystem are completely isolated. This restriction is not necessary for preserving the equilibrium properties of the system. The RE method derives from the observation that adding extra types of MC moves that exchange coordinates between multiple replicas can dramatically increase the relaxation rates of the system. The new type of MC move that is added in the RE method is to randomly choose two replicas i_1 and i_2 in the supersystem and exchange all their coordinates. The change in energy from this MC move is

$$\Delta U = U_{i_2}(x_{i_1}) + U_{i_1}(x_{i_2}) - U_{i_2}(x_{i_2}) - U_{i_1}(x_{i_1}) \tag{1}$$

To preserve detailed balance this exchange is normally made with probability

$$\min\{1, \exp(-\Delta U/RT)\} \tag{2}$$

at a temperature T. R is the gas constant in units of kcal/mol/K if U has units of kcal/mol and T has units of degrees Kelvin.

We have described how multiple simulations at different temperatures can be understood as umbrella simulations. In this special case, replica i has an effective potential $E_i(x_i) = U(x_i)/RT_i$. Substituting this into Eq. (1) gives us the expression for the change in effective energy for a temperature exchange move between replicas i_1 and i_2:

$$\Delta E = U(x_{i_1})/RT_{i_2} + U(x_{i_2})/RT_{i_1} - U(x_{i_2})/RT_{i_2} - U(x_{i_1})/RT_{i_1} \tag{3}$$

$$= -(U_{i_1} - U_{i_2}) \times (1/RT_{i_1} - 1/RT_{i_2}) \tag{4}$$

$$= -\Delta U \times \Delta\beta \tag{5}$$

where ΔU is the difference in the real energy of replicas i_1 and i_2 and $\Delta\beta$ is the difference in the inverse temperatures. Detailed balance can be preserved if exchange moves are accepted with probability

$$\min\{1, \exp(\Delta U \times \Delta\beta)\} \tag{6}$$

For pedagogical reasons we have presented the RE as occurring via the exchange of all the coordinates of system i_1 with all the coordinates of system i_2 but with fixed temperatures T_{i_1} and T_{i_2}. In practice it is much simpler to fix the coordinates of systems i_1 and i_2 and exchange the temperatures instead. This requires the exchange of only one number, the temperature, instead of $3N$ numbers, the particle coordinates. It is in this sense then that we speak of replicas exchanging temperatures. Likewise, a general RE method may correctly be thought of as the exchange of potential functions rather than coordinates.

The RE method has been adapted for use with molecular dynamics.[26] Molecular dynamics is generally easier to implement in classic simulations than in MC and samples basins more efficiently. The essence of the REMD method is to use molecular dynamics to generate a suitable canonical ensemble in each replica rather than MC.

REMD normally occurs in coordinate and momentum space instead of just coordinate space. The total energy for a replica with particle masses $\mathbf{m}$, coordinates $\mathbf{x}$, and velocities $\mathbf{v}$ is now $H = \frac{1}{2}\mathbf{m} \cdot \mathbf{v}^2 + U(\mathbf{x})$. Similar to the RE method, exchanging temperatures T is equivalent to an exchange of coordinates and momenta. For this exchange, the acceptance probability would be

$$\min\{1, \exp(\Delta H \times \Delta\beta)\} \tag{7}$$

where the exchange depends on the difference in total energy ΔH between replicas rather than on just the potential energy U. If done in this way no velocity scaling is done.

This method for REMD is inefficient and should not be used. The inefficiency occurs because the total energy changes more rapidly with temperature than just the potential energy, so the acceptance ratio for RE moves decreases unless more replicas are used in the same temperature range. A better method for REMD was proposed by Sugita and Okamoto[26] and is in common use. The idea is to do a combined move: first exchange temperatures and then scale the momenta by $(T_{\text{new}}/T_{\text{old}})^{1/2}$. This scale transformation has a unit Jacobian, because one replica has its momenta scaled up by some amount and another replica has its momenta scaled down by the same amount. Consequently, this combination move is unbiased, and detailed balance may be preserved by accepting this attempted exchange between replicas i_1 and i_2 with probability

$$\min\{1, \exp(-\Delta E)\} \tag{8}$$

where ΔE is the change in effective energy:

$$
\begin{aligned}
\Delta E &= E_{\text{after}} - E_{\text{before}} \\
&= \left\{ \frac{1}{2}\mathbf{m} \cdot \left[\sqrt{T_2/T_1}\,\mathbf{v}_1\right]^2 /RT_2 + U(\mathbf{x}_1)/RT_2 \right. \\
&\qquad \left. + \frac{1}{2}\mathbf{m} \cdot \left[\sqrt{T_1/T_2}\,\mathbf{v}_2\right]^2 /RT_1 + U(\mathbf{x}_2)/RT_1 \right\} \\
&\qquad - \left\{ \frac{1}{2}\mathbf{m}\cdot\mathbf{v}_2^2/RT_2 + U(\mathbf{x}_2)/RT_2 + \frac{1}{2}\mathbf{m} \cdot \mathbf{v}_1^2/RT_1 + U(\mathbf{x}_1)/RT_1 \right\} \\
&= \{U(\mathbf{x}_1)/RT_2 + U(\mathbf{x}_2)/RT_1\} - \{U(\mathbf{x}_2)/RT_2 + U(\mathbf{x}_1)/RT_1\} \\
&= -\Delta U \times \Delta \beta
\end{aligned}
\tag{9}
$$

Notice that scaling the velocities exactly cancels the change in the effective energy due to the momentum part of phase space; consequently, the probability of accepting this combined temperature exchange and velocity scaling move depends as before only on the difference in potential energy between the two replicas, not their total energy.

Because (1) the temperatures of the replicas are fixed, (2) each replica is identical in all respects except the temperature (for REs in temperature), and (3) there is always one replica at each set temperature, we know that each replica must spend on average an equal amount of time at each temperature. Thus, the RE method is a way to flatten the potential of mean force (PMF) along the direction of temperature. Although temperature is most often exchanged in biological simulations, it may be useful to flatten the PMF along other coordinates (single coordinates or multiple coordinates simultaneously). In particular, the RE method may be used to approximate a multicanonical method by using umbrella restraints that restrain each replica to small overlapping windows in the potential energy. Replicas can then exchange restraints, move easily up and down in potential energy, and will spend on average an equal amount of time in each restraining window of energy.

Because each replica has a temperature that is varying in time, a direct calculation of transition or relaxation rates from the simulations cannot be made. Thermodynamic averages may be computed directly or with methods such as the weighted histogram averaging method.[8,9] To do this, the replicas must be sorted according to temperature after the simulations are complete.

The variation in temperature of any particular replica resembles that seen in the simulated annealing method[43]; however, instead of following a predetermined heating and cooling schedule, this process is self-regulated in RE methods. Replicas that are at high temperature randomly search conformational space until they find a low energy minimum. If the potential energy of this minimum is less than the potential energy of another low-temperature replica, then these two replicas will exchange temperature. Because of this, large temperature changes are slaved to conformational changes in the system. The typical temperature fluctuations and their relation to conformational changes is demonstrated from an actual REMD simulation of a solvated peptide in Fig. 1.

It is the ability of high-temperature replicas to easily cross barriers and search conformational space that makes REMD so effective. In essence, any replica can go around barriers by going up in temperature where the barriers are smaller when measured in units of RT. The flattening of barriers on the free energy surface can be illustrated by plotting the PMF along a structural coordinate and temperature simultaneously for a fully solvated peptide, GB1.[29] This type of PMF is shown in Fig. 2.

Rate enhancement can reach 10-fold or more, depending on the size of the barriers and the choice for the lowest and highest temperatures. Figure 3 shows how helical content as a function of time evolves in the same solvated helix-forming system. The increase in helical content as a function of simulation time fits well to single exponential curves with times in the 1- to 2.5-ns range. We used the last 4.0 ns of the simulation to calculate ensemble averages.

We emphasize that the temperature T that occurs in Eq. (9) is a parameter and hence constant. The temperature is not the so-called instantaneous temperature, which measures the instantaneous kinetic energy of the system and has fluctuations of magnitude $T(2/N_f)^{1/2}$, where N_f is the number of momentum degrees of freedom in the system.

The REMD method is predicated on the fact that the velocity distribution is a Maxwell–Boltzmann distribution. There are some thermostatting

[43] S. Kirkpatrick, C. D. Gelatt, and M. P. Vecchi, *Science* **220**, 671 (1983).

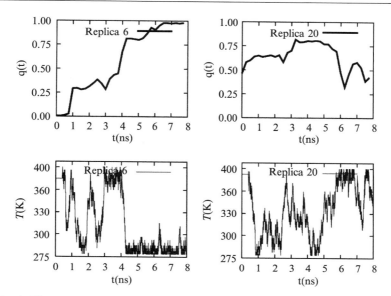

Fig. 1. The variation in helical content, $q(t)$, and temperature for two replicas from a 48-replica REMD simulation. The system is a fully solvated 21-residue peptide (Ala_{21}) that forms a single α helix at low temperature. Notice that temperature appears to undergo a random walk for each replica. Usually, large temperature changes in the replicas are connected to structural changes, where the helical content increases (replica 6) or decreases (replica 20). In replica 6, when helical content increases, the average energy decreases, and this causes the temperature of the replica to drop.

methods that may produce a nearly canonical ensemble in coordinate space, but not in momentum space. These methods should be used with caution. If the momentum space distribution at temperature T_{low} does not map onto the momentum space distribution at temperature T_{high} under scaling all velocities by an amount $(T_{high}/T_{low})^{1/2}$, then scaling the velocities after a temperature exchange can destroy the canonical distribution in coordinate space. Thermostatting methods such as the so-called weak coupling method of Berendsen et al.[44] produce a momentum distribution that is correlated with the coordinate space distribution and that does not change with temperature via this scale transformation. Thus this thermostatting method when used with REMD may produce artifacts. This is

---

[44] H. J. C. Berendsen, J. P. M. Postma, W. F. van Gunsteren, A. DiNola, and J. R. Haak, *J. Chem. Phys.* **81,** 3684 (1984).

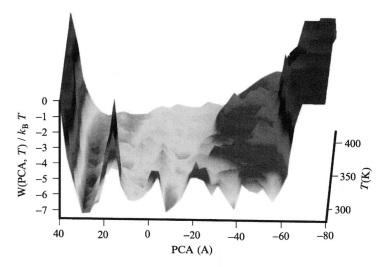

FIG. 2. Illustration of the free energy surface sampled by the REMD method as a function of a structural reaction coordinate (PCA) and temperature (T).[29] At a constant low temperature the energy landscape is rugged, with high-energy barriers separating local minima. However, replicas can move in temperature space and thereby avoid kinetic traps by moving around energetic barriers. (See color insert.)

especially worrisome in systems with implicit solvent, which do not have a large thermal bath.

In our opinion there has not been presented in the literature a proper demonstration or "proof" that combining constant temperature molecular dynamics (MD) with MC temperature exchanges will produce a canonical ensemble over long periods of time. In the absence of such a proof, we feel that it is best to let the MD system equilibrate for some time after making replica exchange moves.

To use REMD one must specify the number of replicas, the temperature of these replicas (i.e., the temperature schedule), and the frequency for attempting temperature exchanges. The effective use of REMD depends critically on the choice of these parameters. In the following section we discuss how to choose appropriate values for these parameters.

Practical Issues

In both RE and REMD methods, there are two types of moves. There are local moves, which move each replica forward by one time step, and RE moves, which exchange temperatures or potentials between replicas.

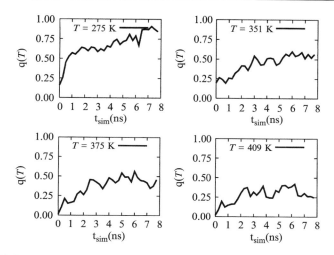

FIG. 3. Helical content (q) as a function of the replica exchange simulation time (in ns) at $T = 275, 351, 375,$ and 409 K. Averages are calculated over a 0.25-ns time window (1000 configurations). Helical content at each temperature approaches equilibrium exponentially with equilibration times between 1 and 2.5 ns.

Although one may randomly alternate between local updates and replica exchange moves, this is not an efficient way to implement the REMD algorithm. Usually, local moves for each replica are made on a different processor. Thus, a temperature exchange between replicas requires the two processors to be synchronized with respect to where in the integration cycle they are. To keep the processors synchronized, it is better to make a fixed number of local updates for each replica followed by one or more attempted RE moves. For example, one might choose to advance each replica forward by 250 integration steps and follow this by one or more attempted temperature exchanges. Making a large fixed number of local moves between RE attempts keeps the synchronization cost to a minimum. If, however, all the replicas are running on a single processor, there is not such a large advantage with this update scheme.

The number of local moves between attempted replica moves should be fairly large. First, because RE moves require synchronization among two or more replicas, these moves are more "expensive" to make than local update moves. Making frequent RE attempts will drastically slow most simulations. Second, RE moves should be made infrequently because there are not normally large barriers to equilibration along the direction of the exchange (usually temperature), so only a few attempted moves are needed to attain a temperature equilibrium. Third, large temperature changes are slaved to conformational changes in the replicas; consequently, temperature changes are not normally rate limited by the exchange attempt frequency but by an intrinsic relaxation process in the system. Fourth, in the REMD method it has not been clearly demonstrated that molecular dynamics with all types of thermostatting will correctly generate a canonical ensemble when combined with RE moves. It is possible that for some thermostats a period of local equilibration must set in before exchanges are attempted. In our estimation, for typical molecular dynamics simulations with 1- to 2-fs time steps, making about 250 integration steps between attempted RE moves is appropriate.

An important issue for RE is the number of replicas and the choice for their potentials. In the special case of temperature this amounts to choosing the number of temperatures (number of replicas), the temperature spacing between replicas, and the minimum and maximum temperatures. These are obviously not all independent choices. The minimum temperature is usually set to be near the temperature of interest biologically, with the knowledge that sampling difficulties and simulation time will increase quickly as the minimum temperature is decreased. The maximum temperature must be set high enough so that replicas at this temperature quickly cross their largest energetic barriers and lose memory of their initial condition. For most small biological simulations this is more than 500 K or about two times larger in temperature than room temperature. The number of replicas and their temperature spacing must then be adjusted to effectively span this temperature range.

Let us discuss generally how the number of replicas needed increases with system size. In large systems the average energy is an extensive quantity, scaling linearly with system size. Also, the energy fluctuations of distant parts of large systems are uncorrelated, so the mean squared fluctuation in energy is also an extensive quantity. Consequently, doubling the system size or number of particles N ($\sim$ Volume) doubles the average energy spacing between adjacent temperatures, but only increases the energy fluctuations at a given temperature by $N^{1/2}$. Because the size of the energy fluctuations must be similar to the energy spacing between replicas for efficient exchanges to occur, the total number of replicas necessary to span a temperature range will in general increase as $N^{1/2}$, the square root of the size of the system, or Volume$^{1/2}$. This dependence ultimately limits the application of the RE method with temperature exchanges.

Simulations of systems with explicit solvent are more costly than simulations with implicit solvent because the explicitly solvated systems have a much greater number of particles. A rough estimate for the temperature spacing for replicas at room temperature that are explicitly solvated can be obtained by assuming that they are mostly water. If the temperature spacing between replicas is ΔT, then the energy spacing between replicas is approximately

$$\Delta E = C_v m \Delta T \tag{10}$$

where C_v is the coordinate space contribution to the specific heat at constant volume (and the mean temperature of the replicas) and m is the mass in grams. The root mean squared fluctuations are approximately

$$\sqrt{\Delta E^2} = \sqrt{C_v m k_B T^2} \tag{11}$$

where $k_B \approx 1.38 \times 10^{-23}$ J/K is the Boltzmann constant. We use k_B in this formula instead of R, the gas constant, because we are working in energy units of joules instead of kcal/mol. This ratio of these quantities should be of order 1, so the temperature spacing should be about

$$\Delta T = \sqrt{k_B T^2 / m C_v} \tag{12}$$

Using $C_v \approx 2.8$ J/g/K $= 4.2$ J/g/K $- 1.4$ J/g/K (approximate C_v from Ref. 45 minus an estimated momentum space contribution of $6/2k_B$ per water molecule because vibrational modes are not significantly excited at room temperature) and assuming a cubic box 30 Å on a side with a water density of 1000 g/liter gives a mass of 27×10^{-21} g and a temperature spacing at 295 K of $\Delta T \approx 4.0$ K. This is in agreement with actual spacings which have worked successfully in solvated systems using REMD (see Table I[26–28, 46–48] and Fig. 4).

Equation (12) indicates several other useful things. First, it confirms our previous statement that the temperature spacing in a homogeneous system should decrease as $1/m^{1/2}$, where m is the mass, so the number of replicas should increase like the square root of the number of particles. Second, if a system has an approximately constant heat capacity over the span of replica temperatures, then the ideal temperature distribution (i.e., the distribution for which exchanges between neighboring replicas is approximately constant) is an exponential distribution. For example, replica $i \in [1, N]$

[45] D. R. Lide (ed.), "CRC Handbook of Chemistry and Physics," 81st Ed. CRC Press, Boca Raton, FL, 2000.

[46] A. E. García and K. Y. Sanbonmatsu, *Proc. Natl. Acad. Sci. USA* **99**, 2782 (2002).

[47] H. Nymeyer and A. E. García, *Proc. Natl. Acad. Sci. USA* **100**, 13934 (2003).

[48] A. E. García and J. N. Onuchic, *Proc. Natl. Acad. Sci. USA* **100**, 13898 (2003).

TABLE I

PARAMETERS USED IN SOME REMD SIMULATIONS CARRIED OUT ON
PROTEINS AND PEPTIDES OF VARIOUS SIZES[a]

System	No. of protein atoms	No. of water molecules	ΔT at 300 K	No. of replicas	T range (K)	Ref.
Met-enkephalin, TIP3P	84	587	9	16	275–419	27
Met-enkephalin, implicit SA	84	0	56	8	200–700	26
SH3 divergent turn, TIP3P	132	903	9	24	276–469	28
A21, TIP3P	222	2640	4	48	278–500	47
A21, GB/SA	222	0	9	16	275–419	48
Fs peptide, TIP3P	282	2660	5	46	275–551	47
Fs peptide, GB/SA	282	0	9	16	200–624	48
Protein A, TIP3P	734	5107	2.2	82	277–548	49

Definitions: The explicit TIP3P water model and the implicit surface area (SA) and generalized born/surface area (GB/SA) models are described in reference 48.

[a] The systems are Met-enkephalin in explicit and implicit (surface area model) solvent, the SH3 divergent turn peptide, A21, and Fs peptides in explicit and GB/SA implicit solvent, and a 46-amino acid fragment of protein A. ΔT is the approximate temperature spacing of neighboring replicas at 300 K.

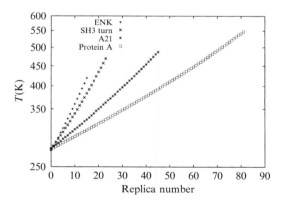

FIG. 4. A semilog plot of the temperature schedules used for various explicitly solvated systems with REMD: Met-enkephalin (ENK), the 7-residue SH3 diverging turn, a 21-residue polyalanine sequence, and protein A. In these systems, individual temperature spacings were adjusted to produce equal temperature exchange rates between neighboring replicas. This optimization produces a nearly exponential distribution of temperatures, which is optimal for systems with constant heat capacities as predicted from Eq. (12). The exponent of these distributions varies as $1/N^{1/2}$.

should have a temperature $T_i = T_1(\Delta T)^{(i-1)}$, where $\Delta T = T_2 - T_1$ is the temperature spacing of the lowest two temperatures. Third, in regions of temperature where the specific heat is larger, the spacing in temperature must be reduced. In particular, this indicates that at a first-order phase transition with a divergent heat capacity the spacing must become infinitely close. This in general indicates that the RE and REMD methods (with exchanges in temperature) do not work well across first-order transitions. However, most applications are to small systems in which finite size broadening smears out sharp specific heat peaks sufficiently that the RE and REMD methods are effective.

In the absence of detailed knowledge about the underlying specific heat, it is best to make some initial guesses about the form of the specific heat and subsequently adjust the temperature spacing of the replicas to equalize replica exchange (RE) move acceptance ratios. For explicitly solvated systems, an exponential temperature distribution is sufficient to start with; however, implicit solvent simulations tend to have a much stronger variation of specific heat with temperature. For implicit solvent simulations the higher temperatures must be spaced much closer. In small systems a uniform temperature spacing may be sufficient.

In Table I we give examples of some biological simulations that used the RE and REMD methods with both implicit and explicit solvent. Notice that for implicit solvent calculations the number of replicas is small, and the temperature steps between replicas are large. For explicit solvent models, the majority of atoms in the system are in the solvent (water). The largest REMD simulation performed to date is for a 46-amino acid fragment of protein A. This system has more than 16,000 atoms. The temperature steps at low temperature are small (2.2 K), and the number of replicas is large (82). Figure 4 shows a semilog plot of the replica temperatures as a function of replica number for the explicit solvent systems listed in Table I. The replica temperatures shown here were adjusted to give an almost constant 20% RE acceptance rate. The replica temperatures approximates an exponential distribution, but for larger systems it increase slightly faster than exponential. Exponential fits to these curves give exponents that are proportional to $1/N^{1/2}$, where N is the number of atoms in the system.

Various applications of the REMD method have been cited in text and in Table I. Among the most important results obtained from explicit solvent simulations have been the use of REMD to study the thermodynamics of helix–coil transitions in peptides that are experimentally characterized. These calculations used two different AMBER force fields (Amber 94 and Amber 96), and found that both force fields fail to reproduce the experimental observations. Use of a perturbation approach showed that a modified force field in which the bias in the backbone dihedral angles was eliminated reproduced the experimental results reasonably well.[49] These calculations showed that the use of exhaustive sampling of well-character- ized systems, combined with the use of multiple force fields, can identify deficiencies in the existing force fields. In addition, it demonstrated that existing force fields are not wildly inaccurate, but that small modifications can bring them into agreement with experimental observations.

The REMD simulations have some limitations. The most obvious is the limitation in system size for models using explicit solvent. Systems with a large number of atoms will require a large number of replicas, and longer simulation times. One advantage of the trivial parallelization features of the REMD is that the method can be implemented in a dual parallelization scheme, in which each replica uses multiple processors, but such implemen- tations may require many hundreds of processors for a large system. Other limitations encountered in common implementations of the REMD are the use of constant volume conditions. Constant volume simulations will exag- gerate the hydrophobic effect at high temperatures and can lead to artifi- cially high temperature transitions and exaggerate the amount of chain collapse.[50,51] This could be overcome by performing constant pressure cal- culations. However, simple water models commonly used in biomolecular simulations do not reproduce the water phase diagram far away from 300 K. Another option is to do the calculations at densities dictated by the water–vapor coexistence curve.

Use of the REMD method has been limited so far to a few systems. We expect that this method, and variations on it, will be widely used in the near future. The accessibility of low-cost computer clusters in particular will spur the use of this method.

Appendix

We describe how one may practically implement REMD, using existing molecular dynamics codes.

Most molecular dynamics codes are essentially large loops over the number of integration steps. Each integration step advances the time for- ward one step. After a certain number of integration steps quantities such as energy or pressure are output. To implement REMD in an existing MD code, the update steps and output steps should be followed at certain intervals by an attempted RE.

[49] S. Gnanakaran and A. E. García, *J. Phys. Chem. B* **107,** 12555 (2003).
[50] G. Hummer, S. Garde, A. E. García, A. Phorille, and L. R. Pratt, *Proc. Natl. Acad. Sci. USA* **93,** 8951 (1996).
[51] G. Hummer, L. W. Pratt, and A. E. García, *J. Phys. Chem. A* **102,** 7885 (1998).

Most MD codes today utilize parallelism through message-passing directives. It can be cumbersome to insert RE moves into such an MD code. First of all, this requires the use of different inputs. Unfortunately, data input/output (I/O) is often a nonstandard feature of message-passing standards. Second, codes often have hard-coded communication that is difficult to locate and change.

To avoid these problems, many people have come up with the idea of using a master program that goes into a loop, within which it creates multiple input files for a system that are identical except for initial conditions and temperature, spawns multiple MD programs to advance the system with these different inputs, and then extracts from the output of these programs the data needed for an RE. This method may run slowly because stopping and restarting an MD program often carries with it considerable overhead such as computing initial cell lists, generating pair exclusion lists, tabulating data for interpolating a complicated function, and doing the initial I/O.

A better method is to use a client/server arrangement. A single server receives data from multiple MD clients, makes any RE moves, and sends the new temperatures to the MD clients. Any modification to the existing MD programs is minimal, usually a few lines of code within the primary integration loop and after the I/O statements to send the potential energy and temperature to the server, receive from the server the new temperature, and then rescale the velocities and reset any temperature-dependent parameters. To avoid difficulties in adapting existing message-passing code, one can do the communication with sockets. Socket codes are robust, working on a range of UNIX and non-UNIX operating systems.

We present the source code in the C language for a rudimentary server and for a subroutine to make an existing MD code into a client able to communicate with that server. This code is provided without any warranties. The user runs it at his or her own risk.

The code consists of several subroutines with the dependencies shown as in Fig. 5. *arbiter* is the server. *arbitrate* is a subroutine called from a MC or MD program. An example MC program for a particle in a bistable potential that utilizes this code is also included. In practice, one copy of *arbiter* is started first, followed by several copies of the MC or MD program, which are informed of the Internet address and port number of the server code.

util.h

```
#ifndef __util_h_
#define __util_h_
```

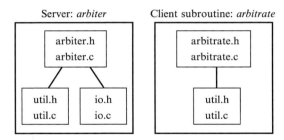

FIG. 5. Diagram of code dependencies for a server and client subroutine that implement the RE or REMD method. For example, *arbiter* can be compiled via a command: cc arbiter.c util.c io.c -o arbiter -lm. The client subroutine should be compiled to object code via a command: cc -c arbitrate.c util.c.

```
double ntohd(double x);/* convert double from network to host
    order */
double htond(double x);/* convert double from host to network
    order */
int big_endian(void);/* is this machine big_endian? */

#endif
```

util.c

```
#include "util.h"

double
ntohd(double x)
{
    static char p;
    static int i;
    static int len;

    if (big_endian())        /* do nothing, if big_endian */
        return x;

    len = sizeof(double);    /* otherwise reverse the byte order */
    for (i = 0 ; i < len/2 ; i++) {
        p = ( (char *)&x) [i];
        ( (char *)&x) [i] = ( (char *)&x) [len - i - 1];
        ( (char *)&x) [len - i - 1] = p;
    }
    return x;
}
```

```
double
htond(double x)
{
   static char p;
   static int i;
   static int len;

   if (big_endian() )/* do nothing, if big_endian */
      return x;

   len = sizeof(double);/* otherwise reverse the byte order */
   for (i = 0;i < len/2;i++) {
      p = ( (char *)&x) [i];
      ( (char *)&x) [i] = ( (char *)&x) [len - i - 1];
      ( (char *)&x) [len - i - 1] = p;
   }
   return x;
}

int
big_endian(void)
{
   static unsigned short s = 1;
   return ( (unsigned char *)&s) [1];
}
```

io.h

```
#ifndef _ _io_h_
#define _ _io_h_
/*
   open_connections opens an array of incoming socket
   connections at the specified port number in timeout number
   of seconds
*/
int
open_connections(int *sockets,
                 int connects_wanted,
                 int port_number,
                 int timeout);
int
close_connections(int *sockets,
                  int connections_open);
#endif
```

io.c

```
#include <stdio.h>
#include <string.h>
#include <stdlib.h>
#include <unistd.h>
#include <fcntl.h>
#include <sys/types.h>
#include <sys/socket.h>
#include <netinet/in.h>
#include <arpa/inet.h>
#include <time.h>
#include "io.h"
#include "arbiter.h"

int
open_connections(int *sockets,
                 int connects_wanted,
                 int port_number,
                 int timeout)
{
   int sin_size;
   int sock_master;
   time_t current_time;
   int connects_made=0;
   struct sockaddr_in my_address, *their_address;

   if ( (their_address =
         calloc(connects_wanted, sizeof(struct sockaddr_in) ) )
         == 0 ) {
     perror("open_connections");
     exit(1);
   }
   if ((sock_master = socket(AF_INET,SOCK_STREAM,0)) == -1) {
     perror ("socket");
     exit(1);
   }

   fcntl(sock_master, F_SETFL, O_NONBLOCK);

   my_address.sin_family = AF_INET;
   my_address.sin_port = htons(port_number);
   my_address.sin_addr.s_addr = INADDR_ANY;
```

```
bzero(&(my_address.sin_zero),8);

if ( bind( sock_master,
           (struct sockaddr *)&my_address,
           sizeof(struct sockaddr) ) == -1 ) {
  perror ("bind");
  exit(1);
}

if (listen(sock_master,BACKLOG) == -1) {
  perror ("listen");
  exit(1);
}

current_time = time(0);
sin_size = sizeof(struct sockaddr_in);
do {

   if ((sockets[connects_made] = \
        accept(
               sock_master,
               (struct sockaddr *) &(their_address
                 [connects_made]),
               &sin_size
               )
        ) != -1)
     ++connects_made;
} while ( (time(0) < current_time + timeout) &&
          (connects_made < connects_wanted) );
free(their_address);
return connects_made;
}

int
close_connections(int *sockets, int connections_open)
{
  int connects_close = 0;
  int i;

  for (i = 0;i < connections_open;i++)
    connects_closed += !close(sockets[i]);
  return connects_closed=connections_open;
}
```

arbiter.h

```
#ifndef --arbiter_h
#define --arbiter_h

#define BACKLOG 10   /* buffer for waiting connections */
#define TIMEOUT 400 /* exit if conect's aren't made in this #
  seconds */
#define sq(x) ((x)*(x))

void
arbitrate(int *sockets,
          int number_of_connects);

void
collect(int *sockets,
        double *Epot,
        double *T,
        int number_of_connects);

void
rearrange(double *Epot,
          double *T,
          int number_of_connects);

void
swap(double *x, double *y);

void
distribute(int *sockets,
           double *T,
           int number_of_connects);
```

arbiter.c

```
#include <stdio.h>
#include <stdlib.h>
#include <string.h>
#include <sys/types.h>
#include <sys/socket.h>
#include <netinet/in.h>
#include <math.h>
#include <time.h>
#include "io.h"
#include "util.h"
#include "arbiter.h"
```

```
/*
   This is the server code which communicates through the
   specified port number (essentially arbitrary except
   low numbers are reserved) and expects a certain number
   of clients to connect.
*/
int
main(int argc, char *argv[])
{
   int *sockets;
   int port_number;
   int numconnects;
   int connects_made;
   long seed;

   if (argc!=4) {
      fprintf(stderr,"!usage: %s port_number #_of_connections
      seed\n",argv[0]);
      exit(0);
   };

   sscanf(argv[1],"%d",&port_number);
   sscanf(argv[2],"%d",&numconnects);
   sscanf(argv[3],"%d",&seed);

   srand48(seed);

   sockets = calloc(numconnects, sizeof(int));
   if (!sockets) {
      perror("calloc");
      exit(1);
   }

   if (numconnects != (connects_made =
      open_connections(sockets, numconnects,port_number,
         TIMEOUT))) {
      perror("open_connections");
      fprintf(stderr,
         "! Only %d connections made instead of %d\n",
         connects_made, numconnects);
      exit(1);
   }

   arbitrate(sockets,numconnects);
```

```
        return 0;
    }

    void
    arbitrate(int *sockets,
              int number_of_connects)
    {
    double *Epot, *T;
    if ((Epot = calloc(number_of_connects,sizeof(double))) == 0) {
      perror("calloc");
      exit(1);
    }
    if ((T = calloc(number_of_connects,sizeof(double))) == 0) {
      perror("calloc");
      exit(1);
    }
    collect(sockets.Epot,T,number_of_connects);
    while (T[0]! = 0.0) {
      rearrange(Epot, T, number_of_connects);
      distribute(sockets, T, number_of_connects);
      collect(sockets, Epot, T, number_of_connects);
    }
    close_connections(sockets,number_of_connects);
    free(Epot);
    free(T);
    }

    void
    collect(int *sockets,
            double *Epot,
            double *T,
            int number_of_connects)
    {
      static int i;
      static int recv_size;

      for (i = 0;i < number_of_connects;i++) {
        recv_size = recv(sockets[i],Epot + i,sizeof(double),0);
        if (recv_size! = sizeof(double)) {
          perror("recv");
          exit(1);
```

```
    }
    Epot[i] = ntohd(Epot[i]);
  }
  for (i = 0;i < number_of_connects;i++) {
    recv_size = recv(sockets[i],T + i,sizeof(double),0);
    if (recv_size! = sizeof(double)) {
        perror("recv");
        exit(1);
    }
    T[i] = ntohd(T[i]);
  }
}

void
rearrange(double *Epot,
          double *T,
          int number_of_connects)
{
  static int i;
  static int first,second;
  static double dE;

  for (i = 0;i < number_of_connects*number_of_connects;i++) {
    first = drand48()*number_of_connects;
    second = drand48()*number_of_connects;
    dE = Epot[first] / T[second] + \
        Epot[second] / T[first] - \
        Epot[first] / T[first] - \
        Epot[second] / T[second];
    if (dE< = 0.0 || drand48() < exp(-dE*200.0/0.397441))
        swap(T + first,T + second);
  }
}

void
swap(double *x, double *y)
{
  static double z;
  z = *x;
  *x = *y;
  *y = z;
}
```

```c
void
distribute(int *sockets,
           double *T,
           int number_of_connects)
{
    static int i;
    for (i = 0; i < number_of_connects; i++) {
    T[i] = htond(T[i]);
    if (send(sockets[i],T + i,sizeof(double),0) != sizeof
      (double)) {
      perror("send");
      exit(1);
    }
  }
}
```

arbitrate.h

```c
#ifndef __arbitrate_h_
#define __arbitrate_h_

int
client_open(char* addr_string,
            int port_number);

/*
Subroutine arbitrate makes an attempted exchange from
the current replica with energy *E and temperature
*T through the server at the internet address
*inet1.*inet2.*inet3.*inet4 and port number given by
*port_number
*/

void
arbitrate(double *E,
          double *T,
          int *inet1,
          int *inet2,
          int *inet3,
          int *inet4,
          int *port_number);

#endif
```

arbitrate.c

```c
#include <stdio.h>
#include <string.h>
#include <stdlib.h>
#include <sys/types.h>
#include <sys/socket.h>
#include <arpa/inet.h>          /* inet_addr function */
#include <netinet/in.h>
#include <unistd.h>             /* close function */
#include "util.h"
#include "arbitrate.h"

int
client_open(char* addr_string,
            int port_number)
{
   static int sock;
   static struct sockaddr_in dest_addr;
   static int error_flag;
   /*- - - - - - - - - - - - - - - - - - - - - - - - -*
     open an IPv4 socket of type SOCK_STREAM
   *- - - - - - - - - - - - - - - - - - - - - - - - - -*/
   sock = socket(AF_INET,SOCK_STREAM,0);
   if (sock == -1) {
     perror("socket");
     exit(1);
   }
   /*- - - - - - - - - - - - - - - - - - - - - - - - -*
          set the destination address
   *- - - - - - - - - - - - - - - - - - - - - - - - - -*/
   dest_addr.sin_family      = AF_INET;
   dest_addr.sin_port        = htons(port_number);
   dest_addr.sin_addr.s_addr = inet_addr(addr_string);
   bzero(&(dest_addr.sin_zero), 8);
   /*- - - - - - - - - - - - - - - - - - - - - - - -*
              connect to server
   *- - - - - - - - - - - - - - - - - - - - - - - - -*/
   error_flag = connect(sock,
                        (struct sockaddr*)&dest_addr,
                        sizeof(struct sockaddr));
   if (error_flag == -1) {
     perror("connect");
```

```
        exit(1);
    }
    /*- - - - - - - - - - - - - - - - - -*
        return the open socket
    *- - - - - - - - - - - - --- - - - -*/
    return sock;
}
void
arbitrate(double *E,
          double *T,
          int *inet1,
          int *inet2,
          int *inet3,
          int *inet4,
          int *port_number)
{
    static int is_initialized=0;
    static int sock;
    static double energy, temperature;
    static char addr_string[16]; /* max inet address + 1 */
    static int len;

    /*
        This subroutine makes an attempted exchange from
        the current replica with energy *E and temperature
        *T through the server at the internet address
        *inet1.*inet2.*inet3.*inet4 and port number
        given by *port_number

        Pointers are used to facilitate use in Fortran code.
        If this subroutine is placed in Fortran code, it may
        need a trailing underscore in the name.
    */
    /*- - - - - - - - - - - - - - - - - - - - - - - - - - - -*
        First time called? Then open a socket
    *- - - - - - - - - - - - - - - - - - - - - - - - - - - -*/
    if (!is_initialized) {
        snprintf(addr_string,16,"%d.%d.%d.%d",
            *inet1,*inet2,*inet3,*inet4);
        sock = client_open(addr_string,*port_number);
        is_initialized=1;
```

```
    }
    /*- - - - - - - - - - - - - - - - - - - - - - - - - -*
         communicate with the server: send
    *- - - - - - - - - - - - - - - - - - - - - - - - -*/
    energy = htond((double)*E);
    temperature = htond((double)*T);

    len = send(sock, &energy, sizeof(double),0);
    if (len! = sizeof(double)) {
       perror("send");
       exit(1);
    }
    len = send(sock, & temperature, sizeof(double),0);
    if (len!= sizeof(double)) {
       perror("send");
       exit(1);
    }

    /*- - - - - - - - - - - - - - - - - - - - - - - - - - -*
        if energy=temperature=0 close socket
    *- - - - - - - - - - - - - - - - - - - - - - - - - - -*/
    if (*E == 0 && *T == 0) {close(sock); return;};

    /*- - - - - - - - - - - - - - - - - - - - - - - - - - -*
            receive new temperature
    *- - - - - - - - - - - - - - - - - - - - - - - - - - -*/
    len = recv(sock, &temperature, sizeof(double),0);
    if (len! = sizeof(double) ) {
       perror("recv");
       exit(1);
    }

    *T = ntohd((double)temperature);
}
```

MC.c

```
#include <stdio.h>
#include <stdlib.h>
#include <math.h>
#include "arbitrate.h"

#define NSTEP_EQUIL 100
```

```
#define NSTEP_PROD 1000
#define NSTEP_SWAP 10
#define NSTEP_OUTPUT 100
double energy(double x);

/*
    A toy Monte Carlo code used to illustrate the replica
    exchange program. "inetA.inetB.inetC.inetD" must be
    the internet address of the machine on which the server is
    running. port must match the port number set by the server.
    Compile via: cc MC.c arbitrate.c util.c -o MC -1m
*/

int main(int argc, char* argv[])
{
   int i;
   doubel T,x,dx,E,dE;
   long seed;
   int inetA = 127, inetB = 0, inetC = 0, inetD = 1;
   int port = 123457;

   if (argc! = 4) {
      fprintf(stderr, "!usage: %s temperature seed
         start-pos\n",argv[0]);
      exit(1);
}

   sscanf(argv[1],"%1f",&T);
   sscanf(argv[2],"%1d",&seed); srand48(seed);
   sscanf(argv[3],"%1f",&x); E = energy(x);

   for (i = 0;i < NSTEP_EQUIL;i++) {
      do {dx = drand48()-0.5;} while (dx == -0.5);
      dE = energy(x + dx)-E;
      if (dE< = 0.0 || drand48()< = exp(-dE/T)) {x += dx; E += dE;};
   }

   for (i = 0; i<NSTEP_PROD; i++) {
      do {dx = drand48()-0.5;} while (dx == -0.5);
      dE = energy(x + dx)-E;
      if (dE <= 0.0 || drand48()<= exp(-dE/T)) {x += dx; E += dE;};
      if (i%NSTEP_SWAP == 0)
          arbitrate(&E,&T,&inetA,& inetB,&inetC,&inetD,
        &port);
      if (i%NSTEP_OUTPUT == 0)
```

```
    printf("%10.41f %10.41f\n",T,x);
  }
  E=T=0.0;
  arbitrate(&E,&T,&inetA,&inetB,& inetC,&inetD,&port);

return 0;
  }

double energy(double x)
{
#define X_MIN 2.0

   return (x*x-X_MIN*X_MIN) * (x*x-X_MIN*X_MIN);
}
```

[7]   DNA Microarray Time Series Analysis: Automated Statistical Assessment of Circadian Rhythms in Gene Expression Patterning

By MARTIN STRAUME

Introduction

DNA microarray technology has emerged as a novel and powerful means for investigating the network architecture underlying transcriptional regulation of biological systems.[1-3] The unique advancement afforded by microarrays is the capacity to conveniently experimentally interrogate genome-wide system responsiveness in a temporally parallel manner.[4-7] This technology is not only expanding our understanding in basic scientific studies of cell and developmental biology,[8] but is also being developed to apply in clinical settings for purposes of diagnosis, prognosis, and

[1] M. Chee, R. Yang, E. Hubbell, A. Berno, X. C. Huang, D. Stern, J. Winkler, D. J. Lockhart, M. S. Morris, and S. P. Fodor, *Science* **274,** 610 (1996).

[2] A. Schulze and J. Downward, *Nat. Cell Biol.* **3,** E190 (2001).

[3] J. Ziauddin and D. M. Sabatini, *Nature* **411,** 107 (2001).

[4] M. N. Arbeitman, E. E. Furlong, F. Imam, E. Johnson, B. H. Null, B. S. Baker, M. A. Krasnow, M. P. Scott, R. W. Davis, and K. P. White, *Science* **297,** 2270 (2002).

[5] R. J. Cho, M. Fromont-Racine, L. Wodicka, B. Feierbach, R. Stearns, P. Legrain, D. J. Lockhart, and R. W. Davis, *Proc. Natl. Acad. Sci. USA* **95,** 3752 (1998).

[6] B. Futcher, *Curr. Opin. Cell Biol.* **12,** 710 (2000).

[7] I. Simon, J. Barnett, N. Hannett, C. T. Harbison, N. J. Rinaldi, T. L. Volkert, J. J. Wyrick, J. Zeitlinger, D. K. Gifford, T. S. Jaakkola, and R. A. Young, *Cell* **106,** 697 (2001).

Copyright 2004, Elsevier Inc.
All rights reserved.
0076-6879/04 $35.00

assessment of treatment outcome of a variety of human diseases and disorders.[9-14]

The specific context in which DNA microarrays are considered in the present discussion is in elucidation of network dynamics of circadian gene regulation.[15] Application of this technology to circadian biology is particularly apropos given the molecular basis of intracellular circadian regulation, a system of transcriptional–translational feedback control.[16-19] The emergence of microarray technology now permits research efforts to go beyond previously technically limited studies employing functional screening, differential display, and other related methods.[20-22]

Examinations characterizing gene expression dynamics in relation to diurnal and circadian rhythmicity have contributed to current understanding of how core clock oscillator mechanisms may functionally interact to orchestrate physiological and behavioral outputs.[23-33] The eventual goals of such studies are to develop an accurate systems-level understanding of

[8] C. Grundschober, F. Delaunay, A. Puhlhofer, G. Triqueneaux, V. Laudet, T. Bartfai, and P. Nef, *J. Biol. Chem.* **276,** 46751 (2001).

[9] S. A. Armstrong, J. E. Staunton, L. B. Silverman, R. Pieters, M. L. den Boer, M. D. Minden, S. E. Sallan, E. S. Lander, T. R. Golub, and S. J. Korsmeyer, *Nat. Genet.* **30,** 41 (2002).

[10] T. R. Golub, D. K. Slonim, P. Tamayo, C. Huard, M. Gaasenbeek, J. P. Mesirov, H. Coller, M. L. Loh, J. R. Downing, M. A. Caligiuri, C. D. Bloomfield, and E. S. Lander, *Science* **286,** 531 (1999).

[11] L. Liotta and E. Petricoin, *Nat. Rev. Genet.* **1,** 48 (2000).

[12] R. S. Thomas, D. R. Rank, S. G. Penn, G. M. Zastrow, K. R. Hayes, K. Pande, E. Glover, T. Silander, M. W. Craven, J. K. Reddy, S. B. Jovanovich, and C. A. Bradfield, *Mol. Pharmacol.* **60,** 1189 (2001).

[13] R. Ulrich and S. H. Friend, *Nat. Rev. Drug Discov.* **1,** 84 (2002).

[14] L. J. van't Veer, H. Dai, M. J. van de Vijver, Y. D. He, A. A. Hart, M. Mao, H. L. Peterse, K. van der Kooy, M. J. Marton, A. T. Witteveen, G. J. Schreiber, R. M. Kerkhoven, C. Roberts, P. S. Linsley, R. Bernards, and S. H. Friend, *Nature* **415,** 530 (2002).

[15] T. K. Sato, S. Panda, S. A. Kay, and J. B. Hogenesch, *J. Biol. Rhythms* **18,** 96 (2003).

[16] J. C. Dunlap, *Cell* **96,** 271 (1999).

[17] S. Panda, J. B. Hogenesch, and S. A. Kay, *Nature* **417,** 329 (2002).

[18] S. M. Reppert and D. R. Weaver, *Nature* **418,** 935 (2002).

[19] M. W. Young and S. A. Kay, *Nat. Rev. Genet.* **2,** 702 (2001).

[20] B. Kornmann, N. Preitner, D. Rifat, F. Fleury-Olela, and U. Schibler, *Nucleic Acids Res.* **29,** E51 (2001).

[21] A. Kramer, F. C. Yang, P. Snodgrass, X. Li, T. E. Scammell, F. C. Davis, and C. J. Weitz, *Science* **294,** 2511 (2001).

[22] R. N. van Gelder and M. A. Krasnow, *EMBO J.* **15,** 1625 (1996).

[23] R. A. Akhtar, A. B. Reddy, E. S. Maywood, J. D. Clayton, V. M. King, A. G. Smith, T. W. Gant, M. H. Hastings, and C. P. Kyriacou, *Curr. Biol.* **12,** 540 (2002).

[24] M. F. Ceriani, J. B. Hogenesch, M. Yanovsky, S. Panda, M. Straume, and S. A. Kay, *J. Neurosci.* **22,** 9305 (2002).

[25] A. Claridge-Chang, H. Wijnen, F. Naef, C. Boothroyd, N. Rajewsky, and M. W. Young, *Neuron* **32,** 657 (2001).

the dynamic interacting regulatory networks that underlie molecular, behavioral, and physiological circadian control, across and between cell and
tissue types of plants, insects, and mammals.[16–19,34]

Whereas it is biological questions that constitute specific aims of in-
quiry, quantitative analytical solution strategies are emerging as critically
important, to interpret the large volumes of data generated by gene array
experiments, and particularly time series thereof.[35–39] To assess whether or
not a diurnal/circadian rhythm is present in gene expression data, "general
purpose" software often fails to offer an ideal solution; rather, idiosyncratic
algorithms, implemented as highly automated, efficient software specifically designed to statistically address this question, have been developed
and applied to microarray time series of *Arabidopsis, Drosophila,* and
mammalian systems.[17,24,26,27]

Statistical Assessment of Daily Rhythms in Microarray Data

COSOPT

The COSOPT algorithm evolved as an extension of the CORRCOS
algorithm (applied to *Arabidopsis* in Harmer *et al.*[27]) and has been applied

[26] G. E. Duffield, J. D. Best, B. H. Meurers, A. Bittner, J. J. Loros, and J. C. Dunlap, *Curr. Biol.* **12,** 551 (2002).
[27] S. L. Harmer, J. B. Hogenesch, M. Straume, H. S. Chang, B. Han, T. Zhu, X. Wang, J. A. Kreps, and S. A. Kay, *Science* **290,** 2110 (2000).
[28] Y. Lin, M. Han, B. Shimada, L. Wang, T. M. Gibler, A. Amarakone, T. A. Awad, G. D. Stormo, R. N. van Gelder, and P. H. Taghert, *Proc. Natl. Acad. Sci. USA* **99,** 9562 (2002).
[29] M. J. McDonald and M. Rosbash, *Cell* **107,** 567 (2001).
[30] S. Panda, M. P. Antoch, B. H. Miller, A. I. Su, A. B. Schook, M. Straume, P. G. Schultz, S. A. Kay, J. S. Takahashi, and J. B. Hogenesch, *Cell* **109,** 307 (2002).
[31] R. Schaffer, J. Landgraf, M. Accerbi, V. V. Simon, M. Larson, and E. Wisman, *Plant Cell* **13,** 113 (2001).
[32] K. F. Storch, O. Lipan, I. Leykin, N. Viswanathan, F. C. Davis, W. H. Wong, and C. J. Weitz, *Nature* **417,** 78 (2002).
[33] H. R. Ueda, A. Matsumoto, M. Kawamura, N. Iino, T. Tanimura, and S. Hashimoto, *J. Biol. Chem.* **277,** 14048 (2002).
[34] T. Ideker, V. Thorsson, J. A. Ranish, R. Christmas, J. Buhler, J. K. Eng, R. Bumgarner, D. R. Goodlett, R. Aebersold, and L. Hood, *Science* **292,** 929 (2001).
[35] M. Eisen, P. T. Spellman, D. Botstein, and P. O. Brown, *Proc. Natl. Acad. Sci. USA* **95,** 14863 (1998).
[36] M. K. Kerr and G. A. Churchill, *Genet. Res.* **77,** 123 (2001).
[37] C. J. Langmead, A. K. Yan, C. R. McClung, and B. R. Donald, *in* "Proceedings of the Sixth Annual International Conference on Computational Molecular Biology (RECOMB)," pp. 1–11. Washington, DC.
[38] M. Ringner, C. Peterson, and J. Khan, *Pharmacogenomics* **3,** 403 (2002).
[39] A. Tefferi, M. E. Bolander, S. M. Ansell, E. D. Wieben, and T. C. Spelsberg, *Mayo Clin. Proc.* **77,** 927 (2002).

to search for diurnal/circadian gene expression profiling in mammalian[17] and *Drosophila* microarray time series.[24] COSOPT is a potentially more useful analytical strategy than CORRCOS because, whereas CORRCOS provides information about period, phase, and statistical significance, it less directly offers information regarding cycling amplitude (only in terms of the variance of the linear-regression detrended original time series). COSOPT does so more directly, in two ways: (1) by directly parameterizing oscillatory amplitude as a derived analytical property and, additionally, (2) by assessing statistical significance directly in terms of the magnitude of this oscillatory amplitude parameter (by way of an empirical resampling statistical strategy, as does CORRCOS, also). COSOPT also reports information about mean expression level and can use this information as an exclusionary criterion (i.e., if expression levels fall below some threshold, then the time series is considered not to contain valid data for interpretive consideration).

COSOPT imports data and calculates the mean expression intensity and its corresponding standard deviation (SD). It then performs an arithmetic linear-regression detrend of the original time series. The mean and SD of the detrended time series are then calculated. COSOPT does not standardize the linear-regression detrended time series to standard normal deviates, as did CORRCOS, thus allowing COSOPT to quantitatively assess oscillatory amplitude much more directly. Variable-weighting of individual time points (as in SEMs from replicate measurements, e.g., or as errors derived from preprocessing with dChip) can be accommodated during analysis for the presence of rhythms in terms of a user-specified number and range of periods (test periods are spaced uniformly in period-space).

For each test period, 101 test cosine basis functions (of unit amplitude) are considered, varying over a range of phase values from (plus one-half the period) to (minus one-half the period) (i.e., such that phase is considered in increments of 1% of each test period). COSOPT calculates, for each test cosine basis function, the least-squares optimized linear correspondence between the linear-regression-detrended data, $y_{lr}(t)$, and the unit-amplitude test cosine basis function, $y_b(t)$, as a function of t [i.e., such that the approximation of $y_{lr}(t)$ by the test cosine basis function, $y_b(t)$, is optimized across all values, t, in terms of two parameters, α and β, whereby $y_{lr}(t) \approx \alpha + \beta^* y_b(t)$]. The quality of optimization possible by the test cosine basis function is quantitatively characterized by the sum of squared residuals (SSR) between $y_{lr}(t)$ and the approximation given by $\alpha + \beta^* y_b(t)$ (SSR is hereafter referred to as χ^2). The values of χ^2 are used to identify the phase at which the optimal correspondence between $y_{lr}(t)$ and $y_b(t)$ is obtained for each test period (i.e., the phase giving the smallest χ^2 value corresponds to the optimal phase). Thus, for each test period are assessed these values of α, β, and χ^2 at the optimal phase. Note that, interpretively, β now represents an optimized, parameterized measure of the magnitude of the oscillatory amplitude expressed by $y_{lr}(t)$ in relation to, or as modeled by, a cosine wave of the corresponding period and optimal phase.

Empirical resampling methods are then employed to assess statistical probabilities of significance directly in terms of this parameter β at each test period and corresponding optimal phase, thus assessing statistically the probability that a significant rhythm is present in $y_{lr}(t)$ (in relation to, or as modeled by, a cosine functional form of the corresponding period and optimal phase). One thousand Monte Carlo cycles are carried out in which surrogate realizations of $y_{lr}(t)$ are generated by both (1) randomly shuffling temporal sequence and (2) adding pseudo–Gaussian-distributed noise to each surrogate point in proportion to the corresponding value of point uncertainty (i.e., replicate SEM, for example). In this way, specifically accounted for in the surrogate realizations are both (1) the influence of temporal patterning and (2) the magnitude of point wise experimental uncertainty. Then, as with the original $y_{lr}(t)$ sequence, optimal values of α and β are determined, along with a corresponding χ^2, and retained in memory for each surrogate at each test period/optimal phase. For each test period/ optimal phase are then calculated the mean and standard deviation of the surrogate β values. These values, in relation to the β value obtained for the original $y_{lr}(t)$ series, are then used to calculate a one-sided significance probability (multiple measures corrected for the number of original data points comprising the time series under consideration) based on a normality assumption (which is in fact satisfied by the distribution of β values obtained from the 1000 randomized surrogates). A summary of the analytical session is then produced for each time series, composed of entries for only those test periods that correspond to χ^2 minima.

MC-FFTSD32

Fast Fourier transform (FFT) methods have long been available for fast, efficient spectral decomposition of time series data when searching for embedded rhythmicities.[40,41] The particular FFT strategy employed in the present work was a full 32-bit implementation (hence "32" in the

[40] W. H. Press, B. P. Flannery, S. A. Teukolsky, and W. T. Vetterling, in "Numerical Recipes: The Art of Scientific Computing." FORTRAN version. Cambridge University Press, Cambridge, 1989.
[41] J. S. Bendat and A. G. Piersol, in "Random Data: Analysis and Measurement Procedures," 3rd Ed. John Wiley & Sons, New York, 2000.

name) in which empirical resampling was employed for assessing statistical significance (in a Monte Carlo procedure, as in COSOPT; hence "MC" in the name), from which power significance probabilities are assessed at each test frequency/period relative to the mean and standard deviation obtained from analyses of 1000 empirically resampled surrogate realizations (hence "SD" in the name) (probabilities multiple measures corrected for the number of original data points in the time series being analyzed).

MC-FFTSD32 imports data and performs an arithmetic linear-regression detrend of the original time series (to make null the zero-frequency power and minimize the otherwise potentially obscuring contribution of the lowest frequency components) after which the data are zero-padded to 1024 total points (thus eliminating artifacts from circular correlation[41] and considerably enhancing the intrinsic frequency resolution of the analysis). Not employed are any windowing strategies to reduce side-lobe leakage.[41]

COSIN2NL

COSIN2NL is an algorithm to assess the period, phase, and amplitude of a one-component cosine function of the form

$$y_{lr}(t) = c_0 + c_1 t + \alpha \cos \left[\frac{2\pi(t + \phi)}{\tau} \right]$$

in which $y_{lr}(t)$ is the linear regression detrended time series to which analysis is being performed, c_0 is a constant offset term, c_1 is a slope term, t is time, and α, ϕ, and τ are the amplitude, phase, and period, respectively, of the cosine function. The parameters of this function are then estimated by nonlinear least squares minimization by a modified Gauss–Newton nonlinear least squares algorithm.[42,43] On convergence, approximate nonlinear asymmetric joint confidence limits can be estimated, if desired, for all parameters (period, phase, amplitude, constant offset, and slope term) at 68.26% (the confidence probability associated with one SD) or any other user-specified confidence probability.[43] An amplitude term significantly different from zero indicates a statistically significant rhythm at the specified level of confidence probability. If the confidence limits of the amplitude term encompass zero, however, the rhythm is not statistically significant at the specified level of confidence probability. Notably, COSIN2NL

[42] M. L. Johnson and S. G. Frasier, *Methods Enzymol.* **117**, 301 (1985).
[43] M. Straume, S. G. Frasier-Cadoret, and M. L. Johnson, *in* "Topics in Fluorescence Spectroscopy," Vol. 2: "Principles" (J. R. Lakowicz, ed.), pp. 117–240. Plenum, New York, 1991.

requires user-specified initialization (i.e., initial guesses for the values of the parameters of the cosine model) and thus is not a fully objective analytical strategy (i.e., it is potentially susceptible to the influence of user-introduced bias).

CIRCCORR

CIRCCORR is an algorithm to assess the extent of correlation, r_0 (in the multiplicative Pearson sense), between an original time series of interest, $y_o(t)$, and that of an equivalently distributed (across the time axis) unit-magnitude cosine test basis function, $y_b(t)$, defined at some user-specified period (i.e., 24 h, in this example of circadian rhythms) and examined over the full range of possible phase angles. If both the original time series and the unit-magnitude cosine test basis functions are linear-regression detrended $[y_o^{LR}(t), y_o^{LR}(t)]$ prior to calculation of Pearson correlation coefficients, r_0^{LR}, the values of r_0^{LR} become measures of similarity between $y_o^{LR}(t)$ and each of the tested $y_b^{LR}(t)$ (in which $-1 \leq r_0^{LR} \leq +1$). In addition, the root mean squared deviation of the linear-regression detrended original time-series values, $\sigma(y_o^{LR})$, is a measure of the dispersion characteristic of the original data set. The phase angle that produces a maximum in the attainable correlation coefficient value, $r_0^{LR}(\phi_{max}, \tau_{circ})$, then represents the optimal phase estimate for the original time series (at the specified test period, τ_{circ}). In this case, $r_0^{LR}(\phi_{max}, \tau_{circ})$ itself is a measure of the maximal rhythmic determination characteristic of the original time series, and $\sigma(y_o^{LR})$ is a measure of overall variability in $y_o^{LR}(t)$. As such, the product $\sigma(y_o^{LR})r_0^{LR}(\phi_{max}, \tau_{circ})$ represents a measure of optimal rhythmic amplitude expressed by $y_o^{LR}(t)$ at $(\phi_{max}, \tau_{circ})$.

Simulation Procedure

Simulated data sets were prepared to approximate previously encountered gene chip profiles from experimental examinations of expression time series.[17,24,27] One thousand surrogate data sets were prepared at each condition considered (see below), in which time series possessed 13 data points, representing 48 h of observation obtained at 4-h sampling intervals. All time series were surrogate realizations of a 24-h-period cosine wave ranging in representational time from -24 h to $+24$ h, at which acrophase (i.e., maximum) occurred at time zero. All data sets were composed of $N(0, 1)$ noise (i.e., normally distributed noise of zero mean and unit variance) to which 24-h cosine profiles were added to produce data with signal-to-noise ratios of either 0, 1, or 2 (i.e., by adding nothing, or a unit-amplitude cosine wave, or an amplitude-2 cosine wave, respectively). At each signal-to-noise ratio, replicate sampling also was varied, ranging from one, two, three, four, or five replicate observations being averaged per data point. A final variable considered in analyses was whether or not replicate pointwise uncertainties were explicitly considered by the analytical strategies or not [i.e., employing either (1) pointwise standard errors of the mean in variable weighting for statistical considerations, or (2) no weighting whatsoever, thus, in effect, assuming each data point to be known with infinite precision (i.e., with associated pointwise SEMs of zero)].

Comparisons of Analytical Results

Tables I–V summarize the results obtained from simulations with each of the aforementioned analytical strategies as a function of simulation condition.

Summary

The challenges for circadian analysis of gene chip-derived time series are indeed considerable, as data sets presenting for analysis are typically characterized by (1) extremely sparse determination (i.e., only 13 points at 4-h sampling frequency for 48 h, for example), (2) extremely high dimensionality (on the order of 10^4 GeneIDs per chip in current Affymetrix implementations, thus making automated, intelligent analytical strategies an imperative), and (3) low replicate numbers (thus limiting pointwise reliability, due primarily to the considerable financial costs of multiple chips per experimental time point). The present report presents a performance comparison and summary for a number of algorithmic approaches that have found application in this setting, employing as a basis for simulation conditions that have been typically encountered in contemporary studies.

Fast Fourier transform methods can reasonably be expected to be employed in the current context, and, as seen from the simulation results, the FFT (by MC-FFTSD32) performs quite admirably, albeit producing somewhat biased period estimates (consistently a bit on the low side), and, under conditions in which pointwise uncertainties are not considered, producing the highest false-positive results of the four algorithms examined.

The straightforward, direct approximation by nonlinear least squares optimization of a single-frequency cosine function to the data in question (by COSIN2NL) produced admirable performance under conditions in which pointwise uncertainties are not considered, but produced the highest false-positive rate of all the methods when pointwise uncertainties were considered. In addition, this method suffers from two drawbacks: (1) it

TABLE I
ANALYSIS WITH COSOPT[a]

±SEM	S/N	%	MeanExpLev	Period	Phase	Ampl	Prob	Prob(MMC)
					$N_{reps} = 5$			
Yes	2	100.0	0.150 (0.127)	24.019 (0.708)	−0.001 (0.341)	2.015 (0.166)	6.57×10^{-4} (2.23×10^{-4})	8.51×10^{-3} (2.87×10^{-3})
No	2	100.0	0.150 (0.127)	24.019 (0.708)	−0.001 (0.341)	2.015 (0.166)	4.10×10^{-4} (1.44×10^{-4})	5.32×10^{-3} (1.86×10^{-3})
Yes	1	66.7	0.079 (0.128)	24.027 (1.365)	0.025 (0.667)	1.110 (0.131)	2.29×10^{-3} (8.16×10^{-4})	2.94×10^{-2} (1.03×10^{-2})
No	1	96.6	0.074 (0.127)	24.056 (1.383)	−0.004 (0.677)	1.045 (0.159)	1.09×10^{-3} (6.77×10^{-4})	1.41×10^{-2} (8.69×10^{-3})
Yes	0	0.0	—	—	—	—	—	—
No	0	2.2	0.040 (0.137)	23.514 (2.585)	−0.289 (7.797)	0.463 (0.109)	2.91×10^{-3} (8.24×10^{-4})	3.72×10^{-3} (1.03×10^{-2})
					$N_{reps} = 4$			
Yes	2	99.9	0.150 (0.141)	23.986 (0.741)	0.017 (0.377)	2.030 (0.195)	7.82×10^{-4} (2.90×10^{-4})	1.01×10^{-2} (3.74×10^{-3})
No	2	100.0	0.150 (0.141)	23.986 (0.741)	0.017 (0.377)	2.030 (0.196)	4.53×10^{-4} (1.91×10^{-4})	5.88×10^{-3} (2.47×10^{-3})
Yes	1	50.5	0.081 (0.137)	23.905 (1.285)	0.042 (0.690)	1.175 (0.151)	2.49×10^{-3} (8.07×10^{-4})	3.19×10^{-2} (1.02×10^{-2})
No	1	92.6	0.073 (0.141)	23.933 (1.405)	0.043 (0.751)	1.074 (0.180)	1.26×10^{-3} (7.60×10^{-4})	1.62×10^{-2} (9.73×10^{-3})
Yes	0	0.0	—	—	—	—	—	—
No	0	2.3	−0.003 (0.125)	22.970 (2.363)	−1.789 (7.025)	0.567 (0.135)	2.45×10^{-3} (9.22×10^{-4})	3.14×10^{-2} (1.16×10^{-2})
					$N_{reps} = 3$			
Yes	2	99.8	0.151 (0.159)	24.037 (0.865)	0.017 (0.467)	2.033 (0.221)	1.08×10^{-3} (4.83×10^{-4})	1.40×10^{-2} (6.20×10^{-3})
No	2	100.0	0.151 (0.159)	24.035 (0.865)	0.016 (0.467)	2.032 (0.222)	5.35×10^{-4} (2.35×10^{-4})	6.93×10^{-3} (3.04×10^{-3})
Yes	1	30.0	0.081 (0.152)	24.099 (1.457)	0.040 (0.844)	1.264 (0.164)	2.73×10^{-3} (7.88×10^{-4})	3.49×10^{-2} (9.91×10^{-3})
No	1	83.1	0.078 (0.157)	24.024 (1.574)	0.047 (0.919)	1.113 (0.195)	1.50×10^{-3} (8.48×10^{-4})	1.93×10^{-2} (1.08×10^{-2})
Yes	0	0.0	—	—	—	—	—	—
No	0	3.4	0.027 (0.164)	23.018 (2.365)	0.928 (5.959)	0.668 (0.148)	2.79×10^{-3} (8.77×10^{-4})	3.57×10^{-2} (1.10×10^{-2})
					$N_{reps} = 2$			
Yes	2	94.8	0.157 (0.191)	24.016 (1.068)	−0.008 (0.561)	2.070 (0.254)	1.59×10^{-3} (7.72×10^{-4})	2.05×10^{-2} (9.85×10^{-3})
No	2	99.7	0.157 (0.191)	24.007 (1.080)	−0.008 (0.561)	2.052 (0.264)	7.10×10^{-4} (3.66×10^{-4})	9.19×10^{-3} (4.72×10^{-3})
Yes	1	12.2	0.073 (0.220)	24.014 (1.588)	−0.205 (0.938)	1.427 (0.188)	2.67×10^{-3} (8.09×10^{-4})	3.42×10^{-2} (1.02×10^{-2})
No	1	64.2	0.076 (0.197)	23.923 (1.762)	−0.045 (1.100)	1.205 (0.220)	1.83×10^{-3} (9.63×10^{-4})	2.35×10^{-2} (1.22×10^{-2})

(continued)

TABLE I (continued)

±SEM	S/N	%	MeanExpLev	Period	Phase	Ampl	Prob	Prob(MMC)
Yes	0	0.0	—	—	—	—	—	—
No	0	2.5	-0.047 (0.192)	23.516 (2.174)	-1.049 (7.219)	0.704 (0.148)	2.43×10^{-3} (8.63×10^{-4})	3.11×10^{-2} (1.09×10^{-2})
				$N_{reps} = 1$				
Yes	2	N/D	N/D	N/D	N/D	N/D	N/D	N/D
No	2	92.0	0.155 (0.278)	24.058 (1.494)	0.048 (0.753)	2.149 (0.354)	1.28×10^{-3} (7.94×10^{-4})	1.65×10^{-2} (1.02×10^{-2})
Yes	1	N/D	N/D	N/D	N/D	N/D	N/D	N/D
No	1	33.7	0.098 (0.286)	23.935 (2.054)	0.142 (1.296)	1.490 (0.280)	2.12×10^{-3} (9.43×10^{-4})	2.72×10^{-2} (1.20×10^{-2})
Yes	0	N/D	N/D	N/D	N/D	N/D	N/D	N/D
No	0	1.5	-0.024 (0.377)	23.313 (2.307)	1.258 (6.676)	1.035 (0.234)	2.49×10^{-3} (8.15×10^{-4})	3.19×10^{-2} (1.03×10^{-2})

[a] ±SEM refers to whether (Yes) or not (No) individual pointwise replicate uncertainties were considered during statistical analysis; S/N, signal-to-noise ratio; %, percentage of files identified as circadianly rhythmic (i.e., with periods such that 20 h < τ_{circ} < 28 h); MeanExpLev, mean expression level of the identified time series (theoretically zero); Period, mean period of the identified time series (theoretically 24); Phase, mean time of acrophase of the identified time series (theoretically zero); Ampl, mean oscillatory amplitude of the identified time series (theoretically either 2, 1, or 0); Prob, mean uncorrected significance probability of the identified time series; Prob(MMC), mean multiple-measures corrected significance probability of the identified time series [Prob(MMC) = 1 − (1 − Prob)^N, N = 13]; circadianly rhythmic if Prob(MMC) < 0.05 (i.e., Prob < 3.9379×10^{-3}); N/D, not determined; values in parentheses are SDs.

TABLE II
ANALYSIS WITH MC-FFTSD32[a]

±SEM	S/N	%	\<Period\>	FracSigPower
		$N_{reps} = 5$		
Yes	2	100.0	23.647 (0.270)	0.953 (0.037)
No	2	100.0	23.573 (0.218)	0.925 (0.035)
Yes	1	81.4	23.765 (0.788)	0.988 (0.040)
No	1	99.8	23.661 (0.593)	0.929 (0.073)
Yes	0	0.0	—	—
No	0	5.9	23.251 (2.490)	0.652 (0.357)
		$N_{reps} = 4$		
Yes	2	100.0	23.665 (0.310)	0.963 (0.038)
No	2	100.0	23.575 (0.242)	0.929 (0.038)
Yes	1	67.1	23.690 (0.778)	0.993 (0.030)
No	1	97.8	23.656 (0.663)	0.937 (0.079)
Yes	0	0.0	—	—
No	0	6.4	22.528 (2.070)	0.690 (0.311)
		$N_{reps} = 3$		
Yes	2	100.0	23.704 (0.397)	0.968 (0.040)
No	2	100.0	23.599 (0.288)	0.927 (0.045)
Yes	1	43.5	23.745 (0.845)	0.995 (0.022)
No	1	93.3	23.683 (0.827)	0.930 (0.095)
Yes	0	0.0	—	—
No	0	7.6	23.108 (2.425)	0.648 (0.346)
		$N_{reps} = 2$		
Yes	2	98.2	23.730 (0.559)	0.977 (0.042)
No	2	100.0	23.608 (0.397)	0.931 (0.058)
Yes	1	22.5	23.665 (0.926)	0.991 (0.032)
No	1	81.4	23.688 (1.088)	0.917 (0.140)
Yes	0	0.0	—	—
No	0	6.4	22.912 (2.165)	0.686 (0.337)
		$N_{reps} = 1$		
Yes	N/D	N/D	N/D	N/D
No	2	98.9	23.667 (0.681)	0.929 (0.084)
Yes	N/D	N/D	N/D	N/D
No	1	51.2	23.732 (1.342)	0.886 (0.177)
Yes	N/D	N/D	N/D	N/D
No	0	5.9	23.213 (2.448)	0.588 (0.354)

[a] ±SEM refers to whether (Yes) or not (No) individual pointwise replicate uncertainties were considered during statistical analysis; S/N, signal-to-noise ratio; %, percentage of files identified as circadianly rhythmic; \<Period\>, mean period recovered for the identified time series (theoretically 24); FracSigPower, fraction of statistically significant spectral power recovered between 20 h $<\tau_{circ} < 28$ h; N/D, not determined; values in parentheses are SDs.

TABLE III
ANALYSIS WITH COSIN2NL[a]

±SEM	S/N	%	Ampl	SD(Ampl)	RAE(68)	Period	Phase	Mean	Slope	$(WSSR/N)^{1/2}$
						$N_{reps} = 5$				
Yes	2	99.9	2.031 (0.196)	0.391 (0.121)	0.194 (0.062)	24.030 (0.807)	−0.004 (0.433)	0.001 (0.164)	0.000 (0.010)	0.885 (0.284)
No	2	100.0	2.017 (0.166)	0.442 (0.114)	0.220 (0.060)	24.028 (0.706)	0.001 (0.368)	0.001 (0.137)	0.000 (0.008)	0.343 (0.088)
Yes	1	78.8	1.098 (0.177)	0.357 (0.096)	0.329 (0.086)	24.019 (1.477)	0.010 (0.842)	−0.000 (0.159)	0.000 (0.010)	0.809 (0.236)
No	1	70.2	1.084 (0.151)	0.396 (0.088)	0.369 (0.079)	24.033 (1.342)	0.001 (0.723)	0.000 (0.137)	0.000 (0.009)	0.305 (0.068)
Yes	0	2.0	0.571 (0.152)	0.240 (0.069)	0.424 (0.068)	24.185 (2.444)	−1.377 (6.652)	0.012 (0.160)	0.003 (0.011)	0.637 (0.247)
No	0	0.2	0.625 (0.108)	0.269 (0.029)	0.433 (0.029)	24.892 (0.584)	−1.132 (8.384)	0.028 (0.044)	0.005 (0.009)	0.199 (0.022)
						$N_{reps} = 4$				
Yes	2	99.7	2.046 (0.243)	0.421 (0.133)	0.208 (0.070)	24.016 (0.910)	0.017 (0.514)	0.001 (0.180)	−0.000 (0.011)	0.929 (0.304)
No	2	99.7	2.033 (0.195)	0.491 (0.120)	0.244 (0.064)	23.997 (0.742)	0.019 (0.416)	−0.003 (0.149)	−0.000 (0.010)	0.381 (0.093)
Yes	1	74.9	1.147 (0.214)	0.382 (0.109)	0.337 (0.088)	23.979 (1.585)	0.037 (0.982)	−0.003 (0.173)	−0.000 (0.011)	0.850 (0.263)
No	1	57.6	1.140 (0.172)	0.430 (0.096)	0.380 (0.075)	23.982 (1.358)	0.029 (0.777)	−0.008 (0.145)	−0.000 (0.010)	0.331 (0.074)
Yes	0	4.1	0.670 (0.159)	0.261 (0.079)	0.392 (0.073)	23.394 (2.085)	−0.427 (6.698)	0.039 (0.159)	0.002 (0.012)	0.760 (0.307)
No	0	0.4	0.652 (0.077)	0.301 (0.027)	0.463 (0.033)	23.303 (2.705)	2.065 (5.393)	−0.071 (0.067)	−0.001 (0.003)	0.223 (0.020)
						$N_{reps} = 3$				
Yes	2	97.6	2.073 (0.298)	0.434 (0.156)	0.213 (0.079)	24.015 (1.126)	−0.018 (0.671)	0.006 (0.227)	−0.000 (0.014)	1.024 (0.400)
No	2	98.6	2.041 (0.219)	0.561 (0.136)	0.278 (0.073)	24.058 (0.863)	0.020 (0.513)	0.004 (0.172)	−0.000 (0.011)	0.435 (0.105)
Yes	1	69.3	1.215 (0.256)	0.382 (0.130)	0.319 (0.095)	23.989 (1.830)	−0.018 (1.272)	0.007 (0.214)	−0.000 (0.015)	0.940 (0.358)
No	1	43.6	1.189 (0.191)	0.469 (0.096)	0.398 (0.069)	24.045 (1.532)	0.036 (0.976)	−0.003 (0.166)	−0.001 (0.012)	0.360 (0.073)
Yes	0	7.6	0.728 (0.208)	0.270 (0.098)	0.376 (0.097)	23.489 (2.331)	−1.045 (7.039)	0.001 (0.211)	0.001 (0.015)	0.852 (0.327)
No	0	0.3	0.770 (0.107)	0.329 (0.025)	0.432 (0.059)	23.920 (2.617)	7.404 (2.810)	−0.196 (0.229)	0.004 (0.009)	0.243 (0.015)
						$N_{reps} = 2$				
Yes	2	89.2	2.128 (0.429)	0.434 (0.200)	0.210 (0.100)	24.018 (1.482)	−0.041 (0.969)	0.016 (0.314)	0.001 (0.023)	1.464 (0.818)
No	2	92.4	2.077 (0.256)	0.675 (0.155)	0.329 (0.081)	24.046 (1.081)	−0.018 (0.624)	0.011 (0.208)	0.000 (0.014)	0.522 (0.120)
Yes	1	61.5	1.323 (0.372)	0.362 (0.161)	0.280 (0.108)	23.817 (1.985)	−0.044 (1.759)	0.002 (0.300)	0.000 (0.021)	1.338 (0.710)
No	1	24.9	1.315 (0.216)	0.534 (0.117)	0.408 (0.069)	23.980 (1.738)	−0.044 (1.098)	0.001 (0.219)	0.001 (0.014)	0.409 (0.090)

±SEM	S/N	%	Ampl	SD(Ampl)	RAE(68)	Period	Phase	Mean	Slope	(WSSR/N)$^{1/2}$
Yes	0	22.2	0.897 (0.336)	0.295 (0.149)	0.331 (0.110)	23.648 (2.161)	−0.388 (6.844)	0.031 (0.277)	−0.000 (0.024)	1.403 (0.715)
No	0	0.4	0.783 (0.173)	0.319 (0.051)	0.416 (0.070)	24.328 (2.410)	−2.606 (6.704)	0.126 (0.132)	0.003 (0.015)	0.237 (0.041)
$N_{reps} = 1$										
Yes	2	N/D	N/D	N/D	N/D	N/D	N/D	N/D	N/D	N/D
No	2	59.4	2.227 (0.339)	0.861 (0.185)	0.382 (0.078)	24.156 (1.436)	0.086 (0.799)	0.023 (0.301)	−0.002 (0.020)	0.662 (0.143)
Yes	1	N/D	N/D	N/D	N/D	N/D	N/D	N/D	N/D	N/D
No	1	10.1	1.614 (0.295)	0.678 (0.149)	0.423 (0.067)	24.034 (2.123)	0.112 (1.342)	−0.001 (0.287)	−0.001 (0.022)	0.517 (0.114)
Yes	0	N/D	N/D	N/D	N/D	N/D	N/D	N/D	N/D	N/D
No	0	0.4	1.154 (0.193)	0.480 (0.136)	0.423 (0.120)	23.693 (3.439)	4.170 (6.206)	−0.068 (0.344)	−0.019 (0.025)	0.362 (0.102)

[a] ±SEM refers to whether (Yes) or not (No) individual pointwise replicate uncertainties were considered during statistical analysis; S/N, signal-to-noise ratio; %, percentage of files identified as circadianly rhythmic; Ampl, mean amplitude recovered for the identified time series (theoretically either 2, 1, or 0); SD(Ampl), mean approximate amplitude SD recovered for the identified time series (actually, the mean of the half-difference between the upper and lower approximate nonlinear asymmetric joint confidence limits of the derived amplitude parameter at 68.26% confidence probability); RAE(68), mean 68.26% relative amplitude error of the identified time series [where RAE(68) is a measure of rhythmic determination (at 68.26% confidence probability) and is defined as the ratio of SD(Ampl)/Ampl; RAE = 0, perfect or infinitely good rhythmic determination; RAE = 1, statistically minimal rhythmic determination]; Period, mean period of the identified time series (theoretically 24); Phase, mean time of acrophase of the identified time series (theoretically zero); Mean, mean expression level of the identified time series (theoretically zero); Slope, mean slope of the identified time series (theoretically zero); (WSSR/N)$^{1/2}$, mean root of the weighted sum of squared residuals of fit divided by the number of time series data points ($N = 13$, in this case) [theoretically, for variably weighted fits (i.e., ±SEM = Yes), this value should approximate unity, whereas for unweighted fits (i.e., ±SEM = No), this value would be expected to be approximately proportional to the inverse root of the number of replicates per time point]; N/D, not determined; values in parentheses are SDs. COSIN2NL is an algorithm defined by five parameters (Ampl, Period, Phase, Mean, and Slope), and must be initialized prior to minimization. Analyses were performed in which COSIN2NL was initialized such that Ampl = 0, Period = 24, Phase = 0, Mean = 0.1, and Slope = 0.1. The first round of iterative minimization involved holding Ampl, Period, and Phase fixed while minimizing to Mean and Slope. After convergence, the second round of iterative minimization involved holding Ampl and Period fixed while minimizing to Phase, Mean, and Slope. After convergence, the third round of iterative minimization involved holding only Period fixed while minimizing to Ampl, Phase, Mean, and Slope. After convergence, the fourth round of iterative minimization involved minimizing to all five parameters, Ampl, Period, Phase, Mean, and Slope, simultaneously, followed by calculation of approximate nonlinear asymmetric joint 68.26% parameter confidence limits. The convergence criterion employed in all COSIN2NL minimizations was a fractional change in variance of fit of 10^{-6} or less. Confidence probability for calculation of approximate nonlinear asymmetric joint parameter confidence limits was 68.26% (the probability associated with one standard deviation) in all calculations reported here.

TABLE IV
ANALYSIS WITH CIRCCORR[a]

±SEM	S/N	%	Mean	SD	Phase	CC	CCRan	CCRanSD	Z-score	Probability
					N_{reps} = 5					
Yes	2	100.0	0.150	1.508	−0.002	0.967	−0.000	0.301	3.219	6.72×10^{-4}
			(0.127)	(0.120)	(0.331)	(0.016)	(0.010)	(0.006)	(0.086)	(2.05×10^{-4})
No	2	100.0	0.150	1.508	−0.002	0.967	0.000	0.301	3.213	6.84×10^{-4}
			(0.127)	(0.120)	(0.331)	(0.016)	(0.009)	(0.006)	(0.086)	(2.07×10^{-4})
Yes	1	100.0	0.073	0.828	−0.000	0.885	−0.000	0.301	2.945	1.97×10^{-3}
			(0.127)	(0.115)	(0.670)	(0.055)	(0.010)	(0.006)	(0.196)	(1.71×10^{-3})
No	1	100.0	0.073	0.828	−0.000	0.885	0.000	0.301	2.943	1.97×10^{-3}
			(0.127)	(0.115)	(0.670)	(0.055)	(0.009)	(0.006)	(0.195)	(1.64×10^{-3})
Yes	0	26.4	0.012	0.404	−0.117	0.593	−0.000	0.294	2.018	0.026
			(0.128)	(0.088)	(7.121)	(0.082)	(0.010)	(0.008)	(0.278)	(0.013)
No	0	27.1	0.010	0.400	0.209	0.590	−0.001	0.294	2.009	0.026
			(0.127)	(0.088)	(6.942)	(0.083)	(0.009)	(0.008)	(0.278)	(0.014)
					N_{reps} = 4					
Yes	2	100.0	0.150	1.528	0.017	0.961	−0.000	0.301	3.196	7.34×10^{-4}
			(0.141)	(0.141)	(0.373)	(0.019)	(0.010)	(0.006)	(0.094)	(2.74×10^{-4})
No	2	100.0	0.150	1.528	0.017	0.961	0.000	0.301	3.192	7.40×10^{-4}
			(0.141)	(0.141)	(0.373)	(0.019)	(0.010)	(0.006)	(0.092)	(2.62×10^{-4})
Yes	1	99.9	0.073	0.857	0.031	0.867	0.000	0.301	2.883	2.56×10^{-3}
			(0.141)	(0.133)	(0.761)	(0.065)	(0.010)	(0.006)	(0.227)	(2.68×10^{-3})
No	1	99.9	0.073	0.857	0.031	0.867	0.000	0.301	2.883	2.54×10^{-3}
			(0.141)	(0.133)	(0.761)	(0.065)	(0.010)	(0.006)	(0.223)	(2.61×10^{-3})
Yes	0	30.3	0.004	0.456	0.195	0.593	−0.001	0.294	2.020	0.026
			(0.139)	(0.105)	(6.832)	(0.082)	(0.009)	(0.008)	(0.281)	(0.013)
No	0	29.8	0.003	0.457	0.125	0.595	0.001	0.294	2.023	0.025
			(0.136)	(0.106)	(6.854)	(0.081)	(0.009)	(0.008)	(0.273)	(0.013)

$N_{\text{reps}} = 3$

Yes	2	100.0	0.151 (0.159)	1.544 (0.157)	0.015 (0.464)	0.949 (0.025)	−0.000 (0.009)	0.301 (0.006)	3.155 (0.108)	8.58×10^{-4} (3.68×10^{-4})
No	2	100.0	0.151 (0.159)	1.544 (0.157)	0.015 (0.464)	0.949 (0.025)	−0.000 (0.010)	0.301 (0.006)	3.152 (0.112)	8.71×10^{-4} (3.70×10^{-4})
Yes	1	99.7	0.074 (0.159)	0.891 (0.146)	0.024 (0.950)	0.834 (0.082)	−0.000 (0.009)	0.301 (0.006)	2.773 (0.280)	4.03×10^{-3} (5.17×10^{-3})
No	1	99.7	0.074 (0.159)	0.891 (0.146)	0.024 (0.950)	0.834 (0.082)	−0.000 (0.010)	0.301 (0.006)	2.774 (0.280)	4.01×10^{-3} (5.02×10^{-3})
Yes	0	30.7	−0.005 (0.153)	0.535 (0.114)	0.265 (6.742)	0.600 (0.088)	−0.001 (0.010)	0.294 (0.008)	2.046 (0.293)	0.024 (0.013)
No	0	30.6	−0.012 (0.155)	0.533 (0.116)	0.237 (6.823)	0.601 (0.088)	−0.000 (0.009)	0.293 (0.008)	2.052 (0.300)	0.024 (0.014)

$N_{\text{reps}} = 2$

Yes	2	100.0	0.157 (0.191)	1.588 (0.186)	−0.010 (0.550)	0.925 (0.036)	−0.000 (0.010)	0.301 (0.006)	3.076 (0.138)	1.16×10^{-3} (6.47×10^{-3})
No	2	100.0	0.157 (0.191)	1.588 (0.186)	−0.010 (0.550)	0.925 (0.036)	0.000 (0.010)	0.301 (0.006)	3.072 (0.137)	1.18×10^{-3} (6.41×10^{-3})
Yes	1	97.8	0.080 (0.192)	0.969 (0.168)	−0.022 (1.118)	0.782 (0.098)	−0.001 (0.010)	0.301 (0.006)	2.604 (0.328)	6.91×10^{-3} (7.78×10^{-3})
No	1	98.3	0.080 (0.191)	0.968 (0.168)	−0.019 (1.120)	0.780 (0.100)	0.000 (0.010)	0.301 (0.006)	2.598 (0.335)	7.17×10^{-3} (8.37×10^{-3})
Yes	0	29.5	0.019 (0.202)	0.652 (0.135)	−0.343 (6.864)	0.596 (0.088)	−0.000 (0.009)	0.293 (0.008)	2.032 (0.302)	0.025 (0.014)
No	0	29.8	0.016 (0.200)	0.651 (0.136)	−0.385 (6.835)	0.594 (0.088)	−0.001 (0.009)	0.293 (0.008)	2.031 (0.299)	0.025 (0.014)

$N_{\text{reps}} = 1$

Yes	2	N/D	N/D	N/D	N/D	N/D	N/D	N/D	N/D	N/D
No	2	100.0	0.155 (0.275)	1.712 (0.254)	0.032 (0.743)	0.865 (0.065)	−0.000 (0.010)	0.301 (0.006)	2.875 (0.227)	2.61×10^{-3} (2.58×10^{-3})
Yes	1	N/D	N/D	N/D	N/D	N/D	N/D	N/D	N/D	N/D

(continued)

TABLE IV (continued)

±SEM	S/N	%	Mean	SD	Phase	CC	CCRan	CCRanSD	Z-score	Probability
No	1	86.6	0.082	1.180	0.058	0.709	−0.000	0.300	2.364	0.013
			(0.278)	(0.222)	(1.476)	(0.111)	(0.010)	(0.006)	(0.373)	(0.012)
Yes	0	N/D	N/D	N/D	N/D	N/D	N/D	N/D	N/D	N/D
No	0	27.7	0.018	0.902	0.063	0.596	−0.000	0.294	2.032	0.025
			(0.269)	(0.190)	(6.685)	(0.085)	(0.009)	(0.008)	(0.287)	(0.014)

[a] ±SEM refers to whether (Yes) or not (No) individual pointwise replicate uncertainties were considered during statistical analysis; S/N, signal-to-noise ratio; %, percentage of files identified as circadianly rhythmic; Mean, mean expression level of the identified time series (theoretically zero); SD, mean standard deviation of points in identified time series; Phase, mean time of acrophase of the identified time series (theoretically zero); CC, mean (Pearson) correlation coefficient calculated for the identified (linear-regression-detrended) original time series correlated with a 24-h-period unit-amplitude cosine basis function; CCRan, mean (Pearson) correlation coefficient calculated for the identified (linear-regression-detrended) original time series after (1) temporal randomization and (2) pointwise noise perturbation (when ±SEM = Yes) correlated with a 24-h-period unit-amplitude cosine basis function; CCRanSD, mean standard deviation of the (Pearson) correlation coefficient calculated for the identified (linear-regression-detrended) original time series after (1) temporal randomization and (2) pointwise noise perturbation (when ±SEM = Yes) correlated with a 24-h-period unit-amplitude cosine basis function; Z-score = (CC−CCRan)/CCRanSD; Probability, mean uncorrected significance probability corresponding to the associated Z-scores of each of the identified time series (considered circadianly rhythmic if Probability <0.05); N/D, not determined; values in parentheses are SDs.

TABLE V

SUMMARY OF PERCENTAGE OF FILES IDENTIFIED AS CIRCADIANLY RHYTHMIC BY EACH OF THE RESPECTIVE ANALYSES UNDER THE VARIOUS CONDITIONS OF SIMULATION[a]

N_{reps}	S/N	COSOPT	MCFFTSD32	COSIN2NL	CIRCCORR $p < 0.05$	CIRCCORR $p < 3.9379 \times 10^{-3}$
			% Circadian (±SEM = Yes)			
5	2	100.0	100.0	99.9	100.0	100.0
	1	66.7	81.4	78.8	100.0	91.3
	0	0.0	0.0	2.0	26.4	0.7
4	2	99.9	100.0	99.7	100.0	99.9
	1	50.5	67.1	74.9	99.9	85.1
	0	0.0	0.0	4.1	30.3	0.6
3	2	99.8	100.0	97.6	100.0	99.9
	1	30.0	43.5	69.3	99.7	70.1
	0	0.0	0.0	7.6	30.7	1.1
2	2	94.8	98.2	89.2	100.0	99.2
	1	12.2	22.5	61.5	97.8	46.7
	0	0.0	0.0	22.2	29.5	1.3
			% Circadian (±SEM = No)			
5	2	100.0	100.0	100.0	100.0	100.0
	1	96.6	99.8	70.2	100.0	91.5
	0	2.2	5.9	0.2	27.1	0.6
4	2	100.0	100.0	99.7	100.0	100.0
	1	92.6	97.8	57.6	99.9	85.8
	0	2.3	6.4	0.4	29.8	0.7
3	2	100.0	100.0	98.6	100.0	100.0
	1	83.1	93.3	43.6	99.7	71.0
	0	3.4	7.6	0.3	30.6	1.3
2	2	99.7	100.0	92.4	100.0	99.5
	1	64.2	81.4	24.9	98.3	47.3
	0	2.5	6.4	0.4	29.8	1.0
1	2	92.0	98.9	59.4	100.0	83.0
	1	33.7	51.2	10.1	86.6	20.5
	0	1.5	5.9	0.4	27.7	0.6

[a] For CIRCCORR, an additional column is reported in which a threshold probability of 3.9379×10^{-3} has been employed [i.e., equivalent to implementing a 13-point multiple-measures correction as $p_{MMC} = 1 - (1 - p)^N$, where $p_{MMC} = 3.9379 \times 10^{-3}$ when $p = 0.05$ and $N = 13$].

requires parameter initialization prior to optimization (thus making it potentially susceptible to user-introduced bias) and (2) it is relatively less computationally stable than the other methods, particularly in situations of poor data determination (i.e., relatively few data points to which to fit).

Analysis by (Pearson) correlation (by CIRCCORR) of observed time series with a variable-phase, period-defined cosine basis function also performs quite admirably (at least after the 13-point multiple-measures significance probability correction is applied), producing quite low false-positive detection rates (largely independent of whether or not pointwise replicate uncertainties are considered in analysis). It was this method, in fact, that was the basis for the original implementation of the algorithm CORRCOS, as applied in Harmer et al.[27] for analysis of circadian gene expression in Arabidopsis.

However, the COSOPT algorithm, to which the approach originally applied by CORRCOS eventually evolved (as applied in Panda et al.[17] to mammalian gene expression and Ceriani et al.[24] to gene expression in Drosophila), appears to perform in a manner superior to any of the approaches considered in this chapter. It extracts reliable estimates of period, phase, and oscillatory amplitude, albeit in manner most conservative of all (i.e., perhaps minimizing false positives at the expense of false negatives). It requires no user initialization, is stable, appears not to produce biased parameter value estimates, is readily amenable to completely automated implementation, and is computationally sufficiently rapid to complete analysis of 20,000 or more GeneIDs in a matter of only a few hours.

[8] Molecular Simulations of Diffusion and Association in Multimacromolecular Systems

By ADRIAN H. ELCOCK

Introduction

Brownian dynamics (BD) is a computer simulation technique used to model the diffusion and association of molecules in solution. Until recently, most BD applications that have employed atomically detailed structures have focused on relatively simple systems, consisting of at most two diffusing macromolecules. However, it has become computationally feasible to consider simulations of many (i.e., hundreds) of diffusing, interacting proteins. This is an important development because it means that it is now possible to investigate theoretically the complicated interdependent physical interactions that proteins encounter in true biological situations—that is, inside the cell. The basis of the method, together with some of its potential applications, is the focus of this chapter.

It is important to state at the outset that this is not explicitly intended to be a "how to" guide to performing BD simulations. Although it might

Copyright 2004, Elsevier Inc.
All rights reserved.
0076-6879/04 $35.00

technically be possible to describe all steps of the process in the allotted space, it would not be possible to cover all of the problems that might actually arise in a practical setting, any one of which might be sufficient to leave the reader at a loss for how to proceed. This is not to say that BD is "rocket science" that is understandable only to a select few; it is rather that the causes of the problems that are encountered are often so obscure (or tiresome) that an easy route to their solution may not be at all obvious. Another reason for not attempting to make this a technical manual is that the software for conducting these kinds of simulations is still very much in development. In fact, the changes that are now being made in the way that simulations are actually performed are significant enough that if a set of detailed instructions were provided here it would certainly be out of date before this volume appeared on the best-seller shelves of the nearest bookstore. I have therefore opted here for more modest objectives: that readers develop a "feel" for the methods, sufficient for them to either consider potential projects, or so that they will no longer feel excluded when the conversation at dinner parties turns (as it invariably does) to "BD methods and their uses."

In addition to tilting this chapter toward readers with little familiarity with molecular simulations, I will also make an attempt to focus the discussion on the kinds of issues that might arise specifically in simulations of systems comprising many macromolecules. Other reviews have focused primarily on the use of BD methods for computing bimolecular association rate constants.[1-3] As noted above, such simulations consider only two diffusing molecules; readers looking for a detailed step-by-step guide to performing these kinds of BD simulations are encouraged to read the excellent technical outline written by Gabdoulline and Wade.[1]

The Basic Idea

The purpose of BD is, like any other dynamic simulation method, to model the time-dependent behavior of a system of molecules. Undoubtedly the most well-known molecular simulation method is molecular dynamics (MD), which in its usual form is used to simulate the internal dynamics (including both small-scale and large-scale conformational changes) of macromolecules such as proteins and nucleic acids in water.[4] The time scale over which MD is currently routinely used ranges from picoseconds to tens or

[1] R. R. Gabdoulline and R. C. Wade, *Methods* **14,** 329 (1998).
[2] A. H. Elcock, D. Sept, and J. A. McCammon, *J. Phys. Chem. B* **105,** 1504 (2001).
[3] R. R. Gabdoulline and R. C. Wade, *Curr. Opin. Struct. Biol.* **12,** 204 (2002).
[4] M. Karplus and J. A. McCammon, *Nat. Struct. Biol.* **9,** 646 (2002).

hundreds of nanoseconds, with the upper limit meaning that some of the more interesting functional motions of macromolecules—which may occur on microsecond or longer time scales—cannot currently be addressed directly.[5] In contrast, in BD the emphasis is on simulating the diffusive properties of molecules—that is, their overall translational and rotational motion and their interactions with other molecules—and the time scale of interest ranges from nanoseconds to milliseconds. To be able to reach these kinds of time scales, we shall see that BD makes a number of assumptions and approximations that are not made in MD; a critical decision that must therefore be made prior to using BD is whether these assumptions are likely to be valid for the system of interest.

Okay, But What Do BD Simulations Look Like?

A discussion of a simulation technique will always seem esoteric and lifeless on paper. A good way to get a more intuitive feel for a method— and to find out if it might yield the kind of information that one is looking for—is to see it in action. To this end, readers might like to take a look at some animations of BD simulations at our website (http://dadiddly.biochem.uiowa.edu); these cover a variety of applications, including concentrated protein solutions, substrates (at their experimental concentrations) diffusing around enzymes, and macromolecular crowding agents diffusing around the GroEL chaperonin. The first feature that will be noticed when one views any of these simulations is how disappointingly random everything is. The word "Brownian" should perhaps have prepared us for this, but it may still be surprising to see that even association events that are subject to long-range attractive attractions (e.g., those between some substrates and enzymes) are subject to huge amounts of noise. This is an important point to keep in mind, however: intracellular environments are not quiet and serene, and any cellular process "worth its salt" must be robust enough to function even when subject to fluctuations imposed on it by external forces. In fact, biological "noise" is an aspect that has begun to attract considerable attention from those interested in simulating (and rewiring) processes such as gene transcription.[6] It is, however, easy to lose sight of this in molecular biology seminars that depict the assembly of multimacromolecular complexes as glorious, orderly processions of spherical and ellipsoidal molecules. The view that emerges from BD simulations is undoubtedly less beautiful, but probably more realistic.

---

[5] V. Daggett, *Curr. Opin. Struct. Biol.* **10,** 160 (2001).
[6] J. Hasty, D. McMillen, F. Isaacs, and J. J. Collins, *Nat. Rev. Genet.* **2,** 268 (2001).

Why Would We Want to Perform BD Simulations?

Before justifying the use of BD simulations to model macromolecular diffusion, it is first worth noting some more general advantages of molecular simulations. The most important advantage will probably be obvious to anyone who has viewed a molecular animation: simulations allow us to observe processes at a level of structural detail that can be impossible to obtain with conventional experimental methods. Of course, the value placed on such a view depends on whether it has been demonstrated that experimentally verifiable aspects of the simulations are consistent with available data; in other words, it is not sufficient to be able to see molecules in action, one must have some reason for believing what one is seeing.[7] Assuming this to be the case, an especially important advantage of simulations is the ability to routinely monitor the behavior of individual molecules; this can be immensely valuable for framing researchers' thinking and for formulating new, testable hypotheses.

A second reason for using molecular simulations is their usefulness for investigating the relative contributions made by different physical forces in determining observed behavior. For example, to assess any role played by electrostatic interactions in controlling, say, the rates at which two molecules associate, all that is required is the running of two simulations: one with the electrostatic interactions modeled (as accurately as possible), and one with the electrostatic interactions switched off, with the latter being achieved simply by skipping the relevant lines of code in the simulation program. This is an advantage that is not to be overlooked: experimentally there are few analogous situations in which one can isolate certain types of interactions with such specificity. For example, the usual way of investigating electrostatic interactions experimentally is to change the ionic strength of the solution, something that is easily achieved by the addition of salts such as KCl. However, salts can exert effects additional to the desired effect of screening electrostatic forces; for example, ions can bind specifically to the macromolecules under study, and at high concentrations they can affect the strength of hydrophobic interactions and the dielectric constant of the solution. Simulations are not subject to these kinds of difficulties of interpretation and are therefore capable (in principle) of yielding unambiguous insights. Unfortunately, this feature is not without a drawback: the very fact that simulations are not subject to the same limitations as experiments means that it is not always possible to directly

---

[7] A. H. Elcock, *Curr. Opin. Struct. Biol.* **12,** 154 (2002).

compare results from the two approaches. This in turn can have two serious consequences. First, it offers those who are pathologically skeptical of theoretical methods an easy opportunity to ridicule valid models: "discrepancies" that might arise between theory and experiment may in fact result only from an attempt to equate "apples with oranges." Second, it allows theoreticians to put up a convenient smokescreen that enables them to repeatedly avoid direct comparison with experiment by performing simulations that cannot be verified experimentally.[7] The last sentence should not be taken too seriously: some of the more powerful insights gained from theory are so precisely because they could not have been arrived at experimentally. This, however, is not an excuse to deliberately avoid opportunities to reconcile the two methods.

Having offended half the readership with the previous paragraph, the next question to address is why specifically we should think about using BD? As noted above, we would usually look to use BD in situations in which the diffusion of molecules is the property of primary interest. In keeping with this, the "traditional" use of atomically detailed BD simulations has been in the modeling of enzyme–substrate or protein–protein encounter events.[2,3] These kinds of simulations have undoubtedly provided important information complementary to that obtainable by experimental methods, although up to now they have been largely limited to systems for which diffusion is known to be the rate-limiting step of the association process. Many applications of BD will continue to be to cases for which the time scales or pathways of diffusion are of central importance, but there will also be opportunities to use the methods in quite different ways. These latter applications will make use of the fact that BD is a method that performs Boltzmann sampling of molecular systems, that is, snapshots of the system are observed with a probability proportional to the negative exponential of their energy (i.e., $e^{-E/kT}$, where E is the energy). In words this simply means that energetically favorable arrangements of the molecules in a system will naturally occur in the simulations more frequently than unfavorable arrangements, and in relative proportions that are physically correct. This means in turn that BD simulations can in principle be used to obtain thermodynamic information such as free energies of association (ΔG_{assoc}); in such situations one is less concerned with the exact pathways followed by diffusing molecules than one is with their time-averaged behavior. This alternative usage of BD again has a parallel in MD, which has more often been used to sample the types of conformations that can be adopted by macromolecules than to observe the time dependence of their behavior. Some of the potential uses of BD in these areas are discussed in a later section.

Theoretical Aspects

Most of the theoretical issues that lie at the heart of BD methods have been covered extensively elsewhere. Here I outline only its most important basic aspects in order to provide appropriate context to the following sections, and highlight the additional issues that have been raised by our extension of the simulation method to systems of many macromolecules.

The Big Assumption

The original Brownian motion, observed by Robert Brown, was the apparently random movement of pollen grains in water, caused by repeated collisions with the much smaller (invisible) solvent molecules. The central assumption of BD simulations is that a similar situation occurs on a more microscopic level, in the sense that biological macromolecules (and even small molecules) perform Brownian motion that can be adequately described by simulation methodology that completely neglects the molecular details of the solvent, water. Making this assumption has several consequences. First, since we do not have to spend any time computing how individual water molecules behave—a step that is one of the major bottlenecks of MD simulations—we can begin to think about carrying out simulations of some large macromolecular systems over long time scales. A second consequence, however, is that we must take steps to ensure that omitting the water does not drastically alter any behavior of interest, and this requires that we make corrections to account for the omission by modeling the effects of water implicitly. Clearly water plays a number of important roles, and all of these must either be correctly described or some convoluted argument that minimizes their importance must be proposed. Foremost among these effects are the dielectric screening of electrostatic forces and the hydrophobic effect. The former has been covered in detail in previous work and, as is outlined below, computationally fast methods for describing such effects are already in existence. The role played by the hydrophobic effect in modulating the dynamics of protein–protein association events has received comparatively little attention (although see Camacho et al.[8] for a nice exception); this situation will certainly change when simulations begin to be applied to large, complex macromolecular solutions.

The Algorithm

All simulation methods involve the use of an algorithm, that is, a series of instructions that the computer is asked to perform repeatedly; in the case of BD, the purpose of the algorithm is to simulate the changes in

[8] C. J. Camacho, S. R. Kimura, C. DeLisi, and S. Vajda, *Biophys. J.* **78,** 1094 (2000).

translation and rotation of molecules as a function of time. The *de facto* standard algorithm for BD is that due to Ermak and McCammon (E–M),[9] the translational part of which may be written

$$\mathbf{r}(t + \Delta t) = \mathbf{r}(t) + D\mathbf{F}(t)/k_{\mathrm{B}}T + \mathbf{R}(t) \tag{1}$$

where $\mathbf{r}$ is the position of a molecule, D is its translational diffusion coefficient (at infinite dilution), $\mathbf{F}$ is the force acting on the molecule due to interactions with other solute molecules, and $\mathbf{R}$ is a random displacement—random in both size and direction—that models the Brownian movement imparted by the solvent. In words, the algorithm simply states that if we know where a molecule is at some time (t), then we can predict where it might be at some time later ($t + \Delta t$), if we know the forces acting on it and its diffusion coefficient (see below). A similar algorithm is used to determine the rotational motion of the molecules, with force replaced by torque, and translational diffusion coefficient replaced by rotational diffusion coefficient. Notice that a molecule that is not interacting with anything else (so that $\mathbf{F} = 0$) will simply undergo a series of purely random steps, the magnitudes of which are chosen so that large molecules diffuse slower than small molecules. Note also that aside from the fact that the dynamic effects of water molecules are modeled implicitly in the BD algorithm (by the presence of D and $\mathbf{R}$), the overall form is similar to that of MD algorithms, the only other exception being that the E–M algorithm makes no mention of the velocities of the molecules (all we need to know is their positions).

As with all cases of numerical integration, a key issue is the magnitude of the timestep, Δt. Obviously we would like to be able to use a value that is as large as possible, so that with a given number of iterations of the algorithm we could simulate as complete a view of the system's behavior as possible. We would, for example, be ecstatic if a time step of 1 ms could reasonably be used: a mere 1000 iterations would then allows us to model an entire second in the life of a molecular system. Unfortunately, this is out of the question; the duration of the time step is limited by our ability to see accurately into the future. If we know where a molecule is at some point in time, we would be hard pressed to accurately predict—in one step—where that same molecule might be 1 ms later because in the intervening time period it is more or less certain to encounter other molecules that will affect its behavior. In fact, for BD, time steps of around 1 ps are more usual, and although this predictive ability might seem rather pathetic, it is still three orders of magnitude greater than that obtaining in MD, where

[9] D. L. Ermak and J. A. McCammon, *J. Chem. Phys.* **69,** 1352 (1978).

time steps of 1 or 2 fs are typical. The reason for this disparity in time steps becomes apparent when we realize that the upper limit on acceptable values is set by the requirement that the forces acting on the atoms or molecules remain approximately constant during the step. Although there is no technical reason why flexible molecules cannot be simulated with BD, applications of the methodology to proteins almost always involve the assumption that they are rigid bodies. For these situations, the forces that will fluctuate the most rapidly will be those acting between the contacting surfaces of different molecules; this is especially true when one of the contacting molecules is small (i.e., has a high value of D) since these are capable of making comparatively large displacements in a single time step. In MD simulations—in which molecules are not rigid bodies—the fastest fluctuations are to be found in the stretching of covalent bonds, where a single period of vibration can be on the order of 10 fs. Interestingly, there is also a theoretical lower limit on the time step that can be used with the E–M BD algorithm,[9] stemming from the fact that the derivation of the algorithm assumes that momentum is rapidly lost relative to the duration of the time step (hence the absence of velocities in the algorithm); this need not concern us further.

At this point, we have not specified what we mean by forces "due to interactions with other solute molecules." Those readers who have a passing familiarity with the method will know that the emphasis of most reported BD studies up to now has been on the role played by electrostatic interactions in promoting molecular associations. It is by now well known, for example, that BD can accurately describe the rapid association kinetics of substrates with diffusion-limited enzymes, as well as those of certain protein–protein systems such as barnase–barstar.[10] Less well known is the fact that BD simulations can also be used to study association processes that are not driven by electrostatics,[8] but the E–M algorithm is completely general in the sense that any intermolecular force that one suspects might play an important role in controlling diffusion can be included. In fact, the only reason that electrostatics have become so intimately associated with BD has been the idea that the long-range nature of these interactions makes them the most obvious candidates for affecting diffusional behavior. There is obviously some truth to this idea. For one thing, diffusion-limited association events such as those mentioned above are more often than not driven by favorable electrostatic interactions between binding partners. For another, the diffusion coefficients of proteins in concentrated solutions are sensitive to both the pH and ionic strength of the solution, indicating that

[10] R. R. Gabdoulline and R. C. Wade, *Biophys. J.* **72**, 1917 (1997).

even (presumably) nonspecific electrostatic interactions can strongly influence diffusional behavior. But those same diffusion coefficients are also strongly sensitive to the protein concentration—probably more so—and can become especially sluggish at the high concentrations likely to be encountered in physiological settings. Clearly, this suggests a role for steric (excluded volume) interactions in modulating diffusional behavior. As is seen below, many of the systems that will in the near future be the focus of BD simulations are likely to be controlled not by electrostatics at all, but instead by these steric (excluded volume) effects. Much of our more recent work has therefore been aimed at considering appropriate methods for dealing with a more complete range of interactions than pure electrostatics.

Electrostatic Interactions

The treatment of electrostatic interactions in BD simulations has been covered in detail in several articles[10,11]; instead of rehashing the technical details of this discussion, my purpose here is to outline to a more general audience why such a model is used in the first place. In MD simulations one usually calculates electrostatic interactions by adding up—using Coulomb's law—contributions from all appropriate pairs of partially charged atoms. In simulations that include water molecules, this will include not only protein–protein atom pairs, but also protein–water and water–water atom pairs. The sheer number of the latter interactions, coupled with the fact that their Coulombic interactions must be calculated even for pairs that are separated by a long distance (e.g., >12 Å) is the major reason that MD simulations are so slow to compute. However, on the positive side it should be noted that by computing all electrostatic interactions in this way, most of the dielectric screening behavior of an aqueous environment should be captured automatically in the simulations, without any additional fiddling or tweaking—such as assigning dielectric constants—being necessary. In fact, as any MD simulator will tell you, water molecules will naturally orient themselves around charged groups and adjust their orientations in a fairly realistic way during the course of a simulation. Because of this, the contributions to dielectric behavior that result from reorientation of molecular dipoles should be described relatively well, and in fact the computed dielectric constants of typical water models such as SPC, although by no means of quantitative accuracy, are nevertheless reasonable.[12] Having said that, conventional MD simulations—which assign fixed partial charges

[11] R. R. Gabdoulline and R. C. Wade, *J. Phys. Chem.* **100,** 3868 (1996).
[12] P. E. Smith and W. F. van Gunsteren, *J. Chem. Phys.* **100,** 3169 (1994).

to all atoms—will not capture effects due to any changes in electron distributions that might wish to occur; this is one of the major reasons why research interest continues into the development of polarizable force fields.

Now, we have already noted that in BD simulations water molecules are not present, and we therefore cannot rely on a simple summation of Coulombic terms between the charged atoms of two proteins to accurately represent their electrostatic interactions as they occur in water. A simple way to model the screening effects of water would be to scale the electrostatic interactions between charges on interacting proteins downward by a factor of 78, corresponding to the dielectric constant of water at 25°. Unfortunately, this has been shown often to be a poor approximation[13]: the much weaker dielectric properties of proteins (compared with water), coupled with their irregular shapes, can dramatically alter the strengths of electrostatic interactions in ways that are impossible to capture in a single dielectric constant. Instead, a more satisfactory approach can be taken, that has the added bonus that it is, from a computational perspective, extremely fast to calculate once an initial overhead has been settled.

The idea is to calculate the electrostatic properties of each molecule prior to BD simulations being performed and to assume that these properties, once calculated, do not alter as molecules diffuse and associate with each other; clearly, we can begin to think about making this assumption only if the molecules themselves are rigid bodies. The property that is computed is actually the electrostatic potential; its value is calculated and stored on a three-dimensional grid that covers all points in space close to the molecule. Importantly, this electrostatic potential grid is not obtained using Coulomb's law, but instead is obtained by solving the more elaborate Poisson–Boltzmann (PB) equation[13] with software programs such as DelPhi,[14] UHBD,[15] or APBS.[16] There are two major reasons for using the PB equation. First, it allows us to take proper account of the fact that proteins and water have different abilities to screen electrostatic interactions (i.e., we do not have to use a single "one size fits all" dielectric constant to describe what may be complicated screening behavior). Second, the PB approach allows the additional electrostatic screening that results from dissolved monovalent ions (such as Na^+ or Cl^-) to be described at

[13] B. Honig and A. J. Nicholls, *Science* **268,** 1144 (1995).
[14] A. Nicholls and B. Honig, *J. Comput. Chem.* **12,** 435 (1991).
[15] J. D. Madura, J. M. Briggs, R. C. Wade, M. E. Davis, B. A. Luty, A. Ilin, J. Antosiewicz, M. K. Gilson, B. Bagheri, L. R. Scott, and J. A. McCammon, *Comput. Phys. Commun.* **91,** 57 (1995).
[16] M. Holst, N. Baker, and F. Wang, *J. Comput. Chem.* **21,** 1319 (2000).

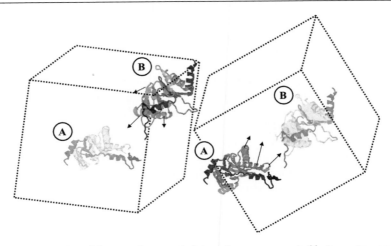

FIG. 1. Illustration of the way electrostatic interactions are computed between two proteins in typical BD simulations. *Left:* The electrostatic forces acting on protein B are computed using the electrostatic potential grid generated by protein A. *Right:* Forces on protein A are computed using the electrostatic potential grid generated by protein B. Note that both grids rotate and translate with their "parent" molecules.

least semiquantitatively, which means that the behavior of macromolecules in salt (physiological) solutions can be simulated without the need to explicitly model the diffusion of perhaps thousands of salt ions.

Although advances have allowed the PB equation to be solved for molecules as large as the ribosome,[17] it is important to keep in mind that it is not a computationally trivial undertaking (the practicalities of how this is actually done are discussed below). Once solved, however, it allows us to calculate the electrostatic interaction of a molecule (A) with other molecules extremely rapidly. In fact, to approximate the force that this potential exerts on a second molecule (B), we need only calculate the slope of molecule A's electrostatic potential at the positions of molecule B's charged atoms. This is easily done by interpolating values from the potential grids, and by making sure that such grids stay attached to their parent molecules as they translate and rotate during the course of a BD simulation (see Fig. 1). As noted above, in adopting this approach there is clearly an assumption being made, namely that the electrostatic potential in the immediate vicinity of each molecule (e.g., A) is unaffected by the approach

[17] N. A. Baker, D. Sept, S. Joseph, and J. A. McCammon, *Proc. Natl. Acad. Sci. USA* **98,** 10037 (2001).

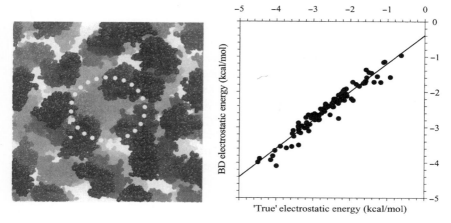

FIG. 2. Comparison of electrostatic interaction energies calculated by the fast approximate method employed in BD and the most rigorous method currently possible. To conduct this comparison, 100 snapshots from a BD simulation *(left)* of aldolase in a 200-mg/ml solution of ribonuclease A were generated, and the interaction energy of the aldolase (circled) with the ribonuclease solution was computed.

of another molecule (B). This would seem to be a drastic approximation and, not surprisingly, it does become progressively less tenable as two molecules approach contact.[11] In principle, one could solve this problem by continually recalculating electrostatic potentials during the course of a BD simulation, but this is currently infeasible and is likely to remain so for some time to come. Despite this, the fixed potential approach can work surprisingly well, even when a protein is surrounded by many others; we can show, for example, that the computed electrostatic interaction energies of proteins in highly concentrated (200 mg/ml) solutions are more or less identical with those obtained by the more rigorous approach of recomputing the potential (Fig. 2).

van der Waals Interactions

In most previous applications of atomically detailed BD methods steric interactions between molecules have been treated rather crudely—in large part because the emphasis of previous studies has so often been on the electrostatic interactions. In fact, the most common approach has been to apply steric considerations only after a simulation step has already been attempted[1]: steps that bring atoms of two molecules into direct contact are rejected, and another move attempted—with a different random displacement **R** applied to the molecules (in the hope that eventually they will choose to move in a direction that does not result in a steric clash). This is not a wholly unreasonable approach to take, but a technical question it raises is whether we should consider time to have elapsed during these "failed" steps. Depending on the answer to this question—and the frequency with which such failed steps occur—we could in principle obtain different estimates for properties, such as diffusion coefficients, that explicitly depend on the time elapsed during the simulation. To clear up this uncertainty, we have taken steps to incorporate van der Waals interactions (which of course include an attractive component in addition to steric exclusion forces) in a more rigorous manner into the simulations. In our current work, forces acting between the surface atoms of proteins are calculated using standard Lennard–Jones potentials in the same way that they are computed in conventional MD simulations. However, an important difference to note is that in (explicit solvent) MD simulations, the association of protein molecules would require that van der Waals interactions between the proteins and water molecules be broken first. In (implicit solvent) BD simulations on the other hand, proteins would not pay any such penalty for associating because there are no van der Waals interactions with water to disrupt. As a consequence, Lennard–Jones parameters that have been shown to provide accurate results for molecular associations in explicit MD simulations are not likely to do so for BD simulations. In fact, they will certainly lead to significant overestimates of the strength of protein–protein interactions. A simple solution to this problem that we are currently exploring is to simply scale down Lennard–Jones interactions by some empirical factor (e.g., 0.5); although crude, this is probably a more tenable solution than it is for electrostatic interactions.

Regardless of the actual parameters used to describe the energetics of van der Waals interactions, it is worth noting that we can again exploit the fact that the proteins are treated as rigid bodies to calculate the interactions extremely rapidly. For one thing, if the interactions are parameterized properly, we should be able to omit contributions from atoms that are not at the protein surface; since the atoms in rigid body molecules are not free to move independently, we would not need to continually check which atoms are at the surface and which are buried—the decision would only need to be made once, at the beginning of the simulation. For another, much of the computationally expensive process of measuring distances between surface atoms—to determine whether they are close enough to interact—can be avoided. Instead, we can construct for each type of molecule in the simulation, a nonbonded list grid; this three-dimensional grid stores for each point in space near to the protein (A), a list of all the atoms in the protein that would interact with an approaching atom of protein B occupying that position in space (Fig. 3). During a BD simulation, all we must do to

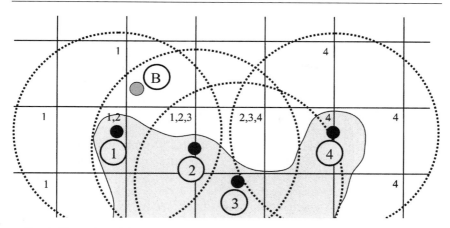

FIG. 3. Illustration of the way a nonbonded list grid is constructed and used in calculations of van der Waals interactions. Here, the numbered atoms (1–4) correspond to surface atoms of protein A (shaded region). Circles centered on each of these atoms determine the furthest distance at which they will interact with atoms of other molecules. A grid is built around protein A, with each grid point storing the names of the atoms on the protein that are within range of interaction. When an atom of another molecule approaches, we look to the nearest grid point to find the list of atoms with which it will interact. In the example shown, atom B will interact with atoms 1 and 2, but not with atoms 3 or 4 of protein A.

find which atoms of A interact with an atom of B is to place B's atom on A's nonbonded list grid and read the names of the atoms listed at that point in space. Of course, we must then still calculate the distances between these atoms, but at least we can be certain that we are calculating only interactions that will be worth calculating; typically this might be all atoms that are within ~12 Å of each other.

Hydrophobic Interactions

An enormous amount of theoretical research has been targeted at understanding the hydrophobic effect. As with the dielectric behavior of water discussed above, the hydrophobic effect is something that should naturally arise in MD simulations that include explicit water molecules; sure enough, simple hydrocarbons such as methane or ethane show a pronounced affinity for one another in aqueous phase MD simulations. Since the effect is a consequence of the molecular nature of the solvent—the hydrogen-bonding abilities of water molecules—any attempt to describe it without actually putting water molecules into the simulations will always be a crude approximation. A common approach is to model the magnitude of hydrophobic interactions as being proportional to the amount of solvent-accessible surface area (SASA) that becomes buried when molecules contact each other[2,8]; it should be noted, however, that a variety of cleverer solutions also exist (see, e.g., Hummer[18]). A key concern is whether these models can be calculated sufficiently rapidly to be useful in BD simulations. This is an as yet unanswered question, but a still simpler (and therefore faster) approach presents itself when we note that the amount of buried SASA in protein–protein complexes scales linearly with the number of contacting atoms in the interface[19]: it should therefore be possible to adapt van der Waals interactions to provide an *ad hoc* description of hydrophobic interactions. Such an approach would have an additional aspect in its favor: a $1/r^6$ dependence (or something similar) would naturally capture the limited range over which such interactions typically operate. Although the details have yet to be worked out—and will certainly require a considerable amount of experimentation to ensure realistic behavior—this may involve as simple a solution as assigning a more favorable energy to van der Waals interactions between nonpolar atoms than polar atoms.

The First Solvation Shell

One aspect of BD simulations that appears to continually offend some researchers is their omission of all water molecules. There is no doubt that waters in the first solvation shells of macromolecules behave differently from those in bulk solution; the more relevant question is whether ignoring their anomalous behavior renders BD simulations completely meaningless. There are several points worth mentioning at this juncture. First, in terms of describing the thermodynamics of associations, it is important to note that the approximations employed in BD are not dramatically worse than those used in other simulation applications: for example, methods for predicting protein tertiary structure from sequence or for computing the thermodynamic effects of mutations in protein–protein interfaces routinely use solvation models of equal or greater crudity without apparently eliciting loud objections. A more legitimate worry is that omitting a first solvation shell may adversely affect the description of the kinetics of association or diffusion and, given the emphasis of this work, it is worth outlining the reasons for this concern. The most important is the fact that the apparent radii of proteins obtained from experiments that measure hydrodynamic properties are significantly greater than the actual radii known from their high-resolution structures: this provides clear evidence that

[18] G. Hummer, *J. Am. Chem. Soc.* **121,** 6299 (1999).
[19] L. Lo Conte, C. Chothia, and J. Janin, *J. Mol. Biol.* **285,** 2177 (1999).

proteins diffuse with one or more hydration shells effectively attached.[20] The fact that proteins diffuse slower than expected on the basis of their structure need not by itself be a disaster—experimental values of diffusion coefficients are preferred for use in the E–M algorithm anyway—but it does become problematic when we stop to consider how the solvation shells of proteins reorganize when association occurs, and how this reorganization process might affect the kinetics of association. Given that it is difficult to imagine how a purely implicit solvation model might account for these rearrangements, it is clearly desirable that some sort of semirigorous method for including a first solvation shell into BD simulations be developed.

No such method exists at the moment, however, so the answer to the question raised above becomes a stark choice. Either one decides that the solvation shell issue is so serious that all BD simulations are a complete waste of time, or one accepts the limitation and ensures that the applications one pursues are not strongly affected by the approximation. Those who select the former choice should recognize that its consequence is to accept that treatment of such phenomena as protein–protein association events will remain beyond molecular simulations until such time as explicit solvent MD simulations of such systems become feasible. Selecting the latter choice implies that the new knowledge one gains from BD simulations outweighs the concerns about its strict validity. It will, of course, be clear that I have decided to make the second of these choices; with luck, some of the applications discussed below will convince readers that this was not an insane choice to make.

Practical Aspects

It is one thing to outline the theoretical issues at the heart of a method, it is quite another to consider how those and other issues actually arise in practice; what follows is intended to cover many, although not all, of the dreary practicalities of setting up and performing BD simulations.

Finding Structures

The first step in performing any BD simulation is to find suitable structural models for the molecules that are of interest. One might think that this would be a trivial step: the primary repository of structural information on macromolecules, the PDB (http://www.rcsb.org), is by now well known to almost all biochemists and many biologists. Surely, then, all that we

---

[20] T. E. Creighton, *in* "Proteins." W. H. Freeman, New York, 1993.

would need to do is search the PDB for our protein and download its structure. As it turns out, things are often not that simple. For a start, searching the PDB can be something of a joke, as anyone who has tried to find a structure of wild-type lysozyme, for example, will attest. Second, even when a structure is present—and can be successfully located—it is not always complete, and a decision is therefore often required as to whether the omission of parts of the protein might affect the validity of simulations.

Obvious examples of incomplete structures are those that cover only one of several domains of a multidomain protein; less obvious but equally serious cases are provided by the legions of macromolecules that undergo weak oligomerization. The latter can be problematic for a number of reasons. First, since oligomerization states are often sensitive to protein concentration, pH, ionic strength, and temperature, care is required to ensure that the structures used in simulations are consistent with the environmental conditions intended to be simulated. Second, even if the oligomeric state of a protein is known, its exact structure may still be in dispute. In crystal structures, for example, there may be multiple contacts between neighboring proteins, so one must decide which of the potential interfaces is the most likely candidate. One way to do this is to examine the size and/or physicochemical nature of the potential interfaces[21]; an alternative and complementary way is to look at the degree of conservation of residues in the interface, with the assumption being that any functionally important interface should be subject to some degree of evolutionary pressure.[22,23] An excellent source of potential structures for oligomeric proteins is the PQS Web server[24] (http://pqs.ebi.ac.uk), which contains the results of automated assessments of potential oligomeric contacts and, when appropriate, stores the structure of the complete protein—something that may not be directly obtainable at the PDB. Although the reconstructed oligomers are occasionally incorrect, the site is such a useful resource that one can often save considerable time by visiting this site in preference to the actual PDB site. One incidental point that may help first-time users is that the structure files in the PQS database have the suffix ".mmol"; their format is essentially identical to the ".pdb" files found at the PDB.

If one cannot find a protein of interest at the PDB or PQS, what does one do? A realistic option for BD simulations is to consider homology modeled structures,[25] that is, structures built using a sequence-similar protein structure as a template. A homology model is not a perfect

[21] J. Janin, *Nat. Struct. Biol.* **4,** 973 (1997).

[22] A. H. Elcock and J. A. McCammon, *Proc. Natl. Acad. Sci. USA* **98,** 2290 (2001).

[23] W. S. J. Valdar and J. M. Thornton, *J. Mol. Biol.* **313,** 399 (2001).

[24] K. Henrick and J. M. Thornton, *Trends Biochem. Sci.* **23,** 358 (1998).

[25] A. Fiser, M. Feig, C. L. Brooks, and A. Sali, *Acc. Chem. Res.* **35,** 413 (2002).

substitute for a high-resolution crystal or NMR-derived structure: the exact positions of atoms will not be correctly predicted and this is usually a sufficiently serious problem that MD simulations, which are often intended to provide a highly detailed view of the behavior of individual atoms, are simply not worth attempting. However, for typical BD simulations, in which we are already making a raft of other serious approximations—for example, that proteins are rigid bodies interacting in a structureless solvent—uncertainties about exact atomic positions should not be seen as a significant problem. In fact, as long as the structure is at least approximately correct, a homology model should be adequate for BD purposes, provided that one can convince oneself that loop regions, where model and template can seriously diverge, are not crucially important. A scenario in which homology models should be particularly appropriate is for studying orthologous proteins from different organisms. For example, structural comparisons of proteins from mesophilic organisms with their counterparts from extremophiles have shown that differences are usually manifested at surface residues, in such a way that the surface amino acid compositions can be different even if the underlying folds are more or less identical.[26] Homology models should be eminently capable of capturing these rather gross compositional differences, and since these in turn can have major consequences for surface properties such as hydrophobicity or electrostatic potential, they offer a simple route to computationally studying differences in protein–protein interactions between organisms. An excellent free resource for building homology models is the SWISS-MODEL[27] Web server (http://www.expasy.ch/swissmod).

A final note relating to structural issues is aimed at those interested in performing simulations of diffusion of small molecules, such as drugs or metabolites. If a high-resolution structure cannot be found (e.g., at the PDB), a resource that is certainly worth investigating is the amazing Web server PRODRG[28] (http://davapc1.bioch.dundee.ac.uk/programs/prodrg) which builds accurate three-dimensional models of molecules given only a two-dimensional representation provided by the user.

Calculate Electrostatic Potentials

There are several articles that discuss the technical details of PB calculations of electrostatic potentials in depth,[1,15] but there are several points that can be especially frustrating to newcomers to this field that I think

[26] C. Cambillau and J. M. Claverie, *J. Biol. Chem.* **275,** 32383 (2000).
[27] N. Guex, A. Diemand, and M. C. Peitsch, *Trends Biochem. Sci.* **24,** 364 (1999).
[28] D. M. F. van Aalten, R. Bywater, J. B. C. Findlay, M. Hendlich, R. W. W. Hooft, and G. Vriend, *J. Comput. Aided Mol. Design* **10,** 255 (1996).

are important to address. To begin, it is first worth being clear about what information is required to perform a PB calculation, and what information is returned by a calculation. Prior to calculation, one needs to know the structure of the protein and the partial charge and atomic radius of every atom in the structure. Partial charges are obviously important for letting the computer know how charge is distributed within the molecule under investigation. Atomic radii, on the other hand, are used to determine what regions of space are to be considered "outside" the molecule and which are to be considered "inside": this information allows the program to assign different dielectric constants to the protein and the aqueous solution and to determine what regions are accessible to dissolved salt. Together, sets of partial atomic charges and atomic radii are termed "parameter sets"; a number of such parameter sets have already been developed for describing the electrostatic properties of common molecules, such as nucleic acids or amino acids (e.g., Refs. 29 and 30).

The first stage in a PB calculation is the correct assignment of these parameters to the atoms of the protein—an innocuous-sounding process that can nevertheless induce tears of frustration in the most wizened simulation practitioners. Those wishing to conduct these kinds of calculations should prepare themselves ahead of time for spending maddening hours in front of a computer monitor ensuring that the names of atoms and residues appearing in a .pdb file match exactly with those in the parameter file (i.e., the file containing the list of charges and radii to be given to atoms). Exaggeration aside, the process is usually relatively straightforward for the vast majority of atoms in a protein. Known trouble spots are the residues at the N and C termini of proteins, which will contain atoms additional to those present in residues located within the chain. Other problematic residues include histidine—which can be protonated at two different (or both) positions, each of which results in a different partial charge distribution, and oddly, isoleucine, about which there appears to be occasional confusion over whether it contains an atom named "CD" or "CD1." An issue for those interested in performing highly detailed PB studies (not usually BD practitioners) is the spatial arrangement of hydrogen atoms: where freely rotating O—H bonds are involved, for example, there can be many potential arrangements or networks of hydrogen bonds that could be formed.[31] Again, there is an excellent Web-based resource that can take care of this: part of the WhatIf Web server[32] (http://www.cmbi.kun.nl/bv/servers/WIWWWI)

[29] D. Sitkoff, K. A. Sharp, and B. Honig, *J. Phys. Chem.* **98,** 1978 (1994).
[30] A. H. Elcock and J. A. McCammon, *J. Phys. Chem. B* **101,** 9624 (1997).
[31] R. W. Hooft, C. Sander, and G. Vriend, *Proteins* **26,** 363 (1996).
[32] G. Vriend, *J. Mol. Graphics* **8,** 52 (1990).

will optimize the positions of hydrogen atoms, making intelligent decisions about the correct protonation state of histidines, and the orientation of asparagine and glutamine side chains (the latter being an issue because the electron density of C=O can be difficult to distinguish from that of C—NH$_2$ in crystal structures). Finally, one should be especially careful when considering nonstandard residues or molecules; for such cases, one will probably have to develop one's own set of parameters. This need not be a terrifying undertaking. The preferred way of doing this is to adopt exactly the same approach used in the derivation of the original parameters. Depending on the parameter set in use, this may involve the use of quantum mechanical calculations to obtain partial charges—something that may not be accessible to all research groups—so an alternative, much simpler approach (usually adopted by my research group) is to derive approximate parameters "by eye" by using one's chemical intuition to identify analogous functional groups that have already been parameterized.

Given suitable input information, the result of a PB calculation will be a three-dimensional grid that stores the value of the electrostatic potential at all grid points—both inside and outside the molecule of interest. The only other common problem to consider is to ensure that this grid is large enough to encompass all regions of space around the protein where the potential is appreciably nonzero. This is usually not a problem when the potential is calculated in the presence of ionic strengths of 50 mM and greater, since the electrostatic potential typically dies off quite rapidly as the distance from the protein increases. However, in low salt concentrations—or worse still, zero salt—the electrostatic potential can extend a surprisingly long way into space, despite the fact that water is such an effective medium for screening electrostatic interactions. It is important to be sure that the potential is zero (or significantly less than the thermal energy, kT) at the grid edges, because otherwise the force experienced by another protein will drop precipitously as the boundary of the grid is crossed; as a result, when the computer comes to calculate electrostatic forces exerted on a protein it may find that some fraction of its atoms (i.e., those lying within the confines of another protein's potential grid) are subject to electrostatic forces, while the others feel no force whatsoever. This abrupt truncation of electrostatic interactions is unlikely to have the same kinds of disastrous effects that it once had on MD simulations of highly charged molecules such as DNA—which would rapidly (and hilariously) fall apart in full view of the paying public—but it is nevertheless undesirable and should be avoided where possible. Fortunately, it is easy to determine whether this is likely to be a problem, since PB electrostatic programs such as UHBD can readily report the value of the electrostatic potential at given distances from the center of the protein.

Perform the Simulation

Having selected structures and computed electrostatic potentials one is almost now in a position to start simulations. The next immediate question is to decide how the molecules in the system are to be spatially arranged at the beginning of the simulation. For studies aimed at simulating freely mixing solutions of macromolecules, a reasonable approach might be to assign initial positions for the molecules randomly, by applying a random translation and rotation to the structure of each molecule in the simulation. Alternatively, one could assign positions on a more energetic basis, by placing each molecule in a large number of randomly chosen initial locations and orientations and then selecting the one with the most favorable interaction energy with those molecules that have already been placed in the system. However, for high protein concentrations (such as that shown at the left of Fig. 2), both approaches require more thought. Finding a position that does not cause a steric clash with other molecules through random selection becomes an increasingly tricky business as more and more molecules are added to the system. In fact, given the high degree of volume excluded in such situations it is conceivable that one might arrive at a point at which it is impossible to add further molecules. A simple way around this is to first perform a short BD simulation in which each molecule is replaced by a single pseudo-atom of similar radius to the molecule it replaces. These pseudo-atoms, which can be made to interact with each other solely through Lennard–Jones interactions, rapidly push each other apart to relieve any nasty steric interactions that might be present. Once this has happened (something that is easily determined by watching how the energy of the system decreases), the original molecules can be reintroduced into the system, into the cavities carved out by their pseudo-atomic "stunt doubles." Regardless of how one initially assigns positions to molecules, it should be remembered that the arrangement with which the simulation is started may not be representative of the actual arrangements that the system will adopt if left to its own devices. This should be not surprising, of course—if one knew exactly how molecules should be arranged without performing BD simulations, there might be no reason to perform the simulations—but a necessary consequence is that an extended period of simulation may be required just to let the system "settle down." The length of this initial period of simulation, which is usually termed "equilibration," will have to be determined on a case-by-case basis since it will depend on the extent to which molecules can freely explore their surroundings; at the absolute minimum it should be sufficient that the total energy of the system has reached some approximately stable value.

One final issue to decide is how to treat the boundaries of the simulation system. If one is interested in the behavior of molecules at infinite dilution, there is clearly no need to set any limit on the space open to the diffusing molecules. If, on the other hand, one is primarily concerned with finite concentrations of molecules, one can make use of "periodic boundary conditions" (PBCs), a commonly used computational trick[33] for ensuring that the system behaves more like one embedded within a solution environment than one at a surface (i.e., at the gas–liquid boundary). PBCs simply mean that the simulation cell—which is usually cubic, but need not be—is replicated in three dimensions in such a way that as a molecule leaves the central cell, a copy of the molecule reappears on the opposite side of the cell. Using this approach allows one to adjust the dimensions of the central simulation cell (or the numbers of molecules) so that proteins are present at any desired concentration.

Stop the Simulation and Analyze the Results

What dictates when one stops performing simulations? Ideally, this should be when one has accrued enough data to obtain an unambiguous answer to the hypothesis that the simulations were intended to address (the issue of what kinds of data to collect is covered in the next section). More often, the upper length of molecular simulation times is determined instead by limitations on the available computational power, and projects are terminated—and the results written up and submitted for publication—regardless of whether any clear conclusions have been obtained. Clearly, this is something that should be avoided if at all possible, and the best way to do this is to be extremely careful in the framing of project goals. This is easy to state, but the formulation of hypotheses that can be properly tested with computer simulations is actually a huge challenge: the most interesting directions to follow somehow always appear to be at or beyond the limits of existing computer power.[7] The next section describes some of the types of applications that we believe can be reliably addressed by BD simulations currently.

There will, of course, be occasions when one cannot know whether a question is answerable with molecular simulations, without actually performing the simulations—no matter how much one scribbles on napkins, it is not always possible to estimate how much simulation time will be required for a proper sampling of the behavior of interest. Taking protein diffusion as an example, one might easily speculate that diffusion will be

---

[33] M. P. Allen and D. Tildesley, *in* "Computer Simulation of Liquids." Clarendon Press, Oxford, 1987.

slowed in the simulation conditions, but by how much may be impossible to guess, and an inability to make an accurate guess means that one cannot reliably estimate how much simulation time will be required to obtain adequate statistics. Does this mean that one should abandon simulations of such systems? Of course not: some of the most exciting simulations are those run to "see what happens," since this is often an effective way to stumble upon interesting behavior. That said, since computer power remains—for most groups—a limited resource, it is probably best invested in projects that are most likely to yield clearly defined results.

In addition to words of caution over the framing of projects, one other aspect that requires close attention in simulation projects is the monitoring of progress. Complacency is a real problem when conducting simulations that require days or weeks to complete: it is easy to leave the laboratory with a deceptive feeling of comfort born of the belief that the computers will continue working overnight. When one returns the following day, one may be tempted to only check that the simulation is still running—that is, consuming CPU time—and did not crash (or fill up the hard disk) during the small hours of the night. This, however, is an imperfect way of assessing progress: a computer program that does nothing but count from one to a bazillion will consume CPU time effectively, but it will not return anything particularly useful. One simple way to make sure that life in the simulation world has not taken a bizarre turn is to check that the energies and spatial coordinates of the molecules, which are usually written out at a user-specified frequency, have not suddenly assumed crazy values. The latter are more easily checked by either examining a snapshot of the system using a free molecular viewer such as RasMol[34] (http://www.umass.edu/microbio/rasmol), or a movie of the simulation (e.g., using the excellent free program PyMol[35]; http://www.pymol.org). In a field of research that (rightly) seeks to place everything on a quantitative basis, this may seem an unnecessarily subjective process, but human eyes are still among the best means of identifying behavior that is clearly ludicrous—assuming, of course, that those eyes are attached to a brain that is making a serious attempt to closely examine the system.

Some Example Applications

The use of BD simulations for understanding the association kinetics of protein–protein and enzyme–substrate pairs has been the subject of several reviews.[2,3,36] The projects briefly outlined here concern more recent

[34] R. Sayle and E. J. Milner-White, *Trends Biochem. Sci.* **20,** 374 (1995).
[35] W. L. DeLano, *in* "DeLano Scientific." San Carlos, CA, 2002.

applications for which simulation of multiple diffusing molecules has been an important development. It is important to note that all of these projects were performed using single-processor PCs identical to the kind that can be bought "off-the-shelf": one does not require access to super-computer time to be able to perform meaningful BD simulations of complex systems.

Substrate Channeling

"Channeling" is the name given to the phenomenon of substrates being transferred directly—that is, without first diffusing through bulk solution—between the active sites of enzymes that catalyze sequential reactions.[37] Thanks to combined crystallographic and kinetic studies, it is now accepted that channeling can occur by molecules diffusing along tunnels running through the interior of proteins[38]; the classic examples of this are trypto-phan synthase and carbamoyl phosphate synthetase. More controversial is the idea that charged substrates can also be channeled by diffusing along the external surface of an enzyme if that surface has a significant favorable electrostatic potential; examples for which this kind of channeling has been proposed include dihydrofolate reductase-thymidylate synthase[39] and cit-rate synthase-malate dehydrogenase[40] (CS-MDH). The latter, which amus-ingly enough is actually an artificially fused construct—and is therefore of uncertain physiological relevance—has been the subject of several experi-mental and theoretical studies, which have arrived at a variety of different conclusions.[41] Without going into the gory details of the debate, the rele-vance of this system for the present work is that it was the subject of the first atomically detailed BD studies of channeling with multiple diffusing substrates.[42] The goal of the simulations was to determine the probability that an oxaloacetate molecule produced at the MDH active site would dif-fuse to a CS active site in preference to diffusing away from the enzyme (Fig. 4). The property monitored during these simulations was therefore simply the position of the oxaloacetate as it diffused from its birthplace; if the oxaloacetate was found to come within a specified distance (e.g., 8 Å) of the CS active site, we assumed that it had been successfully

[36] G. Schreiber, *Curr. Opin. Struct. Biol.* **12,** 41 (2002).

[37] H. O. Spivey and J. Ovadi, *Methods* **19,** 306 (1999).

[38] X. Y. Huang, H. M. Holden, and F. M. Raushel, *Annu. Rev. Biochem.* **70,** 149 (2001).

[39] D. R. Knighton, C. C. Kan, E. Howland, C. A. Janson, Z. Hostomska, K. M. Welsch, and D. A. Matthews, *Nat. Struct. Biol.* **1,** 186 (1994).

[40] A. H. Elcock and J. A. McCammon, *Biochemistry* **35,** 12652 (1996).

[41] H. Pettersson, P. Olsson, L. Bulow, and G. Pettersson, *Eur. J. Biochem.* **267,** 5041 (2000).

[42] A. H. Elcock, *Biophys. J.* **82,** 2326 (2002).

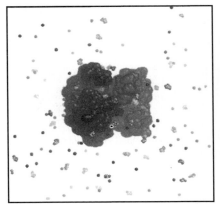

FIG. 4. *Left:* Illustration of the CS-MDH fusion protein surrounded by a 10 mM sodium malate solution. Small spheres represent sodium ions, and small molecules are malates. *Right:* Trajectory followed by an oxaloacetate molecule that successfully channels into the CS active site in a simulation of a 2.5 mM sodium malate solution.

channeled. The central result of these simulations was the demonstration that the answer to the question "Does oxaloacetate channel from MDH to CS?" depends critically on the concentration of malate (the initial substrate of the MDH reaction) present in the solution. This results because the electrostatic properties of oxaloacetate and malate are more or less identical: electrostatic forces exerted by the enzyme on the former can therefore operate equally well on the latter, and since the concentration of malate used in the experiments is so much higher than that of the oxaloacetate, channeling becomes more or less completely suppressed.

There are several implications of this study. First and most importantly, it demonstrates the potential utility of multimolecular BD simulations for guiding experimental strategies: without the ability to directly observe the physical process of metabolite diffusion, we might never have known that the experimental design could unintentionally abolish channeling. Second, the ability of the simulations to unambiguously dissect the contributions of different forces showed that competition between malate and oxaloacetate was actually exerted through direct electrostatic repulsion rather than steric exclusion. Finally, in a broader sense, the fact that a simple coupled enzyme system is so vulnerable to substrate competition effects— even when the competing molecules do not bind strongly—has as yet unexplored implications for metabolic pathways operating in more physiological settings.

Protein Diffusion in Concentrated Solutions

The question of how phenomena observed *in vitro* carry over to situations *in vivo* will become of increasing concern in the coming years for those conducting molecular simulations, just as it already is (or should be) for those carrying out experiments. The ultimate goal in this direction, of course, will be to conduct meaningful simulations of entire cells, so that cellular responses—for example, to perturbations such as the administration of a drug—can be reliably predicted. There have already been some exciting developments in the direction of cellular modeling; and methods are available that operate at a wide range of spatial and temporal resolutions.[43,44] None, however, begin to treat cells at a molecular level of detail, for good reason given that computational resources capable of treating such situations are many years away from being available. But ultimately, we may require methods that are capable of describing the effects of individual, or at least small numbers, of molecules: for example, the number of mRNA copies for a particular gene may be fewer than 10 per cell.[45] Meeting this requirement will almost certainly necessitate some imaginative theoretical developments, since even the most optimistic proponent of BD simulations (i.e., myself) would concede that computational advances alone will not bring cellular simulations into range any time soon. It is, of course, easier to state that theoretical advances are required than it is to fulfill the need. One possible way to proceed is to use atomically detailed methods such as those discussed here to provide precise views of molecular diffusion and association events and then use their results to prompt the development of simpler—and therefore faster—models that can be used to simulate much larger (subcellular) systems.

Our current (preliminary) steps in this direction demonstrate what is probably the state-of-the-art in terms of the accessible time scales and length scales of atomically detailed BD simulations. We are using BD simulations to attempt to reproduce known experimental data[46] on the diffusion coefficients of "tracer" proteins as the concentration of a second protein, present in excess, is increased (Fig. 5); these experimental data cover a variety of different systems, at concentrations ranging up to 200 mg/ml. The property monitored during these kinds of BD simulations is the position of the center of mass of the tracer proteins. If one conducts

[43] M. Tomita, *Trends Biotechnol.* **19**, 205 (2001).
[44] L. M. Loew and J. C. Schaff, *Trends Biotechnol.* **19**, 401 (2001).
[45] J. Stenman, S. Lintula, O. Rissanen, P. Finne, J. Hedstrom, A. Palotie, and A. Orpana, *Biotechniques* **34**, 172 (2003).
[46] N. Muramatsu and A. P. Minton, *Proc. Natl. Acad. Sci. USA* **85**, 2984 (1988).

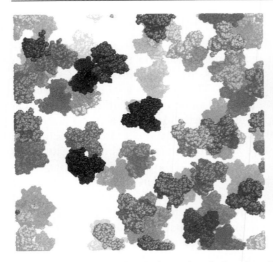
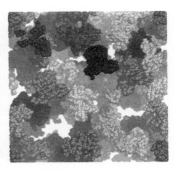

FIG. 5. Illustration of typical protein solutions for which experimental diffusion data have been derived.[46] Shown are simulated solutions of pure aldolase at concentrations of *(left)* 50 mg/ml and *(right)* 200 mg/ml. Both simulation systems shown contain 50 aldolase molecules.

sufficiently long simulations that one fully samples the behavior of the tracer proteins (perhaps 10–100 μs), their effective diffusion coefficients can be calculated straightforwardly using Einstein's diffusion formula:

$$\langle r^2 \rangle = 6D_{\text{eff}}\Delta t$$

where $\langle r^2 \rangle$ is the mean square distance traveled in time Δt and D_{eff} is the effective translational diffusion coefficient (which may be different from the infinite-dilution value assigned to the protein in the E–M algorithm). In addition to monitoring the translational diffusion of the proteins, the orientations of the molecular axes can also be monitored to obtain information on rotational diffusion; given that the latter have not yet been measured for the sets of proteins under study, they provide a good opportunity to test the predictive abilities of the simulation methods.

Macromolecular Crowding Effects

The above-described studies involve monitoring the diffusive behavior of molecules during the course of a simulation, and thus are natural extensions of previous applications of BD methods. As was stated earlier, however, BD simulations may also be used solely for the purpose of sampling possible configurations of a molecular system, so that measurements of thermodynamic properties can be performed. As an example of this kind of application, we have used BD to study the effects of a macromolecular crowding agent on the dissociation kinetics of a protein from the GroEL chaperonin.[47] The potential role played by "crowding" effects has attracted considerable attention because macromolecular concentrations in intracellular conditions are known to be extremely high.[48,49] A considerable degree of qualitative insight into the likely consequences of these effects has already been obtained from theoretical studies, but since they have focused primarily on the derivation of analytical equations, they have of necessity involved a number of often quite serious assumptions. One of our current interests is therefore to ask to what extent (numerical) BD simulations—for which we need not make assumptions such as proteins being spherical particles—might allow more quantitative information to be obtained.

As a step in this direction, the specific system that we studied was one investigated experimentally in 1997.[50] A series of elegant experiments had demonstrated that highly crowded conditions (formed by addition of high concentrations of agents such as Ficoll) made folding of the protein rhodanese much more efficient when high concentrations of a mutant trapping form of the chaperonin were present (Fig. 6). A straightforward interpretation of this result is that partly folded forms of the protein are more likely to rebind to the same GroEL molecule at the end of each round of folding when crowding agents are present. The result is of broader interest, however, because of what it tells us about how physiologically important processes (in this case chaperonin-mediated protein folding) might be affected by the highly crowded cellular environment. It turns out that the experimental results can be quantitatively reproduced by BD simulations that measure the energetics of the dissociation process. To do this, we performed a large number of simulations in which a rhodanese molecule was gradually pulled from the interior of the GroEL "cage" and out into a solution containing varying concentrations of crowder molecules. These simulations were performed using standard free energy perturbation (FEP) methods—methods that have been successfully used in MD simulations—and were repeated >50 times to obtain statistically robust estimates of the free energy changes. To compare the calculated numbers with the experimental numbers, an elaborate theoretical analysis of the experimental data was required; only once this was done was it possible to be sure that the two sets of results were being compared on an equal footing.

[47] A. H. Elcock, *Proc. Natl. Acad. Sci. USA* **100,** 2340 (2003).
[48] A. P. Minton, *J. Biol. Chem.* **276,** 10577 (2001).
[49] R. J. Ellis, *Trends Biochem. Sci.* **10,** 597 (2001).
[50] J. Martin and F.-U. Hartl, *Proc. Natl. Acad. Sci. USA* **94,** 1107 (1997).

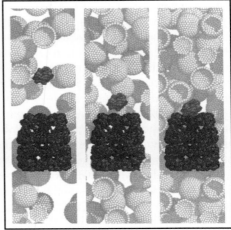

FIG. 6. *Left:* Illustration of the simulated system used in studies of macromolecular crowding effects on the GroEL chaperonin. *Right:* Illustration of the point at which the energy of interaction of the protein rhodanese with the crowding molecules reaches thermal energy ($1\ kT$) for (from left to right, respectively) 10, 20, and 30% w/v crowder solutions.

Conclusion

The use of BD for simulating diffusion and association of multiple macromolecules is a relatively new undertaking, and although it is by no means a technological revolution, it is sufficiently novel that it is difficult to be certain where it will ultimately lead. What we can say with some confidence is that even without any further development, the methods will be useful for studying the effects of excluded volume (i.e., sterics) on the thermodynamics and kinetics of macromolecular association processes; a preliminary demonstration of this is provided by the successful application of the methods to the GroEL system. As noted above, we expect the use of atomically detailed BD to be of particular importance for those interested in understanding macromolecular crowding effects due to the current reliance of that research community on far more structurally simplified models. It has already been shown in at least one other application (the calculation of second virial coefficients) that large quantitative differences can result when atomically detailed protein structures are used in place of spherical models of equal volume.[51]

[51] B. L. Neal and A. M. Lenhoff, *AIChE J.* **41,** 1010 (1995).

To extend the use of BD methods beyond the exploration of excluded volume effects it will first be important to address the following question: can a model that makes such gross approximations capture reality sufficiently faithfully to be of use? The successful past uses of the method for describing association kinetics should partly answer this question,[2,3,8,10] but in any case, it should be recognized that this concern applies equally to all simulation methods—they all neglect something—so there is nothing unique to BD in saying that it requires careful parameterization. Before outlining how this might be carried out, it is worth noting that what we really mean by the term "parameterization" is the adjustment of those energetic terms that are represented in the model so that they also account implicitly for the energetic terms that are not explicitly represented in the model. A good illustration of the consequences of this process can be seen in the common pairwise (i.e., nonpolarizable) water models used in MD simulations: the dipole moment of popular water models, for example, is enhanced by around 25% over the experimental gas phase value to account implicitly for the polarizing effects of other molecules that will typically be encountered in condensed phase situations.[52] In the case of BD it is clear that we have many more terms that must be accounted for implicitly—the varied effects of water, the potential conformational flexibility of solutes, and so on—and it is therefore legitimate to be concerned whether they can all be successfully captured in effective terms. We can, however, cite at least two potential reasons for optimism. First, pure SASA-based energetic models—which are even cruder than the models used in BD simulations—can be made to work surprisingly well for describing the energetics of protein–protein complexes.[53] Second, our own attempts to develop energetic models to describe weak interactions[54] between proteins—by fitting to second virial coefficient data—showed that a combination of a detailed electrostatics term and a surface area–based model worked well over a range of pH and salt concentrations. Although parameterization is often viewed as an excruciatingly boring pursuit—and there is no denying the truth of this view—it can also provide a crucial step forward in understanding the underlying physics of the problem under investigation. This is most clearly demonstrated when it proves impossible to successfully parameterize a model to fit experimental data: it suggests strongly that something essential is missing from the model. The two cases cited above provide good examples of this. In the first,[53] it was demonstrated (albeit perhaps not surprisingly) that the energetic contributions

[52] M. W. Mahoney and W. L. Jorgensen, *J. Chem. Phys.* **112,** 8910 (2000).
[53] N. Horton and M. Lewis, *Protein Sci.* **1,** 169 (1992).
[54] A. H. Elcock and J. A. McCammon, *Biophys. J.* **80,** 613 (2001).

to stability from polar–polar contacts differed substantially from those of nonpolar–nonpolar contacts, to the extent that it was not possible to use a single compromise value to describe the combined effects of the two types of contact. In the second,[54] unsuccessful attempts to fit second virial coefficient data at low pH provided compelling evidence that a general quantitative description of weak interactions would require an explicit treatment of protonation state changes to be included.

There is a variety of data that can be used to parameterize energetic models for use in treating protein association and diffusion. In addition to the obvious candidates for providing thermodynamic information on strong associations—isothermal calorimetry, fluorescence spectroscopy, and so on—there are also sources that provide useful information on relatively weak interactions. One of these is second virial coefficient data, toward which a variety of simulation work has already been directed[55]; a second, more indirect source mentioned earlier—and that actually has the clearest implications for diffusion—is provided by the diffusion coefficient data themselves. Much of our current work is aimed at developing a transferable energetic model (i.e., one that can be applied to new, unstudied proteins) that is applicable to a range of solution conditions.

Regardless of how parameterization is conducted, the key point to be emphasized here is that there is a wide variety of potential uses for BD methodology. The most straightforward are likely to be those seeking to describe processes of macromolecular association, since these have the most in common with the previous successful uses of the methodology. Such studies benefit enormously from the fact that the three-dimensional structure of the macromolecular complex under investigation is known already, and therefore is used as input information for the simulations, so that the additional, tremendous complication of first successfully predicting the structure need not be solved first. This means that simulations of some extremely complex assembly processes, such as those involved in constructing viral capsids or protein crystals, can be performed. Now, whether these are interesting processes to study is, of course, a matter for the individual researcher, and it may be reasonably argued that in most cases we do not care what paths molecules follow as they associate, we care only where they end up. But assembly pathways can be important and just as the resurgence of interest in the protein-folding problem can be attributed in large part to the realization that misfolding is associated with a variety of diseases,[56] so it is also likely that misassembly could have disastrous consequences. By

[55] B. L. Neal, D. Asthagiri, and A. M. Lenhoff, *Biophys. J.* **75,** 2469 (1998).
[56] R. J. Ellis and T. J. T. Pinheiro, *Nature* **416,** 483 (2002).

allowing association pathways to be observed directly, BD methods may shed important light on these processes.

Although the above-described applications take as their starting point knowledge of the final structure, it may also ultimately prove possible to use BD methods to predict the structures of complexes. Since there is a variety of methods already available for addressing the "protein-docking problem,"[57,58] it is important to note what special advantages a BD-based strategy might have. The key benefit stems from the use in BD of forces to determine its sampling of potential structures: using force information ensures that time is not wasted in exploring structures that involve horrible steric (or electrostatic) conflicts. This is to be contrasted with the approach followed by most docking methods, which is to exhaustively construct and evaluate all (i.e., billions of) possible geometries. This difference in approach is not likely to have obvious consequences for predicting the structures of binary complexes, since current computational power makes exhaustive scoring of all possible structures feasible. However, the difference will become apparent when the prediction of structures for ternary, quaternary, or higher-order complexes is attempted. Exhaustive searching of all possible arrangements of three or more proteins is simply not computationally feasible, and is likely to remain so for some time. But simulating the diffusion and association of three or more proteins is feasible with BD and, in principle, would be as simple as placing all the proteins in a box and letting them diffuse until they form stable complexes. This is a seductively simple solution to state, but in practice will probably require the inclusion of an efficient and realistic treatment of conformational flexibility, at least for the interacting surface residues.

The last issue would probably rate as the first item on any "wish list" that one may wish to construct for future technical developments for BD; one step that our group has taken in this direction is the development of methods for simulating flexible polypeptide linkers connecting rigid domains, a step that makes the simulation of multidomain proteins a feasible undertaking. Readers will have their own ideas about what other items should be on the wish list, depending on the importance that they attach to the other shortcomings of the BD methodology. Obvious other items include a treatment of first-solvation shell effects and, eventually, provision for a treatment of hydrodynamic effects more sophisticated than can be described by a single diffusion constant, D. The extent to which the current neglect of these issues limits the applicability of BD methods is at the

[57] C. J. Camacho and S. Vajda, *Proc. Natl. Acad. Sci. USA* **98,** 10636 (2001).
[58] J. Janin, K. Henrick, J. Moult, L. Ten Eyck, M. J. E. Sternberg, S. Vajda, I. Vakser, and S. J. Wodak, *Proteins* **52,** 2 (2003).

moment not clear; parameterization and application of BD to different problems should in the next several years bring the picture into sharper focus.

Acknowledgments

The author wishes to thank Sean R. McGuffee for his many contributions to this research and Drs. Razif R. Gabdoulline and Rebecca C. Wade for the generous gift of source code for the two-molecule BD program SDA. The author is grateful for financial support from the Carver Trust, and the University of Iowa; work in the author's laboratory is not supported by the NIH.

[9] Modeling Lipid–Sterol Bilayers: Applications to Structural Evolution, Lateral Diffusion, and Rafts

By Martin J. Zuckermann, John H. Ipsen,
Ling Miao, Ole G. Mouritsen, Morten Nielsen, James Polson,
Jenifer Thewalt, Ilpo Vattulainen, and Hong Zhu

Introduction

The traditional view of biological membrane organization is that the lipid molecules form a featureless two-dimensional fluid in which proteins are kept in place by hydrophobic interactions with the lipid acyl (fatty acid) chains, hydrophilic interactions with the lipid polar heads, and perhaps anchoring interactions with the cytoskeleton.[1] In the absence of anchoring, the proteins can diffuse laterally in the plane of the membrane.

The advent of the raft hypothesis for vertebrate and fungal plasma membranes has helped to modify this view.[2,3] Research into the properties of detergent-resistant membrane fractions has led to the result that sorting of membrane proteins and signal transduction require colocalization of the related proteins in domains of length scale 70–200 nm, which are called rafts. The lipid compositions of these domains are far from featureless and rich in sphingolipids, cholesterol (vertebrates), or ergosterol (fungi). Moreover, the sterol is vital as its absence impedes raft formation. Finally, it has been conjectured that, in most cases, rafts form a more compact phase than that of the surrounding lipid environment, while still remaining

[1] S. I. Singer and G. L. Nicolson, **175**, 720 (1772).

[2] K. Simons and E. Ikonen, *Nature* **387**, 569 (1997); D. Brown and E. London, *J. Biol. Chem.* **275**, 17221 (2000).

[3] F. R. Maxfield, *Curr. Opin. Cell Biol.* **14**, 483 (2002).

Copyright 2004, Elsevier Inc.
All rights reserved.
0076-6879/04 $35.00

fluid. Thus, the research on rafts suggests strongly the active roles of lipids, especially sterols, in biological functions of cells.

Model Membranes

It is difficult to investigate the properties of biological membranes at a molecular level because of the enormous chemical diversity of lipids and proteins involved. Both experimental and theoretical studies have, therefore, been carried out for many years on model membranes, which are composed of a few lipid and/or protein species. Most of such model membranes can be easily produced *in vitro* as a result of the self-assembly process of their molecular components dispersed in aqueous media.[4]

Despite their relative chemical simplicities the model membranes exhibit sufficient complexity to mimic some of the physical characteristics of biological membranes. For example, ternary mixtures of two lipid species with cholesterol have been used by many researchers as model systems for studying raft formation.[2] In these model membranes, domains in which one of the lipid species forms a compact physical phase—known as the liquid-ordered (**lo**) phase[5]—together with a large concentration of cholesterol molecules appear within a less compact background fluid formed by the other lipid component with little cholesterol. The nature of the background phase is typical of single-component lipid bilayers in the liquid crystalline phase, also known as the liquid-disordered (**ld**) phase.[5] We give a description of the phases below.

We have been particularly interested in investigating the thermodynamic phase behavior of such model membranes with the view to understand the essential microscopic interactions underlying the phase behavior. Our approach has been to carry out theoretical studies based on computer simulations of microscopic interaction models in parallel with experimental investigations. The microscopic interaction models that our group has used or developed have varied focuses, depending on the specific physical questions under investigation.[6,7] Our earlier work involved the use of lattice models.[5,7] Our more recent investigations, however, focused on the development and study of off-lattice models that are able to provide

[4] G. W. Feigenson and J. T. Buboltz, *Biophys. J.* **80,** 2775 (2001).

[5] J. H. Ipsen, G. Karlström, O. G. Mouritsen, H. Wennerström, and M. J. Zuckermann, *Biochim. Biophys. Acta* **905,** 162 (1987); J. H. Ipsen, M. J. Zuckermann, and O. G. Mouritsen, *in* "Cholesterol in Model Membranes" (L. X. Finegold, ed.), pp. 223–257. CRC Press, Boca Raton, FL, 1993.

[6] S. Doniach, *J. Chem. Phys.* **68,** 4912 (1978).

[7] O. G. Mouritsen, B. Dammann, H. C. Fogedby, J. H. Ipsen, C. Jeppersen, K. Jørgensen, J. Risbo, M. C. Sabra, M. M. Sperotto, and M. J. Zuckermann, *Biophys. Chem.* **55,** 55 (1995).

a genuine description of the fluid characteristics of both the **ld** and the **lo** phases.[8] In this chapter, we present our work on the application of off-lattice models to the physical properties of lipid–sterol bilayers.[9,10]

The methodological backbone of our computer simulations is the Metropolis Monte Carlo (MMC) method. This method guarantees that a simulated system will reach its state of thermodynamic equilibrium. A tutorial on the use of MMC methods for the simulation of lattice and off-lattice models for lipid bilayers and lateral organization has already been published in a previous review chapter in this series by Sabra and one of the current authors (O.G.M.).[11] We provide in this chapter a description of the use of the method together with many other simulational techniques within the specific context of our modeling.

It is important to note at the outset that, given the available computational capacity, our simulations are limited in terms of the size of the sample, which is restricted to systems on the order of 10^4 lipid chains for our lattice models and 10^3 lipid chains for the off-lattice models. This enables us to study the nature of distinct phases and phase transitions using dedicated techniques, but not systems with a distribution of domains of the sizes predicted for rafts.

Single-Component Bilayers

Our approach has been to study model systems with the least number of molecular components first and then proceed to systems of multicomponents.[6–8] The initial systems are single-component lipid bilayers that themselves exhibit several phase transitions. We concentrate on one particular phase transition known as the main phase transition, which gives much insight into the interactions between lipid molecules. The transition takes the bilayer from the gel phase to the liquid crystalline phase as the temperature is increased and is abrupt, highly entropic. The gel phase is a two-dimensional (2D) crystalline phase in which conformationally ordered lipid chains form a structural lattice. Here conformationally ordered lipid chains are almost fully extended with few gauche bonds toward their free ends. To describe the nature of the gel phase in more detail, we have renamed it the solid-ordered (**so**) phase. In contrast, the liquid crystalline phase is a fluid

[8] M. Nielsen, L. Miao, J. H. Ipsen, O. G. Mouritsen, and M. J. Zuckermann, *Phys. Rev. E* **54,** 6889 (1996).
[9] M. Nielsen, L. Miao, J. H. Ipsen, M. J. Zuckermann, and O. G. Mouritsen, *Phys. Rev. E* **59,** 5790 (1999).
[10] L. Miao, M. Nielsen, J. Thewalt, J. H. Ipsen, M. Bloom, M. J. Zuckermann, and O. G. Mouritsen, *Biophys. J.* **82,** 1429 (2002).
[11] M. C. Sabra and O. G. Mouritsen, *Methods Enzymol.* **321,** 263 (2000).

phase in which conformationally disordered lipid chains can diffuse laterally in the bilayer plane. Here the conformationally disordered lipid chains contain many gauche bonds, leading to a reduced apparent length, but an increased cross-sectional area. The quantitative properties of the main phase transition depend both on the intrinsic chain length and unsaturation and on the type of polar head of the lipid molecule involved, as well as on the ionic character of the aqueous medium hydrating the polar heads.

Liquid-Ordered Phase and Sterol Evolution

The concept of the **lo** phase emerged from both experimental and theoretical studies of the phase behavior of dipalmitoyl phosphatidylcholine (DPPC)–cholesterol bilayers. DPPC is a glycerophospholipid with a zwitterionic polar head and two saturated (16:0) acyl chains and appears as a molecular component of many biological membranes. The experimental phase diagram of DPPC–cholesterol multibilayer liposomes was first deduced by Vist and Davis[12] from both deuterium nuclear magnetic resonance (^{2}H NMR) and differential scanning calorimetry (DSC) data. The same topology was found for the phase diagram of PPetPC–cholesterol bilayers[10] as shown in Fig. 1A, even though PPetPC differs in structure from DPPC in that one *cis* double bond replaces a single C–C bond in DPPC at C-67. Qualitatively similar phase diagrams were also found for several other systems of binary mixtures containing cholesterol and lipids with PC polar heads by Thewalt *et al.*[13] The most important feature of these phase diagrams is that an abrupt main-phase transition is absent for cholesterol concentrations greater than 20 mol%, and a new phase appears in this region of the phase diagrams. This is a fluid phase in which the lipid chains are reasonably ordered conformationally as demonstrated by the ^{2}H NMR data. Vist and Davis,[12] who first identified this new phase, named it the β phase. In the first theory for the experimental phase diagram, Ipsen *et al.*[5] renamed this phase the **lo** phase since this phase is a compact liquid phase with conformationally ordered chains.[5] This concept has gained a prominent position in raft research, because several experimental studies on rafts suggest that rafts have the physical characteristics of the **lo** phase.[2,3] It is worth pointing out that the only **lo/ld** phase separation observed for single lipid–cholesterol bilayers occurs close to the melting temperature of the pure lipid system. Thus it is reasonable to assume that mixtures of lipids

[12] M. Vist and J. H. Davis, *Biochemistry* **29**, 451 (1990).
[13] J. L. Thewalt and M. Bloom, *Biophys. J.* **63**, 1176 (1992); J. L. Thewalt, C. E. Hanert, F. M. Liseisen, A. J. Farrall, and M. Bloom, *Acta Pharm.* **42**, 9 (1992).

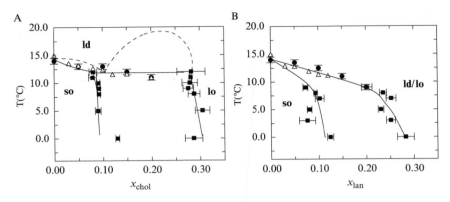

FIG. 1. Experimental phase diagrams as determined by differential scanning calorimetry (△) and by NMR spectroscopy (■ and ●). (A) PPetPC–cholesterol systems; (B) PPetPC–lanosterol systems. The lines connecting the points are guides to the eyes. The phase boundaries indicated by the dashed lines in (A) are not derived from any experimental data and are shown only to illustrate the qualitative structure of the phase diagram, which is consistent with the thermodynamic phase rules.

and cholesterol would require at least some high-T_m components in order to display the **lo** phase at body temperature.

Given the existing understanding of the equilibrium phase behavior of lipid–cholesterol bilayers and the underlying microscopics, and inspired by the work of Bloch on sterol evolution,[14] we have carried out a comparative study of lipid–cholesterol and lipid–lanosterol bilayers, both experimentally and theoretically, by computer simulations. Lanosterol is the first cyclic precursor of cholesterol in the biosynthetic pathway and is regarded as a molecular fossil by Bloch, who proposed that the evolutionary sequence for cholesterol is preserved in its biosynthetic pathway.[14]

The differences in the phase behavior of the two types of lipid–sterol systems are illustrated in the two phase diagrams in Fig. 1.[10] The phase diagram in Fig. 1A shows that the **lo** phase is a well-defined thermodynamic phase distinctly separate from the **ld** phase, but that it can coexist with the **ld** phase under appropriate temperatures and concentrations. In contrast, the phase diagram in Fig. 1B shows that for PPetPC–lanosterol multi-bilayer systems the **lo** and **ld** phases are no longer thermodynamically distinct; that is, coexistence of the two phases is absent.

Moreover, our experimental data for PPetPC–lanosterol and PPetPC–cholesterol show that cholesterol induces greater conformational order in PPetPC–lipid chains than does lanosterol.[10] It may be reasonable to expect

[14] K. Bloch, *Science* **150**, 19 (1965); K. Bloch, *CRC Crit. Rev. Biochem.* **14**, 47 (1983).

such a relative effect to hold in DPPC–cholesterol and DPPC–lanosterol bilayers. Hence a possible hypothesis is that one reason for the evolutionary optimization of cholesterol is that it is able to induce the formation of rafts better than lanosterol. There is in fact some evidence that lanosterol cannot induce raft formation.[15]

Organization of This Chapter

We have so far described both the importance of cholesterol in the lateral organization of vertebrate cell membranes, in particular, the role of cholesterol in raft formation and the issue of sterol evolution. We have also commented on the use of model membranes in the physical study of the relevant phenomena. The rest of the chapter is organized as follows. First, we present our theoretical models for the examination of the physical properties of various types of model membrane. Next we discuss the particular MMC algorithms that have been developed for computer simulations of the theoretical models, as well as several complementary statistical–mechanical techniques that have been applied in order to determine the phase behavior of the model systems. Finally, we present the results and conclude with a brief discussion on possible future extensions of the research.

Theoretical Models

The first step in a statistical mechanical study of a model membrane system is to construct a model describing the microscopic physics of the membrane. The model should contain two basic ingredients: (1) a specific description of the microscopic states of the membrane, that is, the relevant microscopic degrees of freedom; and (2) a specific expression of the total interaction potential energy of the system as a function of the relevant degrees of freedom. In both parts, approximations or simplifications are necessary, mainly because of the limited computational capacity available and the need to underline among the almost formidably complicated microscopic details the general mechanisms that are essential to the macroscopic phenomena under investigation. The models to be described below may in fact be called "minimal" in terms of the approximations and simplifications involved.[8–10,16]

The basic molecular degrees of freedom of a lipid bilayer membrane consist of both the positional (molecular center of mass) and internal

[15] X. Xu and E. London, *Biochemistry* **39,** 843 (2000).
[16] M. Nielsen, Ph.D., thesis. McGill University, Montreal, Canada, 1999.

(conformational) degrees of freedom, and both types are relevant to the phase behavior of the membrane. The description of the positional degrees of freedom is straightforward, given by specifying the spatial coordinates of the molecules. Our description of the conformational isomerism of the lipid chains is a minimal one. Only two internal states are used to approximate the large spectrum of conformations possible for a phospholipid chain: one state, the "ordered" state, has zero internal energy and is nondegenerate, characteristic of the conformational state of a lipid chain in the gel or **so** phase. The other state, the "disordered" state, is characteristic of the average conformational state of a lipid chain in the liquid–crystalline or **ld** phase. This state has a high internal energy, reflecting that energy is required for conformational excitations, and a high degeneracy, effectively representing the large number of conformational excitations that a phospholipid chain can assume in the **ld** phase because of the presence of many gauche bonds. This idea was first proposed and used by Doniach.[6] Both cholesterol and lanosterol are rigid molecules in comparison with the phospholipid chains and we assume that they have no conformational degrees of freedom.

In constructing the function describing the interaction potential energy, several approximations have also been made. First, the interactions between the amphiphilic molecules (phospholipid and sterol) and the water molecules, which are predominantly responsible for stabilizing a bilayer membrane, are approximated by a surface pressure parameter Π, as first suggested by Marčelja.[17] Second, the noncovalent interactions between the amphiphilic molecules may, to a first approximation, be considered as being pairwise, of the van der Waals type, and effective only over a short range of molecular length scale. Finally, only one monolayer is modeled on the basis of the assumption that the two monolayers constituting a lipid bilayer are independent of each other. In essence, our model system consists of microscopic "particles"—distinct lipid chains and sterol molecules—that move within a two-dimensional plane and that interact in a pairwise manner, as described in this section. In the rest of the chapter, the term "particle" may refer either to a lipid chain or to a sterol molecule.

The total potential energy function that models molecular interactions in the lipid–sterol bilayer systems in our simulations is then given by

$$H = H_0 + H_{\text{o}-\text{s}} + H_{\text{d}-\text{s}} + H_{\text{s}-\text{s}} \tag{1}$$

H_0 is the potential energy function describing the interactions between the phospholipid molecules, the specific form of which is as follows:

[17] S. Marčelja, *Biochim. Biophys. Acta* **367,** 165 (1974).

$$H_0 = \sum_i E_d \mathfrak{L}_{id} + \sum_{\langle i<j \rangle} V_{o-o}(R_{ij})\mathfrak{L}_{io}\mathfrak{L}_{jo} + \sum_{\langle i<j \rangle} V_{o-d}(R_{ij})\{\mathfrak{L}_{io}\mathfrak{L}_{jd} + \mathfrak{L}_{jo}\mathfrak{L}_{id}\} + \Pi \cdot A$$

$$(2)$$

Here i is an index labeling the "particles" (lipid chains and sterol molecules) in the system. Correspondingly, $\mathfrak{L}_{io}$ and $\mathfrak{L}_{id}$ are occupation variables that are unity when the ith particle is a lipid chain in the ordered state and the disordered state, respectively, and which are zero otherwise. E_d is the excitation energy of the conformationally disordered state, and $V_{o-o}(R)$ and $V_{o-d}(R)$ are distance-dependent and chain conformation-dependent interactions between two lipid chains. $\langle i<j \rangle$ denotes a summation over nearest neighbors, corresponding to the approximation of short-range interactions. The energy of interaction between two chains that are both in the disordered state is approximated to be zero, which sets the reference point for energy. This potential energy function, H_0, together with the configurational degeneracy of the disordered state, provides a minimal model for the main phase transition of single-component bilayers.

H_{o-s}, H_{d-s}, and H_{s-s}, as indicated by the various subscripts, represent the pairwise interactions between an ordered chain and a sterol molecule, a disordered chain and a sterol molecule, and two sterol molecules, respectively. They are written explicitly as follows:

$$H_{o-s} = \sum_{\langle i<j \rangle} V_{o-s}(R_{ij})\{\mathfrak{L}_{io}\mathfrak{L}_{js} + \mathfrak{L}_{jo}\mathfrak{L}_{is}\}$$

$$H_{d-s} = \sum_{\langle i<j \rangle} V_{d-s}(R_{ij})\{\mathfrak{L}_{id}\mathfrak{L}_{js} + \mathfrak{L}_{jd}\mathfrak{L}_{is}\}$$

$$H_{s-s} = \sum_{\langle i<j \rangle} V_{s-s}(R_{ij})\{\mathfrak{L}_{is}\mathfrak{L}_{js}\}$$

$$(3)$$

Again $\mathfrak{L}_{is}$ is an occupation variable that is unity when the ith particle is a sterol molecule and is zero otherwise. Clearly, $\mathfrak{L}_{is} + \mathfrak{L}_{io} + \mathfrak{L}_{id} = 1$. $V_{o-s}(R)$ and $V_{d-s}(R)$ are the distance-dependent, chain conformation–dependent interaction potentials between a sterol molecule and a lipid chain. Similarly, $V_{s-s}(R)$ is the interaction between two sterol molecules.

All of the pairwise interaction potentials are approximated by a sum of a hard-core repulsive potential of range d, a short-range square-well potential of range R_0, $V^s(R)$, and a longer range attractive square-well potential of range l_{max}, $V^l(R)$. $V^s(R)$ and $V^l(R)$ are given by

$$V^s(R) = \begin{cases} -V^s, & d < R \leq R_0 \\ 0, & \text{otherwise} \end{cases}$$

$$(4)$$

$$V^l(R) = \begin{cases} -V^l, & d < R \leq l_{max} \\ 0, & \text{otherwise} \end{cases} \tag{5}$$

where the specific values of V^l and V^s depend on both the molecular type and conformational state of the interacting species. Here l_{max} is a cutoff distance that will be defined when the simulation algorithm is discussed in the next subsection.

The pairwise interaction potentials are illustrated in Fig. 2, and their construction is based on the hypothesis that cholesterol interacts with lipid molecules as follows. Cholesterol, with its streamlined, rigid hydrophobic backbone, prefers the lipid chains in its immediate neighborhood to be in the conformationally ordered (rigid) state by suppressing the formation of gauche bonds in those chains. At the same time cholesterol tends to disrupt laterally ordered packing of conformationally ordered chains if it is in their midst. A comparison between Fig. 2A and B illustrates the latter, "crystal-breaker" mechanism, as the interactions involved imply that a cholesterol molecule dissolved in an ordered chain environment tends to have a larger surface area than that of a lipid chain, thereby breaking the lateral order of ordered chains. Similarly, a comparison between Fig. 2B and C shows that the former, "chain-rigidifier" mechanism is modeled by a strong interaction between cholesterol and a neighboring, conformationally ordered lipid chain and a weaker interaction with a conformationally disordered lipid chain.

The values of all the relevant parameters in the model are chosen as follows. The unit of length scales in the model is for convenience set at the hard-core diameter, d, of the interaction potentials and the unit of energy is defined to be a quantity J_0. To convert the energy and length scales into units relevant for lipid bilayer systems, J_0 should be on the order 10^{-20} J and d should be on the order 5 Å. The lateral pressure, Π, is fixed at $\Pi d^2/J_0 = 3.0$. The radius of the short-range potential, Eq. (4), is set at $R_0/d = 1.3$. The values of Π and R_0 are chosen such that the change in surface area across the main transition is comparable to that of a pure PC (DPPC) bilayer system. The excitation energy of the disordered state of lipid chains is chosen to be $E_d = 2.78J_0$, and the degeneracy, D_d, of the disordered state, is taken to be $\ln D_d = 12.78$. These values are the same as for the 10th state of the Pink model for the main phase transition of single-component PC bilayers with saturated chains.[18] Other parameters appearing in the definition of the interaction potentials are summarized in Table IA. They are chosen such that the theoretical phase diagram of the lipid–cholesterol model system is similar to that of the DPPC–cholesterol system.

---

[18] D. A. Pink, T. J. Green, and D. Chapman, *Biochemistry* **19**, 349 (1980).

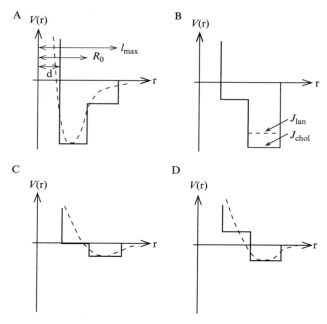

FIG. 2. Schematic illustration of the model interaction potentials. (A) $V_{o-o}(R)$, where subscript "o" refers to a lipid chain in the conformationally ordered state; (B) $V_{o-s}(R)$, where the depth of the potential corresponding to s = chol, is represented by a solid line and a parameter J_{chol} and that corresponding to s = lan is represented by a dotted line and a parameter J_{lan}; (C) $V_{d-s}(R)$, where subscript "d" refers to a lipid chain in the disordered state; (D) $V_{s-s}(R)$. d is the hard-core radius, R_0 is the range of the short-range potentials, and l_{max} is the effective range of the long-range potentials. The R_0/d ratio is chosen so that the average surface area of a sterol molecule is approximately 30% larger than that of a lipid chain in the ordered state. The dashed lines illustrate the more realistic interaction potentials that the model potentials approximate.

The parameters given in Table IB reflect our—again, minimal—modeling of differences in lipid–cholesterol and lipid–lanosterol interactions. The model arises from a comparative analysis of the molecular structures of the two sterols, which are shown in terms of space-filling models in Fig. 3. Figure 3 indicates that the major structural difference is on their respective α faces. In detail, cholesterol has a smooth α face, whereas the α face of lanosterol has an extra methyl group, which forms a rather significant protrusion. Together with two other additional methyl groups this α-face protrusion would decrease the ability of lanosterol to order neighboring lipid chains. It is interesting to note that the α-face methyl group is the first

TABLE I

INTERACTION PARAMETERS FOR MODEL POTENTIALS FOR TWO TYPES OF
LIPID–STEROL MEMBRANES[a]

A

	V_{o-o}	V_{o-d}	V_{d-d}	V_{s-s}	V_{s-d}
Long range	0.40	−0.15	0.20	0.20	0.00
Short range	0.45	0.40	−0.20	−0.15	−0.065

B

	Cholesterol	Lanosterol
V_{o-s}^{l}	0.85	0.75
V_{o-s}^{s}	−0.625	−0.525

[a] (A) The parameter values for the interaction potential, V_{o-o}, between two lipid chains in the ordered state, the interaction potential, V_{o-d}, between a lipid chain in the ordered state and a lipid chain in the disordered state, the interaction potential, V_{d-d}, between two lipid chains in the disordered state, the interaction potential, V_{s-s}, between two sterol molecules, and the interaction potential, V_{s-d}, between a lipid in the disordered state and a sterol molecule. Note that these values are identical for both the lipid–cholesterol and the lipid–lanosterol systems. (B) The parameter values for the interaction potential, V_{o-s}, between a lipid chain in the ordered state and a sterol molecule, where the subscript "s" corresponds to either cholesterol or lanosterol, respectively. All the parameters are given in units of J_0 (see text).

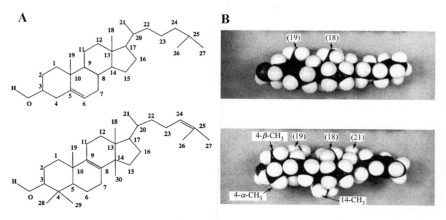

FIG. 3. Molecular structures of cholesterol *(top)* and lanosterol *(bottom)*. (A) Chemical structures; (B) space-filling models. The three additional methyl groups on lanosterol are indicated in (B) as 14-CH₃, 4-α-CH₃, and 4-β-CH₃.

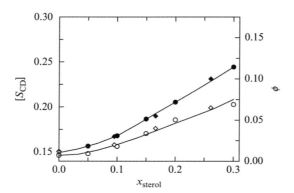

FIG. 4. Fitting of the theoretical (ϕ) order parameter to the equilibrium averages of the experimental order parameter [S_{CD}, as given in Eq. (14)], which leads to the specific value of V^l_{o-s} used for lipid–lanosterol interactions in the model. The experimental data are obtained at $T = 40°$. The theoretical data are calculated for a system containing $N = 1600$ particles at a temperature of $T = 1.0359 T_m$. Corresponding to cholesterol and lanosterol, the experimental data are shown by filled and open circles, respectively, and the theoretical data are given by filled and open diamonds, respectively. The lines connecting the points are guides to the eye.

one to be removed in the biosynthetic pathway at considerable cost in metabolic energy.

Hence, in our theoretical model, the microscopic lipid–cholesterol and lipid–lanosterol interactions differ only in the part that describes the interaction between a sterol molecule and an ordered lipid chain, V_{o-s}. The values listed in Table IB correspond to a decreased strength of the cohesive lipid–lanosterol interaction relative to that of the cohesive lipid–cholesterol interaction. They were determined by fitting the theoretically calculated value of the lipid-chain order parameter, ϕ (see later), to the acyl-chain order parameter experimentally derived from the ^2H NMR data, over a significant range of sterol concentrations for both types of lipid–sterol systems. The fitting was carried out at a single temperature, which was relatively high so that no complications associated with phase transitions arose. The fit is shown in Fig. 4.

Simulation Methods

Metropolis Monte Carlo Procedure

In our computer simulations of the thermodynamic phase behavior of the models described in the previous section, the MMC method provides the backbone method. Generally speaking, the MMC procedure in each simulation samples stochastically and ergodically the complete configuration space spanned by all possible microscopic states of the system and generates a Markovian sequence of microscopic states whose distribution over the configuration space approximates the equilibrium distribution. Explicitly, each simulation consists of many repeated cycles of procedures, and each cycle, labeled by an integer n, is implemented as follows.

1. An arbitrary microscopic state, Ω_n, is chosen.
2. Next, a trial microscopic state, Ω_t, is established by a random, ergodic sampling of the complete configuration space.
3. A quantity, $\Delta\tilde{H} \equiv \tilde{H}(\Omega_t) - \tilde{H}(\Omega_n)$, is calculated, where $\tilde{H}$ depends both on the potential energy function of the system and on the choice of thermodynamic ensemble.
4. A random number, r, is generated.
5. If $r < \exp(-\Delta\tilde{H}/k_B T)$, then the trial microscopic state, Ω_t, is accepted as the new configuration of the system, that is, $\Omega_{n+1} = \Omega_t$. Otherwise the system remains in the same state, that is, $\Omega_{n+1} = \Omega_n$.
6. Quantities of interest for ensemble averaging are calculated and recorded.
7. A new trial microstate, Ω_t, is chosen again randomly and ergodically, and the $n + 1$ cycle is started by going back to step 3.

It is implied in the MMC procedure that a specific choice of thermodynamic ensemble has been made, which affects both the definition of the complete configuration space and the definition of $\Delta\tilde{H}$. In addition to choosing temperature T and surface pressure Π as thermodynamic control variables, we are still left with two different choices regarding the number of "lipid-chain" particles, N_l and the number of "sterol" particles, N_s: (1) fixing both N_l and N_s; and (2) fixing the total number of particles, $N_l + N_s$, and letting the chemical composition of the system fluctuate under the control of an effective chemical potential, μ_Δ. We take μ_Δ to represent the difference between the chemical potentials of the lipid and the sterol particles. We have used either choice 1 or 2 in our simulations, depending on the specific questions that we have asked about the system. In the following, we call choice 1 the "canonical ensemble," and choice 2 the "semigrand canonical ensemble."

In simulations formulated in the canonical ensemble, $\Delta\tilde{H}$ in each MMC cycle is simply given by

$$\Delta\tilde{H} = H(\Omega_t) - H(\Omega_n) - k_B TN \ln[A(\Omega_t)/A(\Omega_n)] \qquad (6)$$

where H is just the potential energy function defined in Eq. (1). The logarithmic term arises from any possible changes in the area of the system (see later). In simulations formulated in the semigrand canonical ensemble, however,

$$\Delta \tilde{H} = H(\Omega_t) - H(\Omega_n) - k_B TN \ln[A(\Omega_t)/A(\Omega_n)] + \mu_\Delta[N_1(\Omega_t) - N_1(\Omega_n)]$$

(7)

Of course, in the two ensembles, the procedures of sampling their corresponding complete configuration spaces differ also. The differences will be pointed out wherever appropriate in the following specific discussions of sampling step 2 in the general procedure.

Sampling step 2 is a combination of different types of "moves" that generate changes in the microscopic state of the system that are associated with the different types of degrees of freedom. We discuss these moves in turn in detail.

Dealing with Translational Degrees of Freedom: Off-Lattice Algorithm. The positions of the particles in a system form a part of the definition of a microscopic state and are essential to calculations of the potential energy arising from position-dependent interactions between the molecular constituents. In dealing with interactions that are pairwise and short-ranged, the most important information required is the local environment of each individual particle, such as the distribution of other particles in its neighborhood and their distances to it. In conventional off-lattice simulations, it is usually one of the most time-consuming steps to obtain this information from each given configuration of molecular positions. We now describe a simulational algorithm that handles structural information in a manner that is distinctly different from conventional algorithms, and that at the same time achieves high computational efficiency.

Our algorithm is an adapted version of the dynamic triangulation algorithm used for modeling random surfaces. It performs two essential tasks: (1) the generation of the (sub)configuration space associated with the translational degrees of freedom; and (2) the generation and the retention of a compact data structure that allows efficient access to structural information contained in each configuration of particle positions. The data structure is based on triangulation of each spatial configuration of the particles. The triangulation itself is performed as follows: an initial configuration in which the particles are positioned with their centers on a regular triangular lattice is used and each particle is linked to its six nearest neighbors by "tethers." The lattice configuration is then represented by a network of tethers forming triangles; the term "triangulation" refers to this representation. The tether network can then be altered randomly, through the moves described below, to provide the subspace of microscopic states

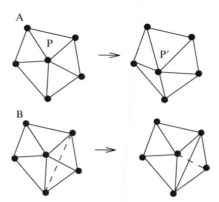

Fɪɢ. 5. (A) Particle displacement. The particle at position P is moved to position P'. (B) Link flip. A tether (shown as a thick line) is replaced by another tether along the unoccupied diagonal (once the first tether is removed) provided that the new tether length does not exceed l_{max}.

that are associated with particle positions. The interacting particle pairs are those that are connected by the tethers.

Particle Displacement. This move is illustrated in Fig. 5A. A particle is chosen at random and its center is subject to a random displacement (δx, δy) where

$$\begin{aligned} \delta x &= (2\zeta_x - 1)\delta r_{MAX} \\ \delta y &= (2\zeta_y - 1)\delta r_{MAX} \end{aligned} \qquad (8)$$

ζ_x and ζ_y are random numbers, $0 \le \zeta_{x(y)} \le 1$. The value of δr_{MAX} is adjusted during the simulations so that approximately 25% of trials are accepted. Consistent with the hard-core repulsion between two interacting particles, moves that would result in an overlap of the hard cores of the particle are always rejected. Another constraint is that the length of every tether is not allowed to exceed a maximum value l_{max}.

Link Flip. The second move is referred to as "link flip," which alters the local connectivity in the tether network. It is shown schematically in Fig. 5B. In each configuration of the tether network, each tether forms one diagonal of a quadrilateral formed by the two adjacent triangles. In the link flip, a tether is chosen at random and then removed; a new tether is placed along the other diagonal of the quadrilateral, provided that its length does not exceed l_{max}.

Change of System Size. Consistent with our choice of thermodynamic ensemble, in which the surface pressure is a control parameter, fluctuations in the total area of the system must be sampled. In our simulations this is achieved via a third move: a random uniform expansion or contraction of the whole system. In this step, a random change in the size of the system is generated as

$$\delta L = (2\zeta - 1)\delta L_{MAX} \qquad (9)$$

where ζ is a random number, $0 \leq \zeta \leq 1$; and the coordinates of all particles in the system are rescaled accordingly. If the distance between any two particles after the rescaling is smaller than the hard-core diameter the change is always rejected. The maximum possible size change, δL_{MAX}, is adjusted during the simulation to give an acceptance ratio of about 50%. The logarithmic term in Eqs. (6) or (7) will necessarily be nonzero following these types of moves. A detailed discussion of these types of sampling moves is given in Frenkel and Smit,[19] pp. 116–122.

Dealing with Conformational and Compositional Degrees of Freedom. In simulations in the canonical ensemble, sampling the part of the configuration space that is associated with the conformational degrees of freedom of the lipid chains is straightforward. A lipid chain is chosen at random and its conformational state is always changed in the trial state. For example, if the chain is in the ordered state, it will be in the disordered state in the trial state and vice versa.

In the case of the semigrand canonical ensemble, a different updating procedure is used, in which the conformational degrees of freedom and the compositional degrees of freedom are handled on the same footing. Explicitly, each particle in the system is assigned three "internal" states: the conformationally ordered and disordered states of a lipid chain, and a third, "sterol" state. Then a random number is used to choose among the three internal states now available.

To ensure detailed balance in the simulations, or the symmetry of the Markov chain generated by the algorithm, all of the different moves discussed above are combined in a random manner. In other words, the simulation algorithm does not impose any preferred sequential order in the sampling procedures that update different types of degrees of freedom.[19] Monte Carlo step (MCS), a time unit for the simulations, is defined to be the time needed to perform on average one complete pass through all the different types of moves.

[19] An excellent textbook for both MMC and molecular dynamics is the following: D. Frenkel and B. Smit, "Understanding Molecular Dynamics," 2nd Ed., Vol. 1. Computational Science Series. Academic Press, San Diego, CA, 2002.

Umbrella Sampling

Up to this point, we have given a detailed recipe for performing an MMC simulation for a system described by the potential energy function of Eq. (1). We now discuss a method for determining phase boundaries specific to our study of the phase behavior of lipid–sterol bilayer membranes. This method is a version of umbrella sampling[20] and it is particularly useful for determining phase transitions the kinetics of which are strongly hysteretic. The transitions between the **so** and the **ld** phases in our systems are precisely such transitions. The high free energy barriers separating the two coexisiting phases make it difficult to achieve an adequate sampling of both phases.

This difficulty was overcome by the use of a simulation technique, which exploits the idea of developing an "artificial" potential energy function that yields a considerably diminished free energy barrier. The simulation carried out for the artificial potential energy function is able to yield an adequate sampling of the two coexisting phases. The equilibrium distribution functions (defined below) for the original potential energy function can then be established from simulations based on the modified potential energy function through a simple reweighting relation. For a complete discussion of the method and related references, the reader is referred to Risbo[20] for details.

The upshot of our simulations performed within the semigrand canonical ensemble was to provide accurate numerical data, from which an equilibrium distribution function, $\mathfrak{P}(T, \mu_\Delta; \varepsilon, x_s)$, could be derived for each given set of values of T and μ_Δ. With $\varepsilon = E/N$ representing the internal interaction energy per particle and x_s the sterol concentration, $\mathfrak{P}(T, \mu_\Delta; \varepsilon, x_s)$ is proportional to the probability of finding the system in states characterized commonly by ε and x_s, and has the property that if T and μ_Δ correspond to a point on a phase coexistence boundary, then it will exhibit two degenerate local minima, representing the two coexisting phases. Thus, by analyzing the change in the "landscape" of $\mathfrak{P}(T, \mu_\Delta; \varepsilon, x_s)$ as values of T and μ_Δ are tuned, the phase boundary can be determined.

The umbrella sampling method was employed to this end and its implementation is described as follows. At values of T and μ_Δ estimated to be close to true coexistence conditions, a single but long simulation is performed for a system of relatively small size L to yield an approximation of the equilibrium distribution function, $\mathfrak{P}_L(T, \mu_\Delta; \varepsilon, x_s)$. When summed over x_s $\mathfrak{P}_L(T, \mu_\Delta; \varepsilon, x_s)$ yields the equilibrium distribution function $\mathfrak{P}_L(T,$

[20] J. Risbo, Ph.D. thesis. Technical University of Denmark, Lyngby, Denmark, 1997; J. Risbo, G. Besold, and O. G. Mouritsen, *Comp. Mater. Sci.* **15,** 311 (1999).

μ_Δ; ε). A spectral free energy function, defined by $\mathfrak{F}_L(T, \mu_\Delta; \varepsilon) = -k_B T$ ln $\mathfrak{P}_L(T, \mu_\Delta; \varepsilon)$, displays a free energy barrier between the two local minima. An iterative cycle is then set up.

1. On the basis of $\mathfrak{F}_L(\varepsilon)$ an extrapolation based on the size dependence of the free energy barrier is used to approximate the barrier of a system of a larger size, L', and in turn, the following function:

$$f(T, \mu_\Delta; \varepsilon) = -\mathfrak{F}_{L'}(T, \mu_\Delta; \varepsilon) = -\frac{L'}{L} \mathfrak{F}_L(T, \mu_\Delta; \varepsilon) \qquad (10)$$

when ε lies in the barrier region. $f(T, \mu_\Delta; \varepsilon)$, known as the shape function, defines the modified potential energy function as $\bar{H} = H + f(T, \mu_\Delta; \varepsilon)$.

2. $\bar{H}$ is used in a second simulation of a system of size L'. From this simulation, a modified probability distribution function, $\mathfrak{P}_{L'}(T, \mu_\Delta; \varepsilon, x_s)$, is obtained. The spectral free energy function corresponding to the modified potential energy function should not show a significant barrier.

3. The required distribution function, $\mathfrak{P}_{L'}(T, \mu_\Delta; \varepsilon, x_s)$, is easily reconstructed from $\mathfrak{P}_{L'}(T, \mu_\Delta; \varepsilon, x_s)$ (see Risbo[20] for details).

4. On the basis of $\mathfrak{P}_{L'}(T, \mu_\Delta; \varepsilon, x_s)$, a method known as Ferrenberg–Swendsen reweighting technique[21] is applied in order to obtain a better estimate of the coexistence condition, T^* and μ_Δ^* and an improved approximation of $\mathfrak{F}(T^*, \mu_\Delta^*; \varepsilon)$ at coexistence from $\mathfrak{F}_{L'}(T^*, \mu_\Delta^*; \varepsilon)$.

5. If desired, another iteration is started from step 1 with $\mathfrak{F}_{L'}(T^*, \mu_\Delta^*; \varepsilon)$, either to obtain an improved statistical sampling of $\mathfrak{P}_{L'}(T^*, \mu_\Delta^*; \varepsilon, x_s)$ for the same system size or to simulate a larger system.

This method enables us to simulate systems of relatively large sizes. Repeated applications of the above-described procedures for systems with systematically varying sizes can yield a series of size-dependent $\mathfrak{P}_L(T^*, \mu_\Delta^*; \varepsilon, x_s)$. A finite-size analysis of the series of data based on the method of Lee and Kosterlitz[22] can provide information on the specific nature of the corresponding transition.

Nevertheless, the method is still quite time consuming. In the iterations, the system size can be increased in only small steps; the statistics required to obtain the initial estimate of the spectral free energy function $\mathfrak{F}$ as described in step 1 are already considerable, being typically on the order of 50×10^6 MCS per particle for a system of size $L = 10$.

[21] A. M. Ferrenberg and R. H. Swendsen, *Phys. Rev. Lett.* **61,** 2635 (1988); R. M. Ferrenberg and R. H. Swendsen, **63,** 1195 (1991).
[22] J. Lee and J. L. Kosterlitz, *Phys. Rev. Lett.* **65,** 137 (1990); J. Lee and J. L. Kosterlitz, *Phys. Rev. B* **43,** 3625 (1991).

Calculation of Physical Quantities

Given that a simulation based on the MMC procedure described above does in fact generate an ensemble of microscopic states that are distributed in the configuration space according to the equilibrium distribution, the macroscopic thermal average of a microscopic physical quantity, $\mathfrak{D}$, can be approximated by its average over the ensemble of states, $\{\Omega_n, n = 1, \ldots, N_{\text{state}}\}$:

$$\langle \mathfrak{D} \rangle = \frac{1}{N_{\text{state}}} \sum_{n=1}^{N_{\text{state}}} \mathfrak{D}(\Omega_n) \tag{11}$$

where N_{state} is the total number of microscopic states in the ensemble. It is important to note that the precise value of N_{state} required for Eq. (11) to be a good approximation for the true thermodynamic averages depends on the nature and size of the system as well as the specific physical observable under investigation.

In general, the macroscopic thermodynamic quantities that are of interest for the systems that we have studied are response functions such as specific heat and isothermal area compressibility as well as a macroscopic conformational order parameter characterizing the conformational ordering of the lipid chains. The response functions can be calculated directly from the simulation data by using the fluctuation-dissipation theorems:

$$C_\Pi(T) = \frac{1}{N k_B T^2} \left[\langle H^2 \rangle - (\langle H \rangle)^2 \right] \tag{12}$$

$$K(T) = \frac{1}{k_B T \langle A \rangle} \left[\langle A^2 \rangle - (\langle A \rangle)^2 \right] \tag{13}$$

where H is the potential energy function defined in Eq. (1), $C_\Pi(T)$ is the molecular specific heat at constant Π, and $K(T)$ is the isothermal area compressibility. The average conformational order parameter of the lipid chains is calculated as

$$\phi = \frac{1}{2} \left(\left\langle \frac{\sum_{i=1}^{N} (\mathfrak{L}_{io} - \mathfrak{L}_{id})}{\sum_{i=1}^{N} (\mathfrak{L}_{io} + \mathfrak{L}_{id})} \right\rangle + 1 \right) \tag{14}$$

In addition to the physical quantities mentioned above, the so-called structure factors, which provide information on the lateral organization of the constituent molecules in the system, have also been calculated, on the basis of the simulations performed within the canonical ensemble. Two of them are listed here, for which numerical data will be provided later:

$$S_T(\mathbf{q}) = \frac{1}{N} \left\{ \langle \rho_T(\mathbf{q}) \rho_T(-\mathbf{q}) \rangle - \langle \rho_T(\mathbf{q}) \rangle^2 \delta_{\mathbf{q},0} \right\}$$
$$S_S(\mathbf{q}) = \frac{1}{N} \left\{ \langle \rho_S(\mathbf{q}) \rho_S(-\mathbf{q}) \rangle - \langle \rho_S(\mathbf{q}) \rangle^2 \delta_{\mathbf{q},0} \right\} \tag{15}$$

Here $\rho_T(\mathbf{q})$ is the Fourier transform of the total density, $\rho_T(\mathbf{r}) \equiv \Sigma_i \, \delta(\mathbf{r} - \mathbf{r}_i)$, and $\rho_S(\mathbf{q})$ the Fourier transforms of the partial density of the sterol molecules, $\rho_S(\mathbf{r}) \equiv \Sigma_i \, \delta(\mathbf{r} - \mathbf{r}_i) \mathfrak{L}_{S_i}$.

Finally, some simulations performed in the canonical ensemble were also used to derive the tracer diffusion coefficient D. According to the Einstein relation, the diffusion constant can be expressed as

$$D = \lim_{t \to \infty} \frac{1}{2dt} \langle |\mathbf{r}(t) - \mathbf{r}(0)|^2 \rangle \tag{16}$$

where $d = 2$ is the number of spatial dimensions of membranes, t is time, and the factor in the angular brackets is the mean square displacement of the diffusing particle over time interval t. In practice, the simulation data $\langle |\mathbf{r}(t) - \mathbf{r}(0)|^2 \rangle$ were plotted as a function of t and the plot displayed linearity for sufficiently large t. D was then determined from the slope of the linear part of the plot. Since the simulations were carried out using Monte Carlo methods, the time scales are meaningful only in a relative sense, as opposed to the absolute physical time scales in the system.

A practical remark on the simulations may be useful. In each simulation run, the system was equilibrated over a period of 200,000 MCS, and the various physical quantities were averaged over a period of 5–20 $\times$ 10^6 MCS.

Results

In this section we present a selection of the results of the Monte Carlo simulations, in terms of the equilibrium phase diagrams, collective conformational ordering of the lipid chains, and structural characterizations of the lipid–sterol bilayer membranes. All the results in this section were obtained using both the potential energy functions and the simulation methods presented in the previous section.

Phase Diagrams

Pure Lipid Bilayers. The phase diagram obtained from our simulation study for model lipid bilayers in the absence of cholesterol[8] is shown in Fig. 6. The phase diagram was calculated using the potential of Fig. 2A and is represented in terms of temperature and the parameter, V_0/J_0, which is a measure of the relative strength of the two square-well attractions. The

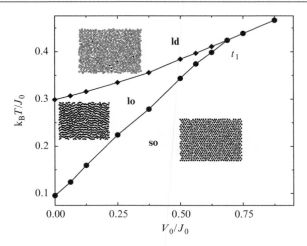

FIG. 6. Phase diagram for the model of the pure lipid bilayer system. All three phase boundaries are first-order phase boundaries. The insets show snapshots of typical microconfigurations for the three different phases labeled **so** (solid-ordered), **ld** (liquid-disordered), and **lo** (liquid-ordered). Chains in the disordered state are plotted as open circles and chains in the ordered chain state as solid circles. The three snapshots are not given to scale. t_1 is the triple point described in text.

point of key importance in this phase diagram is the appearance of two distinct regimes, separated by a triple point, of different types of macroscopic interplay between the translational and conformational degrees of freedom. All the phase lines represent first-order phase transitions, and a triple point is a phase point where two first-order phase lines join to form a single first-order phase line. The two degrees of freedom are uncoupled for values of V_0/J_0 smaller than the triple-point value, where two distinct ordering transitions from the **so** to the **lo** phases and from the **lo** to the **ld** phase take place successively. The first of these transitions is a solid–fluid transition, in which the conformational degrees of freedom do not change, whereas the second transition is between two fluids, in which the conformation degrees of freedom change. However, the two degrees of freedom are macroscopically coupled for values of V_0/J_0 greater than the triple-point value and the intermediate **lo** phase vanishes. This is exactly the case of the main phase transition in single-component lipid bilayers where one passes directly from an **so** phase to an **lo** phase. We conclude, therefore, that although there is no fundamental reason why the intermediate **lo** phase cannot appear as a physical phase of a single-component lipid bilayer, the physical reality is such that the main phase transitions of single-component lipid bilayers take place always above but close to the triple point. For our simulations for lipid–sterol systems, the values of the parameters in the model were then set such that the intermediate phase did not appear as a distinct thermodynamic phase of the corresponding system of pure lipid bilayers.

Lipid–Sterol Bilayers. The simulated equilibrium phase diagrams for the lipid–cholesterol and the lipid–lanosterol membranes[9,10] are shown in Fig. 7. They were obtained from simulations in the semigrand canonical ensemble and are presented in terms of the sterol concentration x_s (where $x_s = x_{chol}$ or x_{lan}) and a reduced temperature, T/T_m, where T_m is the main transition temperature for the system of the pure lipid membrane. In terms of microscopic interaction parameters, the lipid–cholesterol systems are distinguished from the lipid–lanosterol systems by a rather modest (about 10%) increase in the strength of the microscopic interactions between a sterol molecule and a lipid chain in its conformationally ordered state (see Table I). This has been illustrated in Fig. 2B.

It is clear from Fig. 7 that this small modification in the lipid–sterol interaction strength leads to considerable differences in the overall topologies of the two-phase diagrams. In the case of lipid–cholesterol systems (Fig. 7A), the most significant characteristic of the phase diagram is a stable region of coexistence between the **ld** and **lo** phases. Associated with this coexistence are necessarily a stable critical point and the existence of a three-phase line. The critical point is found to be located close to $T \approx 1.0075T_m$, $x_{chol} \approx 0.298$ for the interaction parameters used in the simulations. The temperature of the three-phase coexistence is estimated to be $T = 0.9977T_m$, and the concentrations of cholesterol in the three coexisting

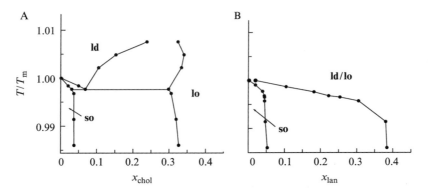

FIG. 7. Theoretical phase diagrams determined from Monte Carlo simulations of the microscopic model of interactions for lipid–sterol membranes. (A) Lipid–cholesterol membranes; (B) lipid–lanosterol membranes.

phases are $x_{chol,so} = 0.030$, $x_{chol,ld} = 0.068$, and $x_{chol,lo} = 0.298$, respectively. By contrast, Fig. 7B shows that no **ld–lo** coexistence can be identified in the phase diagrams for the lipid–lanosterol systems. Correspondingly, only a metastable critical point exists and there is no three-phase line.

The systematic development of the stable **ld–lo** coexistence as lanosterol "evolves" to cholesterol is a macroscopic signature of an increase in the capacity of the sterols to stabilize the **lo** phase. Another consistent signature is that with the sterol "evolution," the **lo** phase boundary of the low-temperature **so–lo** coexistence moves toward lower sterol concentrations, indicating a broadening of the region of stability of the **lo** phase in lipid–cholesterol bilayer membranes. A comparison between Fig. 1 and Fig. 7 shows the close correspondence between experimental and simulation results for both phase diagrams.

Umbrella Sampling and Finite-Size Analysis

It may be instructive at this point to demonstrate, with specific reference to the phase diagrams, how the umbrella sampling technique and finite-size analysis were used to determine the phase diagrams.[9,16]

First, Fig. 8 shows the results from the different steps in the iteration procedure of the umbrella sampling described previously, which was

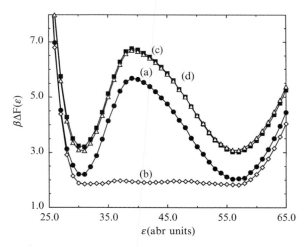

FIG. 8. The different spectral free energies calculated using the modified potential energy function. (a) Initial estimate of the shape function, $f(\varepsilon)$; (b) spectral free energy function for the modified potential energy function; (c) spectral free energy function for the original system; (d; open triangles) spectral free energy function at phase coexistence.

applied in the simulations of the theoretical model for single-component lipid bilayers. In this particular example, the transition between the **so** and the **ld** phases was investigated. At $T = 0.969T_m$, a temperature just slightly below the transition temperature, an initial estimate for the spectral shape function was made for a small system $L = 14$. This is shown in curve (a) of Fig. 8 and has a distinct barrier between the two minima representing different phases. The shape function for a larger system, $L = 16$, was then obtained by the rescaling given in Eq. (10), and the spectral free energy for the modified potential energy function was derived from the new simulation based on the modified energy function, as shown in curve (b) of Fig. 8. The spectral free energy corresponding to the original potential energy function [shown in curve (c), Fig. 8] was reconstructed. Finally, use of the reweighting technique yielded the spectral free energy at coexistence $T = T_m$, now shown in curve (d) of Fig. 8.

Next, finite-size analysis was used to deal with the simulation data obtained for the two types of lipid–sterol systems in order to determine whether phase coexistence occurred in the thermodynamic limit or whether a critical point existed as a terminal point for the **ld–lo** coexistence. Figure 9 gives examples of such a finite-size analysis for the lipid–cholesterol system (refer to the phase diagram in Fig. 7A). Figure 9 shows the spectral free energies, $\mathfrak{F}_L(x_s)$, calculated as a function of the system size L using umbrella sampling, for the **so–lo** coexistence at $T = 0.9860T_m$ (Fig. 9A), the **ld–lo** coexistence at $T = 1.0035T_m$ (Fig. 9B) coexistence, and the **ld–lo** coexistence at $T = 1.0075T_m$ (Fig. 9C). Shown in the insets are the corresponding barrier heights as a function of L for the three cases. In the first two cases, the barrier height increases linearly with L when L is sufficiently large. This finite-size behavior is a characteristic of a first-order transition.[22] In the third case, however, the barrier height approaches a constant value as L is increased, indicating that $T = 1.0075T_m$ is close to the critical point terminating the **ld–lo** coexistence.

Conformational Ordering of the Lipid Molecules

To characterize quantitatively the differential effects of the two sterols on the physical properties of lipid–sterol bilayer membranes, the calculated conformational order parameter, ϕ, as defined in Eq. (14), is presented in Fig. 10.[10] This thermal average represents the collective conformational ordering of the lipid molecules. Figure 10A shows ϕ as a function of sterol concentration for both types of lipid–sterol system at a fixed temperature $T = 1.0129T_m$. This temperature is above that corresponding to the critical point of the **ld–lo** coexistence region of the lipid–cholesterol system. At this

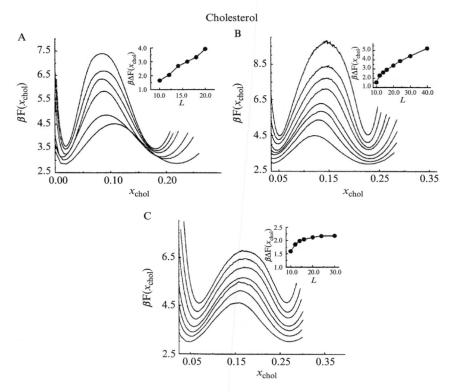

FIG. 9. Finite-size scaling plots of both $\mathfrak{F}_L(x_s)$ and the barrier height $\Delta\mathfrak{F}_L(x_s)$ *(insets)* at (A) $T = 0.9968T_m$ (**so–lo** phase coexistence); (B) $T = 1.0035T_m$ (**ld–lo** phase coexistence); (C) $T = 1.0075T_m$ (close to the critical point). The system sizes are $L = 8, 10, 12, 14, 16,$ and 20.

temperature, neither of the two types of lipid–sterol system undergo any phase transitions as the sterol concentration is changed. Figure 10B gives ϕ as a function of temperature for a fixed sterol concentration $x_s = 0.367$. Clearly both cholesterol and lanosterol order the lipid chains, but the ordering effect of cholesterol is much stronger, qualitatively similar to what has been shown by the experimental data of Fig. 4. For example, Fig. 10A shows that cholesterol at $x_{chol} = 0.40$ is able to rigidify close to 55% of the lipid chains, whereas lanosterol at the same concentration can rigidify only roughly 30%. This differential effect is also clearly illustrated in Fig. 11. Figure 11A and B are, respectively, snapshots of microscopic states of the lipid–cholesterol and lipid–lanosterol systems in the **lo** phase.

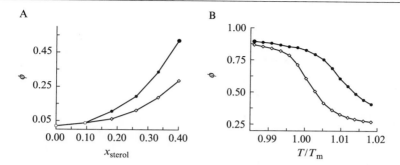

FIG. 10. Theoretically calculated average of lipid-chain order parameter, ϕ, for lipid–sterol membranes containing $N = 1600$ particles. Solid circles correspond to lipid–cholesterol membranes, and open diamonds correspond to lipid–lanosterol membranes. (A) ϕ as a function of sterol concentration at a fixed temperature, $T = 1.0129T_m$; (B) ϕ as a function of temperature for a fixed sterol concentration, $x_s = 0.367$.

Molecular Organization

MMC simulations of the microscopic model allow us to analyze quantitatively the lateral distribution of molecules in the lipid–sterol systems[9,10] in terms of the structure factors defined in Eq. (15). Figure 12 exhibits examples of the calculated structure factors calculated at $T = 0.9806T_m$ and $x_s = 0.367$, where the bilayers are in an **lo** state. Figure 12 shows circular averages over the directions of the Fourier wave vectors, $\mathbf{q}$, of the sterol structure factor $S_{sterol}(|\mathbf{q}|)$, characterizing the distributions of the sterol molecules. Shown in the inset are the circular averages of the total structure factor $S_T(|\mathbf{q}|)$. The total structure factor has the typical features common to all liquid systems. The partial sterol structure factor, however, reveals more interesting structural information. Specifically, there appears a distinct, low-$|\mathbf{q}|$ peak at $|q| \simeq 0.35 \cdot 2\pi/d$, where d is the hard-core radius as discussed in the section describing the theoretical model. This particular q value corresponds to a real-space length scale that covers several molecules, if a representative value of 5 Å for d may be used. The signal almost disappears in $S_{lan}(|\mathbf{q}|)$ for the lipid–lanosterol systems.

Analysis of microscopic configurations such as those shown in Fig. 13 suggests that the special peak in the cholesterol structure factor is related to a microstructure consisting of aligned "threads"[10] of the cholesterol molecules, interspersed with threads of lipid molecules with conformationally ordered chains (see Fig. 14). A qualitative analysis of the microscopic interactions in the model predicts that the stronger the interaction between a sterol molecule and an ordered lipid chain is, the more likely such microdomains are to appear. This explains our observation through simulations

A

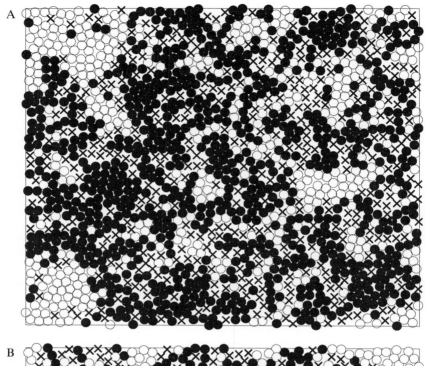

B

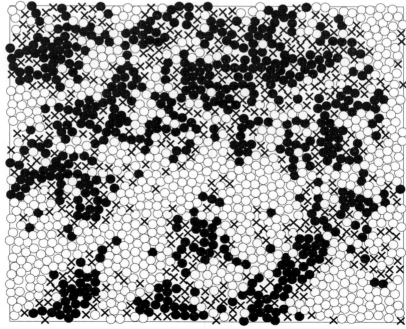

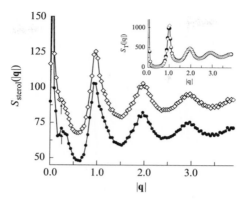

Fig. 12. Partial structure factor describing the distribution of sterol molecules in membranes, $S_{sterol}(|\mathbf{q}|)$, calculated at $T = 0.9806T_m$ and $x_s = 0.367$. $S_{chol}(|\mathbf{q}|)$ is shown by filled circles and $S_{lan}(|\mathbf{q}|)$ by open circles. For clarity, the curve for $S_{lan}(|\mathbf{q}|)$ has been shifted along the y axis. The values of $|\mathbf{q}|$ are given in units of $2\pi/d$, where d is the hard-core diameter assigned to the particles. The arrows indicate an unusual structural signal in addition to the usual peaks characteristic of liquid structure. *Inset:* Total structure factor, $S_T(|\mathbf{q}|)$, for the two lipid–sterol systems.

that such microstructures disappear almost entirely in the lipid–lanosterol systems, where the lipid–sterol interaction becomes weaker.

Lateral Diffusion

The lipid tracer diffusion coefficient in a model lipid–cholesterol binary mixture was calculated as a function of cholesterol concentration and temperature.[23] The model system that was studied was identical to that used in the calculation of the phase diagram of Fig. 7A. The results are summarized in Fig. 15. The diffusion coefficient D increases monotonically with increasing temperature and this qualitative trend is independent of cholesterol concentration. The cholesterol concentration dependence of D, however, shows more interesting behavior. At higher temperatures, D decreases with increasing x_{chol}, while at temperatures below T_m, D increases monotonically

---

[23] J. M. Polson, I. Vattulainen, H. Zhu, and M. J. Zuckermann, *Eur. Phys. J.* **5,** 485 (2001).

Fig. 11. Snapshots of microconfigurations for lipid–sterol systems in the **lo** phase close to T_m when the sterol is cholesterol (A) and when the sterol is lanosterol (B). In the snapshots, a lipid chain in the ordered state is shown as solid circles, a lipid chain in the disordered state is shown as open circles, and a cholesterol molecule is shown as crosses.

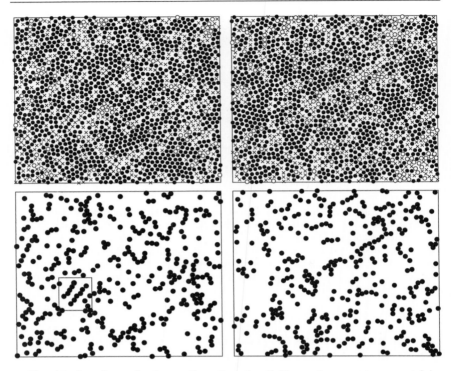

FIG. 13. Snapshots of microconfigurations for lipid membrane systems containing cholesterol *(left)* and lanosterol *(right)*, calculated at $T = 0.9806T_m$ and $x_s = 0.367$. *Top:* All particles, with lipid chains in the ordered state shown by solid circles, lipid chains in the disordered state shown by open circles, and sterol molecules shown by crosses. *Bottom:* Only the corresponding sterol molecules. The part highlighted by the box in the lower left snapshot is shown in detail in Fig. 14.

with x_{chol}. At temperatures slightly greater than T_m, D first decreases, then increases slightly with increasing x_{chol}, as can be seen more clearly in the inset to Fig. 15. A qualitative interpretation of the physical origin of this behavior is provided by the free volume theory of diffusion. An increase in D with x_{chol} is due to an increase in the free area per molecule with increasing x_{chol}. Decreases in D arise from the fact that cholesterol also promotes conformational ordering of the lipid chains, which in turn causes the chains to interact more effectively, and thus increase the effective activation energy for particle movement. These two effects compete with one another and give rise to the nonmonotonic variation of D with x_{chol} at intermediate temperatures. The diffusion results of Fig. 15 are qualitatively consistent with those of a fluorescence recovery after

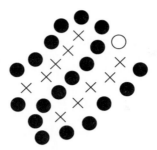

FIG. 14. Enlarged version of the box highlighted in Fig. 13, showing a local "threadlike" distribution of lipid and cholesterol molecules. Lipid chains in the ordered state are shown by filled circles, lipid chains in the disordered state are shown by open circles, and cholesterol molecules are shown by crosses.

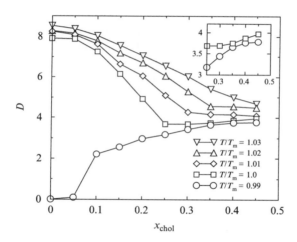

FIG. 15. Diffusion coefficient D versus cholesterol concentration fraction x_s. *Inset*: Data for $T = 0.99T_m$ and $1.0T_m$ in an expanded scale to illustrate the increase in D with x_{chol} in the **lo** phase. The diffusion results for D are given in units of $d^2/MCS \times 10^6$, where d is the hard-core diameter.

photobleaching (FRAP) experimental study of lateral diffusion in dimyristoylphosphatidylcholine (DMPC)–cholesterol binary mixture model membranes.[24] Thus, the minimal model designed to reproduce the phase behavior of lipid–sterol mixtures also yields a correct description of the equilibrium dynamics.

[24] P. F. F. Almeida, W. L. C. Vaz, and T. E. Thomson, *Biochemistry* **31,** 6739 (1992).

Summary and Perspectives

In the preceding sections we gave a detailed description of our minimal, off-lattice models for lipid–cholesterol and lipid–lanosterol bilayers. The difference in behavior between these two sterols was modeled, on the basis of their specific molecular characteristics, in terms of their differential interactions with lipid molecules, namely, that cholesterol tends to rigidify neighboring lipid chains more strongly than lanosterol because of the relative smoothness of its α face compared with that of lanosterol. We also discussed the MMC method as well as the algorithm used in the MMC simulations of the model. We also presented applications of various modern statistical–mechanical methods used to obtain the phase behavior of both lipid–cholesterol and lipid–lanosterol bilayers.

We were able to establish the occurrence/absence of the **lo** phase as a distinct, thermodynamically stable phase in the lipid–cholesterol system and lipid–lanosterol system, respectively. We pointed out that this might have significant implications for understanding both sterol evolution within the context of the structural evolution of membranes and formation of rafts. In particular, our results correlate with the current hypothesis that cholesterol is an important factor in the formation of rafts and their stability and that lanosterol is unable to stabilize rafts.

We were also able to use our model to predict the behavior of the diffusion constant in lipid–cholesterol systems as a function of both temperature and cholesterol concentration. No additional parameters or parameter values were used in these simulations and the results were found to agree qualitatively with experimental results. It is significant that the minimal theoretical models have both descriptive and predictive capabilities, as this shows that the essential microscopic physical mechanisms have been included correctly in the models.

It would be of interest to develop and study extensions of our models with several new projects. These include an examination of *"in vitro"* raft systems such as ternary mixtures composed of two lipids with a low main transition temperature and a high melting temperature, respectively, and cholesterol. The model could also be used to examine lipid–sterol systems containing other sterols, both "raft formers" such as ergosterol and "antiraft" sterols.[15] Another project would involve the inclusion of membrane proteins in the formalism in order to study the effect of cholesterol on protein sorting in binary lipid mixtures. Finally, the effect of sterols and lytic peptides on lipid–membrane lysis can be investigated using our model. Some preliminary results for the last project have been presented in Nielsen,[16] but more studies are required.

Acknowledgments

This work was supported by the Danish Natural Science Research Council, le FCAR du Quebec (Canada) via center and team grants, and the NSERC of Canada via operating and equipment grants. MEMPHYS–Center for Biomembrane Physics is supported by the Danish National Research Foundation. Both M.N. and L.M. acknowledge support from the Danish Research Academy.

[10]   Idealization and Simulation of Single Ion Channel Data

By Antonius M. J. VanDongen

Introduction

The complex behavior of individual ion channels promises to hold the key to a detailed understanding of how they function at the molecular level. Single channel analysis allows functional models to be generated, which can be confronted with available structural information. This chapter discusses methodological aspects of idealization methods for single channel data, and presents a software environment which integrates data analysis and simulation. Ion channels play critical roles in myriad cell physiological processes, including nerve and muscle excitability, synaptic transmission, secretion of salts and hormones, cell–cell contact, and cell cycle progression. Enormous progress has been made in our understanding of how ion channels function at the molecular level. This is in no small part due to a unique property of this class of proteins, namely that they can be studied with high resolution at the level of a single molecule. Development of the patch clamp technology has allowed the behavior of individual ion channels to be experimentally observed with relatively high time resolution. Single channel recordings reveal that ion channels alternate stochastically between discrete functional states that differ in their ionic conductance. A universal aspect of single channel behavior is that it appears to be dominated by two main states, open and closed. In the closed state there is no measurable ion permeation, while in the open state the permeation rate is defined by the single channel conductance, which is a specific characteristic of the channel. Transitions between the open and closed states are too fast to be resolved by current techniques.[1] As a result, single channel recordings have the appearance of a random binary signal, with the current fluctuating between

[1] J. Miodownik and W. Nonner, *Biophys. J.* **68,** A30 (1995).

Copyright 2004, Elsevier Inc.
All rights reserved.
0076-6879/04 $35.00

two well-defined levels. The two functional states must represent different molecular conformations of the channel protein, although the complexity of single channel kinetics implies that the channel can move between many more than two distinct conformations. An important objective of single channel analysis is to uncover how many closed and open states can be distinguished, and to define the kinetic properties and connectivity patterns of these states. The ultimately goal is to map the identified functional states to realistic protein conformations, based on crystallographic channel protein structures.

Noise

The actual behavior of single ion channels is obscured in patch clamp recordings by a substantial amount of noise, which originates from several sources, including (1) the capacitive properties of the glass electrode and its holder, (2) the seal resistance, (3) the electronics used, and (4) the stochastic nature of the ion permeation process. Although the first contribution is usually the largest, it is amendable to substantial experimental minimization (for a detailed description see http://web.ukonline.co.uk/a.hughes/patchwork/patchwork.htm). Dirt in the electrode holder is a common source of extraneous noise, which can be remedied by careful cleaning followed by extensive drying. Noise from the glass electrode can be minimized by several steps. Coating the electrode tip with Sylgard to reduce its capacitance is a standard procedure. Less well known is the advantage of keeping the glass pipette as short as possible, typically around 15 mm. This will likely require cutting down the silver wire extending from the electrode holder and the use of forceps to mount the pipette into the holder. The noise from the pipette can be further reduced by filling only the very tip of the electrode (where the pipette starts to narrow) with buffer solution, and backfilling the rest of the pipette with light mineral oil. It will be necessary to sharpen the end of the silver wire (and rechlorinate it) so that it will be able to reach into the tip and contact the buffer. Together, these techniques will substantially reduce the noise present in single channel recordings. No amount of sophistication in the analysis software can compensate for this.

Filtering

Before analysis, single channel data will have to be converted from an analog signal to a digital representation. This A-to-D conversion requires the use of a low-pass filter (typically a 4- or 8-pole Bessel filter) to prevent introduction of aliasing artifacts. Additional low-pass filtering may be required to increase the signal-to-noise (S/N) ratio to a level suitable for analysis. Analysis algorithms differ in their requirements for an acceptable S/N ratio. The widely used 50% amplitude threshold algorithm (discussed below) has the most stringent S/N requirement, while the Transit idealization algorithm,[2] which is the subject of this chapter, performs much better in this respect. Algorithms employing hidden Markov models (HMMs) can handle data with low S/N ratios.[3-5] In fact, the HMM approach even allows the effect of the antialiasing filtering to be undone by employing an inverse filter that restores the full bandwidth of the signal.[6] HMM methods therefore are expected to have a superior time resolution.

Missed Events

The main negative consequence of low-pass filtering of single channel data is that short-lived events will become distorted or even undetectable. The issue of "missed events" and how to deal with them is an important topic in single channel analysis, which has received much attention (see Ball and Rice[7] for a detailed discussion). Magleby and Weiss[8] have proposed an important general approach for evaluating how noise, filtering, and missed events impact the kinetic models and parameter estimates that are derived from single channel data. In this method, noisy, filtered single channel currents are simulated using the models considered appropriate, and analyzed in the same way as the actual data. Such an approach, which allows a comparison of realistically simulated single channel data with actual experimental data, is also advocated in this chapter.

Subconductance Levels

Another issue complicating single channel analysis is the potential presence of so-called subconductance states. Whereas many single channel recordings appear to be dominated by a relatively simple binary open–close behavior, some ion channels can be seen to occasionally visit current levels whose amplitude is intermediate between the closed and open state. Channels that have been reported to exhibit subconductance levels include inward rectifying K channels, voltage-gated Na and Ca channels, various anion channels, cyclic nucleotide–gated channels, and glutamate receptors. More importantly, careful inspection of large amounts of high-quality single data will invariably turn up examples of subconductance levels, even

[2] A. M. J. VanDongen, *Biophys. J.* **70,** 1303 (1996).
[3] F. Qin, A. Auerbach, and F. Sachs, *Biophys. J.* **79,** 1915 (2000).
[4] F. Qin, A. Auerbach, and F. Sachs, *Biophys. J.* **79,** 1928 (2000).
[5] L. Venkataramanan and F. J. Sigworth, *Biophys. J.* **82,** 1930 (2002).
[6] J. Zheng, L. Vankataramanan, and F. J. Sigworth, *J. Gen. Physiol.* **118,** 547 (2001).
[7] F. G. Ball and J. A. Rice, *Math. Biosci.* **112,** 189 (1992).
[8] K. L. Magleby and D. S. Weiss, *Biophys. J.* **58,** 1411 (1990).

for channels that are considered to be "binary." In the latter case, subconductance levels are usually short-lived and therefore easy to miss. The fact that most of the channel literature ignores the existence of subconductance levels may have more to do with the inability of the prevalent analysis software to correctly handle these events and the uncertainty of their origin, than their actual scarcity. If subconductance levels indeed play a critical role in the mechanism by which channels open and close, as we have proposed,[9-11] then it is paramount that the analysis software properly handle this extra complexity. This was the main driving force for the development of the Transit idealization algorithm,[2] which is discussed in detail below.

Models

Modeling single channel behavior is a critical step in achieving the ultimate goal of understanding how ion channel proteins function at the molecular level. For most channels, the statistical distribution of open and closed durations can be accurately described by a sum of exponential components, consistent with the idea that the channel protein moves between a limited set of relatively stable conformations. As a result, models based on discrete state, continuous time Markov chains have emerged as a favorite paradigm for describing single ion channel behavior. These "Markov models" are characterized by a finite number of states, each with a defined conductance. The Markovian aspect that defines this class of models is a lack of memory: after moving to a new state, the system does not remember where it came from. In addition, the probability of leaving a state does not change over time. These properties result in the lifetime of each state being exponentially distributed. To account for the multiexponential nature of the dwell time distributions that characterize single channel behavior, "aggregated" Markov models can be defined that have multiple interconnected states with the same conductance. Transitions between those states cannot be empirically observed. For instance, closed time distributions usually required more than one exponential component to be properly described. Most Markov models therefore employ multiple connected closed states. The number of exponentials that best fits the empirical closed time distribution provides the minimum number of aggregated, interconnected closed states in the Markov model. The same is true for open and subconductance states. Defining a Markov model

[9] M. L. Chapman, H. M. A. VanDongen, and A. M. J. VanDongen, *Biophys. J.* **72,** 708 (1997).

[10] A. M. J. VanDongen, *Commun. Theor. Biol.* **2,** 429 (1992).

[11] A. M. J. VanDongen and A. M. Brown, *J. Gen. Physiol.* **94,** 133a (1989).

involves specifying the number of closed and open states, their connectivity pattern, and the rate constants for each possible transition.

There are two problems associated with using Markov models to describe empirical single channel data. The first problem is that it may not be possible to distinguish competing models, on the basis of the data alone. This complication is not limited to large or complex Markov models. For instance, the two linear three-state Markov models containing two closed and one open state, C1 ↔ O ↔ C2 and C1 ↔ C2 ↔ O, are indistinguishable, since they give rise to the same dwell time statistics, if their rate constants are chosen properly.[7] The second problem, particularly common in cyclic models, is that a favored model may be unidentifiable: it may not be possible to unequivocally estimate the values of all the rate constants in the model.[7,12] Complications arising from these two problems can sometimes be reduced or removed by the introduction of constraints into the model. Examples of such constraints are a detailed balance (microscopic reversibility) requirement for loops, or interdependence of transition rates that arise from comparable conformational changes, such as movements of identical voltage sensors or ligand binding to multiple equivalent binding sites. These constraints reduce the number of free parameters and this may help make a model identifiable and distinguishable from other models.

Analysis Methods

There are two main approaches to single channel analysis, which differ in their use of models. The first approach, developed later in this chapter, makes no assumptions regarding the underlying model, and attempts to create an "idealized" version of the data consisting of a series of successive current levels, each with a defined amplitude and duration. The idealized time series is then amendable to statistical analysis. An alternative approach makes the explicit assumption that a specific model is underlying the data and uses probabilistic properties derived from the model to guide the analysis. Model-based methods can be used to idealize single channel current records, or alternatively to bypass idealization entirely and directly estimate model parameters from the raw data. Because model-based idealization methods use additional information, their performance could be superior to their model-free counterparts, provided the model used is reasonably close to being correct. However, idealization results by model-based algorithms can also be compromised by the model choice, if it poorly represents the function of the channel being studied. For example, idealizing single channel data containing abundant subconductance levels using a model with

[12] M. Wagner, S. Michalek, and J. Timmer, *Proc. R. Soc. Lond. (Biol.)* **266**, 1919 (1999).

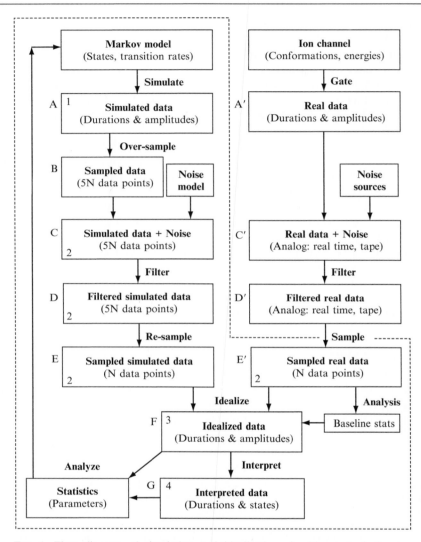

Fig. 1. Flow diagram of simulation and idealization algorithms. A single channel simulation and analysis environment is illustrated. The left side of the diagram shows discrete steps in the simulation process that are designed to mimic events in real single channel behavior and data acquisition, shown on the right. Simulation starts with the definition of a continuous time, discrete state Markov model. The open and closed states in the Markov model correspond to distinct, relatively stable, conformations of the channel. The connectivity between the states reflects the transitions that are possible between channel conformations, while the alterations in free energy associated with conformational changes determine the transition rates in the Markov model. Once a model has been defined, the simulation algorithm generates durations and amplitudes (A), whose statistical properties correspond

only two conductance states, closed and (fully) open, will result in a poor idealization. Model-based analysis methods should therefore be used in those cases only where reasonable models already exist. Channels whose behavior is incompletely characterized should be studied initially, using a model-independent approach. In this first exploratory phase, models are generated on the basis of data obtained, together with additional biophysical or structural information available for the channel. Subsequently, model-based analysis approaches can be applied. Methods based on HMMs have been shown to be particularly powerful for discriminating amongst competing models. They also appear to be superior to many other methods for estimation of Markov model parameters. Of course, these HMM methods are not immune to the identifiability issues discussed above.

Simulation

Simulations of single channel behavior generated by Markov models can serve several purposes. Important applications include testing and comparing the performance of analysis methods and evaluating the effect of missed events resulting from filtering and noise. It is important that the simulations be as realistic as possible and produce single channel data with the same problems as real data. There are a number of design considerations. First, the simulated time series should, initially, not be compromised by limited time resolution. Second, the noise present should be correlated (not "white"), and events that are too fast should be distorted or missing entirely, thereby reflecting the effect of the low-pass Bessel filter. Finally, it would be helpful if the simulated time series could be exactly and repeatedly reproduced, so that the same single channel record can be generated with and without noise, or with different amounts of low-pass filtering.

The flow diagram in Fig. 1 illustrates how such a simulation algorithm was implemented. Different stages of the simulation are shown on the left,

---

accurately to the dwell times and current levels generated by the gating of the channel being modeled (A′). Consecutive closed states are concatenated, since transitions between closed states cannot be experimentally observed. Next, a time series is generated from the amplitudes and durations by sampling the data at a frequency that is five times higher than the final desired rate (B). Real single channel data are contaminated by noise, originating from the amplifier headstage, the patch pipette, and the ionic diffusion process (C′). In the simulation, noise with a Gaussian amplitude distribution is added to the oversampled data (C), which is then subjected to low-pass filtering (D) to emulate the effect of the analog filter used on the real data (D′). The final step in generating simulated data is to resample the filtered noisy data (E), which produces a time series comparable to the sampled real data (E′). Both simulated and real data sets can be idealized (F), optionally interpreted (G), and statistically analyzed. The effect of noise and filtering can be evaluated by comparing the uncompromised data (A) with the sampled noisy, filtered data.

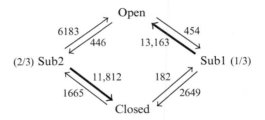

FIG. 2. Cyclic Markov model used for generating test data. The model was first used by Rosales *et al.*[13] to generate noisy and complex single channel data for testing their hidden Markov model method.

with the corresponding aspects of real single channel data on the right. The simulation/analysis software (Transit version 2) can be downloaded from http://vandongen-lab.com. The boxes enclosed by the dotted line in Fig. 1 indicate elements implemented in the software.

The starting point for the simulation is a Markov model, with N states, conductance levels for each state, and rate constants for all allowed transitions. Ideally, the N states of the Markov model correspond to N distinct conformational states of the ion channel protein whose behavior is simulated. The differences in free energy between the conformations determine the transition rates. Data were generated by the various steps of the simulation, using a four-state cyclic Markov model (Fig. 2), originally proposed by Rosales *et al.*[13] to test an HMM-based Bayesian restoration approach. There is one closed state, a fully open state, and two subconductance levels with amplitudes 1/3 and 2/3 of the full conductance. Transition rates vary over two orders of magnitude. Because of the short lifetimes of the two sublevels, the data provide a real challenge for any analysis method.

Figure 3 illustrates how one record is simulated with a duration of 50 ms, sampled at 40 kHz, resulting in 2000 points. The noise standard deviation is 10% of the full conductance, resulting in a S/N ratio between 3 and 10, for the smallest sublevel and the fully open state, respectively. The first step in the simulation is to generate a time series of dwell levels, each with a defined duration and amplitude (Fig. 1A). This is done in an iterative fashion, illustrated by the flow diagram in Fig. 4. By using a seed for the random number generator, the same stochastic behavior can be simulated repeatedly. At this point the data may resemble the actual gating behavior of the channel being simulated (Fig. 1A'). After this first step, the single channel data have been simulated without limiting the time resolution. Extremely short- and long-lived dwell levels are allowed to coexist. The only practical limitations are (1) the total length of the simulation

[13] R. Rosales, A. Stark, W. J. Fitzgerald, and S. B. Hladky, *Biophys. J.* **80,** 1088 (2001).

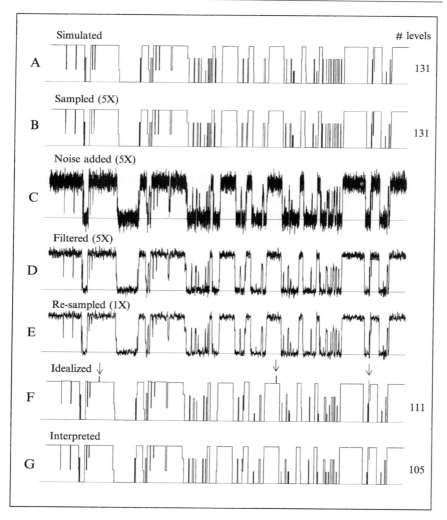

FIG. 3. Simulating, idealizing, and interpreting a single channel record. The individual steps involved in simulating a single channel record are shown. The Markov model shown in Fig. 2 was used. Letters to the left correspond to panels in Fig. 1. Trace F shows the idealization by the Transit algorithm. Arrows indicate three spurious levels that resulted from the noise. Interpretation using the three conductance levels (1/3, 2/3, and 1) present in the Markov model removed all spurious events. Additional details are discussed in text.

(chosen by the user), which may limit the observation of unusually long dwells, and (2) the maximum number of events per simulation that can be stored by the software (chosen by the programmer), which may limit

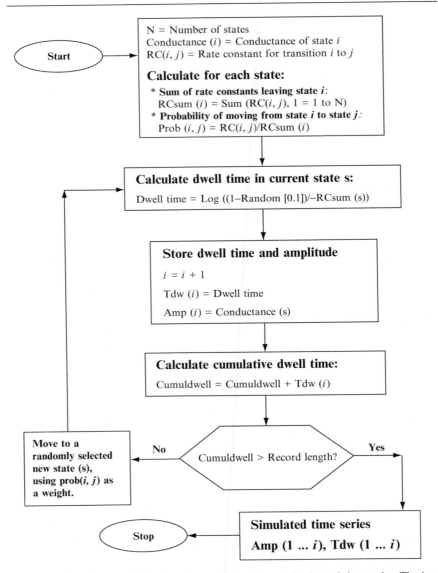

FIG. 4. Flow diagram illustrating the simulation of a single channel time series. The input variables for the algorithm are the desired duration of the simulated time series (RecordLength) and the parameters of the Markov model: the number of states (*N*), an array of length *N* containing the conductance of each state (Conductance), and an *N*-by-*N* matrix (RC) containing the rate constants for each of the connected pairs of states. Before going into the iteration, calculate the following two properties for each state: an array RCsum containing the sum of rate constants leaving state *i* and a matrix Prob(*i*, *j*) containing the probability of moving from state *i* to *j*. To generate the amplitude (Amp) and dwell time (Tdw)

the maximum length of the record to be simulated. The last limitation will cause problems only for models that contain pairs of states that are connected with rate constants much larger than the sample rate, so that the system can visit a large number of states between sample points. This type of gating (sometimes referred to as "buzzing") is extremely difficult to model.[3] The implementation of the simulation software discussed in this chapter does not deal correctly with buzzing gating modes.

To generate a simulated single channel record that resembles real data, the simulated time series will need to be sampled at discrete time points. In the example of Fig. 2, the desired sample frequency is 40 kHz, which implies a sample interval of 25 μs. To realistically simulate the effect of (correlated) noise and filtering, the time series produced in step 1 (Fig. 1A) is over-sampled at a frequency five times the desired final frequency, which is 200 kHz in our example. This produces a sampled, clean version of the time series (Fig. 3B). Because of the sampling, it is possible that short dwells are missed. In the example in Fig. 3 this did not occur: 131 dwell levels are present before and after sampling. To prevent this from becoming a problem, make sure that the oversampled interval (5 μs in Fig. 3) is significantly shorter than the mean lifetime of the most unstable state (56 μs for Sub1).

The next step is addition of white noise to the oversampled record (Figs. 1C and 3C). This is done by adding to each sample a random number, with a zero mean, a Gaussian distribution, and user-specified standard deviation (0.1 in Fig. 3). The random number generator that is used to produce the noise has its own seed associated with it, so that the same noise signal can be generated repeatedly and reproducibly, if desired.

The noisy, oversampled data (Figs. 1C and 3C) resemble empirical single channel data, contaminated by noise from sources discussed above (Fig. 1C'). It is an "analog" (nondigitized) signal that exists in real time at the output of the voltage clamp amplifier or it could be stored on analog tape. Before digitization (analog-to-digital or A/D conversion), it needs to pass through an antialiasing low-pass filter. In the simulation this is mimicked by applying a finite impulse response (FIR) filter with a Bessel characteristic[14] to the noisy, oversampled data, to produce filtered, oversampled data

---

[14] F. J. Sigworth, in "Single-Channel Recording" (B. Sakmann and E. Neher, eds.), pp. 301–321. Plenum Press, New York, 1983.

---

arrays, enter an iterative loop. While in the loop, calculate the dwell time in the current state (s), using a uniformly distributed random number between 0 and 1 as follows: Tdw = log[(1 − Random [0.1]/−RCsum(s)] and store the dwell time with the corresponding amplitude (the conductance of state s). For the next iteration, move to a randomly selected new state, using Prob(i, j) as a weight. Keep iterating until the cumulative dwell time exceeds the desired record length. The output variables of the algorithm are the two arrays Tdw and Amp.

(Figs. 1D and 3D). Digitization (A/D conversion) of the actual single channel signal coming out of the low-pass filter is mimicked by resampling the noisy, filtered, oversampled data (Fig. 3D), using the desired sampled frequency (40 kHz in Fig. 3), to produce the final simulated record (Figs. 1E and 3E). This multistep procedure may appear overly elaborate, but the complexity is required to ensure that the sampled record produced has the same problems associated with it as empirical signal channel data. Both simulated and actual data can now be processed by the idealization algorithm.

Idealization

The objective of an idealization algorithm is to convert a noisy, sampled single channel record into a noise-free series of consecutive current levels with defined amplitudes and durations. The algorithm used in the vast majority of the published literature on single channel analysis is the 50% amplitude threshold method, or variations thereof. The approach is very straightforward. A threshold is placed at 50% of the current amplitude of the (fully) open channel and transitions are detected when the single channel record is seen to cross the threshold in either direction. In between two transitions, the channel is assumed to be in either the open or closed state (depending on the direction of the transitions flanking it), and the distance (in time) between the flanking transitions is estimated as the dwell time. The original algorithm makes the explicit assumption that the channel moves only between two states, open and closed. More elaborate versions have been employed that use multiple thresholds to deal with subconductance levels or multichannel recordings. One big drawback is that in order to place the thresholds correctly, the number of conductance levels and their amplitudes need to be known *a priori*. A second problem is that the S/N ratio needs to be relatively large, to prevent threshold crossings by noise fluctuation to produce spurious dwell levels.

Partly because of these nonideal properties associated with the 50% amplitude threshold methods, an alternative idealization algorithm was developed. The design goal was to maximize objectivity, by not requiring any *a priori* assumptions regarding the underlying model, the number of conductance levels, or their amplitudes. This resulted in the Transit idealization algorithm.[2] In this algorithm, no assumption need be made regarding the behavior of the channel, other than that transitions between conductance levels are fast. Information regarding the number of conductance levels and their amplitudes is not required. The algorithm is relatively insensitive to the complexity of the underlying single channel behavior. Idealization is reliable with relatively low S/N ratios. A detailed description of the algorithm has been previously published.[2] It is based on the idea that single channel behavior, no matter how complex, consists of two components: levels and transitions. If all the transitions could be reliably localized, then the amplitudes and durations of all the levels could be correctly estimated. The Transit idealization algorithm uses a slope detector to localize transitions. Slope detectors are notoriously sensitive to noise fluctuations, and if used with any reasonable sensitivity produce many false signals. The spurious transitions that result from this are removed in Transit by employing a second threshold, a relative amplitude criterion. The idea behind this is that a spurious transition caused by a noise fluctuation divides a genuine conductance level into two shorter levels whose amplitudes are similar. The relative amplitude criterion is used to concatenate any two neighboring levels whose amplitudes are too close.

The slope or first-order derivative of the discrete time series corresponding to the filtered digitized single channel recording can be estimated in various ways. Transit employs two estimators, the central difference and the forward difference, allowing it to combine high noise rejection with the ability to detect fast switching events. The central difference is used as the default slope estimator. Figure 5 illustrates how the central difference can detect a level consisting of a single sample point (Fig. 5, case 1), as long as it is flanked by two longer levels. However, the central difference fails for a series of alternating open and closed levels of one sample point (Fig. 5, case 2), as well as for so-called shoulders of one and two sample points (Fig. 5, cases 3 and 4). The forward difference can properly detect transitions in cases 2 and 4, but it also fails in case 3. Therefore shoulders of one sample point (arrow in Fig. 5) will remain undetected in Transit. Thus far, no acceptable solution for this problem has been found.

The Transit idealization algorithm employs two slope thresholds to detect transitions and a relative amplitude threshold to remove spurious transitions. Optimal performance of the algorithm requires that the thresholds be set properly. To maximize objectivity, the thresholds are set to a level that is based on the noise present in the data. The standard deviations for the amplitude, as well as the forward and central differences, are estimated from the baseline noise. This information can be reliably extracted from the single channel data, even if no empty traces are available.[2] The two slope thresholds (for the forward and central difference) and the relative amplitude threshold are set by multiplying them by a user-selected slope and amplitude factor, respectively. Default values are 3.0 for both factors. Lower values result in more sensitivity, but may result in more spurious levels being generated. A table with probabilities of spurious events surviving both slope and amplitude criterion for different combinations of the slope and amplitude factor has been published.[2]

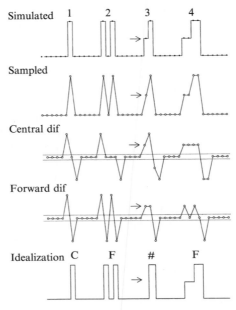

FIG. 5. Use of forward and central differences in the Transit algorithm. Details are explained in text.

Figure 3F illustrates how the Transit idealization algorithm performs on the data simulated for the cyclic model in Fig. 2. The S/N ratio is between 3 and 10. Default values were used for the amplitude and slope factor, respectively. The idealization algorithm detects 111 transitions. Three short-lived spurious levels (six transitions) are indicated by arrows. The idealization algorithm therefore fails to detect $131 - 111 - 6 = 26$ transitions that were present in the simulation in Fig. 3A. Of these, six were single sample point shoulders, which the transit algorithm cannot detect (see Fig. 5). The remaining 20 transitions were removed during filtering and the resampling step, and therefore were not present in the data fed into Transit. This illustrates the importance of having a realistic simulation procedure and the ability to evaluate what happens to the simulated data at each step. The performance of Transit on data from the four-state cyclic model can be significantly improved by increasing the sample frequency (data not shown). In practice, sample frequencies cannot, of course, be arbitrarily chosen but are determined by the noise level present in the data. The procedure outlined here will then allow evaluation of what aspects of the data will be missing, due to filtering, given a model.

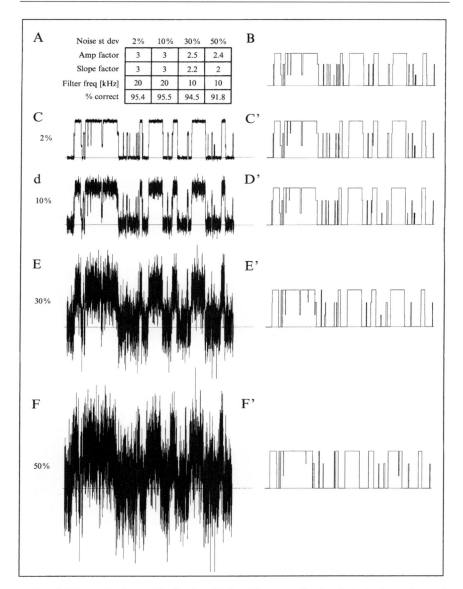

FIG. 6. Effect of noise on idealization fidelity. The same simulated single channel record was idealized and interpreted using four different S/N ratios (50, 10, 3, and 2). (A) summary of the slope and amplitude factors used in the idealization, the frequency of the low-pass filter used, and the percentage of dwell levels correctly identified by the interpretation step.

Interpretation

Idealization results in a consecutive series of amplitude–duration pairs that represent the single channel data. This time series can be subjected to statistical analysis. One important application is the construction of high-resolution amplitude histograms, as previously shown.[2] From such an analysis the number of current levels and their amplitudes can be estimated. This allows a further improvement of the idealized data, which has been termed "interpretation." It involves assigning (conductance) states to each idealized level and if necessary concatenating identical neighboring levels. In addition to improving the quality of the idealization, this interpretation allows a statistical analysis of individual (sub)conductance states.[2] Figure 3G illustrates this for the idealized trace in Fig. 3F. The three spurious levels have been removed and all remaining levels have been assigned to one of the four channel states: the closed state, the fully open state, or one of the two subconductance states. Interpretation has been implemented as a second optional step in Transit, because it adds subjectivity to the algorithm. Also, the same idealization can by interpreted using a different number of states or conductance levels.

Performance

Finally, it will be investigated how Transit performs under conditions of low S/N ratios. The cyclic four-state model in Fig. 2 was simulated with noise levels of 2, 10, 30, and 50% of the full conductance, resulting in a range of S/N ratios between 50 (2% noise and the fully open state) and 0.67 (50% noise and the smallest sublevel). Figure 6 shows the results. On the left are the noise data (before filtering and resampling) and on the right are the interpreted data from Transit. The table in Fig. 6A a gives the filter frequency, and values for the amplitude and slope factors used. The table also provides the percentage of sample points in the interpreted record that was assigned to the correct conductance level of the actual simulated data shown in Fig. 6B. The 30% noise level corresponds to that used in Rosales et al.[13] The main result of the increased noise is an inability to correctly detect the shortest lived subconductance levels, while the overall pattern of activity remains fairly robust. The number of spurious levels increases with the noise level, as is to be expected.

Acknowledgment

This work was supported by NIH Grants NS31557 and MH61506 to A.M.J.V.D.

[11] Statistical Error in Isothermal Titration Calorimetry

By Joel Tellinghuisen

Introduction

The method of isothermal titration calorimetry (ITC) is widely used to obtain thermodynamics information about binding processes in chemical and biochemical systems.[1–3] In a typical application of this technique, one of the reactants (M) is contained in a reaction vessel of small volume (0.2–2.0 ml), and the second reactant (the titrant X) is added stepwise to beyond the end point of the reaction. The instrumental responses following each injection of titrant are analyzed to obtain the heat q associated with the chemical changes from that injection, and the experiment thereby produces a titration curve of q versus extent of reaction. Such titration curves are analyzed by means of a nonlinear least-squares (LS) fit to obtain estimates of the enthalpy change ΔH° and the equilibrium constant K° for the reaction.[1–4]

By repeating the experiment over a range of temperatures, one can determine the T dependence of ΔH° and K°. In fact, when K° is known as a function of T, it becomes possible to estimate ΔH° a second way, from the slope of $\ln K^\circ$ as a function of T^{-1} (the van't Hoff method). Sturtevant's group studied several benchmark reactions by ITC and observed that the level of agreement between the results from the latter method ($\Delta H^\circ_{\text{vH}}$) and the directly measured values ($\Delta H^\circ_{\text{cal}}$) was not convincing.[5–7] These observations prompted a flurry of comments, some of which raised questions about the validity of the van't Hoff relation itself.[8–12] More recent attempts to explain the discrepancies have noted that the van't Hoff estimates are

[1] T. Wiseman, S. Williston, J. F. Brandts, and L.-N. Lin, *Anal. Biochem.* **179**, 131 (1989).
[2] M. El Harrous, S. J. Gill, and A. Parody-Morreale, *Meas. Sci. Technol.* **5**, 1065 (1994).
[3] M. El Harrous, O. L. Mayorga, and A. Parody-Morreale, *Meas. Sci. Technol.* **5**, 1071 (1994).
[4] M. El Harrous and A. Parody-Morreale, *Anal. Biochem.* **254**, 96 (1997).
[5] H. Naghibi, A. Tamura, and J. M. Sturtevant, *Proc. Natl. Acad. Sci. USA* **92**, 5597 (1995).
[6] Y. Liu and J. M. Sturtevant, *Protein Sci.* **4**, 2559 (1995).
[7] Y. Liu and J. M. Sturtevant, *Biophys. Chem.* **64**, 121 (1997).
[8] G. Weber, *J. Phys. Chem.* **99**, 1052, 13051 (1995).
[9] A. Holtzer, *J. Phys. Chem.* **99**, 13048 (1995).
[10] R. Ragone and G. Colonna, *J. Phys. Chem.* **99**, 13050 (1995).
[11] G. Weber, *Proc. Natl. Acad. Sci. USA* **93**, 7452, 14315 (1996).
[12] J. Ross, *Proc. Natl. Acad. Sci. USA* **93**, 14314 (1996).

Copyright 2004, Elsevier Inc.
All rights reserved.
0076-6879/04 $35.00

inherently much less precise than the calorimetric values, and have suggested that when this larger error in ΔH_{vH}° is acknowledged, the two methods are consistent.[13-15] However, when proper error propagation techniques are used to estimate the error in ΔH_{vH}°, inconsistency is still the rule rather than the exception for the available data in the literature.[16] There can be nothing wrong with the van't Hoff approach, since the relevant equation,

$$\left(\frac{\partial \ln K^{\circ}}{\partial T}\right)_{P} = \frac{\Delta H^{\circ}}{RT^{2}} \tag{1}$$

follows directly from the purely mathematical Gibbs–Helmholz equation. Rather, the problems stem from flaws in the procedures for collecting and analyzing ITC data.

One aspect of this problem that has received little attention is the role of statistical error in ITC data and its effect on the determination of ΔH° and K°. In a recent study,[17] I have noted that most ITC data are probably limited by experimental uncertainties in the delivered titrant volume, which means that the estimates of the heat q are subject to proportional rather than constant error. The LS analyses of such data should employ weighted fits instead of the unweighted fits normally done with the standard software in use. Neglect of weights in such situations leads to significant loss of efficiency in the estimation of ΔH° and K°. For treating the effect of uncertainty in the titrant volume, there are two limiting models that yield radically different results. If the error in the injected volume is assumed to be random, the statistical errors of the derived parameters actually increase with increasing number of titration steps. On the other hand, if one assumes that it is the total accumulated volume after i injections that is subject to random error, the incremental volume in the ith injection is the difference between two such independent quantities and is now correlated with these two volumes. This situation requires the use of weighted, correlated, nonlinear LS for analysis and leads to parameter standard errors that decrease with increasing number of steps.

The key tool for implementing the statistical studies in Refs. 16 and 17 was the use of the LS variance–covariance matrix $\mathbf{V}$ to assess parameter confidence limits and to properly propagate statistical errors in functions of the LS parameters. Since the $\mathbf{V}$ matrix is heavily underappreciated for

[13] J. B. Chaires, *Biophys. Chem.* **64**, 15 (1997).
[14] J. R. Horn, D. Russell, E. A. Lewis, and K. P. Murphy, *Biochemistry* **40**, 1774 (2001).
[15] J. R. Horn, J. F. Brandts, and K. P. Murphy, *Biochemistry* **41**, 7501 (2002).
[16] L. S. Mizoue and J. Tellinghuisen, *Biophys. Chem.* (in press).
[17] J. Tellinghuisen, *Anal. Biochem.* **321**, 79 (2003).

this purpose, I have included in the present work some computations designed to illustrate clearly its properties. Chief among these are the following: (1) In linear LS fits with all the usual assumptions,[18] especially normally distributed (Gaussian) random error about the true values of the dependent variable (y), the $\mathbf{V}$ matrix yields exact parameter variances, standard errors, and correlation coefficients when the error structure of the data is known; (2) one can similarly define an "exact" $\mathbf{V}_{nl}$ for nonlinear LS by using exactly fitting data; and (3) although this $\mathbf{V}_{nl}$ does not have the same rigorous validity as in linear LS, previous Monte Carlo (MC) studies have led to a useful "10% rule of thumb": if the predicted standard error for a nonlinear parameter is less than 10% of the parameter's magnitude, the $\mathbf{V}_{nl}$-based prediction is likely to be good within 10% in predicting the confidence limits.[19] Note that in doing MC computations on LS models, one must assume an error structure in order to add the random error to the data. Thus, the simple predictions for linear models can be used the verify the MC algorithms. In checking the extent of the error in $\mathbf{V}_{nl}$ for nonlinear models, I have typically used 10^5 simulated data sets.

In the following sections, I briefly review the essential LS and thermo-dynamics relations relevant to the study by ITC of the simplest binding case, 1:1 complexation $(X + M \rightleftharpoons MX)$. The properties of linear and nonlinear fits are illustrated on a simple model for van't Hoff analysis. The $\mathbf{V}_{nl}$ matrix is then used to assess the statistical errors in ΔH° and K° as estimated from ITC data for various error structures—constant error, proportional error, and both. The use of correlated LS is described in detail for the relevant model of titrant volume delivery. Monte Carlo calculations are used to check the validity of the $\mathbf{V}_{nl}$ matrix in cases of relatively large parameter error, and also to assess the loss of precision when heteroscedastic and correlated data are analyzed by ordinary unweighted nonlinear LS.

Variance–Covariance Matrix in Least Squares

Linear Least Squares

The LS equations are obtained by minimizing the sum S,

$$S = \Sigma w_i \delta_i^2 \tag{2}$$

with respect to a set of adjustable parameters $\boldsymbol{\beta}$, where δ_i is the residual (observed–calculated mismatch) for the ith point and w_i is its weight. In the present matrix notation, $\boldsymbol{\beta}$ is a column vector containing p elements,

[18] M. L. Johnson and L. M. Faunt, *Methods Enzymol.* **210**, 1 (1992).
[19] J. Tellinghuisen, *J. Phys. Chem. A* **104**, 2834 (2000).

one for each adjustable parameter. Thus its transpose is a row vector: $\boldsymbol{\beta}^T = (\beta_1, \beta_2, \ldots, \beta_p)$. The problem is a linear one if the measured values of the dependent variable (y) can be related to those of the independent variable(s) ($x, u, \ldots$) and the adjustable parameters through the matrix equation,[19–21]

$$\mathbf{y} = \mathbf{X}\boldsymbol{\beta} + \boldsymbol{\delta} \tag{3}$$

where $\mathbf{y}$ and $\boldsymbol{\delta}$ are column vectors containing n elements (for the n measured values), and the design matrix $\mathbf{X}$ has n rows and p columns, and depends only on the values of the independent variable(s) (assumed to be error free) and not on the parameters $\boldsymbol{\beta}$ or dependent variables $\mathbf{y}$. For example, a fit to $y = ax + b/x^3 + c\,\exp(2u)$ qualifies as a linear fit, with two independent variables (x, u), three adjustable parameters (a, b, c), and $\mathbf{X}$ elements $X_{i1} = x_i, X_{i2} = x_i^{-3}, X_{i3} = \exp(2u_i)$. On the other hand, the fit becomes nonlinear if, for example, the first term is changed to x/a, or the third to $2\exp(cu)$. It also becomes nonlinear if one or more of the "independent" variables is not error free, hence is treated (along with y) as a dependent variable.

The solution to the minimization problem in the linear case is the set of equations,

$$\mathbf{X}^T\mathbf{W}\mathbf{X}\boldsymbol{\beta} \equiv \mathbf{A}\boldsymbol{\beta} = \mathbf{X}^T\mathbf{W}\mathbf{y} \tag{4}$$

When the data are subject to random error only, the square weight matrix $\mathbf{W}$ is diagonal, with n elements $W_{ii} = w_i$; when the data are correlated, $\mathbf{W}$ contains off-diagonal elements. Equations (4) are solved for the parameters $\boldsymbol{\beta}$, for example,

$$\boldsymbol{\beta} = \mathbf{A}^{-1}\mathbf{X}^T\mathbf{W}\mathbf{y} \tag{5}$$

where $\mathbf{A}^{-1}$ is the inverse of $\mathbf{A}$. Knowledge of the parameters permits calculation of the residuals $\boldsymbol{\delta}$ from Eq. (3) and thence S, which in matrix form is

$$S = \boldsymbol{\delta}^T\mathbf{W}\boldsymbol{\delta} \tag{6}$$

Importantly, the variances in the parameters are the diagonal elements of the variance–covariance matrix $\mathbf{V}$, which is proportional to $\mathbf{A}^{-1}$ (see below).

For these equations to make sense, it is essential that the measurements y_i be drawn from parent distributions of finite variance.[20] (This, e.g., excludes

[20] W. C. Hamilton, "Statistics in Physical Science: Estimation, Hypothesis Testing, and Least Squares." Ronald Press, New York, 1964.

[21] D. L. Albritton, A. L. Schmeltekopf, and R. N. Zare, in "Molecular Spectroscopy: Modern Research II" (K. Narahari Rao, ed.), pp. 1–67. Academic Press, New York, 1976.

Lorentzian distributions.) If, in addition, they are unbiased estimates of the true means, then the LS equations will yield unbiased estimates of the parameters β. If the parent distributions are normal, the parameter estimates will also be normally distributed. For these to be minimum variance estimates as well, it is necessary that the weights be taken as proportional to the inverse variances,[20–22]

$$w_i \propto \sigma_{yi}^{-2} \tag{7}$$

Under these conditions, LS is also a maximum likelihood method. Note that it is possible to have linear LS estimators that are unbiased but not minimum variance, or minimum variance but not unbiased, or even unbiased and minimum variance, but nonnormal.

A Priori Weights. At this point, let us assume that the parent distributions for the data are normal (Gaussian) and that we know the σ_{yi} (as we do any time we run a Monte Carlo LS calculation). With this prior knowledge of the weights, we take the proportionality constant in Eq. (7) to be 1.00. Then S is distributed as a χ^2 variate for $\nu = n - p$ degrees of freedom.[20–23] Correspondingly, the quantity S/ν follows the reduced (χ^2) distribution, given by

$$P(z)dz = Cz^{(\nu-2)/2} \exp(-\nu z/2)dz \tag{8}$$

where $z = \chi_\nu^2$ and C is a normalization constant. It is useful to note that a χ^2 variate has a mean of ν and a variance of 2ν,[24] which means that χ_ν^2 has a mean of unity and a variance of $2/\nu$. In the limit of large ν, $P(z)$ becomes Gaussian.

With the proportionality constant in Eq. (7) taken as unity, the proportionality constant connecting $\mathbf{V}$ and $\mathbf{A}^{-1}$ is likewise unity, giving

$$\mathbf{V} = \mathbf{A}^{-1} \tag{9}$$

Since the parent data distributions are normal, the parameter distributions are also normal, as already noted. Then our prior knowledge of the σ_{yi} renders Eq. (9) exact. This is true even when the number of data points equals the number of adjustable parameters, giving an exact solution for the parameters ($\nu = 0$). For example, the 95% confidence interval on β_1 is $\pm 1.96 V_{11}^{1/2}$, so in MC calculations on linear fit models, 95% of the

[22] A. M. Mood and F. A. Graybill, "Introduction to the Theory of Statistics," 2nd Ed. McGraw-Hill, New York, 1963.
[23] P. R. Bevington, "Data Reduction and Error Analysis for the Physical Sciences." McGraw-Hill, New York, 1969.
[24] M. Abramowitz and I. A. Stegun, "Handbook of Mathematical Functions." Dover, New York, 1965.

estimates of β_1 are expected within $\pm 1.96 V_{11}^{1/2}$ of the true value. Conversely, a significant deviation from this prediction indicates a flaw in the MC procedures.

There is much confusion in the literature regarding these matters. In general the off-diagonal elements in $\mathbf{V}$ (the covariances) are nonzero, for both linear and nonlinear fits. This means that the parameters $\boldsymbol{\beta}$ are correlated. The correlation matrix $\mathbf{C}$ is obtained from $\mathbf{V}$ through

$$C_{ij} = V_{ij}/(V_{ii}V_{jj})^{1/2} \tag{10}$$

and yields elements that range between -1 and 1. However, each of the parameters in a linear fit is distributed normally about its true value, with $\sigma_{\beta j} = V_{jj}^{1/2}$, irrespective of its correlation with the other parameters. The correlation comes into play only when we ask for joint confidence intervals of two or more parameters, in which case the confidence bands become ellipsoids in two or more dimensions.[25] Then the correlation is also correctly predicted by Eq. (10), obviating MC computations for characterizing such joint confidence ellipsoids for linear LS fits.

It is useful to note from the structure of $\mathbf{A}$ that it scales with σ_y^{-2} and with the number of data points n. Accordingly, $\mathbf{V}$ scales with σ_y^2 and with $1/n$. Thus, the parameter standard errors go as σ_y and as $n^{-1/2}$. For example, if all σ_{yi} are increased by the factor f for a given data structure, all $\sigma_{\beta j}$ increase by the same factor f. To observe the $n^{-1/2}$ dependence exactly, it is necessary to preserve the structure of the data set, for example, by using the same 5 x_i values on going from $n = 5$ to 10, 15, 20, and so on. This means that the $\sigma_{\beta j}$ are to be interpreted in the same manner as the standard deviation in the mean in the case of a simple average. (One can readily verify that for a fit of data to $y = a$, the equations do yield for σ_a the usual expressions for the standard deviation in the mean.)

Of course, all of the foregoing does assume prior knowledge of the statistics of the y_i. Unfortunately, from the experimental side we never have perfect *a priori* information about σ_{yi}. However, there are cases, especially with extensive computer logging of data, where the *a priori* information may be good enough to make Eq. (9) the proper choice and the resulting $\mathbf{V}$ virtually exact. A good example is data obtained using counting instruments, which often follow Poisson statistics closely, so that the variance in $y_i(\sigma_{yi}^2)$ can be taken as y_i. (For large y_i Poisson data are also very nearly Gaussian.) An important reason for using prior weighting when it can be justified is that one can then use the χ^2 statistic as an indicator of goodness of fit.

[25] W. H. Press, B. P. Flannery, S. A. Teukolsky, and W. T. Vetterling, "Numerical Recipes." Cambridge University Press, Cambridge, 1986.

A Posteriori Weights. At the other extreme we have the situation where nothing is known in advance about the statistics of the y_i, except that we believe the parent distributions to be normal and to have the same variance, independent of y_i. In this case the weights w_i are all the same constant, which without loss of generality we can take to be 1.00. This is the case of unweighted least squares. The variance in y is then estimated from the fit itself, as

$$\sigma_y^2 \approx s_y^2 = \frac{\Sigma \delta_i^2}{n - p} = \frac{S}{\nu} \tag{11}$$

which is recognized as the usual expression for estimating a variance by sampling. The use of Eq. (11) represents an *a posteriori* assessment of the variance in y_i. (This was designated "external consistency" by Birge[26] and Deming,[27] as opposed to "internal consistency" for the situation where the σ_{yi} are known *a priori*.) The variance–covariance matrix now becomes

$$\mathbf{V} = \frac{S}{\nu} \mathbf{A}^{-1} \tag{12}$$

Under the same conditions as stated before Eq. (7), s_y^2 is distributed as a scaled χ^2 variate. This means, for example, if the s_y^2 values from an MC treatment of unweighted LS are divided by the true value σ_y^2 used to generate the random noise, the resulting ratios are distributed in accord with Eq. (8) for χ_ν^2.

In the case of *a posteriori* assessment, the uncertainty in s_y can greatly limit the reliability of the parameter standard error estimates when the data set is small. Since the variance in χ_ν^2 is $2/\nu$, the relative standard deviation in s_y^2 is $(2/\nu)^{1/2}$. From error propagation (see below), the relative standard deviation in s_y is half that in s_y^2, or $[1/(2\nu)]^{1/2}$. For $\nu = 200$, this translates into a nominal 5% relative standard deviation in s_y and hence also in all the parameter standard error estimates ($V_{jj}^{1/2}$); but it is a whopping 50% when $\nu = 2$, as in the test model for van't Hoff analysis explored below (in which case the distribution of s_y^2 is also far from normal, in fact is exponential). It is for this reason that it is highly desirable to characterize the data error independently from a particular experiment in those situations where it may be difficult to obtain a large number of experimental points in each run. Such information then permits use of the *a priori* $\mathbf{V}$ for confidence limits, and the χ^2 test for the data from the experiment in question.[28]

[26] R. T. Birge, *Phys. Rev.* **40**, 207 (1932).
[27] W. E. Deming, "Statistical Adjustment of Data." Dover, New York, 1964.
[28] Y. Hayashi, R. Matsuda, and R. B. Poe, *Analyst* **121**, 591 (1996).

What about the confidence limits on the parameters in the case of *a posteriori* assessment? The need to rely on the fit itself to estimate s_y means the parameter errors are no longer exact but are uncertain estimates. Accordingly we must employ the t distribution to assess the parameter confidence limits. Under the same conditions that yield a normal distribution for the parameters $\boldsymbol{\beta}$ and scaled χ^2 distributions for s_y^2 and for the V_{jj} from Eq. (12), the quantities $(\beta_j - \beta_{j,\text{true}})/V_{jj}^{1/2}$ belong to the t distribution for ν degrees of freedom,[22] which is given by

$$f(t)dt = C'(1 + t^2/\nu)^{-(\nu+1)/2}dt \qquad (13)$$

with C' another normalizing constant. For small ν the t distribution is narrower in the peak than the Gaussian distribution, with more extensive tails. However, the t distribution converges on the unit variance normal distribution in the limit of large ν, making the distinction between the two distributions unimportant for large data sets.

Intermediate Situations. Sometimes one has *a priori* information about the relative variation of σ_{yi} with y_i but not a good handle on the absolute σ_{yi}. For example, data might be read from a logarithmic scale, or transformed in some way to simplify the LS analysis. As a specific example of the latter, data might be fitted to $y = ax + bx^2$ by first dividing by x to yield $y' \equiv y/x$, then fitting to $y' = a + bx$. If the original y_i have constant standard deviation σ_y, then simple error propagation shows that the standard deviations in the y_i' values are σ_y/x_i, meaning the weights $w_i \propto x_i^2$. One can readily show that the resulting weighted "straight-line" analysis yields equations [Eqs. (4)] that are identical to those for the unweighted fit to $y = ax + bx^2$. This is a general property of linear LS fits to alternative forms relatable by linear variable transformations (which preserve the normal structure of the original data). Also, the results for both $\boldsymbol{\beta}$ and $\mathbf{V}$ [through Eq. (12)] are independent of arbitrary scale factors in the weights. In the present example, if the latter are taken as simply $w_i = x_i^2$, S/ν will be an estimate of σ_y^2.

Another situation is when data come from two or more parent distributions of differing σ_y, but again known in only a relative sense. As before, the results of the calculations are independent of an arbitrary scale factor in the weights. However, to obtain meaningful estimates of the parent variances, it is customary to designate one subset as reference and assign $w_i = 1$ for these data, with all other weights taken as $w_i = s_{\text{ref}}^2/s_i^2$ (hence the need for knowledge of the relative precisions). Then the quantity S/ν $(= s_{\text{ref}}^2)$ obtained from the fit is more properly referred to as the "estimated variance for data of unit weight," and the estimated variance for a general point in the data set is s_{ref}^2/w_i.

Users of commercial data analysis programs should be aware that those programs that provide estimates of the parameter errors do not always make clear which equation—Eq. (9) or Eq. (12)—is used. For example, recent versions of the program KaleidaGraph (Synergy Software) use Eq. (12) in unweighted fits to user-defined functions, but Eq. (9) in all weighted fits.[29] This means that in cases like those just discussed, where the weights are known in only a relative sense, the user must scale the parameter error estimates by the factor $(S/\nu)^{1/2}$ to obtain the correct *a posteriori* values. (In the KaleidaGraph program the quantity called "Chisq" in the output box is just the sum of weighted squared residuals S, which is χ^2 only when the input σ_i values are valid in an absolute sense.)

Nonlinear Least Squares

In nonlinear fitting the quantity minimized is again S, and the LS equations take a form similar to Eq. (4) but must be solved iteratively. The search for the minimum in S can be carried out in a number of different ways[18,23,25]; but sufficiently near this minimum, the corrections $\Delta\beta$ to the current values β_0 of the parameters can be evaluated from[23,25,27]

$$X^{T}WX\Delta\beta \equiv A\Delta\beta = X^{T}W\delta \tag{14}$$

leading to improved values,

$$\beta_1 = \beta_0 + \Delta\beta \tag{15}$$

The quantities W and δ have the same meaning as before; but the elements of X are $X_{ij} = (\partial F_i/\partial\beta_j)$, evaluated at x_i using the current values β_0 of the parameters. The resulting matrix A is now an approximation of the Hessian matrix.[25] The function F expresses the relations among the variables and parameters, and it is convenient to express it in such a way that a perfect fit yields $F_i = 0$. For the commonly occurring case where y can be expressed as an explicit function of x, it can be written

$$F_i = y_{\text{calc}}(x_i) - y_i = -\delta_i \tag{16}$$

In the case of a linear fit, starting with $\beta_0 = 0$, these relations yield for β_1 equations identical to Eqs. (4) and (5) for β; and convergence occurs in a single cycle. In the more general case where y cannot be written explicitly in terms of the other variables, these equations still hold, but with δ in Eq. (14) replaced by $-F_0$, where the subscript indicates that the F_i values are calculated using the current values β_0 of the parameters.

Regardless of how convergence is achieved, the variance–covariance matrix is again given by Eq. (9) in the case of *a priori* weighting and

[29] J. Tellinghuisen, *J. Chem. Educ.* **77**, 1233 (2000).

Eq. (12) for *a posteriori* weighting, with $\mathbf{X}$ as redefined just below Eq. (15). However, there is an important distinction between $\mathbf{V}$ in the general non-linear case versus the linear case: the matrix $\mathbf{A}$ now contains a dependence on the parameters. Also, in general there is no need to distinguish between dependent and independent variables in nonlinear fitting, as all variables can be taken to be uncertain.[27] In that case $\mathbf{A}$ may also depend on the values of all the variables, not just the (previously) independent variables. Thus even in the case of *a priori* weighting, $\mathbf{V}$ from Eq. (9) will vary from data set to data set. However, one can extract estimates of $\mathbf{V}$ from a per-fectly fitting theoretical curve and use this $\mathbf{V}$ in the same fashion as in the case of linear fitting. This is the "exact" $\mathbf{V}_{nl}$ menioned in the Introduction.

Through $\mathbf{V}_{nl}$, one can estimate parameter confidence limits for a par-ticular nonlinear fit model and data structure almost trivially, often with a few minutes of effort using a program like KaleidaGraph. While it is true that the nonlinear LS parameter distributions are generally not normal, often they are close enough thereto to permit estimation of confidence intervals in this *a priori* fashion with a reliability that exceeds that achiev-able in typical MC calculations. This is because the MC variance estimates are subject to the previously noted statistics of a χ^2 variate, which means for a 1000-set MC calculation a relative standard deviation of about 4.5% in the variances, or half that in the standard errors. And many published studies have employed far fewer than 1000 data sets, with concomitant loss in error precision as $N^{-1/2}$. The 10% rule of thumb for the reliability of $\mathbf{V}_{nl}$, mentioned in the Introduction, actually turns out to be conservative for most of the cases I have examined to date.

Statistical Error Propagation

The textbook expression,

$$\sigma_f^2 = \sum \left(\frac{\partial f}{\partial \beta_j}\right)^2 \sigma_{\beta_j^2} \tag{17}$$

is normally used to compute the propagated error in a function f of the independent variables $\boldsymbol{\beta}$, where the sum runs over all uncertain variables β_j. However, Eq. (17) assumes that these variables are uncorrelated. This assumption seldom holds for a set of parameters $\boldsymbol{\beta}$ returned by an LS fit, and one must use the more general expression,[30]

$$\sigma_f^2 = \mathbf{g}^T \mathbf{V} \mathbf{g} \tag{18}$$

[30] J. Tellinghuisen, *J. Phys. Chem. A* **105,** 3917 (2001).

in which the jth element of the vector $\mathbf{g}$ is $\partial f / \partial \beta_j$. This expression is rigorously correct for functions f that are linear in variables β_j that are themselves normal variates. For nonlinear functions of normal and nonnormal variates, its validity is limited by the same 10% rule of thumb that applies to parameters estimated by nonlinear LS.

In many cases the computation of σ_f can be facilitated by simply redefining the fit function so that f is one of the adjustable parameters of the fit.[30] For example, suppose a set of data is fitted to the quadratic function, $y = a + bz + cz^2/2$, where $z = (x - x_0)$, and it is the errors in this function and its derivatives that are of interest. For $f = y$, $\mathbf{g}^T = (1, z, z^2/2)$, from which it is clear that $\sigma_f = \sigma_a$ for $x = x_0$. Similarly, the statistical errors in the first and second derivatives of f at x_0 are σ_b and σ_c, respectively. Thus, one can bypass Eq. (18) by simply repeating the fit for the several values of x_0 that are of interest.

Monte Carlo Computational Methods

The Monte Carlo LS calculations are done using programs coded in FORTRAN and methods that are detailed elsewhere.[19] To minimize postprocessing of the very large files that would be produced in a run of 10^5 data sets, the distributional information is obtained by binning "on the fly." The statistical averages and higher moments are similarly computed by running accumulation. For most of the computations, a typical run of 10^5 data sets takes less than 1 min.

The statistics for the various quantities from the MC calculations are calculated by accumulating the relevant sums and then dividing by the number of sets N at the conclusion. For example, the estimated variance in a parameter a is $s_a^2 = \langle a^2 \rangle - \langle a \rangle^2$. For assessing the significance of bias, it is necessary to know the precision of the MC parameter estimates, which (at the 68.3% or 1-σ level, for normal data) is their estimated standard error, $s_a/N^{1/2}$. On the other hand the sampling estimates of the parameter standard errors are subject to the previously mentioned properties of the χ^2 distribution, for N degrees of freedom in this case. Thus their relative standard errors are $(2N)^{-1/2} = 0.00224$ for $N = 10^5$.

The histogrammed data are analyzed by fitting to the appropriate models using the user-defined curve-fitting function in KaleidaGraph. The uncertainties in the binned values are taken as their square roots, in keeping with the Poisson nature of the binning process. Bins containing fewer than six counts are normally omitted. For the most part the values are fitted simply as sampled points. However, technically the bin counts represent integrals over the specified intervals, a distinction that can make a difference if the data are not binned on a fine enough scale. For example, in the present case, proper treatment of narrow χ_ν^2 distributions requires breaking each interval into subintervals and integrating.

Van't Hoff Analysis of $K^\circ(T)$: Least-Squares Demonstration

With assumption of a functional form for $\Delta H^\circ(T)$, Eq. (1) can be integrated to yield a form suitable for analysis by LS fitting. In the examples discussed below, ΔH° is assumed to be expressible as quadratic in the temperature T over the range encompassed by the data,

$$\Delta H^\circ = a + b(T - T_0) + c(T - T_0)^2 \tag{19}$$

From this expression, at $T = T_0$, $\Delta H^\circ = a$, $\Delta C_P^\circ = b$, and $d\Delta C_P^\circ/dT = c$. Integration of Eq. (1) then yields

$$R\ln(K^\circ/K_0^\circ) = A\left(\frac{1}{T_0} - \frac{1}{T}\right) + B\ln\left(\frac{T}{T_0}\right) + c(T - T_0) \tag{20}$$

where R is the gas constant,

$$A = a - bT_0 + cT_0^2 \quad \text{and} \quad B = b - 2cT_0 \tag{21}$$

If we take y as $\ln K^\circ$ and define $\ln K_0^\circ$ as one of the fit parameters, the fit to Eq. (20) becomes linear and should obey all the rules for linear fits described above. Alternatively, the fit of K° to the exponential form of Eq. (20) is nonlinear and can be expected to deviate from these rules. These behaviors are illustrated through a series of MC calculations employing the model spelled out in Table I, which is based on observations for the complexation of Ba^{2+} with 18-crown-6 ether, a reaction that is sometimes used to calibrate ITC instruments[31] and that has been examined for consistency between ΔH_{vH}° and ΔH_{cal}°.[6,14] For simplicity, ΔC_P° is assumed to be constant ($c = 0$); however, the data error is intentionally set large enough to ensure that ΔC_P° is not statistically defined for the five-point data set (since cases of large relative error are more likely sources of significant deviations from linear behavior in nonlinear fits). Note also that the assumption of proportional error in K° means that the error in $\ln K^\circ$ is constant, since $\sigma(\ln y) = \sigma_y/y$.

Results of several MC computations on 10^5 data sets are summarized in Table II and illustrated in Figs. 1 and 2. In cases 1 and 2, the random error is assumed to be normal in the fitted quantity ($\ln K^\circ$), and the MC deviations from the true values for the parameters, their standard errors, and χ_ν^2 are reasonable—about equally positive and negative, with 11 of 16 being less than 1σ and none exceeding 2σ. In case 3 $\ln K^\circ$ is still the fitted

[31] L.-E. Briggner and I. Wadsö, *J. Biochem. Biophys. Methods* **22**, 101 (1991).

TABLE I
MODEL USED IN MONTE CARLO TESTS OF VAN'T HOFF ANALYSIS OF $K^\circ(T)$

$$\Delta H^\circ = -32{,}000 \text{ J mol}^{-1} \text{ K}^{-1} + 130 \text{ J mol}^{-1} \text{ K}^{-2}(T - 298.15) \equiv a + b(T - T_0)$$

$$\ln(K^\circ) = \ln(K_0^\circ) + \frac{a - bT_0}{R}\left(\frac{1}{T_0} - \frac{1}{T}\right) + \frac{b}{R}\ln\left(\frac{T}{T_0}\right); \quad K_{298}^\circ = 5.5$$

$$n = 5: \quad t_i(^\circ\text{C}) = 5, 15, 25, 35, 45$$

$$\sigma_{K_i^\circ} = 0.10 \ K_i^\circ; \quad \sigma_{\ln(K_i^\circ)} = 0.10$$

For log fit with $T_0 = 298.15$ K:

$$X_{i1} = 1.0; \quad X_{i2} = R^{-1}(298.15^{-1} - T_i^{-1}); \quad X_{i3} = R^{-1}[\ln(T_i/T_0) + T_0/T_i - 1]$$

$$W_{ii} = 100; \qquad\qquad W_{ij} = 0, i \neq j$$

$$\boldsymbol{\beta} = \begin{pmatrix} 1.70475 \\ -32{,}000 \\ 130 \end{pmatrix}; \quad \mathbf{A} \equiv \mathbf{X}^{\mathsf{T}}\mathbf{W}\mathbf{X} = \begin{pmatrix} 5.00000 \times 10^2 & -4.55537 \times 10^{-4} & 6.80393 \times 10^{-2} \\ -4.55537 \times 10^{-4} & 1.85173 \times 10^{-7} & -2.46001 \times 10^{-7} \\ 6.80393 \times 10^{-2} & -2.46001 \times 10^{-7} & 1.58826 \times 10^{-5} \end{pmatrix}$$

$$\mathbf{V} = \begin{pmatrix} 4.84137 \times 10^{-3} & -1.59714 \times 10^1 & -2.09873 \times 10^1 \\ -1.59714 \times 10^1 & 5.56650 \times 10^6 & 1.54638 \times 10^5 \\ -2.09873 \times 10^1 & 1.54638 \times 10^5 & 1.55264 \times 10^5 \end{pmatrix};$$

$$\mathbf{C} = \begin{pmatrix} 1.00000 & -0.09729 & -0.76548 \\ -0.09729 & 1.00000 & 0.16634 \\ -0.76548 & 0.16634 & 1.00000 \end{pmatrix}$$

For log fit with $T_0 = 278.15$ K:

$$\boldsymbol{\beta} = \begin{pmatrix} 2.67152 \\ -34{,}600 \\ 130 \end{pmatrix}; \quad \mathbf{A} = \begin{pmatrix} 5.00000 \times 10^2 & 1.40472 \times 10^{-2} & 2.00600 \times 10^{-1} \\ 1.40472 \times 10^{-2} & 5.79405 \times 10^{-7} & 9.14687 \times 10^{-6} \\ 2.00600 \times 10^{-1} & 9.14687 \times 10^{-6} & 1.53647 \times 10^{-4} \end{pmatrix}$$

$$\mathbf{V} = \begin{pmatrix} 9.00642 \times 10^{-3} & -5.43633 \times 10^2 & 2.06047 \times 10^1 \\ -5.43633 \times 10^2 & 6.14868 \times 10^7 & -2.95065 \times 10^6 \\ 2.06047 \times 10^1 & -2.95065 \times 10^6 & 1.55264 \times 10^5 \end{pmatrix};$$

$$\mathbf{C} = \begin{pmatrix} 1.00000 & -0.73053 & 0.55100 \\ -0.73053 & 1.00000 & -0.95497 \\ 0.55100 & -0.95497 & 1.00000 \end{pmatrix}$$

TABLE II

MONTE CARLO RESULTS FROM 10^5 DATA SETS ON MODEL IN TABLE I[a]

Parameter	Value	δ^b	s^c	δs^b
Exact values for $T_0 = 298.15$ and $n = 5$:				
$\ln K_0^\circ$	1.70475		0.06959	
K_0°	5.5		0.3827	
ΔH_0°	−32,000		2359.3	
ΔC_P°	130		394.04	
1. Error normal in $\ln K^\circ$; $\ln K^\circ$ fitted:				
$\ln(K_0^\circ)$	1.70485	0.46	0.06958	−0.09
ΔH_0°	−31,990.7	1.25	2357.5	−0.36
ΔC_P°	129.76	−0.19	395.4	1.49
$\chi_\nu^{2\,d}$	0.99709	−0.92	0.99939	−0.27
2. Error normal in $\ln K^\circ$; $\ln K^\circ$ fitted, 20 data points (4 at each T_i):				
$\ln(K_0^\circ)$	1.70476	0.11	0.03484	0.63
ΔH_0°	−31,998.0	0.54	1176.5	−1.21
ΔC_P°	129.66	−0.55	197.8	1.75
$\chi_\nu^{2\,d}$	1.00110	1.01	0.34276	−0.31
3. Error normal in K°; $\ln K^\circ$ fitted:				
$\ln(K_0^\circ)$	1.69979	−22.24	0.07049	5.79
ΔH_0°	−31,991.0	1.19	2387.6	5.35
ΔC_P°	129.66	−0.27	400.4	7.25
$\chi_\nu^{2\,d}$	1.02276	6.83	1.05416	24.22
4. Error normal in K°; K° fitted using theoretical weighting:				
K_0°	5.50457	3.78	0.38281	0.13
ΔH_0°	−31,995.1	0.65	2378.8	3.68
ΔC_P°	119.93	−8.00	398.1	4.65
$\chi_\nu^{2\,d}$	0.99662	−1.07	0.99884	−0.51
5. Error normal in K°; K° fitted using "observed" weighting:				
K_0°	5.43230	−54.50	0.39285	11.86
ΔH_0°	−31,979.8	2.66	2405.5	8.74
ΔC_P°	158.72	22.44	404.8	12.17
$\chi_\nu^{2\,d}$	1.01077	3.28	1.03959	17.71
6. Error normal in K°; K° fitted unweighted[e]:				
K_0°	5.50332	2.38	0.44079	67.87
ΔH_0°	−32,185.9	−21.14	2781.0	79.90
ΔC_P°	103.88	−15.40	536.3	161.50

[a] Units: J mol^{-1} for ΔH°, J mol^{-1} K^{-1} for ΔC_P°. All results for 5-point data sets except where indicated otherwise.

[b] Deviations (observed − true) in units of standard errors, which are $\sigma/(10^5)^{1/2}$ for parameters, $\sigma/(2 \times 10^5)^{1/2}$ for standard deviations.

[c] Ensemble estimates of standard deviations for Monte Carlo data.

[d] Reduced χ^2 has expected value 1.00 and standard error $(2/\nu)^{1/2}$, $= 1.00$ for 5-point data sets, $(2/17)^{1/2}$ for 20-point data sets.

[e] RMS standard errors from *a posteriori* **V** [Eq. (12)] are 0.3391, 3636, and 408.3, respectively.

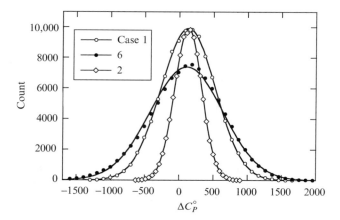

FIG. 1. Histogram data for ΔC_P°, as estimated in the indicated computations summarized in Table II. The results from cases 1 and 2 fit the Gaussian curve with reasonable χ^2 values (32.4 for case 1, 31 points; 36.2 for case 2, 32 points), while the data from the unweighted fit (case 6) are clearly not Gaussian ($\chi^2 = 298$). Similar data from case 3, although biased in s, do fit the Gaussian distribution adequately at this level of scrutiny ($\chi^2 = 36.2$, not shown). Note that the binning interval for case 2 was half that for the other two, resulting in its factor of 2 smaller area in this display.

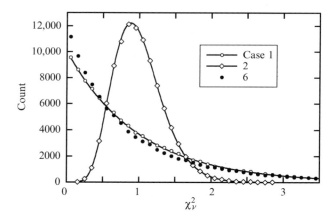

FIG. 2. Histogram data for χ_ν^2 for the indicated computations of Table II. The results from cases 1 and 2 fit the theoretical distribution of Eq. (8), yielding $\nu = 1.997 \pm 0.007$ and 17.085 ± 0.076, respectively. The estimates of s/ν from the unweighted fit (case 6) were scaled to an average value of 1.00 for this plot; they do not follow Eq. (8) for any ν.

quantity, but the error is taken to be normal in $K°$ (i.e., the random error is added to $K°$ before the logarithm is taken). As a result of the log transformation, the fitted data are no longer normal, and this results in significant bias in χ_ν^2, all parameter standard errors, and $\ln K_0°$. Cases 4–6 involve the nonlinear fitting of $K°$, both with (cases 4 and 5) and without weighting (case 6). Since the data error is taken as 10% of $K°$, there are several options for actually assessing the weights: Use the true $K°$ values (theoretical weighting, available only in the MC context), the actual values after the random error has been added (observed weighting), or the adjusted values from the fit itself (not included here).[19] All choices produce bias in most of the tabulated quantities, but the magnitude of the bias varies considerably from case to case.

Even though the biases in cases 3–6 are mathematically significant, they are mostly not practically significant from the standpoint of a single experiment. For example, the disparity of -0.068 in $K_0°$ in case 5 still represents only $\simeq 1/6$ of 1 standard deviation for a single experiment; and remember that this s itself, if estimated *a posteriori,* would be uncertain by 50% ($\nu = 2$). Except for case 6, none of the MC standard error estimates differs by more than 3% from the "exact" predictions, even for the very uncertain $\Delta C_P°$, for which the relative standard error is $\sim 300\%$. The results for the unweighted fit to the $K°$ values (case 6) demonstrate how neglect of weights for heteroscedastic data leads to standard errors that must exceed the minimum variance values—by 15, 18, and 36%, respectively. At the same time, the *a posteriori* estimates from Eq. (12) differ from both the exact values (from a properly weighted fit) and the observed (MC) values; relative to the latter they are off by -23, $+31$, and -24%. This demonstrates that the estimated $\mathbf{V}$ from Eq. (12) is simply wrong when it results from an unweighted fit of data that should properly be analyzed by a weighted fit. Not surprisingly, in this situation, the quantity called χ_ν^2 also fails to follow the true χ_ν^2 distribution (Fig. 2). Interestingly, the unweighted fit does yield the least bias for $K_0°$.

Case 2 and Fig. 1 illustrate how taking four values at each T_i instead of one drops the parameter standard errors by a factor of 2. We might ask what would happen if we simply averaged each set of four points and fitted only the five average values. Clearly each single MC data set will yield exactly the same results if we weight each set of four by four times the original weight for individual values. This is the same as taking each of the five averages of $K°$ to have $\sigma_{\bar{y}i} = \sigma_{yi}/2$, that is, the theoretical standard deviation in the mean. Alternatively, if we use the actual statistics of each group of four, we will not get identical results, and the statistics of this process will not follow the usual rules.[32] In effect, by this procedure we are granting the data the right to determine their own destiny. This disparity emphasizes the importance of having prior knowledge of the data error and using this prior knowledge to assign the weights. If such prior weighting is used in the present case, all of the results will be identical to those obtained for the original 20-point data sets, except one: χ_ν^2 will now follow the distribution for $\nu = 2$ instead of that for $\nu = 17$ (Fig. 2).

Table I includes the **A**, **V**, and **C** matrices for two different choices of T_0, 298.15 K and 278.15 K. Note first that both **V** matrices yield identical V_{33} elements, confirming expectations that the constant ΔC_P° should not have a statistical error that depends on the arbitrary choice of reference temperature. The other two parameters do vary with T, and so do their errors. To assess the error in ΔH° as a function of T, note that $\mathbf{g}^T$ in Eq. (18) is $(0, 1, z)$, where $z = T - T_0$. It is easy to verify that

$$\sigma_{\Delta H^\circ}^2 = V_{22} + 2V_{23}z + V_{33}z^2 \tag{22}$$

and that both **V** matrices thus yield identical estimates of $\sigma_{\Delta H^\circ}^2$ at all T. Of course the two **V** values give this quantity directly at 298.15 K and 278.15 K, confirming that one can obtain the desired standard error by simply repeating the fit for the several T_0 values of interest.

Finally, it is noteworthy that ΔH° and ΔC_P° are highy correlated at 278.15 K but largely uncorrelated at 298.15 K. This means that the naive error propagation formula of Eq. (17) will work fairly well on the results at the latter T_0 but will be badly in error if used with the results obtained for $T_0 = 278.15$ K.

Isothermal Titration Calorimetry

Fit Model for 1:1 Binding

In an ITC experiment, the heat q_i determined for the ith injection of titrant X represents the result of changes in the amount of the complex MX in the active volume V_0 of the reactant vessel. The cells in the most widely used instruments are of the perfusion type, in which a volume ν of solution is expelled each time the same volume of titrant is injected. It is assumed that prior to each injection the system is uniform and at equilibrium, and solution of this composition is expelled on injection, after which the injected titrant mixes and reacts to achieve the new equilibrium. Following the ith injection the total concentrations (free and complexed) of X and M are given by[3]

$$[X]_{0,i} = [X]_0(1 - d^i) \quad \text{and} \quad [M]_{0,i} = [M]_0 d^i \tag{23}$$

---

[32] J. Tellinghuisen, *J. Mol. Spectrosc.* **179**, 299 (1996).

where $[X]_0$ is the concentration of titrant in the syringe, and $[M]_0$ is the starting concentration of M in the reaction vessel. The dilution factor $d = 1 - \nu/V_0$.

At equilibrium the concentrations of reactants and product satisfy the equilibrium expression (using the concentration reference state),

$$\frac{[MX]_i}{([X]_{0,i} - [MX]_i)([M]_{0,i} - [MX]_i)} = K \equiv K° \times (\text{liters mol}^{-1}) \qquad (24)$$

The number of moles of complex produced by the ith injection is thus

$$\Delta n_i = V_0[MX]_i - (V_0 - \nu)[MX]_{i-1} = V_0\{[MX]_i - d[MX]_{i-1}\} \qquad (25)$$

and the associated heat is

$$q_i = \Delta H° \Delta n_i \qquad (26)$$

For notational simplicity, I work with the dimensionless $K°$ below, which is tantamount to taking all activity coefficients to be unity at all times in Eq. (24). I also neglect such experimental complications as the need to estimate heats of dilution for the titrant, and the related concentration dependence of q_i. Within the framework of these assumptions, this model is exact; that is, there is no need for the differential approximation described by Wiseman et al.[1]

This model has two adjustable parameters, $\Delta H°$ and $K°$, and as many data points as injections. The software in general use for analyzing ITC data includes a third parameter, the "site number" n_s. For 1:1 complexes this parameter is typically within ~ 0.05 of 1.0 and should usually be viewed as a concentration correction factor, needed to put the concentrations of X and M on a common footing. Since inclusion of this factor is important for achieving a good fit of typical ITC data, I have also included it in the present model, where I have taken it as a correction factor to $[M]_0$. (The matter of how this factor should be applied is discussed further below.)

Knowledge of the error structure of ITC data is key to estimating the parameter standard errors, as it is also for realistic MC calculations. There are two clear sources of random error in ITC: (1) the extraction of q_i values from the recorded data, and (2) the delivery of the metered volume ν of titrant from the syringe. The first of these is essentially a sensitivity of measurement limitation and is expected to be roughly constant, independent of q_i. The effects of the second depend strongly on the specific assumptions about the volume error. If the incremental volume ν is assumed to possess random error, simple error propagation yields a proportional error, $\sigma_{qi} = q_i(\sigma_\nu/\nu)$. On the other hand, if it is the total *accumulated* volume after i steps that is assumed to possess random error, the incremental volume, being the difference between two such independent quantities, possesses correlated error. These three kinds of error affect the precisions differently and are examined individually and in combination in the calculations described below. When simultaneous contributions from the measurement and volume errors are considered, the variances are assumed to be additive, for example, $\sigma_i^2 = \sigma_q^2 + q_i^2 (\sigma_\nu / \nu)^2$ for the random volume error. Note that random volume error leads to data that are inherently heteroscedastic, requiring a weighted fit for proper analysis. Similarly, correlated volume error requires the use of weighted, correlated LS. For reference, the two uncertainties are estimated as $\sigma_q = 0.28$ μcal and $\sigma_\nu = 0.015$ μl in Ref. 1; for the benchmark reaction of 2'CMP with RNase studied there, the volume error dominates over most of the titration curve in the random volume error model (see below).

The point made after Eq. (10), about the dependence of V on the data error, is worth revisiting here. If the computations for a given model are repeated after simply scaling σ_q and σ_ν by a factor f, the parameter standard errors will scale by the same factor f. Since sufficiently small data errors yield adequately Gaussian parameter distributions, this means that the error structure for the model can always be evaluated from V_{nl}. The only question then is the extent to which this structure applies to the actual situation; in previously examined cases, the 10% rule of thumb has proved a reliable guideline for applicability.

In the LS fitting codes the independent variable is taken as the titration index i, which is rigorously error free. The error in q is taken to be normal at all times, and the solution concentrations are treated as exact. Although uncertainty in the prepared concentrations is significant in most actual experiments, this uncertainty is not manifested as point-to-point random error in a given experiment; and it is anyway partly compensated through the parameter n_s, as was already noted and is discussed further below.

Check on 10% Rule of Thumb

For most of the calculations discussed in this and subsequent sections, $[M]_0$ was fixed at 1.00 mM, V_0 was 1.4 ml (as in the instrument of Ref. 1), and ΔH° was 10.0 kcal/mol. The total titrant volume for m injections was typically 0.1 ml; and the precisions in K°, n_s, and ΔH° were investigated as functions of K° (more properly $K[M]_0$), the number of injections m, and the stoichiometry range of the experiment, $R_m = [X]_{0,m}/[M]_{0,m}$.

I address first the circumstances under which V_{nl} can be trusted to yield reliable estimates of the parameter errors in the analysis of ITC curves. For reference, Fig. 3 illustrates typical titration curves spanning the approximate extremes of analyzable values of the product $K[M]_0$. These curves

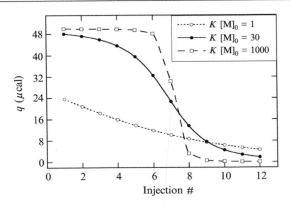

Fig. 3. Computed ITC curves for $K° = 5 \times 10^3$, 1.5×10^5, and 5×10^6 and $[M]_0 = 0.200$ mM. Other conditions: $V_0 = 0.20$ ml, $R_m = 2.08$, $\nu = 10$ μl, $\Delta H° = 10$ kcal/mol. Neglecting the error in ν and taking $\sigma_q = 0.6$ μcal, as in Ref. 4, the predicted standard errors in $K°$ are 1.845×10^3, 1.107×10^4, and 1.008×10^6, respectively, while the corresponding errors in $\Delta H°$ are 2.486, 0.0932, and 0.0533 kcal/mol. The standard errors in the concentration correction parameter n_s (taken to be 1.00) are 0.105, 0.0074, and 0.0030 (same order).

represent an extension of the example, $K[M]_0 = 1000$, explored in Fig. 3A of Ref. 4. Since the latter MC analysis also included an assumed 2% error in the concentrations, the present results cannot be compared quantitatively; however, the results for this case are commensurate with those obtained there. It is interesting that even though $K°$ is uncertain by 20% for $K[M]_0 = 1000$, $\Delta H°$ and n_s are actually quite well defined in this case. On the other hand, all three parameters are much less precise at the other extreme.

Figure 4 illustrates the results of MC calculations for the case $K° = 5 \times 10^6$ in Fig. 3. The results for $K°$ are clearly non-Gaussian, with a bias of +3.4% (even though the peak in the distribution is shifted negatively). However, the MC statistical error in $K°$ exceeds the predicted value by less than 6%, showing that even for this 20% relative error, the predicted value from $\mathbf{V}_{nl}$ would be adequate for many applications. The histograms for $\Delta H°$ and n_s appear to be Gaussian, as expected from their smaller percent standard errors (0.5 and 0.3%, respectively). However, only the former actually fits the Gaussian with an adequately small χ^2—21.6 for 33 points. It is interesting that the reciprocal of $K°$ (or the dissociation constant) is actually much closer to Gaussian than $K°$ itself; similar results were obtained in the large-$K°$ regime in the study of complexation equilibria in Ref. 19.

Monte Carlo calculations have been carried out for a number of other choices of the ITC parameters, for both constant and proportional error, random and correlated. None of the results has shown any problem with the 10% rule for the analysis of typical ITC data. Its validity extends even

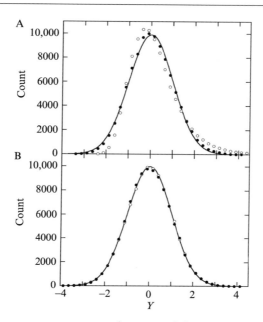

FIG. 4. Histogrammed results for (A) $K^°$ (○) and $K^{°-1}$ (●), and (B) $\Delta H^°$ (○) and n_s (●), from 10^5 MC fits of 12-point data sets like that shown for $K^° = 5 \times 10^6$ in Fig. 3, with superimposed random error having $\sigma_q = 0.6$ μcal. In each case the histogrammed quantity is $Y = (\beta - \beta_{\text{true}})/\sigma_\beta$, with the values for these quantities given in the caption to Fig. 3. The smooth curves are standard Gaussians, scaled into optimal agreement with the data for $K^{°-1}$ in (A) and with $\Delta H^°$ in (B).

to the low limit of three titration increments ($m = 3$), where there are no degrees of freedom and the LS equations yield exact fits at all times. Of course, the $\mathbf{V}_{\text{nl}}$ matrix can say nothing about the extent of nonnormality or the bias, so if these are at issue, MC calculations must be employed for the specific cases in question.

Dependence on Stoichiometry Range and Titration Increments

Having demonstrated the approximate validity of the "exact" nonlinear variance-covariance matrix $\mathbf{V}_{\text{nl}}$ under the relatively extreme conditions of Fig. 4, I now use it to investigate the dependence of the ITC parameter standard errors on the other experimental quantities. Figure 5 illustrates the computed standard errors in $K^°$ and $\Delta H^°$ as functions of the range of titration and the number of titration increments, for the midrange $K[\text{M}]_0$ value of 36 and constant error. Other parameters were chosen to resemble those for the instrument described in Ref. 1 (except the error, which was

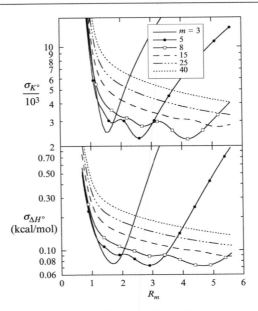

FIG. 5. Calculated standard errors in $K°$ and $\Delta H°$ for $K° = 36,000$ and $[M]_0 = 1.0$ mM, as functions of the stoichiometry range and the number of titration steps m, for constant absolute error $\sigma_q = 0.04$ mcal. Other parameters: $V_0 = 1.4$ ml, $\Delta H° = 10$ kcal/mol, and $vm = 0.10$ ml. For reference, the total q for complete reaction is 14 mcal, so for $m = 15$ and $R_m = 5$, the error on q_1 is ~1%.

artificially large). The most interesting result of these calculations is the observation that better precision is achieved with fewer points and a larger titration range than is customarily employed in such work. For small m, the standard errors exhibit structure, showing that they are sensitive to just where the points fall on the titration curve.

The loss of precision with increasing number of points seems at odds with expectations, but it can be understood as follows. The total heat q is limited by the fixed amount of M in the reaction vessel, so increasing the number of titration increments m decreases each q_i. At first approximation, if $K°$ and n_s are held constant, one obtains m estimates of $\Delta H°$ from $\Delta H° = q_i/\Delta n_i$ [Eq.(26)]. The relative error in each such estimate of $\Delta H°$ is approximately σ_q/q_i, and since q_i decreases with increasing m, the error in $\Delta H°$ increases concomitantly. This is partially offset by the statistical averaging effect, which goes as $m^{-1/2}$. The net result is a standard error that increases roughly as $m^{1/2}$. Similar observations were made some time ago by Doyle et al. in connection with a study of a differential absorption technique for characterizing binding isotherms.[33]

The possible role of correlated error in the titrant volumes will be considered in detail below. When the error in the incremental volume ν is assumed to be random, there are several cases to consider: (1) ν and the titrant concentration $[X]_0$ are fixed, and the operator decides on m, and hence R_m; (2) $[X]_0$ and the stoichiometry range R_m are set in advance, and the operator decides on m; and (3) ν and R_m are set, and the operator varies m. In the first case, an increase in m means simply adding additional points at the end of the titration range, which always leads to a decrease in all of the parameter standard errors (i.e., improved precision). In case 2 the total titrant volume is fixed, so increasing m decreases ν; this leads to an increase in the relative uncertainty σ_ν/ν and hence an m dependence similar to that shown for constant error σ_q in Fig. 5. Case 3 is the least feasible from an experimental standpoint, since it means altering the titrant concentration when either m is changed for fixed R_m, or R_m is changed for fixed m. However, this case does lead to increasing precision with increasing m, since the relative error σ_ν/ν is now fixed. Results for this case show that minimal error in $\Delta H°$ occurs near $R_m = 1.5$, but a much larger titration range of $R_m \approx 4$ is needed for optimal precision in $K°$.[17] Of course, with increasing dilution of titrant, all q_i will decrease as m^{-1}, so that eventually the constant error σ_q in q will dominate.

Figure 6 illustrates the dependence of the relative errors in $K°$ and $\Delta H°$ on $K°$ and R_m for $m = 7$ and constant absolute error. There is a fairly flat minimum in the error surface for $K°$, centered near $K[M]_0 = 10$ and $R_m = 4$. $\Delta H°$ is an order of magnitude more precise, with a large region at large $K°$ where the relative error is less than 1%. For both quantities there is considerable structure in the error surface for the relatively small m value of 7. Note again that the structure of these contour diagrams is unaffected by the actual value used for σ_q, so, for example, reducing σ_q by a factor of 10 would simply result in a relabeling of the contours by the same factor. Also, in regions where the relative errors in $K°$ and $\Delta H°$ exceed ~ 0.2, the actual statistical distributions may be far from Gaussian, as already noted in the discussion of Fig. 4.

Although for efficiency I have used a programming language to generate all the results described in this section, it is worth noting that the "exact" parameter standard errors for a specific data structure can be obtained quite easily using some desktop data presentation and analysis packages. For example, I have used KaleidaGraph to double-check some of the results obtained from the FORTRAN programs. Such calculations are facilitated by defining the key quantities from Eqs. (23–26) as library functions, for example,

[33] M. L. Doyle, J. H. Simmons, and S. J. Gill, *Biopolymers* **29**, 1129 (1990).

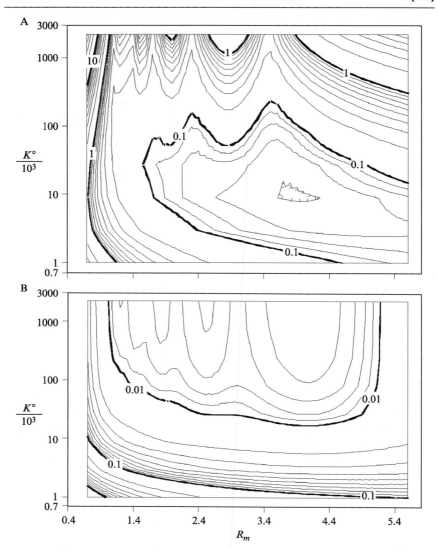

Fig. 6. Contour plots of the relative standard errors σ_β/β as functions of K° and ΔH°, for $m = 7$ and constant error, $\sigma_q = 0.04$ mcal: (A) for K°; (B) for ΔH°. Other parameters as in Fig. 5.

```
di(x) = (df^x);
xOi(x) = (XO * (1. - di(x)));
yOi(x) = (YO * c * di(x));
tsm(x) = (xOi(x) + yOi(x) + 1./a);
tmp(x) = (tsm(x)^2 - 4. * yOi(x) * xOi(x));
mxi(x) = ((x > 0)?(VO/2 * (tsm(x) - sqrt(tmp(x))) * b) : 0);
```
(27)

where x is the (integer) running index for the titration steps, and `di(x)`, `xOi(x)`, and `yOi(x)` are as defined in Eq. (23), with $c = n_s$, $a = K°$, and $b = \Delta H°$. `tsm` and `tmp` are used in the quadratic solution for $[MX]_i$; and `df`, `XO`, `YO`, and `VO` are d, $[X]_0$, $[M]_0$, and V_0, respectively, and can be replaced by their numerical values in these expressions or entered through definition statements preceding these in the library. The fit function to be entered in the user-defined function box (General) is then

$$mxi(x) - df * mxi(x - 1) \qquad (28)$$

Since the first value of x is 1, the definition of `mxi(x)` contains a branching statement to set the second term in Eq. (28) $= 0$ for this first injection of titrant.

When Weights Are Neglected

I now take a closer look at the prototype case of Ref. 1 (Figs. 4A and 5), and ask what happens when such data are analyzed by unweighted LS. As was already noted for case 6 in Table II, when heteroscedastic data are analyzed with neglect of weights, there are two main consequences: (1) the parameter estimators are no longer minimum variance, so the parameter distributions must broaden. (2) The error estimates from the *a posteriori* **V** of Eq. (12) are not reliable and can be either pessimistic or optimistic. The magnitudes of these effects can be determined only through MC computations.

For the aforementioned experiment in Ref. 1, the following values of the key parameters apply: $v = 4$ μl, $\Delta H° = -13.7$ kcal/mol, $[M]_0 = 0.651$ mM, $K° = 4.88 \times 10^4$, $R_m = 2.06$. As was noted earlier, σ_v and σ_q were estimated to be 0.015 μl and 0.28 μcal, respectively. The calculated q_i values range from 1.18 mcal for $i = 1$ to 0.04 mcal for $i = 20$. Since the relative error in v is 0.0038, the relative error will exceed the absolute error (σ_q) until $i = 17$, where $q_i = 0.07$ mcal. With data properly weighted for the combined errors, $\mathbf{V}_{nl}$ yields $\sigma_{K°} = 200$ and $\sigma_{\Delta H°} = 21.0$ cal/mol.

For comparison, MC calculations employing unweighted fits yield the results illustrated in Fig. 7. The neglect of weights has led to increases in

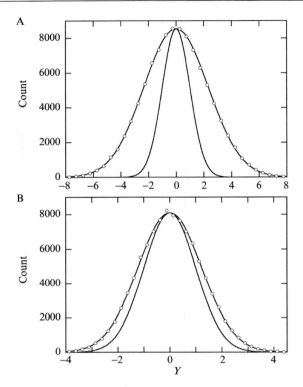

FIG. 7. Histogrammed results from 10^5 unweighted MC fits of 20-point data sets resembling that illustrated in Figs. 4A and 5 of Ref. 1, for K° (A) and ΔH° (B). In each case the histogrammed quantity is $Y = (\beta - \beta_{\text{true}})/\sigma_\beta$, using σ_β values as given in text for a properly weighted fit. The inner curve in each case is the normal curve ($\sigma = 1$) scaled to the peak of the data, while the other curves are fitted Gaussians having $\sigma = 2.334$ (A) and 1.232 (B).

both σ values, as anticipated. The corresponding loss of efficiency for the determination of K° is $2.334^2 \approx 5$, meaning one would need to run five equivalent experiments with unweighted analysis to match the precision achievable through proper weighting of a single experiment. On the other hand, the biases in both K° and ΔH° were not statistically significant. Interestingly, the results obtained from the *a posteriori* **V** (by averaging the appropriate MC V_{jj} elements and taking the square roots) were surprisingly close to correct for σ_{K° (500 versus 472 observed) and only a bit worse for $\sigma_{\Delta H^\circ}$ (17.8 versus 25.7 cal/mol). Still, these results further illustrate how Eq. (12), when applied naively to heteroscedastic data, can both over- and undershoot the actual values.

Correlated Error in Titrant Volume

The motorized syringes used to inject the titrant in ITC apparatuses are programmed to travel a certain distance with each injection. For such devices it may be more appropriate to consider the end points of travel of the plunger as the quantities subject to random error, rather than the incremental volume v. Then the latter, being the difference between two independent quantities, possesses correlated error. The effects of this change in assumption about the error can be seen dramatically in MC computations: With increasing number of steps m, the case 2 model above (fixed $[X]_0$ and R_m) now gives decreasing rather than increasing parameter error. This effect, too, was observed previously by Doyle et al.,[33] in fitting differences extracted from an absorbance titration curve.

In the correlated LS fit needed for proper analysis of such data, the weight matrix $\mathbf{W}$ of Eqs. (4)–(6) and (14) is the inverse of the variance–covariance matrix associated with the incremental volumes.[20,21,32] Let us represent the total delivered titrant volume after i injections as u_i. Then the ith incremental volume is $v_i = u_i - u_{i-1}$. The v_i and u_i are related via a linear transformation,

$$\mathbf{v} = \begin{pmatrix} 1 & 0 & 0 & & 0 & 0 \\ -1 & 1 & 0 & & 0 & 0 \\ 0 & -1 & 1 & \cdots & 0 & 0 \\ & & & \cdots & & \\ 0 & 0 & 0 & & -1 & 1 \end{pmatrix} \mathbf{u} \equiv \mathbf{L}\mathbf{u} \qquad (29)$$

and thus $\mathbf{V}_v$ and $\mathbf{V}_u$ are related by[20,30]

$$\mathbf{V}_v = \mathbf{L}\mathbf{V}_u\mathbf{L}^T \qquad (30)$$

Because the u_i are independent, $\mathbf{V}_u$ is diagonal, with elements σ_u^2 (constant by assumption). Since it is q_i that is fitted here, these σ_u^2 values must be converted to σ_{qi}^2 by error propagation, using numerical differentiation to assess dq_i/du_i. Calling the resulting matrix $\mathbf{V}_u'$, $\mathbf{V}_v' = \mathbf{L}\mathbf{V}_u'\mathbf{L}^T$ and $\mathbf{W} = \mathbf{V}_v'^{-1}$. It is noteworthy that the matrix $\mathbf{V}_v'$ is tridiagonal, with elements $(i,i) = (\sigma_{qi-1}^2 + \sigma_{qi}^2)$, $(i, i+1) = -\sigma_{qi}^2$, and $(i, i-1) = -\sigma_{qi-1}^2$ (with all indices limited to the range $1 - m$ of the data). The diagonal terms are thus seen to be the expected results for subtraction of two uncorrelated quantities.

Use of this $\mathbf{W}$ with the model described and used above yielded formal parameter error estimates (from $\mathbf{V}_{nl}$) in good agreement with the results from the MC computations and well within the framework of the 10% rule of thumb. For example, in computations for $R_m = 3$ on the model described in Fig. 5 but having $\sigma_q = 0$ and $\sigma_v = 0.0015$ ml, the choice $m = 5$ yielded

13% relative standard error in K° and 6.4% in ΔH°, as compared with estimates higher by 5.0 and 4.6%, respectively, from MC computations on 10^5 data sets. However, the reduced χ^2 from the MC computations was 1.034, which deviates more from the expected value of 1.00 than was found in the earlier MC results.

With this demonstrated reliability, the correlated model was used to estimate the parameter errors as a function of m for a more realistic $\sigma_v = 0.015 \ \mu l,$[1] while, for comparison, MC computations were carried out on the same model (i.e., random error in the accumulated titrant volume u_i) using ordinary unweighted LS to extract the parameters and their *a posteriori* estimated standard errors. The results (Fig. 8) show that unweighted LS

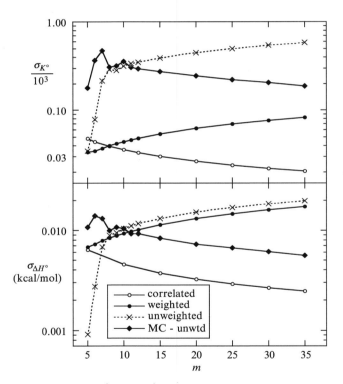

FIG. 8. Statistical errors in K° and ΔH° as a function of m, for model of Fig. 5, but with $\sigma_q = 0$ and random error $\sigma_v = 0.015 \ \mu l$ in the accumulated titrant volume u_i (hence correlated error in v). The results marked "weighted" are obtained when this same random error is in v. The "MC" results are from the statistics of 10^4 data sets analyzed using unweighted LS, while those marked "unweighted" represent the apparent values (RMS) returned by the *a posteriori* $\mathbf{V}$ of Eq. (12), from the same MC computations. The structure at small m in the "unweighted" and "MC" results is real.

correctly tracks the m dependence predicted by the correlated model $\mathbf{V}_{nl}$, but gives standard errors too large by about a factor of 10 for K° and 2 for ΔH°. On the other hand, the *a posteriori* $\mathbf{V}$ (Eq. 12) gives estimates that happen to be roughly correct only for the range $m \approx 8–12$, deviating sharply to overly optimistic errors for smaller m and pessimistic for larger.

Since the error in measuring q_i is presumed to be random, and since this error leads to decreasing precision with increasing m (Fig. 5), we might expect that addition of random error in q_i to this correlated v model should neutralize the increased precision at large m in Fig. 8. That is in fact observed. To generate the data for the MC computations, the effects of the volume error are first computed, as before: the u_i are given random error, v_i is calculated from $v_i = u_i - u_{i-1}$, and the heat q_i is calculated using a variable v version of Eqs. (23) and (25). Then random error is added to each q_i value. This error is correctly accommodated in the correlated fit model by adding σ_q^2 to the diagonal elements of $\mathbf{V}_v'$, as was confirmed in the MC computations. Figure 9 shows that the addition of the random measurement error has rendered K° less precise by a factor of ~ 3, and made σ_K° almost independent of m. A smaller reduction of precision occurs for ΔH°, and its standard error still decreases with increasing m.

Figure 9 also includes results of analyzing these same data with neglect of the correlation, using either unweighted LS or the weighted model that is correct for random error in both v and q_i. The MC statistics show that the weighted model does a good job of extracting the K° values but performs less well for ΔH°. However, without repeating their experiments enough times to obtain ensemble statistics, users of the weighted model would normally rely on the *a posteriori* $\mathbf{V}$ for error estimates. Accordingly, they would report errors for ΔH° that are significantly too large over the full range of m in Fig. 9, and somewhat less pessimistic errors for K° for most m. Not surprisingly, unweighted LS performs worse almost across the board; the one exception is in the assessment of ΔH°, where the unweighted model actually betters the weighted model for $m > 35$.

Calorimetric Versus Van't Hoff ΔH° from ITC

Test Model

I return now to an issue raised in the Introduction, namely the matter of the precision of ΔH° as estimated directly from the ITC titration curves versus that determined indirectly from the T dependence of $K^\circ(T)$. To investigate this problem in the ITC framework, I have devised a model that consists of five ITC experiments run on the same thermodynamics model described in Table I. As was already noted, this model was designed

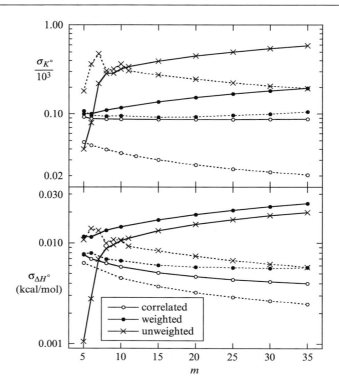

Fig. 9. Results for same model as in Fig. 8, but with the addition of random measurement error $\sigma_q = 0.28$ μcal (as in Ref. 1). The dashed "correlated" curves are from Fig. 8 ($\sigma_q = 0$). The other results are obtained by analyzing the same MC data sets with neglect of correlation, using the weighted fit model and unweighted LS. In both cases, the solid curves represent the apparent standard errors, from the *a posteriori* $\mathbf{V}$, while the dashed curves represent the actual MC statistics from 10^4 data sets for each point.

to resemble the Ba^{2+}/18-crown-6 ether complexation (except that the actual K_{298}° of 5500 is used in the present ITC model). The model was examined for both constant error and proportional error (random v model only). The reaction volume was taken as 1.4 ml and the initial concentration of the crown ether as 0.020 mol/liter. For each titration curve, 25 incremental 10-μl injections of titrant (Ba^{2+}) were taken, covering a stoichiometry range of 2.0 (Ba^{2+}: ether). The same conditions were assumed for all five temperatures. Under these conditions, the magnitude of q_1 was 70 mJ at 5°, dropping to 59 mJ at 45°. For the constant error model, σ_i was 1.0 mJ for all i. In the proportional error model, the relative σ was set at 2%. This choice made the uncertainty in ΔH_{cal}° nearly the same in the two models.

The "site parameter" n_s was taken to be 1.00 at all times, and was defined as a correction to $[M]_0$, as in Eqs. (27). In these model calculations, ΔH°_{cal} and ΔH°_{vH} are identical by definition. In that case, titration data recorded at different temperatures can (and should) be analyzed simultaneously (global analysis) to yield a single reference K° value and a single determination of $\Delta H^\circ(T)$, as defined here by the two parameters a and b. The statistical errors for such a determination were examined both for the assumption of a single n_s value, and for separate n_s values for each titration curve. The latter would be the less presumptuous approach in a set of experiments. In fact, the standard errors in the key thermodynamic parameters (a, b, and K°_0) differed very little for these two approaches.

Results for K° are illustrated in Fig. 10. Since the largest predicted relative error is only $\sim 8\%$ (for K° in the constant error model), no MC confirmations are deemed necessary, and I present just the "exact" ($\mathbf{V}_{nl}$-based) results here. From the results in Fig. 10, two observations are noteworthy: (1) the constant error model yields results that are less precise by almost an order of magnitude; and (2) in both models the relative error in K° is more nearly constant than the absolute error. The latter observation means that van't Hoff analysis through an unweighted fit to the logarithmic form of Eq. (20) is not a bad approximation, which is reassuring, since the unweighted log fit is the usual approach taken for such analyses.

In keeping with all earlier indications, ΔH°_{cal} is determined with much better relative precision ($<0.8\%$) than K° in both error models. Results for these directly estimated values are illustrated in Fig. 11, together with their counterpart ΔH°_{vH} values, as obtained by fitting the five $K^\circ(T)$ values to Eq. (20). Not surprisingly, the poorer precision in K° translates into poorer precision in ΔH°_{vH}. The curves at the bottom of this figure show the reduction of error due to the averaging effect when the five individual ΔH°_{cal}

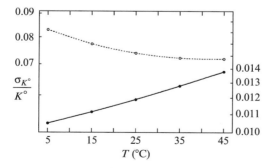

FIG. 10. Relative standard errors in K° from model calculations for van't Hoff analysis, for proportional error (solid, scale to right) and constant error (dashed, scale to left).

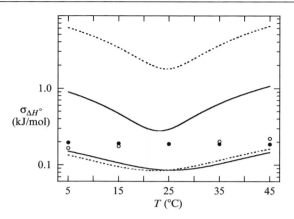

FIG. 11. Standard errors in ΔH° from the model calculations, as obtained from the individual ITC data sets (points), from weighted fits of these points to a linear relation (curves at bottom), and from weighted fits of $K^\circ(T)$ to the exponential form of Eq. (20) (upper curves). The solid points and curves represent the proportional error model; the open points and dashed curves represent the constant error model.

values are fitted to a straight line; this reduction is not uniform across the T range of the data, rather it is most prominent in the midrange. The van't Hoff results behave similarly, although even more dramatically. In both cases the reason is as already discussed in connection with Eq. (22). It is noteworthy that the results from the global fits of all five titration curves to the same linear ΔH° function and a single K_0° reference value yield only slight reduction in the error of the fitted ΔH° relative to the results in the lower curves in Fig. 11. Further, if ΔH° is fixed at the fitted results (i.e., if a and b are taken as known), the fits of the five $K^\circ(T)$ values to Eq. (20) return K° with standard errors insignificantly different from those from the global analyses. Thus, while the global analysis of the ITC data is the "proper" statistical procedure, there is little practical gain in this approach over the customary one of just fitting the ΔH°_{cal} values from the several ITC experiments run at different temperatures to an appropriate function of T. If ΔH°_{vH} and ΔH°_{cal} are deemed consistent (as they of course are in the present model), the subsequent fit of the individual $K^\circ(T)$ values with ΔH° taken as "known" further sharpens the definition of $K^\circ(T)$. Of course, in any effort to determine whether the two estimates of ΔH° are consistent, the ΔH°_{vH} values will be limited by the greatly reduced precision shown in the top two curves in Fig. 11.

The fits of the five $K^\circ(T)$ values to Eq. (20) yield parameter errors that are quite large. For example, in the constant error model, this fit yields $\Delta C^\circ_{P,vH} = 130 \pm 299$ J mol^{-1} K^{-1}. While this parameter thus appears to be undefined, it is still necessary to include it in any attempt to evaluate the consistency between ΔH_{vH}° and ΔH_{cal}°, because it is well defined in the linear fit of the ΔH_{cal}° values. Omitting it would make the two models incompatible. With it included, the error in ΔH_{vH}° is also relatively large ($\sim$20%) at the ends of the T range. However, the MC computations summarized in Tables I and II have already shown that this is not a significant source of error in estimating either the parameters or their standard errors in this case.

Although not treated here, the correlated v model of Fig. 8 yields comparative precisions for K° and ΔH° that resemble those for the present random proportional error model, that is, relative errors a factor of $\sim$2 larger for K° than for ΔH°. With both constant and proportional error present, the comparative precisions fall between the limits of the two error models, for both correlated error (Fig. 9) and random. Thus the results in Figs. 10 and 11 can be considered to bracket the range of actual observations.

Case Study: Ba²⁺ Complexation with Crown Ether

The complexation of Ba^{2+} with 18-crown-6 ether in water (unbuffered) has been studied by at least five groups, but the experimental results hardly show a developing consensus (Fig. 12). The ΔH_{cal}° values from Liu and

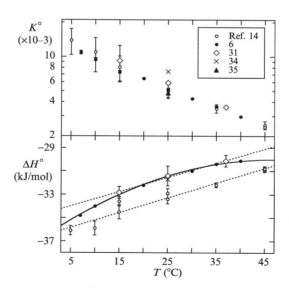

FIG. 12. Literature results for K° and ΔH_{cal}° for the complexation of Ba^{2+} by 18-crown-6 ether in pure water. Note log scale for K°. Error bars represent reported standard errors. Results from Refs. 34 and 35 were reported only at 25° (no ΔH° for the latter).

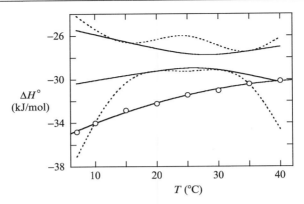

FIG. 13. ΔH°_{cal} estimates from Liu and Sturtevant[6] for Ba^{2+}/ether complexation (points and their quadratic least-squares representation), and 1-σ error bands on van't Hoff estimates using quadratic (dashed curves) and linear (solid) representations for ΔH° in the LS fit function [Eq. (20)].

Sturtevant[6] agree well with those from Briggner and Wadsö[31] in the restricted T region where they overlap, but the former show clear curvature over the full T range. Although statistical errors were not reported for ΔH°_{cal} in Ref. 6, the results from the unweighted fit indicate that the errors are comparable to the size of the displayed points. The most recent results, from Horn et al.,[14] are less precise but are still larger in magnitude by a statistically significant 5%. The K° values from Ref. 14 appear to be much less precise than the other results; however, the quoted errors on K° in that study are overly pessimistic, as is discussed further below.

The directly measured ΔH° values from Liu and Sturtevant are compared with the van't Hoff estimates in Fig. 13. The latter were obtained from weighted fits of their K° values to Eq. (20), taking their tabulated uncertainties as the estimated σ values. Neither the quadratic nor the linear representation of ΔH° gave a statistically significant determination of $\Delta C^\circ_{P,vH}$. Even worse, both fits gave χ^2 values >10 (cf. expected values of 4 and 5, respectively). This suggests that the estimated errors on K° in Ref. 6 were optimistic. With this interpretation, the conservative approach is to use the a posteriori $\mathbf{V}$ of Eq. (12) to estimate the errors, which is what I have done to generate the error bands in Fig. 13. Even with these enlarged errors, the ΔH°_{vH} values are not consistent with the ΔH°_{cal} estimates, in agreement with the conclusions of Liu and Sturtevant. If we take the view

[34] R. M. Izatt, R. E. Terry, B. L. Haymore, L. D. Hansen, N. K. Dalley, A. G. Avondet, and J. J. Christensen, J. Am. Chem. Soc. 98, 7620 (1976).
[35] Y. Hasegawa and H. Date, Solvent Extr. Ion Exch. 6, 431 (1988).

that only a constant ΔH° is justified by the van't Hoff analysis, we obtain $\Delta H^\circ = -28.4(6)$ kJ mol^{-1}, which, for comparison purposes, should be taken as the estimate in the middle of the T range, or around 23°. The inconsistency remains.

Briggner and Wadsö published values for K° and ΔH° at only three temperatures.[31] Fitting their K° values (weighted) to a constant ΔH° yields -31.7 (1.8) kJ mol^{-1}, in good agreement with their ΔH°_{cal} value of -31.42 (20) kJ mol^{-1} at 25°C. Horn et al. reported that their van't Hoff estimates of ΔH° were statistically consistent with their ΔH°_{cal} values.[14] However, if their published K° values are analyzed by weighted LS using their uncertainties, and the error bands on ΔH°_{vH} are calculated using Eq. (18) to properly accommodate the interparameter correlation, the case for consistency is less convincing. In this weighted fit, the χ^2 is 0.41, which is much smaller than the expected 6, indicating that the error estimates on K° were pessimistic. Accordingly, use of the a posteriori $\mathbf{V}$ in the error propagation calculation results in a narrowing of the error bands by a factor of $\sim$4.

Although the χ^2 from the weighted fit of the K° values in Ref. 14 is an order of magnitude too small, a weighted fit of the published ΔH°_{cal} values to a linear function of T gives a reasonable value of 9.7 (for $\nu = 7$). This observation is consistent with the use of unweighted LS to analyze ITC data that are actually dominated by proportional error.[16] The situation is analogous to that illustrated in Fig. 9, where the actual statistical errors in the parameters from the unweighted analysis are significantly smaller than indicated by the a posteriori $\mathbf{V}$. And it is the actual errors in K° and ΔH° that are manifested in subsequent fits of these quantities to functions of T. The proportional error model predicts that the relative error in K° should be a factor of 2–3 larger than that in ΔH°, while the constant error model (implicit in unweighted LS) predicts a factor of $\sim$10. The latter is consistent with the published results in Ref. 14, while the former is closer to observations from the fits of the extracted values to functions of T.

Beyond the complexation of Ba^{2+} with ether, few other reactions have been examined closely for the consistency of ΔH°_{cal} and ΔH°_{vH}. Those that have also fail to yield convincing agreement.[16]

Conclusion

A computational study of statistical error in the nonlinear LS analysis of ITC data for 1:1 complexation reactions yields the following main results: (1) when the data uncertainty is dominated by constant measurement uncertainty in the heat q, better precision is achieved in both K° and ΔH° for fewer titrant injections than are customarily used; as few as five may be optimal in many cases; (2) under usual experimental circumstances, the same result holds for data uncertainty dominated by the relative error in the titrant volume v, if this error is assumed to be random; (3) on the other hand, if volume error dominates and it is the error in the integral titrant volume that is random, the precision increases with increasing number of titration steps; (4) for typical conditions, the relative precision for K° is a factor of 2–10 poorer than that for ΔH°; higher precision in K° is generally favored by larger stoichiometry ranges than are customarily used; the same holds for ΔH° in the constant error model, but $R_m < 2$ is optimal for ΔH° when the volume error dominates; (5) actual ITC data are typically dominated by the relative error in v for large q (early titrant injections i) and by the constant absolute uncertainty in q for small q (large i); for optimal extraction of K° and ΔH°, such data require analysis by weighted LS, or in the case of random error in the integral titrant volume, by correlated LS; and (6) the larger relative error in K° versus ΔH° means that ΔH° estimates obtained from van't Hoff analysis of the T dependence of K° (ΔH°_{vH}) will in turn be inherently less precise than the directly extracted estimates (ΔH°_{cal}).

One of the most intriguing results of this study is the observation that the dependence of the parameter standard errors on the number of titration steps is so completely tied to assumptions about the nature of the error in the titrant volume: If random in the differential volume v, the precision decreases with increasing m; if random in the integrated volume u, it increases with m. The true situation probably lies somewhere between these two extremes, but it can be determined only by properly designed experiments. If the random v error model is found to dominate, experiments should be designed to use much smaller m than is currently the practice.

The present computations have dealt with only the case of 1:1 binding, and even if the random v error is found to dominate, it may be necessary to use $m = 10$ or more to demonstrate that the stoichiometry actually is 1:1. Also, in practice it is wise to work with at least a few degrees of statistical freedom, in order to obtain some indication of the goodness of fit. Other cases, like multiple binding,[36] will have to be examined specifically. However, it is possible that there, too, better precision can be achieved using fewer injections than are commonly employed.

The reduced precision in ΔH°_{vH} means that ΔC°_P and its T derivative may not be statistically defined in a van't Hoff analysis, even though they may be well determined by the ΔH°_{cal} data. However, these parameters may still be necessary as means to an end in the attempt to confirm consistency between ΔH°_{cal} and ΔH°_{vH}. If they are neglected in the van't

[36] C. A. DiTusa, T. Christensen, K. A. McCall, C. A. Fierke, and E. J. Toone, *Biochemistry* **40**, 5338 (2001).

Hoff analysis, then the resulting $\Delta H°$ estimate should be reported as the value for the average T of the data set (or an appropriately weighted average, for weighted data). This will eliminate the bias problem noted by Chaires.[13]

With proper attention to error propagation to generate error bands on $\Delta H°_{vH}$, the case for consistency between $\Delta H°_{cal}$ and $\Delta H°_{vH}$ in ITC is not as sanguine as has been suggested in recent work.[14] A number of factors have been considered in the effort to explain the discrepancies, but one that has largely escaped attention is the role of the ITC site parameter n_s. This parameter is typically determined with greater relative precision than either $K°$ or $\Delta H°$, and it is clearly needed in the analysis of most ITC data to achieve a satisfactory fit. In the 1:1 binding case considered here, n_s is in effect correcting for errors in the stated concentrations of the two solutions. It is normally defined in the stoichiometry sense MX_{ns}, which means that it is serving as a correction factor for M [i.e., $n_s = 1.05$ means the true $[M]_0$ is 5% larger than stated; see Eqs. (27)]. Since M is often a macromolecule and is typically harder to prepare to known concentrations, this is probably appropriate in most cases. However, when the titrant concentration is less certain, it is proper to redefine the correction. The effect of such redefinition is a correction factor of $1/n_s$ to $[X]_0$, and it results in a change in both $K°$ and $\Delta H°$ by the factor n_s. If experiments are repeated with the same solutions over a range of temperatures, the result will be a systematic shift in both sets of results. A constant proportional error in $K°$ will have no effect on $\Delta H°_{vH}$, so the result will be an error in the apparently more precise $\Delta H°_{cal}$ and possible discord between the two estimates. When the site parameter is covering for a mix of macromolecules (e.g., 2.5% having two sites to yield $n_s = 1.05$), the 1:1 fit model is not really correct. Either way, the deviation of n_s from "chemical" stoichiometry for the process under investigation is an indication of systematic error, and a conservative assessment of the parameter errors in such cases should include its consideration.

Throughout this work I have assumed that equal volume aliquots (v) of titrant are added sequentially to generate the titration curve. This appears to be the only mode used by workers in the field, also. However, from the structure evident in Figs. 5 and 6, it seems likely that for small m and a chosen R_m, some sequence of volumes that vary from step to step might yield smaller parameter errors. Indeed, preliminary results from an examination of this problem show that for a 7-step titration to $R_m = 3$, a variable-v algorithm can reduce the statistical error in $K°$ by 40% from the constant-v approach.

Most of the statistical errors reported, plotted, and discussed in this work have been based on the predictions of the "exact" nonlinear variance–covariance matrix $\mathbf{V}_{nl}$, bolstered by MC calculations in selected cases. To use this approach, it is necessary to know the error structure of the data ... which, of course, one always does in an MC calculation. Unfortunately, most experimentalists still take the ignorance approach in their LS analyses, using unweighted fits and the *a posteriori* **V** from Eq. (12) to estimate parameter errors. Much is to be gained from taking the trouble to assess the experimental statistical error apart from the data for a given run.[28,37,38] The obvious advantage is the narrowed confidence bands that attach to the *a priori* **V** versus the *a posteriori* **V** and its concomitant need for the *t* distribution to assess confidence limits.[28] But in addition, one has the χ^2 statistic to assess goodness of fit—a quantitative answer to the question, "Are my fit residuals commensurate with my data error?" [22,23] In view of the persistent discrepancies between ΔH_{cal}° and ΔH_{vH}°, it is possible that the fit models in current use do not adequately reflect the actual physical situation in an ITC experiment. To address this issue, reliable information about the data error is essential.

There is another "downside" to the naive use of unweighted LS and the *a posteriori* **V** in cases like ITC, where the data are inherently heteroscedastic: Eq. (12) always "lies" in such cases. The extent of the lie can only be determined through MC calculations. If proportional error is assumed to dominate over the entire titration curve, the error in $\sigma_{K^{\circ}}$ can be as much as a factor of ∼10 in the case of correlated error (Fig. 8), resulting in a 100-fold loss in efficiency in the estimation of K°.[17] In linear LS, neglect of weights does not bias the estimation of the parameters. However, nonlinear LS parameters are inherently biased to some degree; and since the bias scales with the variance,[19] neglect of weights will exacerbate the bias, possibly converting an insignificant bias into a significant one.

[37] J. Tellinghuisen, *Appl. Spectrosc.* **54**, 431 (2000).
[38] J. Tellinghuisen, *Appl. Spectrosc.* **54**, 1208 (2000).

[12] Analysis of Circular Dichroism Data

By Norma J. Greenfield

Introduction

The optical bands of molecules without a center of symmetry often exhibit the property of circular dichroism (CD), which is the unequal absorption of right-handed and left-handed polarized light. There are many basic reviews of the CD of proteins in the literature,[1–12] so the experimental techniques of obtaining and converting data to molar or mean residue

Copyright 2004, Elsevier Inc.
All rights reserved.
0076-6879/04 $35.00

ellipticity are not restated here. This review, rather, focuses on how one can analyze CD data, once it is obtained, to determine thermodynamic parameters of folding, binding constants, and estimates of secondary structure.

Proteins and polypeptides have CD bands in the far ultraviolet region (178–260 nm) that arise mainly from the amides of the protein backbone and are sensitive to their conformations. In addition, proteins have CD bands in the near ultraviolet (350–260 nm) and visible regions, which arise from aromatic amino acids and prosthetic groups. The CD in these regions depends on the tertiary structure of the protein. The changes in CD in both the near and far UV as a function of temperature, denaturants, and time can be used to estimate the thermodynamic and kinetic parameters of protein and nucleic acid folding (see reviews in Refs. 13–17). In addition, there are often changes in both the near and far UV bands when proteins interact with ligands or other proteins. Since CD is a quantitative technique, changes in CD spectra are directly proportional to the amount of the complexes formed, and these changes can be used to estimate binding constants.

The theory of CD and the numerical methods used for the analysis of CD to estimate structural parameters are discussed in detail in another chapter (see [13] in this volume[18]), so methods of analyzing CD data to obtain structural information are only briefly summarized here without mathematical details. The programs needed to analyze CD data, including the source code in most cases, are widely available (see Appendix I). This

[1] S. Beychok, *Science* **154,** 1288 (1966).

[2] B. L. Vallee, J. F. Riordan, J. T. Johansen, and D. M. Livingston, *Cold Spring Harb. Symp. Quant. Biol.* **36,** 517 (1972).

[3] A. J. Adler, N. J. Greenfield, and G. D. Fasman, *Methods Enzymol.* **27,** 675 (1973).

[4] H. Rosenkranz, *Z. Klin. Chem. Klin. Biochem.* **12,** 415 (1974).

[5] N. J. Greenfield, *CRC Crit. Rev. Biochem.* **3,** 71 (1975).

[6] G. D. Fasman, *Methods Cell. Biol.* **18,** 327 (1978).

[7] W. C. Johnson, Jr., *Methods Biochem. Anal.* **31,** 61 (1985).

[8] W. C. Johnson, Jr., *Annu. Rev. Biophys. Biophys. Chem.* **17,** 145 (1988).

[9] W. C. Johnson, Jr., *Proteins* **7,** 205 (1990).

[10] R. W. Woody, *Methods Enzymol.* **246,** 34 (1995).

[11] N. J. Greenfield, *Anal. Biochem.* **235,** 1 (1996).

[12] S. Y. Venyaminov and J. T. Yang, *in* "Circular Dichroism and the Conformational Analysis of Biomolecules" (G. D. Fasman, ed.), Vol. 69–107. Plenum Press, New York, 1996.

[13] A. M. Labhardt, *Methods Enzymol.* **131,** 126 (1986).

[14] G. D. Ramsay and M. R. Eftink, *Methods Enzymol.* **240,** 615 (1994).

[15] M. R. Eftink, *Methods Enzymol.* **259,** 487 (1995).

[16] N. J. Greenfield, *in* "Encylopedea of Spectroscopy and Spectrometry" (J. C. Lindon, G. E. Tranter, and J. L. Holmes, eds.), p. 117. Academic Press, London, 2000.

[17] N. J. Greenfield, *Trends Anal. Chem.* **18,** 236 (1999).

[18] N. Sreerama and R. W. Woody, *Methods Enzymol.* **383**(13) (2004), this volume.

chapter focuses on numerical methods to obtain binding and folding constants from CD data.

Summary of Methods to Obtain Secondary Structure of Proteins from Circular Dichroism Data

Methods of analyzing protein CD spectra to yield secondary structure have been reviewed in detail.[11,12] All the analytical methods commonly used to analyze CD data assume that the spectrum of a protein can be represented by a linear combination of the spectra of the secondary structural elements plus noise:

$$[\theta]_\lambda = \Sigma F_i S_{\lambda i} + \text{noise} \tag{1}$$

where $[\theta]_\lambda$ is the mean residue ellipticity of the unknown protein as a function of wavelength; F_i is the fraction of each structure; and $S_{\lambda i}$ is each basis spectrum used in the fit. The structural components that are estimated include α helices, parallel and antiparallel β-pleated sheets, turns, poly-L-proline II–like helices, and random conformations. Representative CD curves corresponding to common secondary structures are shown in Fig. 1.

Method of Least Squares

The simplest method of estimating conformation is to fit the spectrum of a protein by a linear combination of basis spectra corresponding to known secondary structural elements. Here it is assumed that $[\theta]_\lambda = c_1 S_1 + c_2 S_2 + c_3 S_3 + \ldots + C$, where c_n is the fraction of conformation n, S_n is the spectrum of conformation n, and C is a constant. In constrained least-squares fits Σc_n is fixed to equal 1 and C is set to 0. One finds the values of each c_n that minimize the error between the actual spectrum and the calculated spectrum using the of method of least squares[19,20] (multilinear regression) first described by Gauss. An alternative method is to systematically create a large set of linear combinations of the basis spectra and to search the set for the combination that gives the lowest standard deviation to the known spectrum.[21,22] Before the structure of many proteins had been determined by X-ray crystallography or nuclear magnetic resonance (NMR) spectroscopy, the spectra of unknown proteins

[19] N. Greenfield and G. D. Fasman, *Biochemistry* **8,** 4108 (1969).
[20] S. Brahms and J. Brahms, *J. Mol. Biol.* **138,** 149 (1980).
[21] N. Greenfield, B. Davidson, and G. D. Fasman, *Biochemistry* **6,** 1630 (1967).
[22] Y. H. Chen, J. T. Yang, and H. M. Martinez, *Biochemistry* **11,** 4120 (1972).

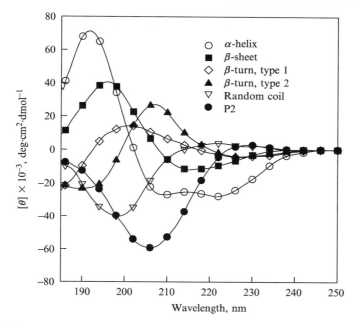

FIG. 1. Representative circular dichroism curves corresponding to common secondary structural elements. The spectra of the α-helix, β-sheet, β-turns, and random coil were drawn from data in Brahms and Brahms.[20] The spectrum of poly-L-proline II (P2) was contributed by B. Brodsky, RobertWood Johnson Medical School, Piscataway, NJ.

were fit by a combination of the spectra of polypeptides with known conformations (Fig. 1). Once the conformation of many proteins had been determined, the CD spectra of several well-characterized proteins were deconvoluted into reference (basis) spectra for the α-helix, antiparallel and parallel β-pleated sheets, β-turn, and random conformations using multilinear regression analysis,[22,23] singular value decomposition,[24–26] or the convex constraint algorithm.[27] Such extracted basis curves are often used now in place of polypeptide standard curves.

The method of least squares works well for predicting the percentages of α helix in a protein, because the spectrum of the α helix has high intensity and a distinctive shape with several nodes (Fig. 1). However, the method is poorer for determining the content of sheets and turns. The

[23] V. P. Saxena and D. B. Wetlaufer, *Proc. Natl. Acad. Sci. USA* **68,** 969 (1971).

[24] J. P. Hennessey, Jr. and W. C. Johnson, Jr., *Biochemistry* **20,** 1085 (1981).

[25] L. A. Compton and W. C. Johnson, Jr., *Anal. Biochem.* **155,** 155 (1986).

[26] P. Manavalan and W. C. Johnson, Jr., *Anal. Biochem.* **167,** 76 (1987).

[27] A. Perczel, M. Hollosi, G. Tusnady, and G. D. Fasman, *Protein Eng.* **4,** 669 (1991).

use either of spectra of polypeptides with known conformations or of spectra of extracted basis curves as standards has its drawbacks. The polypeptide databases, which use the spectra of relatively long peptides as standards, are not suitable models for the CD of the short strands of helix and sheet found in proteins. In addition, there is no way to correct spectra for the contributions of disulfide bonds, aromatic amino acids, prosthetic groups, or bound ligands and these sometimes make large contributions to the CD of proteins in the far UV. It is also difficult to extract good standards for sheets and turns using a protein database because the contributions of the β-sheet and disordered conformations are almost opposite to each other. Thus, proteins with approximately 50% sheet and low helical contents have low ellipticity. Deconvolution methods using the spectra of such proteins do not yield curves that are indicative of the spectra of the pure conformations, although they are useful for analyzing the conformations of similar proteins. The basis spectra extracted from such proteins are not suitable for determining the conformation of polypeptides.

A variation of the method of least squares is ridge regression (CONTIN).[28] In this method the spectrum of the unknown protein is fit by a linear combination of the spectra of a large database of proteins with known conformations. The contribution of each reference spectrum is kept small, unless it contributes to a good agreement between the theoretical best fit curve and the raw data. The method usually gives good estimates of β sheets and turns and gives good fits of the spectra of polypeptides in known conformations.

Singular Value Decomposition: SVD, VARSLC, SELCON

The application of singular value decomposition to determine the secondary structure of proteins was first developed by Hennessey and Johnson.[24] In this method, basis curves, with unique shapes, are extracted from a set of spectra of proteins with known structures. Each basis curve is then related to a mixture of secondary structures, which are then used to analyze the conformation of an unknown protein. The method excels in estimating the α-helical content of proteins, but is poor for sheets and turns unless the data are collected to short wavelengths, at least 184 nm, which is not possible for solutions containing buffers commonly used to stabilize proteins, for example, Tris, HEPES, physiological concentrations of NaCl, dithiothreitol, glycerol, or EDTA.

Several newer programs have improved the method of singular value decomposition by selecting references that have spectra that closely match

---

[28] S. W. Provencher and J. Glockner, *Biochemistry* **20,** 33 (1981).

the protein of interest. They include the variable selection method (VAR-SELEC, or VARSLC)[26] and the self-consistent method (SELCON).[29–33] These give good estimates of β sheet and turns in globular proteins and work with data collected only to 200 nm.[11] However, they also do not have suitable standards to be used for the analysis of long polypeptides, short peptide models, coiled coils, or collagen-like proteins.

Neural Net Analysis

Neural networks are artificial intelligence programs for detecting patterns in data. Two programs widely used to analyze CD data are CDNN[34] and K2D.[35] A neural network is first trained, using a set of proteins with known structures. The CD contribution at each wavelength is weighed, leading to the output of the correct secondary structure. The trained network is then used to analyze unknown proteins. The fits to known proteins are good and seem to be relatively independent of the wavelength range that is analyzed.

Accuracy of Secondary Structural Estimates

It is difficult to know with confidence the probable error in the estimation of secondary structure of an unknown protein, using any of the CD deconvolution methods described above. The biggest factor contributing to poor estimates is inaccurate determination of protein concentration, leading to errors in the estimate of the mean residue ellipticity. The probable accuracy of each curve fitting method is usually evaluated by examining how well a procedure can estimate the conformation from the CD data of a set of proteins whose conformations have been evaluated by X-ray crystallography. One then obtains the correlation coefficient of the agreement between the estimated and X-ray parameters and the mean square error between the calculated and X-ray structures. The agreements for predicted versus observed structures for globular proteins, using the self-consistent methods (the SELCON programs) or a neural net program (CDNN), are good, giving mean square errors in the agreements for α-helix and β-structures of approximately 5%.[11] However, a

[29] N. Sreerama and R. W. Woody, *Anal. Biochem.* **287,** 252 (2000).
[30] N. Sreerama and R. W. Woody, *Biochemistry* **33,** 10022 (1994).
[31] N. Sreerama and R. W. Woody, *J. Mol. Biol.* **242,** 497 (1994).
[32] N. Sreerama and R. W. Woody, *Anal. Biochem.* **209,** 32 (1993).
[33] N. Sreerama, S. Y. Venyaminov, and R. W. Woody, *Anal. Biochem.* **299,** 271 (2001).
[34] G. Bohm, R. Muhr, and R. Jaenicke, *Protein Eng.* **5,** 191 (1992).
[35] M. A. Andrade, P. Chacon, J. J. Merelo, and F. Moran, *Protein Eng.* **6,** 383 (1993).

good correlation for any given set of proteins does not assure that the method will work with all proteins. In addition, obtaining a good agreement between the observed spectrum and calculated spectrum by any of the fitting schemes does not necessarily mean that the structural estimate is accurate. This is especially true if one tries to analyze the conformation of small peptides by methods that utilize protein databases as references.

Recommended Methods

All the computational methods described above are suitable for determining whether mutations or the binding of ligands changes the conformation of a protein. They are also useful for determining whether recombinant proteins are folded and whether proteins with similar sequences have the same conformation. However, because there are so many methods to estimate secondary structure from CD data it often becomes difficult to decide which method to use. The "best" method of analyzing CD spectra depends on the quality of the raw data and the desired information. The following are recommended.

1. For estimating the conformation of globular proteins in solution: SELCON[29,36,37] and CDNN[34]

2. For determining the conformation of model polypeptides: constrained least-squares analysis,[19,20] for example, the program LINCOMB[38] and ridge regression, for example, CONTIN[28]

3. For estimating the conformation of a protein when the precise concentration is not known, or for evaluating of small conformational changes on protein–protein or protein–ligand interactions: nonconstrained least-squares analysis,[20] for example, the program MLR[11]

4. For quantifying changes in secondary structure as a function of varying conditions: constrained least-squares fits with a fixed set of reference spectra

Determination of Thermodynamics of Protein Folding/Unfolding from CD Data

Circular dichroism can be used to determine the enthalpy, entropy and midpoints, and T_m values of unfolding/refolding transitions of a protein if they are reversible as a function of temperature or denaturant.

[36] N. Sreerama, S. Y. Venyaminov, and R. W. Woody, *Anal. Biochem.* **287,** 243 (2000).
[37] N. Sreerama, S. Y. Venyaminov, and R. W. Woody, *Protein Sci.* **8,** 370 (1999).
[38] A. Perczel, K. Park, and G. D. Fasman, *Anal. Biochem.* **203,** 83 (1992).

Definition of Terms

Several thermodynamic parameters are often calculated from CD data and they are briefly described below.

Heat Capacity, C_p. If two bodies have masses m_1 and m_2 and initial temperatures T_1 and T_2, where $T_2 > T_1$ and they are in contact, they will eventually reach the same temperature, T_{final}:

$$m_2 c_2 (T_2 - T_{\text{final}}) = m_1 c_1 (T_{\text{final}} - T_1) = q \tag{2}$$

c_1 and c_2 are constants called the specific heat capacities of the bodies at T_1 and T_2, respectively, and q is the heat of the system. The specific heat capacity of water is 1.00 cal/g at 15° and 1 atm of pressure. The heat capacity of everything else can be defined in terms of water. (One calorie is the quantity of heat needed to raise the temperature of 1 g of water 1° between 14.5 and 15.5°.) The heat capacity at constant pressure is C_p:

$$C_p = \delta q_{\text{pr}} / \delta T \tag{3}$$

where δq_{pr} is the heat added to the system at constant pressure and δT is the change in temperature.

Internal Energy, U. When a suspended mass is above the ground it has potential energy, PE. When it falls the potential energy is converted to kinetic energy, KE. When it hits the ground it becomes hot, because its kinetic energy is converted to internal energy. Internal energy, U, consists of rotational, vibrational, and electronic energies. The total energy of a body = KE + PE + U.

Enthalpy, H. The enthalpy of a system is defined as

$$H = U + PV \tag{4}$$

H has units of energy (calories or joules per mole), P is pressure, and V is volume. ΔH is the change in enthalpy, which equals the heat absorbed at constant pressure between two temperatures:

$$\Delta H = q_p = S_{T_1}^{T_2} C_p(T) \delta T \tag{5}$$

Entropy, S. The entropy of a body is its heat divided by the ambient temperature. For a reversible change of state at constant pressure:

$$\Delta S = \Delta H / T \tag{6}$$

The entropy of a system is at its maximum at equilibrium.

Gibbs Free Energy, G.

$$G = H - TS \tag{7}$$

When a body undergoes a change of state:

$$\Delta G = \Delta H - T\Delta S \tag{8}$$

At chemical equilibrium:

$$\Delta G = -nRT \ln(K) \tag{9}$$

where K is the equilibrium constant and R is the gas constant $= 1.987$ kcal/mol, and n is the number of moles.

$$\ln(K) = -\Delta G/nRT \quad \text{and} \quad K = \exp(-\Delta G/nRT) \tag{10}$$

The equilibrium constant is a function of temperature:

$$\ln K = -\Delta G/RT \tag{11}$$

Differentiating with respect to temperature:

$$\delta \ln (K)/\delta T = \Delta G/RT^2 - 1[\delta(\Delta G)/\delta T]/RT \tag{12}$$

Making the substitutions $\delta G/\delta T = \Delta S$ and $\Delta G = \Delta H - T\Delta S$ Eq. (12) becomes

$$\delta \ln (K)/\delta T = \Delta H/RT^2 \tag{13}$$

At two different temperatures, T_1 and T_2:

$$\ln (K_2/K_1) = \Delta H(1/T_1 - 1/T_2)/R \tag{14}$$

Equation (14) was first derived by van't Hoff (who received the first Nobel Prize for Chemistry in 1901 for discoveries in reaction kinetics and osmotic pressure of solutions) and is known as the van't Hoff equation.

Determination of Thermodynamics of Unfolding/Refolding of Proteins as Function of Temperature

The Gibbs–Helmholtz equation can be used to describe the change in energy when a protein goes from a native to a denatured state. Equation (15) essentially combines all the equations above. ΔC_p for a transition is usually taken to be independent of temperature, but ΔH and ΔS may vary as a function of temperature.

$$\Delta G° = \Delta H°(1 - T/T_M) - \Delta C_p[T_M - T + T \ln(T/T_M)] \tag{15}$$

The unfolded and folded states of proteins usually have different heat capacities, because of the differences in exposure of the hydrophobic groups of the protein to the solvent.[39] Since it can be difficult to estimate heat capacity changes (see below), the thermodynamic parameters are sometimes calculated making the assumption that $\Delta C_p = 0$. This simplification can lead

---

[39] W. J. Becktel and J. A. Schellman, *Biopolymers* **26**, 1859 (1987).

to an underestimate of the enthalpy value and an overestimate of the T_M at $K = 1$, but the values calculated can still be useful for comparative purposes and for estimating binding constants (see below).

Fitting Folding/Unfolding Curves Setting $\Delta C_p = 0$. If it is assumed that the denatured and native states have the same heat capacity, the Gibbs–Helmholtz equation reduces to

$$\Delta G° = \Delta H°(1 - T/T_M) \quad \text{or} \quad \Delta G° = \Delta H°[(T_M - T)/(TT_M)] \quad (16)$$

where $\Delta H°$ is the enthalpy change for the transition at the transition temperature, T_M.

For a protein that undergoes a two-state transitions between a folded and unfolded form, at low temperature it is in state 1, folded or f, and at high temperature it is in state 2, unfolded or u. At any given temperature the amount of protein in state 1 or state 2 can be described by an equilibrium constant, k_F:

$$k_F = f/u \quad (17)$$

If one looks at a protein with a high α-helical or β-sheet content, the ellipticity at 222 and 208 nm (the maxima for an α helix) or at 218 nm (the maximum for β sheet) may change as a function of temperature. The change in ellipticity as a function of temperature will be proportional to the change in concentration of u and f. The ellipticity, θ, as a function of temperature, T (degrees Kelvin), therefore can be fitted in the Gibbs–Helmholtz equation to determine ΔH, the enthalpy associated with the change of state at the T_M. The folding constant, k_F, is a function of the free energy of folding:

$$k_F = \exp(-\Delta G/RT) = \exp\{(\Delta H/RT)[(T/T_M) - 1]\} \quad (18)$$

The fraction folded, α, is the folded protein concentration divided by the total protein concentration. The folding constant $k_F = \alpha P_t/(1 - \alpha)P_t$, where P_t is the total protein concentration. Solving for the fraction folded, α, in terms of the folding constant k_F:

$$\propto = k_F/(1 + k_F) \quad (19)$$

Since CD is a quantitative technique, the total ellipticity, θ, is proportional to the sum of the concentration of unfolded protein times its extinction coefficient plus the folded protein times its extinction coefficient:

$$[\theta] = [(\varepsilon_F - \varepsilon_U)\alpha] + \varepsilon_U \quad (20)$$

Here $[\theta]$ is the ellipticity at any temperature, ε_F is the ellipticity when the protein is fully folded, and ε_U is the ellipticity when the protein is totally unfolded.

To determine the thermodynamic constants from CD data, it is necessary to determine the enthalpy and T_M values that give the best agreement between the experimentally determined ellipticity, $[\theta]$, as a function of temperature, and the Gibbs–Helmholtz equation. Since the equation is nonlinear it must be solved by nonlinear curve fitting techniques such as the Levenberg–Marquardt algorithm[40] to find the parameters that best fit the equation. Nonlinear curve fitting has been reviewed,[41] so the methods and the analysis of errors in the fits are not discussed in detail here, with the caveat that the errors determined by curve fitting are not a good indication of the actual errors in the determinations of the constants, which should be estimated by doing replicate experiments. The Levenberg–Marquardt algorithm, and other similar methods, has been implemented in many commercial graphics and curve fitting programs and is readily available (see Appendix I).

To use the curve-fitting algorithms one inputs initial values for T_M, ΔH, ε_1, and ε_2. The method will change the parameters to minimize the difference between the observed and calculated ellipticities as a function of temperature. ΔS can be evaluated at the T_M, where $K = 1$ and therefore $\Delta G = 0$, using Eq. (21):

$$\Delta S = \Delta H / T_M \tag{21}$$

For the folding/unfolding of a dimer the equations are

$$k_F = \exp\{[\Delta H/(RT)](T/T_M - 1) - \ln(P_t)\} \tag{22}$$

$$q = (4P_t k_F + 1) \tag{23}$$

$$e = (8P_t k_F + 1)^{1/2} \tag{24}$$

$$d = 4P_t k_F \tag{25}$$

$$\alpha = (q - e)/d \tag{26}$$

$$[\theta] = [(\varepsilon_F - \varepsilon_U)\alpha] + \varepsilon_U \tag{27}$$

Here α is the fraction folded and P_t is the total protein concentration. T_M is the temperature at which $\alpha = 0.5$ and it changes as a function of the protein concentration.

The thermodynamic parameters of folding of a trimer have been described by Engel.[42] In the case of a trimer–monomer equilibrium:

[40] D. W. Marquardt, *J. Soc. Indust. Appl. Math.* **11**, 431 (1963).
[41] M. L. Johnson and S. G. Frasier, *Methods Enzymol.* **117**, 301 (1985).

$$K = f/u^3 \tag{28}$$

$$K = \exp\{(\Delta H/RT)(T/T_M - 1) - \ln[(3/4)[P_t]^2]\} \tag{29}$$

$$d = \{(9K[P_t]^2 + 1)/(3K[P_t]^2)\} - 3 \tag{30}$$

$$A = \left\{[(-1)(d/2)] + [(d^2/4) + (d^3/27)]^{1/2}\right\}^{1/3} \tag{31}$$

$$B = (-1)\{(d/2) + [(d^2/4) + (d^3/27)]^{1/2}\}^{1/3} \tag{32}$$

$$\alpha = A + B + 1 \tag{33}$$

$$[\theta] = (\varepsilon_1 - \varepsilon_2)\alpha + \varepsilon_2 \tag{34}$$

Here T_M is also the apparent T_M observed at $\alpha = 0.5$ and depends on the protein concentration. Similar equations can be used to fit the folding of heterotrimers[43] and tetramers.[44,45]

Breslauer[46] has shown that if one plots $1/T_{M \text{ (observed)}}$ versus $\ln C_{\text{total}}$:

$$m = (n-1)R/\Delta H \tag{35}$$

$$B = [\Delta S - (n-1)R\ln(2) + R\ln(n)]/\Delta H \tag{36}$$

where n is the number of chains in the molecule, m is the slope, and B is intercept. Equations (35) and (36) can be used to confirm the number of chains in a folded peptide or protein. The enthalpies determined utilizing CD data are model dependent, as the calculated enthalpies depend on the number of chains in the fully folded species. However, the values of n determined from the slope and intercept will not be self-consistent if the wrong folding equations are fitted to yield the enthalpies and entropies of folding.

Fitting Folding/Unfolding Curves Assuming $\Delta C_p \neq 0$. Theoretically one can fit CD data obtained as a function of temperature to Eq. (37):

$$k_F = \exp\{\Delta H^\circ(1 - T/T_M) - \Delta C_p[T_M - T + T\ln(T/T_M)]\} \tag{37}$$

and include the ΔC_p term as one of the constants to evaluate. For a monomeric protein, however, there is not enough information in the CD data

[42] G. Engel, *Anal. Biochem.* **61**, 184 (1974).
[43] N. J. Greenfield and V. M. Fowler, *Biophys. J.* **82**, 2580 (2002).
[44] C. R. Johnson, P. E. Morin, C. H. Arrowsmith, and E. Freire, *Biochemistry* **34**, 5309 (1995).
[45] R. Fairman *et al.*, *Protein Sci.* **4**, 1457 (1995).
[46] K. J. Breslauer, *Methods Enzymol.* **259**, 221 (1995).

obtained under a single condition to determine the heat capacity with any accuracy. Indeed, trying to include the ΔC_p term can worsen the agreement between the T_M values determined by CD and scanning calorimetry.[47] To determine ΔC_p, unfolding measurements may be obtained under varying conditions of pH or denaturants. If the structure of the fully folded protein does not change as a function of conditions, but the observed T_M of folding changes,[48] it is possible to determine the heat capacity by plotting the apparent enthalpy determined at several pH or denaturant values as a function of T_M in degrees K. The slope is equal to the heat capacity. If one has determined the ΔC_p value by other methods, such as scanning calorimetry, it is also possible to include the ΔC_p term as a constant when analyzing CD data.

If one has a monomer-to-dimer transition on folding, one can fit the mean residue ellipticity as both a function of temperature and protein concentration simultaneously to determine the ΔC_p directly.[49] Here one uses Eqs. (38)–(44).

$$\Delta G = \Delta H + \{\Delta C_p(T - T_M) - T((\Delta H/T_M) + (\Delta C_p[\ln(T/T_m)]))\} \quad (38)$$

$$k_F = \exp(-\Delta G/RT) \quad (39)$$

$$a = 4k_F P_t^2 \quad (40)$$

$$b = -8k_F P_t^2 - P_t \quad (41)$$

$$c = 4k_F P_t^2 \quad (42)$$

$$\alpha = \left\{(-b) - [(b^2) - (4ac)]^{1/2}\right\}/(2a) \quad (43)$$

$$[\theta] = [(\varepsilon_F - \varepsilon_U)\alpha] + \varepsilon_U \quad (44)$$

The independent variables are the temperature and the concentration. The dependent variable is the ellipticity. One inputs initial estimates of the T_M at $K = 1$, ΔH, and ΔC_p and uses nonlinear least-squares fitting to find the best fits to the folding/unfolding data. (Note that the T_M at $K = 1$ is independent of the protein concentration used to evaluate it.)

Correcting CD Data for Linear Changes in CD of Folded and Unfolded Forms as Function of Temperature. When examining CD data of a polypeptide or protein obtained as a function of temperature while there is only

[47] J. W. Taylor, N. J. Greenfield, B. Wu, and P. L. Privalov, *J. Mol. Biol.* **291,** 965 (1999).
[48] P. L. Privalov and N. N. Khechinashvili, *J. Mol. Biol.* **86,** 665 (1974).
[49] R. M. Ionescu and M. R. Eftink, *Biochemistry* **36,** 1129 (1997).

one major folding transition, there may be a small, almost linear change in the ellipticity of both the folded and unfolded forms with temperature. It should be emphasized that the small changes in the ellipticity of the folded form as a function of temperature reflect true conformational changes and are not simply due to changes in the optical properties of a helix or β strand as a function of temperature. In the case of folded, mainly helical proteins, there are several possible explanations for the "linear" decrease in ellipticity with temperature, which include possible fraying of the ends of the helix before the major cooperative unfolding transition,[50] changes in helix–helix interactions[51] or changes from a well-packed folded state to a more molten conformation.[52,53] Dragan and Privalov[54] have shown that the almost linear change in the CD of a folded, two-chain α-helical coiled-coil GCN4 leucine zipper analog before the major folding transition is accompanied by significant heat absorption, showing that enthalpic conformational changes are occurring during the linear region of the unfolding curve. The linear changes in the CD of unfolded proteins as a function of temperature probably arise from changes in the population of conformers in the unfolded state as the temperature is raised.[55–57]

Figure 2 illustrates the unfolding of a dimeric coiled coil, AcTM1a-Zip,[43,58] as a function of concentration and temperature. Data were collected at four concentrations, ranging from 1.5×10^{-6} to 1.5×10^{-4} M. The data were fit globally to a two-state transition, and the ellipticities of the folded and unfolded forms were corrected for pre- and posttransition "linear" changes of ellipticity. Fitting the data globally to Eqs. (38)–(44) gave a T_M of folding at $K = 1$ of 74.3°, a ΔH of folding of -55.1 kcal/mol, and a ΔC_p of folding of -384 cal/mole·deg.

Comparison of Thermodynamic Parameters Estimated from CD Data with Those Determined from Scanning Calorimetry. Freire[59] has reviewed thermal denaturation methods to study protein folding. Since all spectroscopic methods of determining folding are indirect, the parameters obtained depend on the model used to fit the data. If a protein undergoes a simple two-state transition, the enthalpy determined from fitting the

[50] M. E. Holtzer and A. Holtzer, *Biopolymers* **32**, 1589 (1992).
[51] T. M. Cooper and R. W. Woody, *Biopolymers* **30**, 657 (1990).
[52] N. J. Greenfield and S. E. Hitchcock-DeGregori, *Protein Sci.* **2**, 1263 (1993).
[53] N. J. Greenfield *et al., J. Mol. Biol.* **312**, 833 (2001).
[54] A. I. Dragan and P. L. Privalov, *J. Mol. Biol.* **321**, 891 (2002).
[55] R. W. Woody and A. Koslowski, *Biophys. Chem.* **101–102**, 535 (2002).
[56] T. P. Creamer and M. N. Campbell, *Adv. Protein Chem.* **62**, 263 (2002).
[57] Z. Shi, R. W. Woody, and N. R. Kallenbach, *Adv. Protein Chem.* **62**, 163 (2002).
[58] N. J. Greenfield, G. T. Montelione, R. S. Farid, and S. E. Hitchcock-DeGregori, *Biochemistry* **37**, 7834 (1998).
[59] E. Freire, *Methods Enzymol.* **259**, 144 (1995).

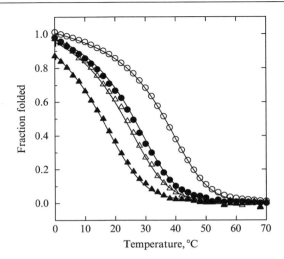

FIG. 2. The unfolding of a chimeric coiled coil containing the first 14 residues of rat striated muscle α-tropomyosin and the last 18 residues of the yeast transcription factor, GCN4, as a function of temperature and concentration followed by the changes in ellipticity at 222 nm: (▲) 1.5 μM; (△) 7.5 μM; (●) 15 μM; (○), 150 μM. The points were fit globally to the Gibb–Helmholtz equation for a two-state transition between a folded dimer and unfolded monomers. The pre- and posttransition ellipticities were corrected for small linear changes as a function of temperature. Redrawn with permission from data in Greenfield et al.[58] Copyright (1998) American Chemical Society.

Gibbs–Helmholtz equation, called the van't Hoff enthalpy, equals the total molar enthalpy, which is the total heat absorbed by a protein when it undergoes a thermal transition determined from scanning calorimetry. In this case, the enthalpy determined from CD data should be equal to the enthalpy determined by scanning calorimetry.

In two examples, Thompson et al.[60] studied a dimeric peptide containing the leucine zipper region of GCN4 and Taylor et al.[47] studied a monomeric peptide with constrained ends using both scanning calorimetry and CD. In both cases there was reasonable agreement between the thermodynamic parameters determined by the two methods. For the coiled-coil dimer, the ΔH and T_M values at $K = 1$ determined by CD were 35.0 ± 1.2 kcal/mol and 94.1 ± 0.2°, respectively, while from scanning calorimetry the values were 34.7 ± 0.3 kcal/mol and 96.9 ± 0.8°, respectively. For the monomeric helix with constrained ends, the respective values from CD measurements were 14.7 ± 0.5 kcal/mol and 30.1 ± 0.9° and from calorimetry the values were 16.4 ± 0.9 kcal/mol and 28.0 ± 1.6°, respectively.

[60] K. S. Thompson, C. R. Vinson, and E. Freire, Biochemistry 32, 5491 (1993).

Thermodynamics of Protein Folding/Unfolding Determined from Changes in CD as Function of Denaturants

When denaturants such as guanidine-HCl or urea are added to a protein it unfolds. It has been shown for many proteins that the free energy of folding of the protein, ΔG, is a linear function of the detergent concentration, [D], and one can fit the change in ellipticity, θ, as a function of detergent to obtain the free energy of folding of a protein in the absence of the denaturant.[39,61–63] The parameters to be fit are ΔG_0, m, ε_F, and ε_U, where ΔG_0 is the free energy of folding in the absence of detergent, m is the constant relating the linear change in free energy to the detergent concentration, and ε_F and ε_U are the ellipticities of the fully folded and unfolded proteins, respectively. One can also include the terms m_1 and m_2, which are the slopes describing any linear changes in ellipticity of the folded and unfolded proteins before and after the folding transition. For a monomer:

$$\Delta G = \Delta G_0 + m[D] \tag{45}$$

$$k_F = \exp(-\Delta G/RT) \tag{46}$$

$$\alpha = k_F/(k_F + 1) \tag{47}$$

$$\theta = \alpha(\varepsilon_F - m_1[D] - \varepsilon_U - m_2[D]) + \varepsilon_U \tag{48}$$

In Eqs. (45)–(48), ΔG is the free energy of folding in the presence of denaturant, k_F is the folding constant, and α is the fraction folded. The difference between the calculated change in ellipticity and the observed change in ellipticity is minimized to give the best values of the fitting parameters, using nonlinear least-squares curve fitting.[40,41] If one knows the values of ε_1, ε_2, m_1, and m_2 they may be input as constants rather than as parameters to be determined. Once ΔG_0 is determined, the folding constant is calculated by the equation $k = \exp(-\Delta G/RT)$.

For a dimeric protein:

$$\alpha = \{2P_t^2 k_F + P_t - [(-2P_t^2 k_F - P_t)^2 - 4P_t^4 k_F^2)]^{1/2}\}/(2P_t^2 k_F) \tag{49}$$

where P_t is the total protein concentration.

Before nonlinear least-squares programs became widely available, the value of k_F at each detergent concentration was evaluated, and the ΔG of folding was plotted as a function of the denaturant concentration. The

[61] J. A. Schellman, *Biopolymers* **15**, 999 (1976).
[62] J. A. Schellman, *Biopolymers* **14**, 999 (1975).
[63] C. N. Pace and T. McGrath, *J. Biol. Chem.* **255**, 3862 (1980).

straight line obtained was extrapolated to zero detergent concentration to determine ΔG_0. If the protein is stable, a high concentration of detergent may be necessary to unfold the protein, and the extrapolation to zero may be prone to large errors. It is better to directly fit the data using non-linear least-squares procedures.[64] Ibarra-Molero and Sanchez-Ruiz[65] have shown for several proteins that better agreement is obtained with the free energy values determined by calorimetry, carrying out the denaturations at several temperatures and not assuming that the effect of the denaturant is linear; there may be higher-order terms.

As well as using denaturants to induced unfolding, salts and osmotic agents can be used to induce folding, and the same treatments of the circular dichroism data can be used to determine the free energy of the unfolded state. Figure 3 illustrates that the osmolyte trimethylamine N-oxide (TMAO) can induce a protein subunit of ribonuclease P from *Bacillus subtilis,* P protein, to fold.[66] Both the direct data and linear transformations of the data to obtain the free energy of folding in the absence of TMAO are shown.

Multiple Transitions

Some proteins do not unfold in a single transition, but instead appear to unfold in segments. For example, the tropomyosins are two-chain coiled-coil proteins that exhibit multiple unfolding transitions both using CD and calorimetry. The unfolding of these proteins can be fit to a model that assumes that the overall folding of the protein is represented by the sum of several independent transitions. The equations used to fit the data are a linear sum of the individual folding transitions.

Folding Intermediates

The folding/unfolding of proteins is often not a two-state transition between a folded and unfolded state and there may be several intermediate states. Often these intermediate states may be observable by CD. If CD spectra are collected as a function of temperature or denaturant intermediate states may be readily apparent if the spectra do not exhibit an isodichroic (also called isosbestic) point at which the CD does not change, showing that at least three basis component spectra are contributing to the spectrum obtained under each condition. However, the lack of an isodichroic point does not mean that only two spectra are contributing to

[64] M. M. Santoro and D. W. Bolen, *Biochemistry* **27,** 8063 (1988).

[65] B. Ibarra-Molero and J. M. Sanchez-Ruiz, *Biochemistry* **35,** 14689 (1996).

[66] C. H. Henkels, J. C. Kurz, C. A. Fierke, and T. G. Oas, *Biochemistry* **40,** 2777 (2001).

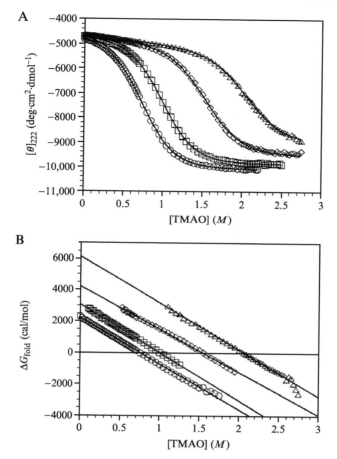

Fig. 3. The effect of an osmolyte, TMAO, on the folding of an RNase T_1 subunit, P protein. (A) CD signal at 222 nm of the P protein (10 μM) in 10 mM Tris-cacodylate (pH 8) monitored at four different temperature: 28° (O), 37° (□), 50° (◇), and 60° (△). Solid lines represent the nonlinear least-squares fit to the data, using a linear extrapolation model assuming two-state folding [Eqs. (45)–(49)]. (B) The same data converted to a free energy of folding-versus-TMAO concentration profile. Reprinted with permission from Henkles *et al.*[66] Copyright (2001) American Chemical Society.

the set of spectra, as more than two basis spectra may have a point in common.

There are several algorithms that allow a set of spectra, obtained under different conditions, to be deconvoluted into a smaller number of basis curves. These spectra can then be used to evaluate the fraction of curve contributing to each spectrum. Note that the basis sets that are obtained do not necessarily reflect the spectrum of a pure conformational state, since many conformational states of a protein may have secondary structural elements in common. Algorithms for deconvoluting sets of CD spectra into basis curves include singular value decomposition (SVD), reviewed by Hendler and Shrager,[67] the convex constraint algorithm (CCA) of Perczel et al.,[38] matrix rank analysis,[68] and multivariant curve analysis.[69,70] These deconvolution methods are described briefly below:

Singular Value Decomposition. Konno[71] has used SVD to determine spectra and quantify conformational intermediates. In this method one starts with an $M \times N$ data matrix, A, which contains a far-UV CD spectrum in each of N columns. M is the number of data points in a single spectrum. In the case $M < N$, M is increased to N by filling zero to the $(N - M)$ additional rows. Singular value decomposition is then used to determine the principal component spectra. These components, do not, however, correspond to the spectra of any physical conformers. Further matrix manipulations are used to determine the spectra corresponding to each conformer and the contribution of each conformer to each of the original spectra. The method is described in detail in the original article and is applied to determining intermediates in the folding of acid denatured cytochrome *c*.

Convex Constraint Algorithm. Perczel et al.[27,38,72] developed an algorithm called convex constraint analysis (CCA), which like SVD deconvolutes a database of spectra into components, but has different criteria for defining the basis curves. In CCA, the sum of the fractional weights of each component spectrum is constrained to be equal to 1. In addition, a constraint called volume minimization is defined, which allows a finite number of component curves to be extracted from a set of spectra. This method directly finds the set of basis curves and the weights of each curve contributing to the original set and the standard deviation between the original set of data and reconstructed data, using the basis curves and weights, and is simpler to use than the SVD methods. The algorithm does not correct for possible noise, and does not *a priori* determine the minimum number of basis curves needed to fit the data. The authors suggest using an increasing number of curves to sequentially fit the data and stopping when there is no further decrease in the standard deviation between the original and reconstructed data sets. They also show that if one tries to obtain too many

[67] R. W. Hendler and R. I. Shrager, *J. Biochem. Biophys. Methods* **28,** 1 (1994).
[68] G. Peintler, I. Nagypál, A. Jancsó, I. R. Epstein, and J. Kustin, *J. Phys. Chem. A* **101,** 8013 (1997).
[69] J. Mendieta, M. S. Diaz-Cruz, M. Esteban, and R. Tauler, *Biophys. J.* **74,** 2876 (1998).
[70] S. Navea, A. de Juan, and R. Tauler, *Anal. Chem.* **74,** 6031 (2002).
[71] T. Konno, *Protein Sci.* **7,** 975 (1998).
[72] A. Perczel, K. Park, and G. D. Fasman, *Proteins* **13,** 57 (1992).

curves to reconstruct the data set, some of the basis spectra will be almost identical. The convex constraint algorithm can distinguish a chiral component only if it represents at least 50% of one of the curves in the data set and the authors suggest including reference spectra with known conformations to resolve the curves fully into components when using the algorithm.

The convex constraint algorithm has been used to show that there are folding intermediates when some highly stable coiled coils are subjected to thermal denaturation.[52,53] Such intermediates in the folding of a stable coiled coil have been confirmed by calorimetry.[54] The use of the algorithm to deconvolute folding curves has been criticized, however, because if one is not careful one can find more spectra than actually contribute to data sets of proteins.[73] Indeed, any spectrum with outlying noise will be resolved by the convex constraint algorithm as a unique basis set. As with any other analytical technique it is necessary to show that the model makes sense in light of all the experimental data.

Figure 4 illustrates the deconvolution of two sets of CD data obtained for the unfolding of two chimeric peptides containing the N termini of two isoforms of tropomyosin. Both peptides are two-stranded coiled coils. The first, AcTM1aZip,[58] unfolds cooperatively in a two-state transition (Fig. 2).

When sets of spectra (Fig. 4A) are deconvoluted into three curves by convex constraint analysis, two of them are almost identical (Fig. 4B). The changes in the fractions of each basis curve as a function of temperature are shown in Fig. 4C. Note that in Fig. 4 the fractions of the two spectra corresponding to the unfolded peptide are added. Although the second peptide, GlyTM1bZip, is also a two-stranded coiled coil, it does not unfold in a single step.[53] NMR peaks are broadened well before there is loss of helical content as measured by circular dichroism. Deconvolution of a set of CD spectra of GlyTM1bZip (Fig. 4D) resolves a folding intermediate (Fig. 4E). The fraction of each basis set as a function of temperature is shown in Fig. 4F. Note that the basis spectra resolved by the CCA algorithm do not necessarily reflect pure conformational states, as the fully unfolded AcTM1aZip peptide still appears to have some helical character, as typical of many unfolded proteins that are helical in their native states.

Matrix Rank Analysis. Matrix rank analysis (MRA), long used to determine the number of independent absorbing species in chemically reacting systems or in equilibrium systems, has been applied to deconvolute sets of CD spectra.[74] A newer algorithm has been developed to detect and remove erroneous rows and/or columns from the matrixes and to monitor the most significant experimental information along the rows and/or columns of the

[73] M. E. Holtzer and A. Holtzer, *Biopolymers* **36**, 365 (1995).
[74] M. F. Brown and T. Schleich, *Biochim. Biophys. Acta* **485**, 37 (1977).

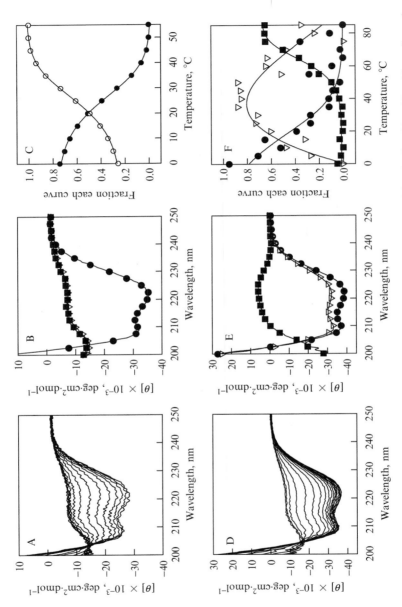

FIG. 4. (A and D) Spectra of two chimeric peptides, AcTM1aZip and GLyTM1bZip, containing the first 14 or 19 residues of muscle and nonmuscle rat α-tropomyosins encoded by exons 1a and 1b, respectively, and the last 18 residues of the yeast transcription factor GCN4 as a function of temperature. (B and E) Deconvolution of the spectra in (A) and (D) into three basis sets, using the convex constraint algorithm.[27,72]

data sets as well as to determine the number of absorbing species.[68] This method has been applied to study the CD of peptide–copper complexes.[75]

Multivariant Curve Analysis. In this technique the number of component spectra contributing to a set of data is initially estimated by SVD and the changes and structure of the experimental data matrix are then analyzed by factor analysis. This approach gives an estimate of the regions where the concentration of the different components is changing or evolving and it also provides an initial estimation of how these concentration profiles change along the experiment. The method is based on evaluation of the magnitude of the singular values (or of the eigenvalues) associated with all the submatrices of a matrix built up by adding successively, one by one, all the rows of the original data matrix. The calculations are performed in two directions: forward (in the same direction of the experiment), starting with the two first spectra, and backward (in the opposite direction of the experiment), starting with the last two spectra. In the forward direction, the detection of a new component is detected by the upsurging of a new singular value; in the backward direction, the disappearance of a component is detected by the upsurging of a new singular value. Singular values related with significant components become larger and clearly distinguished from the singular values associated with noise, in their graphical representation. The method is discussed in detail in several articles.[69,70]

Fitting Thermal Folding/Unfolding of Short Peptides Using Statistical Models

Many proteins unfold in cooperative "two-state" at transitions between the folded and unfolded forms. However, the folding of isolated α-helical peptides, which do not have interhelical interactions, cannot be fit by simple two-state models. Two general theories to explain the unfolding of helical peptides were proposed by Zimm and Bragg[76] and Lifson and Roig.[77] Their treatments have been reviewed by Doig.[78]

The data of the exon 1a peptide can be fit by only two basis sets and two of the sets are therefore almost identical, whereas the exon 1b peptide requires a minimum of three basis sets to reconstruct the data. (C and F). Change in each basis set as a function of temperature. For AcTM1aZip the weight corresponding to the two almost identical spectra in (B) (■, ∇) are added to give the composite curve (○) in (C). Redrawn with permission from data of Greenfield *et al.*[58] [copyright (1998) American Chemical Society] and Greenfield *et al.*[79] [copyright (2001) Elsevier].

[75] B. Gyurcsik, I. Vosekalna, and E. Larsen, *J. Inorg. Biochem.* **85,** 89 (2001).
[76] B. H. Zimm and J. K. Bragg, *J. Chem. Phys.* **31,** 526 (1959).
[77] S. Lifson and A. Roig, *J. Chem. Phys.* **34,** 1963 (1961).
[78] A. J. Doig, *Biophys. Chem.* **101–102,** 281 (2002).
[79] N. J. Greenfield *et al., J. Mol. Biol.* **312,** 833 (2001).

Zimm–Bragg Equation. For long proteins, the predominant interactions determining folding are usually short- or long-range hydrophobic contacts. For α-helical synthetic polypeptides, as opposed to small native proteins, the change in state as a function of temperature often cannot be fit by a simple two-state equilibrium. For long polypeptides the transition going from an α helix to a random coil is sharp, but for short peptides the transition is broad. Zimm and Bragg[76] in 1959 suggested that the difference in sharpness is a consequence of the following: the formation of the first turn of a helix is difficult because it involves a large reduction in entropy or disorder. Once formed, however, this turn acts as a nucleus to which further turns can be added by hydrogen bonding. Thus helix formation could be described by a process of nucleation followed by propagation.

The Zimm–Bragg model of folding assumes that a residue in a helix can be described by whether its oxygen atom is bonded to the hydrogen atom of the third preceding residue. In their model of helix folding, the state of a chain of three segments can be described by a sequence of $n - 3$ symbols, each of which can have one value, either 1 or 0, with a nonbonded oxygen having the symbol 0 and a bonded oxygen having the symbol 1. The first three segments (e.g., amides in a polypeptide) are not bonded because there are no H^N protons to form hydrogen bonds with the backbone carbonyls. The statistical weight of a given state of the chain is assumed to be the product of the following factors.

1. The quantity unity for every 0 that appears (unbonded segment)
2. The quantity s for every 1 that follows a 1 (bonded segment)
3. The quantity σs for every 1 that follows μ or more 0's (boundary between bonded and unbonded regions)
4. The quantity 0 for every 1 that follows a number of 0's less than μ

Sequences less than μ zeros do not appear. For the α helix, μ is considered to be about 3. Sigma, σ, is the initiation factor or the difficulty of forming the first hydrogen bond. The propagation factor is s and is a function of the helix propensity of a given amino acid. Zimm and Bragg calculated that σ usually would be 10^{-2} or less because it is difficult to initiate a helix because of the unfavorable entropy change.

Zimm and Bragg defined the partition function Q as the number of ways of arranging a given number of 0's and 1's in a chain, assuming that the chain always began with three 0's. Using this model they showed that $\delta \ln Q / \delta \ln s$ is the average number of hydrogen bonds formed in a chain at a given value of s. θ is defined as the fraction of possible hydrogen bonds formed, not the ellipticity.

$$\theta = [1/(n - 3)](\delta \ln Q / \delta \ln s) \tag{50}$$

For small peptides that contain only one nucleating helical segment, the following equations can be used to determine ΔH and s: s, the propagation factor (sometimes called the helical propensity), varies as a function of temperature and $\delta \ln s / \delta T = \Delta H / RT^2$. Thus, it is possible to fit θ as a function of temperature to derive ΔH, σ, and s at any given temperature from the change in θ as a function of temperature, using Eqs. (51) and (52).

$$s = \exp\{(\Delta H/1.987)[(T - T_M)/(TT_M)]\} \tag{51}$$

$$\theta = \{(n-3)(s-1) - 2 + [(n-3)(s-1) + 2s](s^{-n+2})\}/$$
$$\{(n-3)(s-1)[1 + ((s-1)^2(s^{n-1})/\alpha) - ((n-3)(s-1) + s)(s^{-n+2})]\} \tag{52}$$

s, σ, and ΔH are assumed to be independent of temperature.

The constant, n, is the length of the peptide chain. If the peptide folds cooperatively, σ will be low. If the peptide folds completely uncooperatively, σ will be equal to 1. When σ is unity, there is no interaction between the states of successive segments. The fraction of hydrogen bonds (i.e., α helix) shows a gradual rise with increasing s according to the formula

$$Q = s/(1+s) \tag{53}$$

According to the Zimm–Bragg treatment there is a critical value of the size n at which substantial helix formation appears. This value is approximately that at which

$$(s-1)^2 s^{-n+1} = \sigma \tag{54}$$

In the Zimm–Bragg model of helix formation, θ, the partition function is related to k_F by the following formula:

$$\alpha/(1-\alpha) = k_F = [(n-3)\theta + 3]/(n-3)(1-\theta) \tag{55}$$

and α, the fraction folded, is $k_F/(1 + k_F)$ as in Eq. (19) and $[\theta]$, the mean residue ellipticity, is $(\varepsilon_F - \varepsilon_U)\alpha + \varepsilon_U$ as in the case of fitting the Gibbs–Helmholtz equation [Eq. (20)]. The difference between α, the fraction folded from CD, and θ, the partition function, is minimal for long peptides but can significantly affect the estimation of s, σ, and ΔH for short peptides.

The parameters σ, s, and ΔH in the Zimm–Bragg equation are dependent on one another, and it is difficult to find unique values for a single polypeptide with a single chain length. If one does a global fit of data obtained with polypeptides with the same repeating sequence of known varying length, it is easier to find a unique value of σ that fits all the data.

Figure 5 shows the ellipticity of short helical polypeptides fit to the Zimm–Bragg model of helix formation.[80]

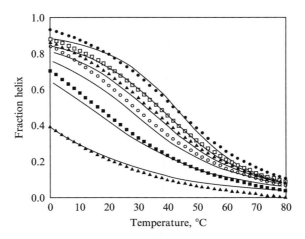

FIG. 5. Comparison of the measured fractional helicity (symbols) with curves calculated using the Zimm–Bragg folding model for short peptides. Curves are shown for peptide chain lengths of 50 (●), 38 (□), 32 (▲), 26 (○), 20 (■), and 14 (▲) residues. Curves were generated using Eqs. (52) and (55) with s at $0° = 1.35, \sigma = 0.0033$, and $\Delta H° = -955$ cal/mol residue. Figure from Scholtz et al.[80] Used with permission [copyright (1991) John Wiley & Sons].

Cantor and Shimmel[81] have solved the Zimm–Bragg equation in terms of the change in the fraction folded as determined by CD, optical rotatory dispersion, difference spectra, and so on. In their treatment, the fractional helicity (partition function) is θ:

$$\theta = [\sigma s/(s-1)^3][ns^{n+2} - (n+2)s^{n-1} + (n+2)s - n]/ \\ n\{1 + [\sigma s/(s-1)^2][s^{n+1} + n - (n+1)s]\} \tag{56}$$

The two treatments of the Zimm–Bragg equation give similar results. The differences in the constants evaluated are within the curve-fitting errors of each other.

As well as using heat to denature peptides, the Zimm–Bragg parameters can be used to fit the unfolding of peptide using denaturants. For example, Scholtz et al.[82] studied the urea unfolding of protein α helices in a homologous series of helical peptides with the repeating sequence Ala-Glu-Ala-Ala-Lys-Ala and chain lengths varying from 14 to 50 residues.

[80] J. M. Scholtz, H. Qian, E. J. York, J. M. Stewart, and R. L. Baldwin, *Biopolymers* **31,** 1463 (1991).

[81] C. R. Cantor and P. R. Shimmel, "Biophysical Chemistry," Part 3: "The Behavior of Biological Macromolecules," p. 1058. W. H. Freeman, 1980.

[82] J. M. Scholtz, D. Barrick, E. J. York, J. M. Stewart, and R. L. Baldwin, *Proc. Natl. Acad. Sci. USA* **92,** 185 (1995).

The dependence of the helix propagation parameter of the Zimm–Bragg model for helix–coil transition on urea molarity was determined at $0°$ with data for the entire set of peptides, and a linear dependence was found for $\ln(s)$ as a function of urea concentration.

For long peptides, with possibly several helical segments, the Zimm–Bragg theory uses matrix methods for determining the partition function. This treatment is discussed in full detail in the original article by Zimm and Bragg[76] and by Cantor and Shimmel.[81]

Lifson–Roig Theory. Lifson–Roig theory[77] is similar to the Zimm–Bragg treatment, but the number of residues is defined as the number of α carbons that are flanked by CONH units on both sides. Lifson–Roig theory predicts that modification of the N-terminal amino group should have no effect on the α-helical content, but Chakrabartty[83] showed experimentally that that was not the case. Doig *et al.*[84] have modified Lifson–Roig theory to take into account N-capping effects. In Lifson–Roig theory, each residue is considered to exist in one of two states: helical, where the ψ and φ angles are near $-57°$ and $-47°$, respectively, or coiled, which is everything else. Lifson–Roig theory gives each residue in a polypeptide chain a statistical weight, depending on the states of the residue on either side. The states are defined as u (coil), v (next to a coil with one or both residues not in a helix), and w (a helix state flanked on both sides by a coil). hhh triplets have a weight of w; hhc, chc, and chh have a weight of v; ccc, cch, hcc, and hch have weights of 1. w has a weight >1. The weight, w, depends on the amino acid composition of the peptide.

A peptide with the sequence ccccchhhhccchhcc has weight *uuuuvwww-vuuvvuu*. N- and C-capping effects can be included by modifying the definition.

Helical states are partitioned this way into v and w weights because a w weight will have helical hydrogen bonds and a v weight will not. Lifson–Roig theory predicts that one needs three consecutive residues with helical ψ and φ angles to form the proper hydrogen bonds. The unfavorable nature of the event of forming a helix shows that the v state must have a weight of less than 1. The c state is the reference and its weight is set to 1. The weight w is greater than 1. The nucleation parameter in Lifson–Roig theory is v^2. Comparing the formalism of Lifson–Roig with Zimm–Bragg gives the equations $s = w/1 + v$ and $\sigma = v^2/(1 + v)^4$.

Rohl *et al.*[85] used the Lifson–Roig model to compare NH exchange and CD for measuring the helix–coil transitions in short polypeptides. In their

[83] A. Chakrabartty, T. Kortemme, and R. L. Baldwin, *Protein Sci.* **3**, 843 (1994).
[84] A. J. Doig, A. Chakrabartty, T. M. Klingler, and R. L. Baldwin, *Biochemistry* **33**, 3396 (1994).
[85] C. A. Rohl, A. Chakrabartty, and R. L. Baldwin, *Protein Sci.* **5**, 2623 (1996).

data analysis they make the assumptions that the ellipticity of the folded helical and unfolded peptides both vary linearly as a function of temperature, although as shown above these changes reflect true conformational changes:

$$[\theta]_{coil} = 2220 - 53T \tag{57}$$

$$[\theta]_{helix} = (-44{,}000 + 250T)(1 - 3/n) \tag{58}$$

where T is the temperature in degrees Celsius, n is the chain length, and $[\theta]$ is the mean residue ellipticity. The mean residue ellipticity is assumed to be a linear function of the helical content.

They postulate that the parameter w depends on temperature with the usual van't Hoff treatment and that

$$\ln(w) = \ln(w_0) - \Delta H/R(1/T - 1/T_0) \tag{59}$$

They use $0°$ (273.15 K) as their reference temperature and w_0 is assumed to be the helical propensity (propagation factor) at $T = 0$.

In the simplest treatment, the equilibrium constant for adding a residue to a helical segment in terms of Lifson–Roig theory is

$$k = w/(1 + v) \tag{60}$$

Therefore the free energy for the change in stability of the helix on adding a residue is

$$\Delta G = -RT \ln[w/(1 - v)] \tag{61}$$

When one fits the unfolding of helical peptides as monitored by changes in ellipticity, w and v are completely dependent on one another. The value one determines for w_0 depends on the value of v. Usually v is chosen as 0.036 and is assumed not to change with amino acid composition or buffer. Relative changes in stability are calculated by assuming that $\Delta\Delta G = -RT \ln(w_{guest}/w_{host})$.

The Lifson–Roig treatment has been extended to study both N- and C-capping effects; the weights of residues N1, N2, and N3 following the N-cap position; and the effects of solvents on the helical propensity of amino acids in short polypeptides.[78,86–90] In these studies values for various parameters are determined by statistical mechanics algorithms. For example, the program N1N2N3 implements a modified Lifson–Roig helix–coil theory

[86] A. J. Doig and R. L. Baldwin, *Protein Sci.* **4**, 1325 (1995).
[87] W. Shalongo and E. Stellwagen, *Protein Sci.* **4**, 1161 (1995).
[88] B. J. Stapley, C. A. Rohl, and A. J. Doig, *Protein Sci.* **4**, 2383 (1995).
[89] D. A. Cochran and A. J. Doig, *Protein Sci.* **10**, 1305 (2001).
[90] D. A. Cochran, S. Penel, and A. J. Doig, *Protein Sci.* **10**, 463 (2001).

to calculate the helix content of a given peptide sequence. The inputs to the program are the peptide sequences and corresponding experimental helix contents; a library of w, N-cap, n_1, n_2, n_3, c_1, and C-cap values for each amino acid; an N-cap value for acetyl and a C-cap value for amide; and a list of parameters to be determined. All n_2, n_3, and c values are initially set to 1. Parameters are then varied until they converge on essentially unchanging values for a set of peptide sequences and ellipticities.[90]

Determination of Binding Constants from CD Data

There are two different techniques that can be used to determine binding affinities from CD data. The same methods can be used to analyze protein–protein, protein–DNA, and protein–small molecule interactions. First, a protein can be directly titrated with a ligand at a fixed temperature if the binding causes a change in the CD spectrum of the protein–ligand complex compared with the unmixed components. These changes could arise from conformational changes caused by ligand binding, which would affect the CD of the backbone amides in the ultraviolet region, or could arise from changes in the environment of the aromatic or prosthetic groups of the protein or the ligand on binding. The change in CD on binding is directly proportional to the amount of complex formed, and thus the binding constant can be directly determined from the CD changes. However, even if there are no changes in the spectrum of a protein on complex formation, if the interactions increase its stability one can often estimate the binding constant. The binding constant can then be determined using the relationship $k_A = \exp(-\Delta\Delta G/RT)$, where $\Delta\Delta G$ is the difference in free energy of folding of the mixture versus the unmixed components, R is the gas constant, T is the temperature, and k_A is the association constant.

Direct Isothermal Titrations: Equivalent Binding Sites

When a ligand binds to a protein it can induce a change in ellipticity relative to the unmixed components. The ligand can be another protein, a small molecule, or a macromolecule such as DNA. For example, assume a ligand with concentration $[L_i]$ binds to a protein with a concentration $[P_t]$, with n equivalent binding sites, where $[LP_i]$ is the concentration of protein–ligand complex formed. The association constant, K, obeys Eq. (62):

$$K = ([LP_i])/([L_i] - n[LP_i])([P_0] - [LP_i]) \qquad (62)$$

The change in ellipticity, $[\theta]$, due to complex formation is directly proportional to $[LP_i]$. When all of the protein has bound ligand, $[\theta] = \theta_{max}$:

$$[LP_i] = ([P_t])([\theta]/\theta_{max}) \qquad (63)$$

$$[q] = \theta_{max}\left[\left(\frac{(1 + K[L_i]/n + K[P_t])}{2K[P_t]}\right)\right) - \left(\left(\frac{([1 + K[L_i]/n + K[P_t])}{2K[P_t]^2}\right)^2 - \left(\frac{[L_i]}{n[P_t]}\right)\right)^{1/2}\right]$$ (64)

This equation was derived by Engel,[91] who originally developed the equation for determining binding constants of enzyme–ligand complexes from fluorescence titrations, as a special case of the Scatchard equation.[92] By fitting Eq. (64) to the raw data, the binding constant of the ligand for the protein and the number of binding sites can be estimated.

Before the availability of high-speed computers it was usual to try to linearize the equation to find the binding constants (reviewed in Greenfield[5]). Now the equations can be solved by the nonlinear least-squares procedures described above.

The change in CD as a function of ligand concentration has been used to study numerous systems (see reviews in Refs. 17 and 93–95). Figure 6 illustrates a study in which changes in the CD spectra of bovine factor V_a on ligand binding were used to measure the binding constants for a soluble C_6 phosphatidylserine (C_6PS).[96]

Direct Isothermal Titrations: Multiple Binding Sites

In many cases proteins, polypeptides, or DNA do not have independent ligand-binding sites. This is especially true where there may be closely spaced binding sites and the binding of a ligand to one site inhibits (negative cooperativity) or facilitates (positive cooperativity) the binding of a second ligand. Two models to describe cooperative or negative cooperative binding are the Hill equation[97] and the McGhee–von Hippel equation[98] and CD data can be fit to either model to estimate binding constants and cooperativity factors.

Hill Equation. The Hill equation was originally developed to describe the binding of oxygen to hemoglobin, which shows positive cooperativity.

[91] G. Engel, *Anal. Biochem.* **61**, 184 (1974).
[92] G. Scatchard, *Ann. N. Y. Acad. Sci.* **51**, 660 (1949).
[93] N. J. Greenfield, *Biochim. Biophys. Acta* **403**, 32 (1975).
[94] D. S. Linthicum, S. Y. Tetin, J. M. Anchin, and T. R. Ioerger, *Comb. Chem. High Throughput Screen* **4**, 439 (2001).
[95] S. M. Kelly and N. C. Price, *Curr. Protein Pept. Sci.* **1**, 349 (2000).
[96] X. Zhai, A. Srivastava, D. C. Drummond, D. Daleke, and B. R. Lentz, *Biochemistry* **41**, 5675 (2002).
[97] A. V. Hill, *J. Physiol. (Lond.)* **40**, iv (1910).
[98] J. D. McGhee and P. H. von Hippel, *J. Mol. Biol.* **86**, 469 (1974).

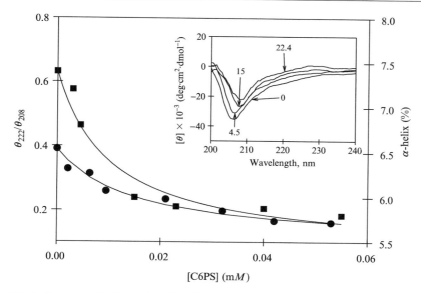

FIG. 6. Changes in the the molar ellipticity ratio ($\theta_{222}/\theta_{208}$) (left-hand scale) and α helicity (right-hand scale) were used to measure the binding constants of bovine factor V_a for a soluble C_6 phosphatidylserine (C_6PS). The data were fit to a single binding-site model, to yield apparent binding constants of 9.3 and 10.2 μM, respectively. *Inset:* CD spectra of factor V_a (1.3 μM) in the presence of 0, 4.5, 15.0, and 22.4 μM C_6PS and 1 mM Ca^{2+}. Redrawn with permission from Zhai *et al.*[96] Copyright (2002) American Chemical Society.

When using the Hill equation, the change in ellipticity as a function of free ligand concentration is fit to Eq. (65):

$$\theta_i = (\varepsilon k_a^H [L_i]^H)/(1 + k_a^H [L_i]^H) \tag{65}$$

where θ_i is the change in ellipticity observed when a ligand binds to a protein, ε is the extinction coefficient for the spectral change, $[L_i]$ is the free ligand concentration, and H is the Hill coefficient, a measure of cooperativity. The values of ε, k_a, and H that best fit the equation are estimated by nonlinear least-squares curve fitting. Equation (65) has often been used to determine the binding constants of magnesium and calcium to calcium-binding proteins such as calmodulin[99–102] and troponin.[103–107] Binding of the ligands causes an increase in helical content.

[99] R. Kobayashi, I. C. Kuo, C. J. Coffee, and J. B. Field, *Metabolism* **28,** 169 (1979).
[100] M. Walsh, F. C. Stevens, K. Oikawa, and C. M. Kay, *Can. J. Biochem.* **57,** 267 (1979).
[101] T. H. Crouch and C. B. Klee, *Biochemistry* **19,** 3692 (1980).
[102] W. D. McCubbin, M. T. Hincke, and C. M. Kay, *Can. J. Biochem.* **58,** 683 (1980).
[103] A. C. Murray and C. M. Kay, *Biochemistry* **11,** 2622 (1972).

McGhee–von Hippel Equation. The McGhee–von Hippel equation[108] was originally derived to describe the binding of ligands to polynucleotides. It is a model for the nonspecific binding of large ligands to one-dimensional homogeneous lattices. The equation is also referred to as a model for solving multiple-contact binding, multivalent binding, or parking problems and its quantitative properties have been analyzed.[109] The equation essentially is a generalized form of the Scatchard binding equation:

$$v/L = KN(1 - lv/N)\{(1 - lv/N)/[1 - (l - 1)v/N]\}^{l-1} \qquad (66)$$

where v is the concentration of bound ligand, L is the concentration of free ligand, K is the intrinsic binding constant, N is the number of binding sites, and l is the length of the binding site (base pairs in the case of DNA binding). Complete saturation occurs when $v = N/l$. Equation (66) can also be used to model the binding of ligands to polypeptides with repeating sequences or interactions of fibrous proteins with repeating units such as actin with their binding partners.

As an illustration, in one study Hayes *et al.*[110] studied the interaction of several small spore proteins with DNA. The proteins by themselves are unfolded but become helical when bound to DNA and the binding is almost stoichiometric, as shown in Fig. 7.

Figure 8 shows the effects of the binding of the proteins on the CD spectrum of the DNA in the near-UV region. The changes in ellipticity in this region as a function of protein concentration are cooperative (Fig. 8A) and can be modeled by the McGhee–von Hippel equation (Fig. 8B).

Serial Dilutions

If one has a 1:1 mixture of a protein and ligand that form a binary complex that stabilizes the conformation of the protein, it is possible to determine the binding constant by measuring the ellipticity of a 1:1 mixture as a function of total concentration. This method has been used to determine the association constant of coiled coils[111] and protein–protein complexes.[112,113] The equation relating the ellipticity to the total protein concentration is

[104] W. D. McCubbin and C. M. Kay, *Biochemistry* **12**, 4228 (1973).
[105] R. S. Mani, W. D. McCubbin, and C. M. Kay, *FEBS Lett.* **52**, 127 (1975).
[106] J. D. Johnson, J. H. Collins, and J. D. Potter, *J. Biol. Chem.* **253**, 6451 (1978).
[107] L. Smith, N. J. Greenfield, and S. E. Hitchcock-DeGregori, *J. Biol. Chem.* **269**, 9857 (1994).
[108] J. D. McGhee and P. H. von Hippel, *Biochemistry* **14**, 1297 (1975).
[109] Y. Kong, *Biophys. Chem.* **95**, 1 (2002).
[110] C. S. Hayes, Z. Y. Peng, and P. Setlow, *J. Biol. Chem.* **275**, 35040 (2000).
[111] H. Wendt *et al.*, *Biochemistry* **36**, 204 (1997).

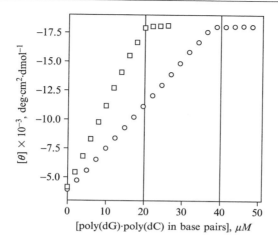

FIG. 7. Stoichiometric reverse titrations of SspCTyr with poly(dG)·poly(dC). Poly (dG)·poly(dC) was added sequentially to stirring solutions of SspCTyr at 5 μM (□) or 10 μM (○). The CD signal at 222 nm was corrected for dilution and buffer and polynucleotide contributions and then used to calculate mean residue ellipticity, $[\theta]_{222}$. Reprinted with permission from Hayes et al.[110] Copyright (2000) American Society for Biochemistry and Molecular Biology.

$$[\theta]_{obs} = [\theta]_F + ([\theta]_U - [\theta]_F)[-k_D + (k_D^2 + 4P_t k_D)^{1/2}]/(2P_t) \qquad (67)$$

where k is the association constant of the two proteins; $[\theta]_{obs}$ is the ellipticity of the mixture of the proteins, at any total concentration, P_t; $[\theta]_F$ is the ellipticity of the fully folded complex; and $[\theta]_U$ is the sum of the ellipticity of the fully unfolded individual proteins or protein chains in the case of homodimers.

Determination of Binding Constants Using Change in Stability of Proteins on Complex Formation

When proteins interact with other proteins or ligands, even if there is no change in the CD of the folded protein on complex formation, there may be a change in the stability if the association occurs only when the protein is folded. This happens because the folded protein is in chemical equilibrium with the unfolded protein, and complex formation removes some of the folded protein, so more of the unfolded protein folds to maintain

[112] N. Kobayashi, S. Honda, and E. Munekata, Biochemistry 38, 3228 (1999).
[113] R. E. Georgescu, E. H. Braswell, D. Zhu, and M. L. Tasayco, Biochemistry 38, 13355 (1999).

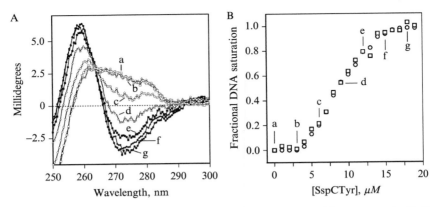

Fig. 8. Induced changes in the near-UV CD spectra of polynucleotides correspond with the fractional DNA saturation calculated from $[\theta]_{222}$. (A) Near-UV CD spectra (250–300 nm) of poly(dA-dT)·poly(dA-dT) with increasing amounts of SspCTyr in 5 mM sodium phosphate (pH 7.5) and 40 mM NaCl at 21° were acquired. The concentrations of added SspCTyr were 0 μM (curve a), 3 μM (curve b), 6 μM (curve c), 9 μM (curve d), 12 μM (curve e), 15 μM (curve f), and 18 μM (curve g). (B) Percent change in the ellipticity of poly(dA-dT)·poly(dA-dT) at 271 nm (□) and the fractional saturation of poly(dA-dT)·poly(dA-dT) calculated from $[\theta]_{222}$ (○) were plotted as a function of total SspCTyr concentration from a forward titration of poly(dA-dT)·poly(dA-dT) in 5 mM sodium phosphate (pH 7.5) and 40 mM NaCl at 21°. The concentrations of added SspCTyr were the same as for (A). Figure reprinted with permission from Hayes *et al.*[110] Copyright (2000) American Society for Biochemistry and Molecular Biology.

the equilibrium.[114] The increase in stability can be used to determine the association constants of the protein–protein or protein–ligand complexes.

Protein–Protein Interactions. If two proteins bind to each other only when they are folded and the protein complex unfolds cooperatively and reversibly to give two unfolded monomers, it is easy to determine the binding constant by determining the thermodynamics of folding of the complex compared with the thermodynamics of folding of the monomers. In the simplest case both proteins (e.g., X and Y) are monomers. The constant of folding of protein X is $k_{FX} = [X_F]/[X_U]$ and the constant of folding of protein Y is $k_{FY} = [Y_F]/[Y_U]$. The constant of folding of the dimer is $k_{FXY} = [XY_F]/[X_U][Y_U]$. The association constant $K_a = [XY_F]/[X_F][Y_F]$. By making the substitutions, it is easy to show that $K_A = k_{FXY}/k_{FX}k_{FY}$. To determine the association constant one determines the thermodynamics of folding of the individual monomers X and Y and the dimer XY by monitoring their ellipticity as a function of temperature or denaturant. The free energy of folding is calculated using the relationship $\Delta G = \Delta H - T\Delta S$.

[114] C. N. Pace *et al.*, *J. Mol. Biol.* **312**, 393 (2001).

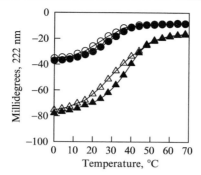

FIG. 9. Increase in ellipticity when two model peptides containing the N and C termini of rat striated α-tropomyosin interact with each other. (○) C-terminal peptide, GlyTM9a$_{251-284}$; (●)N-terminal peptide; (△) addition of the unfolding curves of the individual peptides; (▲) mixture of the two peptides. Binding causes an increase in the T_M and enthalpy of folding relative to the unmixed components. The peptides were each 10 μM in 100 mM NaCl, 10 mM sodium phosphate buffer, pH 6.5. Redrawn with permission from data in Greenfield *et al.*[115] Copyright (2002) Biophysical Society.

The folding constant is then determined from the relationship

$$k_F = \exp(-\Delta G/RT) \qquad (68)$$

One then divides the k_F of folding of the dimer by the k_F values of folding of the two monomers to obtain the binding constant for dimer formation.

Figure 9 shows the folding of two model peptides containing the N and C termini of striated muscle α-tropomyosin.

The N-terminal peptide is a two-chain homodimer.[58] The C-terminal peptide is a two-chain homodimer that is cross-linked by a disulfide bond,[115] so that it folds as a monomer. The two peptides associate to form a heterotrimer, which has slightly increased ellipticity, and greatly enhanced stability. The data were fit to equations of folding of dimer, monomer, and heterotrimer, respectively, to give k_F values at 25° of 2.6 × 10^4, 1.06 × 10^0, and 1.62 × 10^{10}, respectively. These constants were then used to calculate a dissociation constant of the complex, 1.7 μM.

Protein–Ligand Interactions. If a ligand binds to a protein only when it is folded, the increase in stability of the complex compared with the free protein can be used to estimate the binding constant.[61–63] If the changes in enthalpy of unfolding are small, the binding constant can be evaluated by Eq. (69):

$$\Delta T_M = (TT_0P/\Delta H_p)\ln(1 + K_B[L]) \qquad (69)$$

---

[115] N. J. Greenfield, T. Palm, and S. E. Hitchcock-DeGregori, *Biophys. J.* **83,** 2754 (2002).

where T_M is the change in the midpoint of the thermal denaturation curve, T and T_0 are the T_M values in the presence and absence of ligand, ΔH_p is the enthalpy of folding of the protein, K_B is the equilibrium for the binding of the ligand to the folded protein, and [L] is the free concentration of ligand. Schellman cautioned that Eq. (69) works only when ligand binding has only a small effect of the ΔH of the folding.[61]

Pace and McGrath[63] have also shown that one can use the change in stability of the protein to denaturation by detergents by utilizing the change in the extrapolated free energy of folding as a function of ligand concentration to determine the binding constant of the ligand, assuming that all the increase in free energy is due to binding.

Horn et al.[116] have compared the enthalpies of binding obtained from isothermal titration calorimetry and van't Hoff analysis. Their treatment, which can be adapted to analyze CD data, shows that if folding and binding are linked, the observed binding constant includes contributions from both the binding and unfolding equilibria:

$$K_{obs} = K_{int}/(1 + K_U) \tag{70}$$

where K_{int} is the intrinsic binding constant for binding of the ligand to the folded molecule and K_U is the unfolding constant for the protein. The observed enthalpy is described by

$$\Delta H_{obs}^{\circ} = \Delta H_{int}^{\circ} - [K_U/(1 + K_U)]\Delta H_U^{\circ} \tag{71}$$

where ΔH_{int}° is the intrinsic enthalpy for ligand–protein binding, the ratio $K_U/(1 + K_U)$ describes the fraction of unfolded molecule, and ΔH_U° is the enthalpy of macromolecular unfolding. Applying these equations, Horn et al.[116] showed that the same enthalpies of binding were determined using direct titrations and from the van't Hoff analysis of ligand binding.

Jones et al.[117] have described a global method of fitting of CD thermal unfolding curves obtained in the presence of ligands to determine binding constants. Brandts and Lin[118] have discussed in detail the thermodynamic implications in cases in which there is more than one protein–ligand binding site.

Conclusion

Circular dichroism is a useful technique for analyzing protein secondary structure, protein folding, and protein interactions. CD has also been used

[116] J. R. Horn, D. Russell, E. A. Lewis, and K. P. Murphy, *Biochemistry* **40**, 1774 (2001).
[117] C. L. Jones, F. Fish, and D. D. Muccio, *Anal. Biochem.* **302**, 184 (2002).
[118] J. F. Brandts and L. N. Lin, *Biochemistry* **29**, 6927 (1990).

extensively to study the kinetics of protein folding and protein–ligand interactions, but this is beyond the scope of this review. Since CD is an indirect method of measurement of thermodynamics and binding, however, care must be used to ensure that the models used to fit the CD data are based in reality.

Appendix I

Structural Analysis Programs

There are currently no commercial packages for analyzing CD data to determine structural parameters. The following is a list of Web sites where CD analysis programs are currently available.

Structural analysis programs that run in a DOS environment, including SELCON, VARSLC, K2D, MLR, LINCOMB CONTIN, and CCA algorithm, are available at the following Web site: http://www2.umdnj. edu/cdrwjweb

Structural analysis programs, including MLR, VARSLC, SELCON, and CONTIN, that run in a Windows environment are available at the following ftp site: ftp://ftp.ibcp.fr/pub/C_DICROISM/

Web sites containing the latest versions of individual programs:

CCA algorithm: http://www2.chem.elte.hu/protein/programs/cca/

CDNN: http://bioinformatik.biochemtech.uni-halle.de/cd_spec/index.html

CONTIN: http://s-provencher.com/index.shtml

K2D: http://www.embl-heidelberg.de/%7Eandrade/k2d.html

SELCON: http://lamar.colostate.edu/~sreeram/CD

VARSLC: http://oregonstate.edu/dept/biochem/faculty/johnson download.html

Curve Fitting

There are many commercial programs for fitting nonlinear equations. Several popular programs include the following:

KaleidaGraph (Synergy Software, Reading PA)
Origin (OriginLab, Northampton, MA)
PsiPlot (Poly Software, Pearl River, NY)
SigmaPlot (SPSS Science, Chicago, IL)

[13] Computation and Analysis of Protein Circular Dichroism Spectra

By Narasimha Sreerama and Robert W. Woody

Introduction

Circular dichroism (CD) is the most widely used form of chiroptical spectroscopy, spectroscopic techniques that utilize the differential inter-action of molecules with left- and right-circularly polarized light. In the absence of a magnetic field, the molecule must be chiral to give rise to a difference in the interaction with the two types of circularly polarized light. The phenomenon of CD consists of the differential absorption of left- and right-circularly polarized light by a chiral molecule. Circular dichroism due to electronic transitions is generally referred to as CD (also as ECD), and that due to vibrational transitions is referred to as VCD. There are other types of chiroptical spectroscopy, which include circularly polarized luminescence, circular intensity differential scattering, and Raman optical activity.

In this chapter we discuss the origins of electronic CD in proteins, the-oretical methods for computing protein CD, and empirical analysis of CD for estimating structural composition of proteins. A large number of scien-tists have contributed to our understanding of protein CD. Given the wide scope of this chapter and space limitations, we have been forced to be highly selective in our references, citing representative papers and review articles.

Basic Definitions

In this section we summarize the definition and units for CD. We also discuss the origin of CD in proteins and the terminology utilized in reporting and discussing protein CD spectra.

CD and Rotational Strength

The phenomenon of CD involves the absorption of light and it can be considered as a special type of absorption spectroscopy. CD is the differ-ence in the absorption of left- and right-circularly polarized light and is defined as

$$\Delta\varepsilon(\lambda) = \varepsilon_{L}(\lambda) - \varepsilon_{R}(\lambda) \tag{1}$$

Copyright 2004, Elsevier Inc.
All rights reserved.
0076-6879/04 $35.00

where ε_L and ε_R, respectively, are the extinction coefficients for the left- and right-circularly polarized components at wavelength λ. The units for CD, when defined as $\Delta\varepsilon$, are $M^{-1}\cdot cm^{-1}$, where M is the molar concentration.

Molar ellipticity is another measure for CD, and is defined as

$$[\theta] = 100\theta/Cl \tag{2}$$

where C and l are the molar concentration and pathlength (cm) of the sample, respectively. Molar ellipticity is reported either as $deg\cdot cm^2\cdot dmol^{-1}$ or as $deg\cdot M^{-1}\cdot m^{-1}$, and these two units are equivalent. Molar ellipticity and molar CD (as it is called when expressed in $\Delta\varepsilon$) are interconvertible by a factor

$$[\theta] = 3298\Delta\varepsilon \tag{3}$$

The integrated intensity of a CD band gives a measure of the strength of CD, called the rotational strength. Rotational strength has units of $erg\cdot cm^3$, and is experimentally defined as

$$\begin{aligned} R &= (hc/32\pi^3 N_A)\int(\Delta\varepsilon/\lambda)d\lambda \\ &= 2.295 \times 10^{-39}\int(\Delta\varepsilon/\lambda)d\lambda \end{aligned} \tag{4}$$

where h is Planck's constant, c is the velocity of light, and N_A is Avogadro's number.

Rotational strength is defined theoretically, following Rosenfeld's treatment,[1] as the imaginary part of the scalar product of the electric ($\boldsymbol{\mu}$) and magnetic ($\mathbf{m}$) dipole transition moments of an electronic transition

$$R = \text{Im}\{\boldsymbol{\mu}\cdot\mathbf{m}\} \tag{5}$$

This definition suggests the more frequently used units for rotational strength of Debye–Bohr magnetrons (DBM; 1 DBM = 0.9274×10^{-38} $erg\cdot cm^3$).

Using Eq. (5) and quantum mechanical wave functions for the ground and excited states, one can calculate the rotational strength for a given transition as

$$R_{oa} = \text{Im}\{(o|\boldsymbol{\mu}|a)\cdot(a|\mathbf{m}|o)\} \tag{6}$$

where $\boldsymbol{\mu}$ is the electric dipole transition moment operator, a measure of the linear displacement of charge on excitation; and $\mathbf{m}$ is the magnetic dipole transition moment operator, a measure of the circular displacement of electron density on excitation. The superposition of $\boldsymbol{\mu}$ and $\mathbf{m}$ results in a

[1] L. Rosenfeld, *Z. Physik* **52,** 161 (1928).

helical displacement of charge, which interacts differently with left- and right-circularly polarized light.

The expression for rotational strength as given in Eq. (6) is origin dependent because of the origin dependence of the magnetic dipole transition moment operator. An alternate origin-independent formulation of rotational strength, known as the dipole-velocity formulation,[2,3] uses the gradient operator, ∇, and is normally used in the theoretical computation of CD.

$$R_{oa} = -(eh/2\pi m\nu_{oa}) \operatorname{Im} \{(o|\nabla|a) \cdot (a|\mathbf{m}|o)\} \tag{7}$$

where e and m are, respectively, charge and mass of an electron, and ν_{oa} is the frequency of the transition $o \rightarrow a$.

Calculation of the CD spectrum at a given wavelength, λ, is done by assuming Gaussian bands for all transitions and using the relation between molar CD, $\Delta\varepsilon_k$, and rotational strength, R_k, for a given transition k with a half-bandwidth (one-half of the width of the CD band at $1/e$ of its maximum) of Δ_k.

$$\Delta\varepsilon_k = 2.278 R_k \lambda_k / \Delta_k \tag{8}$$

Protein CD

The CD spectra of proteins are generally divided into three wavelength ranges, based on the energy of the electronic transitions that dominate in the given range (Fig. 1). These are (1) the far UV (below 250-nm), where the peptide contributions dominate, (2) the near UV (250–300-nm), where aromatic side chains contribute, and (3) the near UV–visible region (300–700-nm), where extrinsic chromophores contribute. Applications of CD to protein structure and folding have been developed on the basis of

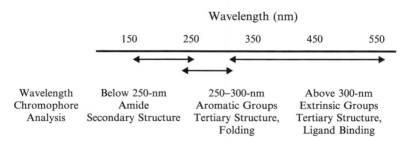

FIG. 1. CD spectral regions and contributing chromophores in proteins.

[2] W. Moffitt, *J. Chem. Phys.* **25,** 467 (1956).
[3] A. Moscowitz, *in* "Modern Quantum Chemistry—Istanbul Lectures" (O. Sinanoğlu, ed.), Part III, p. 31. Academic Press, New York, 1965.

these characteristic origins of protein CD spectra. The protein CD spectrum in the far UV is almost always reported on a residue basis, which corresponds to the concentration expressed in molar concentration of protein times the number of residues. The near-UV CD is generally reported on the basis of residue molar concentration, but sometimes the protein concentration or aromatic residue concentration is used. Extrinsic chromophore CD is reported on the basis of protein concentration.

Protein CD has also been classified into "intrinsic" and "extrinsic," based on the contributing chromophores. Intrinsic CD of proteins is due to chromophores that are part of the protein (peptides, side-chain groups, and disulfides), and extrinsic CD is due to chromophores that are not part of the protein chain (heme, flavin, ligands, etc.). The early literature on protein CD considered the near-UV CD due to aromatic side chains as extrinsic CD, but we consider amino acid side chains as part of the protein and their CD contributions as intrinsic to the protein. Intrinsic protein CD is different from the "intrinsic CD" of a chromophore, which arises from the inherent chirality of the chromophore.[4] Chromophores that contribute to protein CD are generally achiral. The peptide and the aromatic side-chain groups have a plane of symmetry and do not exhibit intrinsic optical activity. Many extrinsic chromophores are also achiral in the isolated state and, when free, do not give CD bands. The "intrinsic CD" contributions of protein chromophores are generally assumed to be nonexistent, except for the disulfide group. The disulfide chromophore is nonplanar ($|\chi_{ss}| \sim 90°$), making it inherently dissymmetric and giving rise to intrinsic CD.

The electronic transitions in an achiral molecule have either the electric (e.g., $n\pi^*$ transitions) or the magnetic (e.g., $\pi\pi^*$ transitions) dipole transition moment equal to zero, or the two kinds of transition moments are perpendicular to each other (e.g., $n\sigma^*$ transitions), which results in zero rotational strength. The interaction between protein chromophores in the chiral field of the protein introduces perturbations leading to optical activity. Moffitt's seminal contributions[2] laid the foundations for the theoretical treatment of protein CD. The mechanisms that generate rotational strength have been worked out by Born, Kuhn, Kirkwood, Moffitt, Condon, Eyring, Tinoco, Schellman, and co-workers. These can be grouped into three basic types of interactions between transitions on different chromophores: (1) the coupled-oscillator mechanism[5] (also known as the μ–μ mechanism) describing the Coulomb interactions between two electrically allowed transitions, such as the $\pi\pi^*$ transitions of the peptide group, and

[4] W. Moffitt and A. Moscowitz, J. Chem. Phys. **30,** 648 (1959).
[5] J. G. Kirkwood, J. Chem. Phys. **5,** 479 (1937).

resulting in composite transitions with nonvanishing electric and magnetic transition moments; (2) the $\mu-m$ mechanism,[6] wherein the electrically allowed transition on one chromophore mixes with the magnetically allowed transition on another; and (3) the one-electron mechanism[7] (also known as static-field mixing), wherein the electrically and magnetically allowed transitions on the same chromophore mix in the electrostatic field of the rest of the molecule. Exciton coupling is a special case of the coupled-oscillator mechanism in which degenerate transitions on identical chromophores mix. Theoretical calculation of CD of proteins is performed by considering each chromophore to be independent of others, thus neglecting overlap and electron exchange between chromophores, and constructing a composite wave function for the protein molecule from the wave functions of the constituent chromophores.

Computation of Protein CD

In this section we discuss the computation of protein CD, using theoretical methods. The basic approach used to compute the CD of complex systems, such as proteins and nucleic acids, is the so-called "divide and conquer" approach.[8] As one can glean from the previous section, the protein is treated as a collection of independent and mostly achiral chromophores, which are perturbed in the chiral environment of the protein. The characterization of the chromophores and their interactions forms the basis of theoretical calculation of CD. Methods based both on classical physics and on quantum mechanics have been developed to compute CD. The classical approach is based on the interaction of electromagnetic radiation with matter and the changes in polarizability, while the characteristics of electronic transitions form the basis of the quantum mechanical approach.

Classical Approaches

The framework for the classical method was developed by DeVoe.[9,10] Here the electronic transitions, either individually or collectively, are considered as oscillators. The dipole moment induced by the electromagnetic field in an oscillator is described by the complex polarizability tensor. The real part of the polarizability tensor, which is important in wavelength

---

[6] J. A. Schellman, *Acc. Chem. Res.* **1,** 144 (1968).

[7] E. U. Condon, W. Altar, and H. Eyring, *J. Chem. Phys.* **5,** 753 (1937).

[8] R. W. Woody, *in* "Circular Dichroism and the Conformational Analysis of Biomolecules" (G. D. Fasman, ed.), p. 25. Plenum Press, New York, 1996.

[9] H. DeVoe, *J. Chem. Phys.* **41,** 393 (1964).

[10] H. DeVoe, *J. Chem. Phys.* **43,** 3199 (1965).

regions away from the absorption band, is related to the index of refraction, and the complex part is proportional to the extinction coefficient; the real and imaginary parts of a complex function of a real variable, such as the complex polarizability, are related to one another by Kramers–Kronig relations.[11]

The induced dipole moment in a given oscillator i has two sources: the electromagnetic field of the incident light, and the field produced by the induced dipole moments of all other oscillators. The expression for the induced dipole moment for oscillator i is given as

$$\boldsymbol{\mu}_i^{\text{ind}} = \alpha_i \mathbf{E}_i^{\text{ext}} - \sum_{j \neq i} \alpha_i \mathbf{T}_{ij} \boldsymbol{\mu}_j^{\text{ind}} \tag{9}$$

where α_i is the complex polarizability, $\mathbf{E}_i^{\text{ext}}$ is the external field, $\mathbf{T}_{ij}$ is the dipole interaction tensor for oscillators i and j, $\boldsymbol{\mu}_j^{\text{ind}}$ is the induced dipole moment for oscillator j. The set of equations for N oscillators can be written as a matrix equation:

$$\boldsymbol{\mu}_i^{\text{ind}} = \mathbf{G}^{-1} \mathbf{E}_i^{\text{ext}} = \mathbf{A} \mathbf{E}_i^{\text{ext}} \tag{10}$$

where $\mathbf{G}$ is a $3N \times 3N$ matrix composed of N^2 3×3 submatrices, α_i^{-1} as diagonal elements, and $\mathbf{T}_{ij}$ as off-diagonal elements. The inverse of matrix $\mathbf{G}$ is matrix $\mathbf{A}$.

The induced dipole moments μ^{ind} are analogous to the electric dipole transition moments given in Eq. (7), and the moments of induced moments with respect to the origin, $\mathbf{r} \times \mu^{\text{ind}}$, are analogous to the magnetic dipole transition moments. The expressions for absorption and CD for electrically allowed transitions are then derived as[9,10]

$$\varepsilon = 8\pi^2 N_A / 6909\lambda \sum_i \sum_j \text{Im}(A_{ij}) \mathbf{R}_{ij} \mathbf{e}_i \cdot \mathbf{e}_j \tag{11}$$

$$\Delta\varepsilon = 16\pi^3 N_A / 6909\lambda^2 \sum_i \sum_j \text{Im}(A_{ij}) \mathbf{R}_{ij} \cdot \mathbf{e}_i \times \mathbf{e}_j \tag{12}$$

where $\mathbf{e}_i$ and $\mathbf{e}_j$ are the unit vectors along $\boldsymbol{\mu}_i^{\text{ind}}$ and $\boldsymbol{\mu}_j^{\text{ind}}$, and $\mathbf{R}_{ij}$ is the vector from i to j. In this approach, the extinction coefficient and CD at a given wavelength are calculated directly as a summation of contributions from all oscillators in a given molecule. These expressions describe the coupled-oscillator interactions between the electrically allowed transitions, which are normally considered in classical polarizability theory applications for organic molecules and biopolymers.

[11] A. Moscowitz, *Adv. Chem. Phys.* **4**, 67 (1962).

Extensive applications of classical polarizability theory for proteins have been performed by Applequist and co-workers.[12–16] They have developed the atom dipole interaction model in which each atom has an isotropic polarizability, independent of wavelength, and the dipole moments induced on the atoms are incorporated in the calculation of matrix **A**. Small organic molecule polarizability data at the NaD line were used to obtain polarizability parameters for C, O, N, H, and the halogens. In addition to the isotropic atomic polarizabilities, the anisotropic polarizability of the first electrically allowed peptide transition ($\pi\pi^*$ at 190-nm) was included in the dipole interaction model for protein CD calculations. Polarizability anisotropy data for simple amides were used for obtaining polarizability parameters. The results for proteins and polypeptides in specific conformations obtained with the dipole-interaction model are satisfactory in the $\pi\pi^*$ region. The major drawback of the dipole-interaction model is the neglect of the electrically forbidden amide $n\pi^*$ transition (at 220 nm), which is important in proteins.

Matrix Method

Direct calculation of rotational strengths of proteins by quantum chemical methods and Eq. (7) is, in principle, possible. However, the size and complexity of a protein make such a computation next to impossible. The matrix method[17,18] performs this task by combining the interactions between the chromophores in the protein, and can be considered as a quantum chemical method depending on the description of chromophores. The matrix method has its origins in the coupled-oscillator model and the exciton model, wherein a secular determinant is constructed on the basis of the energy and interactions between the transitions on chromophores. The secular matrix for a two-chromophore system with one transition on each chromophore is written as

$$\begin{bmatrix} E_1 & V_{12} \\ V_{12} & E_2 \end{bmatrix} \tag{13}$$

where E_1 and E_2 are the energies of the transitions and V_{12} is the interaction energy between the two transition moments. The secular matrix is

[12] J. Applequist, *J. Chem. Phys.* **71**, 1983 (1979).

[13] J. Applequist, *J. Chem. Phys.* **71**, 4332 (1979).

[14] J. Applequist, *Biopolymers* **20**, 2311 (1981).

[15] J. Applequist and K. Bode, *J. Phys. Chem. B* **103**, 1767 (1999).

[16] K. Bode and J. Applequist, *J. Am. Chem. Soc.* **120**, 10938 (1998).

[17] P. M. Bayley, E. B. Nielsen, and J. A. Schellman, *J. Phys. Chem.* **73**, 228 (1969).

[18] P. M. Bayley, *Prog. Biophys. Mol. Biol.* **27**, 1 (1973).

diagonalized, yielding two excited states that are linear combinations of individual transitions.

An analogous matrix can be constructed for a protein, composed of a set of chromophores with localized transitions, with the energies of the transitions forming the diagonal elements and the interactions between different localized transitions on chromophores forming the off-diagonal elements.

$$
\begin{bmatrix}
E_{11} & V_{11,12} & V_{11,13} & V_{11,21} & V_{11,22} & V_{11,23} & \cdot & \cdot & \cdot & V_{11,N1} & V_{11,N2} & V_{11,N3} \\
\cdot & E_{12} & V_{12,13} & V_{12,21} & V_{12,22} & V_{12,23} & & & & V_{12,N1} & V_{12,N2} & V_{12,N3} \\
\cdot & & E_{13} & V_{13,21} & V_{13,22} & V_{13,23} & & & & V_{13,N1} & V_{13,N2} & V_{13,N3} \\
\cdot & & & E_{21} & V_{21,22} & V_{21,23} & & & & V_{21,N1} & V_{21,N2} & V_{21,N3} \\
\cdot & & & & E_{22} & V_{22,23} & & & & V_{22,N1} & V_{22,N2} & V_{22,N3} \\
\cdot & & & & & E_{23} & & & & V_{23,N1} & V_{23,N2} & V_{23,N3} \\
\cdot & & & & & & \cdot & & \cdot & \cdot & & \cdot \\
\cdot & & & & & & & \cdot & & \cdot & \cdot & \cdot \\
\cdot & & & & & & & & & E_{N1} & V_{N1,N2} & V_{N1,N3} \\
\cdot & & & & & & & & & & E_{N2} & V_{N2,N3} \\
\cdot & & & & & & & & & & & E_{N3}
\end{bmatrix}
$$

(14)

For example, one can construct a $3N \times 3N$ interaction energy matrix, **H**, for a protein with N residues with three transitions on each amide chromophore. Here E_{ij} represents the energy of the transition j on chromophore i, and $V_{ij,kl}$ represents the interaction between transition j on chromophore i and the transition l on chromophore k. A transition localized on a given chromophore is represented by an appropriate charge distribution, for example, a set of point charges describing the dipole transition moment. Electrically allowed $\pi\pi^*$ transitions have a nonzero electric dipole transition moment and are represented by dipolar charge distributions. Electrically forbidden $n\pi^*$ transitions are represented by a set of quadrupolar charges centered on relevant atoms. The point charges are determined either from the quantum chemical wave function or from experiment, or a combination of both. The Coulomb interaction between localized transitions j and l on chromophores i and k, respectively, are calculated using the monopole–monopole approximation as

$$
V_{ij,kl} = \sum_m \sum_n q_{ijm} q_{kln} / r_{ijm,kln}
$$

(15)

where indices m and n correspond to the point charges for transitions j and l, respectively, and the distance between point charges is given by r. The monopole–monopole approximation is a better approach to compute the Coulomb interaction between two charge distributions because the dipole–dipole approximation breaks down at short distances, for example, when chromophores are at or near van der Waals distances.

In the example given above, the first $n\pi^*$ ($j = 1$) and the first two $\pi\pi^*$ ($j = 2$ and 3) transitions on the peptide group have been considered. The three types of interactions between different transitions described by the μ–μ, μ–m, and one-electron mechanisms are distributed among the various off-diagonal elements of this matrix. For the terms with $i \neq k$, $j = 2$ or 3, and $l = 2$ or 3, both interacting transitions are electrically allowed, and the terms correspond to the μ–μ mechanism. Terms with $i \neq k$, $j = 1$, and $l = 2$ or 3 describe the interaction between an $n\pi^*$ and a $\pi\pi^*$ transition on different chromophores and correspond to the μ–m mechanism. When $i = k$, both transitions lie on the same chromophore representing the mixing of excited states in the static field of the molecule, corresponding to the one-electron mechanism. An additional term, with $i \neq k$, $j = 1$, and $l = 1$, describes the interaction between quadrupolar charge distributions of two $n\pi^*$ transitions on two different chromophores.

Diagonalization of matrix $\mathbf{H}$ gives the eigenvalues and eigenvectors that describe the excited states of the composite molecule; these can be called group states.

$$\mathbf{H} = \mathbf{U}\mathbf{H_d}\mathbf{U}^T \qquad (16)$$

Eigenvalues, the diagonal elements of matrix $\mathbf{H_d}$, give the energies of the transitions; eigenvectors, elements of the unitary matrix $\mathbf{U}(\mathbf{U}\mathbf{U}^T = \mathbf{I}$, where matrix $\mathbf{U}^T$ is the transpose of matrix $\mathbf{U}$ and $\mathbf{I}$ is the identity matrix), describe the mixing of localized transitions. By combining eigenvectors with the properties of localized transitions, $\boldsymbol{\mu}_k^0$ and $\mathbf{m}_k^0$, one can determine the transition moments for each group state for the composite system, and calculate the rotational strength.

$$\mathbf{m}_i = \sum_k \mathbf{U}_{ki}\mathbf{m}_k^0 \qquad \text{and} \qquad \boldsymbol{\mu}_i = \sum_k \mathbf{U}_{ki}\boldsymbol{\mu}_k^0 \qquad (17)$$

Tinoco's first-order perturbation method,[19] which predates the matrix method, considers doubly excited states (excitation of electrons on two chromophores) in addition to the singly excited states. While this method gives expressions for μ–μ, μ–m, and one-electron terms generating rotational strength, it requires the diagonalization of the Hamiltonian matrix for degenerate or near-degenerate transitions before the perturbative treatment. On the other hand, the matrix method is equivalent to an all-order perturbation approach and is much easier to implement. Nowadays, most protein CD calculations are performed using the matrix method.

[19] I. Tinoco, Jr., *Adv. Chem. Phys.* **4,** 113 (1962).

Chromophores and Transitions

The chromophores that are important in protein CD calculations are the amide, the aromatic side chains, disulfides, and extrinsic groups. Of these, the amide chromophore, which forms the backbone of the polypeptide chain, is the most abundant and dominates the far–UV CD; the contributions from aromatic side chains and extrinsic groups, respectively, dominate the near-UV and visible regions (Fig. 1). Only a few of the amino acid side chains have transitions at wavelengths longer than 180 nm[20] (aromatic side chains, disulfides, histidine, arginine, glutamine, asparagine, glutamic acid, and aspartic acid), and among those only the aromatic transitions and disulfides have large enough electric or magnetic dipole transition moments to make appreciable spectral contributions. Most protein CD calculations, however, include only the amide transitions. Only a limited number of CD calculations have incorporated the aromatic side-chain and disulfide transitions, and those that include extrinsic chromophores (e.g., the heme group in myoglobin and retinal in rhodopsins) are even rarer.

The transition parameters are generally obtained from molecular orbital calculations on structures modeling the chromophores, combined with experimentally available transition dipole moments and wavelengths.[21] The electrically allowed $\pi\pi^*$ transitions, experimental data for which can be obtained, are represented by an electric dipole transition moment. The magnetically allowed $n\pi^*$ transitions, which are difficult to fully characterize experimentally, are represented by a magnetic dipole transition moment and an electric quadrupole transition moment,[22] obtained from quantum chemical calculations.

One or two $\pi\pi^*$ and one or two $n\pi^*$ amide transitions have been used in protein and polypeptide CD calculations, and the parameters were mostly obtained from semiempirical molecular orbital (MO) calculations and available experimental data on secondary amides. The amide group has four π electrons, three π orbitals (π_+, π_0, and π^*) and two lone pairs in the nonbonding orbitals (n and n') of the valence shell. Two $\pi\pi^*$ transitions, $\pi_0 \rightarrow \pi^*$ (denoted NV$_1$, at 190 nm, electric dipole transition moment $|\mu| \approx 3.1$ Debyes, D) and $\pi_+ \rightarrow \pi^*$ (NV$_2$, 139 nm, $|\mu| \approx 1.8$ D), and one $n\pi^*$ ($n \rightarrow \pi^*$, 220 nm) transition in the amide chromophore have been identified. Semiempirical calculations predict the second $n\pi^*$ transition ($n' \rightarrow \pi^*$) near the NV$_1$ transition, but it has not been identified

[20] C. R. Cantor and S. N. Timasheff, in "The Proteins" (H. Neurath, ed.), Vol. V, p. 145. Academic Press, San Diego, CA, 1982.
[21] M. C. Manning and R. W. Woody, Biopolymers 31, 569 (1991).
[22] R. W. Woody, J. Chem. Phys. 49, 4797 (1968).

experimentally and *ab initio* calculations place it at much higher energies.[23] The experimental transition moment directions of NV_1 and NV_2, characterized by single-crystal polarized reflection studies of *N*-acetylglycine,[24] are $-55°$ and $10°$ or $61°$, respectively. The $n \rightarrow \pi^*$ transition has a large magnetic dipole transition moment along the carbonyl bond. The wavelength of the $n \rightarrow \pi^*$ transition in amides is dependent on solvent (230-nm in apolar solvents to 210-nm in strongly polar solvents), while those of $\pi \rightarrow \pi^*$ transitions are much less sensitive to solvent. The parameters for the tertiary amide, formed by the imino acid proline, are slightly different from those for secondary amide; the wavelength of the first $\pi\pi^*$ transition is at $\sim$200-nm, which is $\sim$10 nm longer than that in secondary amide. The polarizations of the $\pi\pi^*$ transitions in tertiary amides are not known, but *ab initio* calculations indicate that they are similar to those for secondary amides.[23]

The transitions of Phe and Tyr side chains that contribute to near- and far-UV protein CD are well understood, as the chromophores are derivatives of benzene; the Phe side chain is an alkyl-substituted benzene, and the Tyr side chain is an alkylated phenol. These benzenoid chromophores have four $\pi\pi^*$ transitions: L_b, L_a, B_b, and B_a, in Platt's notation.[25] Of these, the L transitions are forbidden and the B transitions are allowed. Furthermore, the B transitions are degenerate in benzene. The L transitions, however, are coupled to B transitions vibronically and borrow most of their intensity from the B transitions.[26] In the Phe side chain, the weakly perturbing alkyl substitution does not change the situation; here, the B transitions are nearly degenerate and the L transitions derive most of their intensity through vibronic coupling with the B transitions. Phe side-chain transitions are as follows: L_b, 260-nm, $|\mu| \approx 0.4$ D; L_a, 210-nm, $|\mu| \approx 2.7$ D; B_b and B_a, 185-nm, $|\mu| \approx 6$ D. In the Tyr side chain, stronger perturbation by the phenolic oxygen shifts the L transitions to longer wavelengths, makes the L transitions electrically allowed, and makes the B transitions nondegenerate. In addition, the vibronic coupling between the B and L transitions is weaker than that in the Phe side chain because of larger energy separation. Tyr side-chain transitions are as follows: L_b, 275-nm, $|\mu| \approx 1.2$ D; L_a, 230-nm, $|\mu| \approx 2.8$ D; B_b and B_a, 190-nm, $|\mu| \approx 6$ D.

The Trp side-chain transitions can be approximated by those of indole, a perturbed naphthalene, which in Platt's model are also: L_b, L_a, B_b, and B_a. These $\pi\pi^*$ transitions occur in the near- and far-UV region, and are

[23] L. Serrano-Andrés and M. P. Fülscher, *J. Am. Chem. Soc.* **118**, 12190 (1996).
[24] L. B. Clark, *J. Am. Chem. Soc.* **117**, 7974 (1995).
[25] J. R. Platt, *J. Chem. Phys.* **17**, 484 (1949).
[26] J. N. Murrell and J. A. Pople, *Proc. Phys. Soc. A* **69**, 245 (1956).

more complex than those of the Tyr side chain due to vibronic fine structure and overlapping transitions. The first three transitions have been characterized experimentally.[27] The L_b and L_a transitions are between 270 and 280 nm and the L_b band has well-resolved vibronic components; the L_a transition is quite sensitive to the environment; the B_b transition occurs near 225 nm. There are two to three strong transitions between 180 and 210 nm, as determined by the absorption spectra of indole derivatives. Parameters for six transitions of the Trp side chain were obtained by combining experimental results with Pariser–Parr–Pople calculations.

Two $n \rightarrow \sigma^*$ transitions have been characterized in the disulfide group, which arise from excitation of lone pair orbitals on the sulfurs to the σ^* orbital of the S–S bond. These are electrically forbidden transitions but have strong magnetic dipole transition moments. The transition energies and transition dipole moments depend on the C_β–S–S–C_β dihedral angle, and the excitation energies were determined empirically by fitting experimental transition energies for model disulfides with different C_β–S–S–C_β angles to a cubic polynomial.[28] By extending the Bergson model accounting for the C_β–S–S–C_β angle dependence of the absorption bands, analytical expressions for transition monopoles as a function of the C_β–S–S–C_β angle have been derived.[29]

Protein CD Spectra

The polypeptide chain in a protein is primarily made up of secondary amides. Proline, which accounts for ~5% of residues in proteins, forms tertiary amides. Polypeptides form different secondary structures based on the arrangement of amide groups dictated by the backbone conformation. The interaction between transition moments leads to CD, and the geometric relation between amide groups in different secondary structures determines the characteristic secondary structure CD spectra. Proteins vary in the extent of their secondary structure, and the secondary structure composition and spatial arrangement of amides determines the far-UV CD spectrum. The correspondence between the secondary structure and CD spectrum is exploited in developing applications of CD.

α Helices and β sheets are the two most important secondary structures in proteins and they are stabilized, respectively, by intra- and interchain hydrogen bonds. They are characterized by a set of dihedral angles ϕ and ψ (α: $-57°$, $-47°$; β: $-120°$, $+120°$) that repeat along the polypeptide chain

[27] B. Albinsson and B. Norden, *J. Phys. Chem.* **96**, 6204 (1992).

[28] G. Kurapkat, P. Krüger, A. Wollmer, J. Fleischhauer, B. Kramer, E. Zobel, A. Koslowski, H. Botterweck, and R. W. Woody, *Biopolymers* **41**, 267 (1997).

[29] R. W. Woody, *Tetrahedron* **29**, 1273 (1973).

so that the successive amide groups are oriented identically with respect to the overall direction of chain propagation. Amide groups in α helices form an approximately cylindrical surface with intrachain hydrogen bonds parallel to the helix axis, and those in β sheets span a planar surface. Two types of β sheets exist, having polypeptide strands either parallel or antiparallel to each other. In general, α helices are longer and more rigid than β sheets because of the nature of the hydrogen bonds that stabilize them. Proteins are often characterized by the relative contents of α and β structures as α-rich, β-rich, and $\alpha\beta$ proteins. Another important secondary structure is the β turn, generally formed by three residues and stabilized by a hydrogen bond between the first and the third amide group, which effectively reverses the direction of chain propagation. Turns are characterized by sets of differing dihedral angles for successive residues[30] and need not have a stabilizing hydrogen bond. Another secondary structure that has repeating backbone dihedral angles (ϕ, ψ: $-78°$, $150°$) and is present in proteins, although to limited extent, is the poly(Pro)II type structure (P_2). This structure is the preferred conformation of proline-rich polypeptides, such as collagen, and is formed because of the limited conformational mobility of the proline side chain. Short stretches of P_2 structures involving Pro and non-Pro residues do occur in proteins, and they form an appreciable fraction of the residues that do not form α, β, or turn structures.[31] Amino acid residues that do not form any of these secondary structures exist in proteins, and these are called "unordered" for lack of ordered conformation. In many instances, the P_2 conformation is not explicitly identified and is implicitly treated as part of the unordered structure.

The three regular secondary structures, α, β, and P_2, that have repeating ϕ and ψ angles have characteristic CD spectra (Fig. 2). The α-helical CD spectrum[32] is characterized by two negative bands at 222 and 208 nm, and a positive band at 192 nm, which are normally used in CD analysis. Additional bands, a positive shoulder at 175 nm, a negative band at 160 nm, and a positive band centered below 140 nm, have been observed in the vacuum UV,[33] but the latter two are generally inaccessible for solution studies. In proteins, α helices show variations in their structure and the average geometry is different from the canonical form. Calculated CD spectra of α-helical fragments are consistent with the model α-helical CD. That the intensity of α-helical CD increases with the chain length has been confirmed by theoretical calculations. An empirical expression

[30] C. M. Wilmot and J. M. Thornton, *Protein Eng.* **3,** 479 (1990).
[31] N. Sreerama and R. W. Woody, *Biochemistry* **33,** 10022 (1994).
[32] A. Toumadje, S. W. Alcorn, and W. C. Johnson, Jr., *Anal. Biochem.* **200,** 321 (1992).
[33] W. C. Johnson, Jr. and I. Tinoco, Jr., *J. Am. Chem. Soc.* **94,** 4389 (1972).

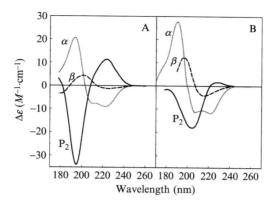

FIG. 2. (A) CD spectra of α-helix, β-sheet, and P_2 structure deconvoluted from a reference protein set of 37 proteins. (B) CD spectra of model polypeptides in α-helical,[32] β-sheet,[36] and poly(Pro)II helix[39] conformations.

$$V_r = V_\infty (r - k)/r \qquad (18)$$

where V is the CD amplitude at a specific wavelength for a helix of r residues, V_∞ is the CD of an infinite helix at the same wavelength, and k is an empirical parameter, describes the chain-length dependence of α-helical CD.[34] Physically, the parameter k corresponds to the number of residues "missing" due to end effects. Experimental data suggest that the value of k is not a constant but decreases with the helix length for short helices.[35] However, for longer helices it can be considered a constant ($\sim$4).

The CD spectrum of the typical β sheet[36] has a negative band near 215-nm and a positive band near 198 nm (Fig. 2); additional short-wavelength bands near 175 nm (negative) and 168 nm (positive) are also observed.[37] The β-sheet CD is difficult to characterize because of the variations in the geometry of β structure in polypeptides and proteins, and the limited solubility of polypeptides in the β structure. Among the variations in β-sheet geometry, twisting of the β sheet has been examined, and it was found that twisting led to increased amplitudes of the CD bands. Weakly twisted β sheets have similar amplitudes for the two bands of opposite sign at 215 and 198 nm, whereas strongly twisted β sheets have a stronger

[34] P. J. Gans, P. C. Lyu, M. C. Manning, R. W. Woody, and N. R. Kallenbach, *Biopolymers* **31**, 1605 (1991).
[35] D.-H. Chin, R. W. Woody, C. A. Rohl, and R. L. Baldwin, *Proc. Natl. Acad. Sci. USA* **99**, 15416 (2002).
[36] S. Brahms, J. Brahms, G. Spach, and A. Brack, *Proc. Natl. Acad. Sci. USA* **74**, 3208 (1977).
[37] B. A. Wallace and R. W. Janes, *Curr. Opin. Chem. Biol.* **5**, 567 (2001).

198 nm band.[38] Theoretical studies of chain-length dependence of β-sheet CD predict an increase in the intensity of both bands with increasing length of the β sheet.

The P_2 conformation is a left-handed helix with three residues per turn, the geometry of which is derived from the poly(Pro)II helix formed by *trans* peptides; the poly(Pro)I helix has *cis* peptides. The CD spectrum of the poly(Pro)II helix is characterized by a weak positive band near 226 nm and a strong negative band near 206 nm (Fig. 2).[39] This spectrum is qualitatively similar to those of unordered (called random coil in older literature) and unfolded proteins, which suggested the existence of P_2 conformation in them.[40] Subsequent analyses of protein crystal structures confirmed the existence of short stretches of P_2 structure in globular proteins.[31,41,42]

The variable nature of turns in proteins, with the characterization of at least eight different kinds of turns,[30] makes its CD characterization difficult. Model cyclic and linear peptide studies combined with theoretical calculations have been used in identifying salient features for the major types of turns in proteins.[43] Type I turns have CD spectra qualitatively resembling that of the α helix. Type II turns have CD spectra similar to that of the β sheet, but with red-shifted extrema. Type II' turns are predicted and observed to have an α-helix–like CD spectrum.

Moffitt's exciton picture[2] and subsequent theoretical developments explain the basic interactions resulting in the CD patterns for various polypeptide secondary structures and proteins. In the α helix the exciton splitting of the first peptide $\pi\pi^*$ transition gives rise to three components. A long-wavelength component polarized along the helix axis is responsible for the negative 208-nm band, and two degenerate components polarized perpendicular to the helix axis result in the positive 192-nm band. The negative band at 222 nm is due to the peptide $n\pi^*$ transition. In addition to the Moffitt exciton bands, theory predicts a "helix band," which is a consequence of helical geometry.[44] The helix band corresponds to two strong, slightly split bands of opposite sign, giving rise to a positive couplet centered at the perpendicularly polarized band position. (A couplet is a pair of closely spaced CD bands of opposite sign and is designated as

[38] M. C. Manning, M. Illangasekare, and R. W. Woody, *Biophys. Chem.* **31,** 77 (1988).

[39] G. D. Jenness, C. Sprecher, and W. C. Johnson, Jr., *Biopolymers* **15,** 513 (1976).

[40] M. L. Tiffany and S. Krimm, *Biopolymers* **6,** 1379 (1968).

[41] A. A. Adzhubei and M. Sternberg, *J. Mol. Biol.* **229,** 472 (1993).

[42] B. J. Stapley and T. P. Creamer, *Protein Sci.* **8,** 587 (1999).

[43] A. Perczel and M. Hollosi, *in* "Circular Dichroism and the Conformational Analysis of Biomolecules" (G. D. Fasman, ed.), p. 285. Plenum Press, New York, 1996.

[44] W. Moffit, D. D. Fitts, and J. G. Kirkwood, *Proc. Natl. Acad. Sci. USA* **43,** 723 (1957).

positive or negative according to the sign of the long-wavelength component). While direct experimental evidence is lacking for the helix band in the α helix, its existence has been established indirectly using multiple spectroscopic techniques.

The exciton interactions in β sheets result in the positive band near 195 nm and a negative band near 175 nm. The negative band near 215 nm is assigned to the $n\pi^*$ transition. The amplitude of the 215-nm band is largely determined by the mixing of $n\pi^*$ and $\pi\pi^*$ via the $\mu-m$ mechanism. Exciton coupling theory predicts a positive $\pi\pi^*$ couplet in both parallel and antiparallel β-sheet structures, but the effect is smaller than that in the α helix, resulting in smaller amplitudes. The helix band is absent in β sheets because the two directions perpendicular to the chain direction are not equivalent (nondegenerate).

The assignments of far-UV CD bands in the P_2 conformation are more complicated because of the nonconservative nature of its CD spectrum, that is, the sum of the rotational strengths does not vanish. The long-wavelength band ($\sim$217 nm) is generally assigned to the $n\pi^*$ transition and the short-wavelength band ($\sim$200 nm) is assigned to the $\pi\pi^*$ transition. Strong mixing with the higher energy transitions and/or asymmetric solvent effects are believed to be responsible for the nonconservative CD spectrum.

Protein CD can be approximated as a linear combination of secondary structure spectra determined by their relative contents in the protein structure. The coupled-oscillator interactions between various $\pi\pi^*$ transitions and the mixing of $\pi\pi^*$ and $n\pi^*$ transitions describe the basic mechanisms. Madison and Schellman[45] calculated protein CD spectra by the matrix method[17] and by combining the CD from different segments. The results for α-helical segments resembled those of ideal α helices. For β structures, agreement with ideal β-sheet CD was only qualitative, which was attributed to variations of β structures in proteins. The nonperiodic chain fragments showed the greatest disagreement between experiment and theory, and even poorer agreement was obtained for poly(Pro)II helix.

Systematic comparisons[16,46–50] of calculated CD with experiment have been carried out for a set of proteins with known crystal structures and different compositions of secondary structures, using both matrix and

[45] V. Madison and J. A. Schellman, *Biopolymers* **11,** 1041 (1972).

[46] J. D. Hirst, *J. Chem. Phys.* **109,** 782 (1998).

[47] N. A. Besley and J. D. Hirst, *J. Phys. Chem. A* **102,** 10791 (1998).

[48] J. D. Hirst, S. Bhattacharjee, and A. V. Onufriev, *Faraday Discuss.* **122,** 253 (2002).

[49] R. W. Woody and N. Sreerama, *J. Chem. Phys.* **111,** 2844 (1999).

[50] A. Koslowski, N. Sreerama, and R. W. Woody, *in* "Circular Dichroism: Principles and Applications" (N. Berova, K. Nakanishi, and R. W. Woody, eds.), 2nd Ed., p. 55. John Wiley & Sons, New York, 2000.

polarizability methods. These calculations were performed using a description of transitions, obtained either by high-level quantum chemical methods or by experimental data on transition moment directions.

The dipole interaction model calculations performed by Bode and Applequist[16] used the empirical polarizabilities of the amide $\pi_o\pi^*$ transition and the aliphatic side chains, and were performed on protein fragments representing different structural components (helix, sheet, and unordered; here, the unordered structure includes turns). The CD spectra of the reassembled proteins constructed from the individual component spectra were comparable to the weighted average component spectra. Bandwidths of 4000 and 6000 cm^{-1} were employed in these calculations and, in general, the larger bandwidth (6000 cm^{-1}, which is ~13 nm at 190 nm) gave better results. The $n\pi^*$ transition was not included in these calculations, and the theoretical predictions for the 190-nm band compared qualitatively with experiment. The inclusion of solvent effects in the dipole-interaction model indicated that the effects of the solvent on the calculated CD spectra are small for proteins. The dipole-interaction model predicts a qualitatively correct CD spectrum for the poly(Pro)II helix,[13] but fails to reproduce the CD spectra of unordered regions.

Protein CD spectra were calculated by Hirst and co-workers,[46,47] using the matrix method[17] and amide transition parameters derived from various quantum chemical methods, ranging from semiempirical CNDO/S to solution-phase *ab initio* CASSCF/SCRF methods, with acetamide and *N*-methyl acetamide as models for the peptide chromophore. Their best results were obtained with transition monopoles that reproduce the *ab initio* electrostatic potential arising from the transition charge densities.[47] A bandwidth of 15.5 nm was used for all transitions. The Spearman rank correlation coefficients between the experimental and calculated values of $[\theta]_{190}$ for 29 proteins improved from 0.44 (CNDO/S parameters) to 0.66, and from 0.48 to 0.90 for $[\theta]_{220}$. The improvement appears to be mainly due to the use of transition charge densities fitted to the electrostatic potential rather than use of the CASSCF parameters per se, since the correlation coefficients obtained when the solution-phase transition monopole charges were placed near atomic centers were poorer or comparable to those obtained with the gas-phase parameters. The results for 15 proteins that were common between this protein set and that used in the dipole interaction model were comparable. The dipole interaction model showed better correlation at 190 nm while the matrix method with improved parameters showed better correlation at 220 nm.

Woody and Sreerama[49] also used the matrix method[17] in their calculation of protein CD spectra, but their transition parameter set was derived from a combination of experimental data and theoretical parameters. For the two amide $\pi\pi^*$ transitions, they used experimental data for secondary amides, and for the $n\pi^*$ transition they used parameters derived from INDO/S wavefunctions. The transition parameters used for protein chromophores are described in the preceding section. Bandwidths used were as follows: 10.5 nm for the $n\pi^*$ transition; 11.3 nm for the $\pi_o\pi^*$ transition; 7.2 nm for the $\pi_+\pi^*$ transition; and 12.8 to 7.1 nm for the various aromatic transitions, calculated from an empirical relationship between bandwidth and transition wavelength. The Spearman rank correlation coefficients obtained for a set of 23 proteins at 190 and 220 nm were 0.66 and 0.84, respectively. Inclusion of side-chain transitions led to inferior correlation coefficients (0.46 and 0.71, respectively). The rank correlation coefficient, which calculates the correlation between the ranking of the proteins determined by the CD value at a given wavelength, is a poor measure. A more detailed comparison indicates larger disagreements between theory and experiment, which is pronounced in the 195 to 208 nm region where the mean absolute error in ellipticity is $2 - 3\Delta\varepsilon$.

Theoretical CD spectra (from Woody and Sreerama[49]) for eight proteins, two each from α-rich, $\alpha\beta$, β_I, and β_{II} classes,[51] are compared with experiment in Fig. 3. These represent typical results one obtains from CD calculations. The theoretical CD spectra of proteins with large α-helix content from all three sets of calculations generally agree with the experiment, as do those for ideal α-helix and β-sheet structures. The results for β sheet–rich proteins are much poorer, predicting positive and larger than experimentally observed amplitudes for the 195-nm band. Agreement is especially poor for some β proteins that have a negative band around 200 nm. The results for proteins with comparable α and β contents are intermediate. Overall, the agreement between theoretical and experimental protein CD spectra could be called semiquantitative, and the β proteins are the major source of disagreement.

The β-rich proteins exhibit two types of CD spectra, one showing features that are reminiscent of model β sheets, a positive $\pi\pi^*$ band ($\sim$195 nm) followed by a negative $n\pi^*$ band ($\sim$215 nm), and the other those of unordered polypeptides, a negative $\pi\pi^*$ band ($\sim$220 nm), leading to the classification as β_I and β_{II} proteins.[51] The existence of P_2 structure in the unordered conformation has been established.[31] An analysis of β sheet and P_2 structure contents in β-rich proteins showed that β_{II} proteins have larger ratios of P_2 to β-sheet contents (>0.4) than β_I proteins (<0.4).[52] The average P_2 structure content in globular proteins is small ($<10\%$), and generally the P_2 structure appears in short stretches.[31,41,42] In proteins

[51] J. Wu, J. T. Yang, and C.-S. C. Wu, *Anal. Biochem.* **200,** 359 (1992).
[52] N. Sreerama and R. W. Woody, *Protein Sci.* **12,** 384 (2003).

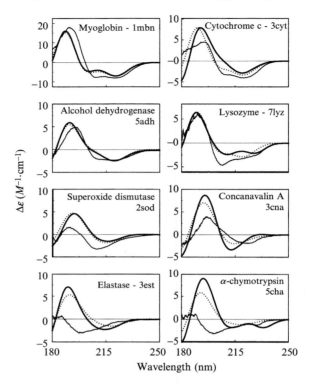

FIG. 3. Comparison of theoretical and experimental CD spectra for selected proteins. The two theoretical CD spectra[49] were obtained by excluding (including only peptide groups, dotted line) or including aromatic chromophores (including all aromatic side chains and peptide groups, thick line) in the calculation. The experimental CD spectra[53] are shown as thin lines. The PDB code of the crystal structure used in the CD calculation is given for each protein.

with moderate to high α content, CD contributions from P_2 structure are probably masked by the strong α-helical contribution. However, in β-rich proteins contributions from P_2 structure seems to be influencing the protein CD, and the inability of the matrix method to correctly calculate the P_2 CD appears to be largely responsible for the greater disagreement between theory and experiment for β-rich proteins. A better description of transition parameters or improvements in the methodology that predict the nonconservative CD spectrum of P_2 structure is needed for a better understanding of protein CD spectra.

Most protein CD calculations use a static structure, generally determined by X-ray diffraction, while the experimental data correspond to the dynamic solution–state structure. Integration of molecular dynamics

with CD calculations provides a means for introducing structural fluctuations in these calculations.[48,54,55] Both equilibrium and folding/unfolding
molecular dynamics trajectories can be used in such studies. The equilibrium dynamics of elastase and concanavalin A indicate that while the
ensemble-averaged CD spectrum of concanavalin A, a β_I protein, was similar to that of the crystal structure, that of elastase, a β_{II} protein, was quite
different.[48] A shift in the population of dihedral angles toward the P_2
region in the Ramachandran plot that was observed in the elastase ensemble was deemed responsible for the changed CD spectrum, which reduced
the disagreement with experiment.[48] The myoglobin unfolding trajectory
was examined for changes in helicity, estimated from the $[\theta]_{220}$ values,
which stayed near 0.3 while that calculated from the structure fluctuated
between 0.2 and 0.4.[48] Equilibrium dynamics of ribonuclease A combined
with CD calculations resulted in somewhat better agreement between
theory and experiment than that obtained with the static crystal structure
above 195 nm and poorer agreement at shorter wavelength[28,55] (Fig. 4).
These studies have demonstrated the feasibility of the combined molecular

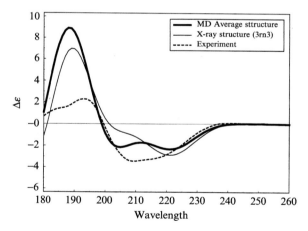

FIG. 4. Comparison of theoretical (solid lines) and experimental (dotted line) CD spectra
of ribonuclease A. One theoretical CD spectrum (thin line) corresponds to the static structure
(PDB code of the crystal structure, 3rn3). The second theoretical spectrum (thick line) was the
result of a combined molecular dynamics/CD study, and was obtained as an average of 75 CD
spectra corresponding to 75 structures, at an interval of 4 ps, along a 300-ps molecular
dynamics trajectory.[55]

[53] N. Sreerama and R. W. Woody, *Anal. Biochem.* **282**, 252 (2000).
[54] J. Fleischhauer, J. Grötzinger, B. Kramer, P. Krüger, A. Wollmer, R. W. Woody, and
E. Zobel, *Biophys. Chem.* **49**, 141 (1994).
[55] A. Koslowski, H. Botterweck, J. Fleischhauer, G. Kurapkat, A. Wollmer, and R. W.
Woody, *Prog. Biophys. Mol. Biol.* **65**(Suppl. 1), 43 (1996).

dynamics/CD approach for proteins, but the inability of the CD calculations to account for the nonconservative CD spectrum of the P_2 conformation still exists. Inclusion of other factors, such as the solvent effects, nonplanarity of the peptide bond, and mixing with high-energy transitions, should lead to further improvements in the calculation of protein CD.

Analysis of Protein CD

In this section we discuss the application of CD spectroscopy to obtain insights on protein structure and conformation. The analysis generally starts with the protein CD spectrum and may be considered the inverse of the computation of protein CD, where one starts with the structure. However, computed CD spectra can also be used in deriving helpful information about the conformation of side-chain and extrinsic chromophores.

The far-UV CD of a protein is generally reflective of the secondary structure content of the protein because it is dominated by contributions from the amide chromophore, which are determined by the backbone conformation. One of the most successful applications of CD, the structural characterization of proteins, is developed from this remarkable sensitivity of far-UV CD to secondary structure. Although the side-chain contributions to protein far-UV CD are generally submerged under amide contributions, some proteins have strong interactions between aromatic chromophores that can lead to significant far-UV CD bands. On the other hand, the near-UV CD is generally due to aromatic and disulfide contributions, providing insights into tertiary interactions. CD bands due to extrinsic chromophores, such as prosthetic groups, metal ions, inhibitors, and substrate analogs, are seen in the visible and near-UV regions and give information about protein–ligand interactions.

Secondary Structure Analysis

The approximation that a given protein CD spectrum (C_λ) can be expressed as a linear combination of secondary structure component spectra, $B_{k\lambda}$, given as

$$C_\lambda = \sum f_k B_{k\lambda} \qquad (19)$$

where f_k is the fraction of secondary structure k, forms the basis of analytical methods for protein CD analysis.[56–61] With the knowledge of $B_{k\lambda}$ one can determine the fractions f_k corresponding to a given C_λ using Eq. (19).

[56] N. J. Greenfield and G. D. Fasman, *Biochemistry* **8,** 4108 (1969).
[57] J. T. Yang, C.-S. C. Wu, and H. M. Martinez, *Methods Enzymol.* **130,** 208 (1986).
[58] W. C. Johnson, Jr., *Annu. Rev. Biophys. Biophys. Chem.* **17,** 145 (1988).

$B_{k\lambda}$ can be obtained from polypeptides in specific conformations, as was done in the earlier methods. Secondary structures in proteins deviate from ideal conformations, and in most current methods $B_{k\lambda}$ that give a better representation of component spectra in proteins are derived from a set of CD spectra of proteins with known secondary structures. Such a set of proteins is called a basis set or reference set.

For a reference set of N proteins with known CD spectra and secondary structures, one can write N linear equations.

$$[C_\lambda]_N = \left[\sum f_k B_{k\lambda}\right]_N \tag{20}$$

The common elements among them are the component spectra $B_{k\lambda}$, which can be obtained by solving the N equations simultaneously and minimizing the difference between the calculated and experimental CD spectra for the reference proteins by the least-squares method.[56,57,62] However, such a simple procedure becomes unstable because the number of equations, N, far exceeds the number of unknowns, f_k. Ridge regression, adopted in the program CONTIN,[63] fits the analyzed CD spectrum (C_λ^{obs}) as a linear combination of the reference CD spectra by minimizing the function:

$$\sum_\lambda (C_\lambda^{\text{calc}} - C_\lambda^{\text{obs}})^2 + \alpha^2 \sum_{j=1}^N (\nu_j - N^{-1})^2 \tag{21}$$

where α is the regularizer and ν_j is the coefficient of the CD spectrum for the jth reference protein used to construct the analyzed spectrum, C_λ^{calc}. The secondary structures corresponding to the analyzed CD spectrum are calculated using the coefficients ν_j with the secondary structures of protein j:

$$f_k^{\text{calc}} = \sum \nu_j f_k^j \tag{22}$$

One can also formulate this problem as a matrix equation by constructing a CD spectral matrix $\mathbf{C}$ of dimension $M \times N$, with M CD data points, and a structural matrix $\mathbf{F}$ of dimension $K \times N$, with K secondary structures.

[59] S. Y. Venyaminov and J. T. Yang, in "Circular Dichroism and the Conformational Analysis of Biomolecules" (G. D. Fasman, ed.), p. 69. Plenum, New York, 1996.

[60] N. J. Greenfield, *Anal. Biochem.* **235**, 1 (1996).

[61] N. Sreerama and R. W. Woody, in "Circular Dichroism: Principles and Applications" (N. Berova, K. Nakanishi, and R. W. Woody, eds.), 2nd Ed., p. 601. John Wiley & Sons, New York, 2000.

[62] Y. H. Chen, J. T. Yang, and H. M. Martinez, *Biochemistry* **11**, 4120 (1972).

[63] S. W. Provencher and J. Glöckner, *Biochemistry* **20**, 33 (1981).

$$\mathbf{C} = \mathbf{BF} \tag{23}$$

The inverse of matrix $\mathbf{B}$, which has a dimension $K \times M$, can be used with any C_λ for estimating f_k (represented as vectors C and f of dimensions M and K, respectively).[64]

$$\mathbf{B}^{-1}\mathbf{C} = \mathbf{F} \qquad \text{or} \qquad \mathbf{B}^{-1}C = f \tag{24}$$

The matrix equation can be solved using singular value decomposition[58] or factor analysis[65] techniques, which are mathematically similar. In the singular value decomposition algorithm, one writes matrix $\mathbf{C}$ as a product of three matrices, $\mathbf{USV}^{\mathrm{T}}$, where $\mathbf{U}$ and $\mathbf{V}$ are unitary matrices ($\mathbf{UU}^{\mathrm{T}} = \mathbf{I}$, where matrix $\mathbf{U}^{\mathrm{T}}$ is the transpose of matrix $\mathbf{U}$ and $\mathbf{I}$ is the identity matrix), and $\mathbf{S}$ is a diagonal matrix. The diagonal elements of matrix $\mathbf{S}$ are the singular values, and their magnitudes determine their importance in the analysis; smaller values generally correspond to experimental noise and are ignored in the analysis. Equation (24) is solved by determining $\mathbf{B}^{-1}$ as a multiplicative inverse of the matrix $\mathbf{B}$, written as a matrix product: $\mathbf{FVS}^{+}\,\mathbf{U}^{\mathrm{T}}$, where $\mathbf{S}^{+}$ is the inverse of matrix $\mathbf{S}$.[66]

Alternatively, one can also consider a given pair of CD spectrum and corresponding secondary structure as a pattern, $C_\lambda \leftrightarrow f_k$, and create N such patterns for an N-protein reference set. Pattern recognition techniques, such as neural networks, are utilized to determine information flow from the spectral pattern to the structure pattern, which can then be used with any given spectral pattern.[67–69]

The protein secondary structures used are determined from their crystal structures, which are available in the Protein Data Bank as atomic coordinates. The secondary structure content is determined by an unbiased assignment of the secondary structures, using certain geometric rules that define a given structure. Of the many algorithms that are in use for assigning secondary structures to a set of coordinates, the DSSP[70] method is the most popular. Structures assigned by DSSP include α, 3_{10}, and π helices, β sheets, β turns, and S bends; α, 3_{10}, and π helices are combined to obtain the α-helical fraction, and β turns and S bends are combined for the turns fraction. One can consider four or two residues, respectively, for each α-helical or β-strand segment as distorted, as an approximation for end effects, and

[64] J. P. Hennessey, Jr. and W. C. Johnson, Jr., *Biochemistry* **20**, 1085 (1981).
[65] P. Pancoska and T. A. Keiderling, *Biochemistry* **30**, 6885 (1991).
[66] L. A. Compton and W. C. Johnson, Jr., *Anal. Biochem.* **155**, 155 (1986).
[67] G. Böhm, R. Muhr, and R. Jaenicke, *Protein Eng.* **5**, 191 (1992).
[68] M. A. Andrade, P. Chacán, J. J. Merolo, and F. Morán, *Protein Eng.* **6**, 383 (1993).
[69] N. Sreerama and R. W. Woody, *J. Mol. Biol.* **242**, 497 (1994).
[70] W. Kabsch and C. Sander, *Biopolymers* **22**, 2577 (1983).

further split the α and β fractions into distorted and regular fractions.[71] Algorithms for assigning the P_2 structure are also available.[31,72] The protein secondary structures that are included in CD analysis are as follows: regular and distorted α helices, regular and distorted β strands, turns, P_2 structure, and unordered or remainder.

The construction of a good reference protein set that represents the spectral and structural variations present in proteins is important for a successful protein CD analysis. The characteristics of protein CD spectra and their relation to secondary structure are discussed in the previous section. The dominating effect of α–helical CD contributions in proteins is well documented. In comparison with α helices, other secondary structures not only have weaker CD but also have greater structural variability. Presently, a 42-protein reference set created by combining different reference protein sets developed independently by three research groups is available for the analysis of protein CD spectra.[53] Inclusion of a large number of proteins, belonging to different tertiary structure classes and with varying secondary structure contents, is expected to provide a good representation of the spectral and structural variability in proteins. A larger, 48-protein reference set that includes 6 denatured proteins is also available[53] and can be used in specific applications requiring the characterization of the unordered state.

Using all proteins in the reference set and a mathematical method based on the formulation of Eq. (19), one obtains a solution for a given CD spectrum C_λ, which gives the corresponding secondary structure fractions f_k. Ideally, the values of f_k thus obtained should be positive, the sum of f_k should be unity, and the reconstructed CD spectrum should match C_λ. These ideal conditions are rarely met. The additivity of individual secondary structure contributions involved in the linear approximation of Eq. (19) assumes the equivalence of ensemble-averaged solution and the time-averaged solid-state structures, and ignores the contributions from nonpeptide chromophores and the effects of geometric variability of secondary structure and of tertiary structure. The nonideality of the situation often results in negative f_k and/or $\sum f_k$ not being unity, and one can use these criteria as selection rules for the validity of the analysis. They are used as constraints in some least-squares methods.[62,63] By and large, the assumptions involved in protein CD analysis are valid, but a poor or inadequate representation of the analyzed spectrum in the reference set results in a failed analysis (nonideal f_k values) or erroneous results. The inadequacies of the assumptions and their effects on protein CD analysis are

[71] N. Sreerama, S. Y. Venyaminov, and R. W. Woody, *Protein Sci.* **8**, 370 (1999).
[72] S. M. King and W. C. Johnson, Jr., *Proteins Struct. Funct. Genet.* **35**, 313 (1999).

overcome by relaxing the physical constraints and/or introducing flexibility in the analysis.

The principle of selecting a limited number of proteins from the reference set specifically for the analysis of a given CD spectrum, thus creating one or more subsets, is embodied in the variable selection method.[73] It effectively creates reference sets that include proteins that are important for the analysis and excludes proteins that adversely affect the analysis. *A priori*, one does not know whether a given protein needs to be included or excluded in the analysis, and the selection is accomplished differently by different algorithms. In the basic variable selection method[73] one or more proteins are deleted from the reference set sequentially; removal of one protein results in N reference protein sets of size $N - 1$; removal of two proteins gives $N \times (N - 1)/2$ reference sets of size $N - 2$; and so on. This results in a large number of solutions, with one solution from each reference set, and performing variable selection exhaustively becomes computationally prohibitive. Proteins in the reference set can also be removed on the basis of the similarity of their CD spectra with the analyzed spectrum, with the least similar protein being removed first and a few most similar proteins always included in the analysis. This implementation is referred to as the locally linearized model[74] and is computationally much less intensive. One can also select a specific number of proteins, randomly chosen from the reference protein set, to create a minimal basis.[75] Since this approach generates a large number of combinations of proteins the process is stopped once an acceptable number of valid solutions is obtained.[76] With the knowledge of the tertiary structure class (α-rich, β-rich, or $\alpha\beta$) of the analyzed protein, one can use tertiary class-specific reference proteins[77] for CD analysis. Proteins in the reference set can also be weighted variably in a least-squares fitting of the analyzed CD spectrum,[63] as is done in the ridge regression technique implemented in the computer program CONTIN. Variable selection has been explicitly introduced[53] in CONTIN via the locally linearized implementation (CONTIN/LL).

The inclusion of the analyzed protein in the reference set improves the quality of $B_{k\lambda}$ and the subsequent analysis, but the corresponding f_k values are unknown. The self-consistent method[78] resolves this problem by making a guess for the unknown structure, the secondary structure of

[73] P. Manavalan and W. C. Johnson, Jr., *Anal. Biochem.* **167,** 76 (1987).

[74] I. H. M. van Stokkum, H. J. W. Spoelder, M. Bloemendal, R. van Grondelle, and F. C. A. Groen, *Anal. Biochem.* **191,** 110 (1990).

[75] B. Dalmas and W. H. Bannister, *Anal. Biochem.* **225,** 39 (1995).

[76] W. C. Johnson, Jr., *Proteins Struct. Funct. Genet.* **35,** 307 (1999).

[77] N. Sreerama, S. Y. Venyaminov, and R. W. Woody, *Anal. Biochem.* **299,** 271 (2001).

[78] N. Sreerama and R. W. Woody, *Anal. Biochem.* **209,** 32 (1993).

the protein in the reference set that is spectrally most similar to the analyzed protein. Solution of the matrix equation gives a new set of f_k values that replace the initial guess and the process is iterated for convergence.

Implementation of variable selection invariably gives a large number of solutions. The validity of a given solution is generally measured by the relaxed selection rules, as applicable[58–61]: Sum rule, $|\sum f_k - 1.0| \leq 0.05$; Fraction rule, $f_k \geq -0.025$; Spectral rule, error between the experimental and reconstructed spectra is $\leq 0.25\Delta\varepsilon$; Helix rule,[76] the difference between the helix fractions obtained with the all-protein reference set and the variably selected reference set is ≤ 0.03. The first two selection rules are derived from the physical constraints of f_k, the third follows Eq. (19), and the fourth is a consequence of the dominance of α–helical CD in proteins. Generally, all valid solutions are averaged to obtain the secondary structure fractions f_k corresponding to a given CD spectrum.

Two approaches for introducing variable selection, locally linearized and minimal basis approaches, and the self-consistent method are combined with the singular value decomposition algorithm for solving matrix equation in two popular computer programs (SELCON3 and CDSSTR) for protein CD analysis. The three computer programs SELCON3, CDSSTR, and CONTIN/LL, provided with multiple reference protein sets with different wavelength ranges, have been made available with the CDPro software package (Internet: http://lamar.colostate.edu/~sreeram/CDPro). These programs and the reference sets are also available for use at the Internet-based CD analysis site DICHROWEB (http://www.cryst.bbk.ac.uk/cdweb/). Computer programs that use neural network methods are also available for protein CD analysis (http://bioinformatik.biochemtech.uni–halle.de/cd_spec/), but these do not include variable selection and the resulting flexibility.

The success of an analytical method is generally measured by the accuracy of the results and reliability of the analysis. The accuracy of protein CD analysis is determined by the performance of the method for the members of a given reference set by cross-validation. CD estimates of the secondary structure fractions of each protein in the reference set are obtained by removing it from the reference set and analyzing its CD spectrum, using the other members of the reference set. The performance of a method is assessed statistically, using RMS deviations (δ) and correlation coefficients (r) between CD and X-ray estimates of secondary structures. Such a comparison of the performance of the three programs and three reference sets from CDPro[53] is given in Table I, where δ and r for each secondary structure, separately and collectively, are given. These performance indices give an estimate of the accuracy of either a given method (e.g., δ and r, for CDSSTR) or a given secondary structure (e.g., δ_T and r_T, for turns). On the whole, CDSSTR performed better than the other two

TABLE I

PERFORMANCE OF SELCON3, CDSSTR, AND CONTIN/LL PROGRAMS IN ANALYZING PROTEIN CD SPECTRA OF 29 PROTEINS IN WAVELENGTH RANGE 190–240 nm[a]

Reference protein	Method	α_R		α_D		β_R		β_D		T		U		Overall[b]	
		δ	r	δ	r	δ	r	δ	r	δ	r	δ	r	δ	r
29	SELCON3	0.052	0.949	0.053	0.689	0.102	0.547	0.036	0.709	0.075	0.302	0.118	0.268	0.078	0.773
	CDSSTR	0.059	0.938	0.052	0.785	0.083	0.655	0.030	0.790	0.074	0.337	0.097	0.491	0.070	0.817
	CONTIN/LL	0.058	0.936	0.055	0.679	0.102	0.486	0.035	0.719	0.074	0.323	0.103	0.317	0.075	0.784
37	SELCON3	0.047	0.960	0.050	0.715	0.094	0.638	0.036	0.704	0.063	0.538	0.116	0.142	0.073	0.795
	CDSSTR	0.059	0.939	0.047	0.811	0.087	0.648	0.030	0.801	0.066	0.452	0.098	0.413	0.069	0.819
	CONTIN/LL	0.054	0.944	0.052	0.706	0.093	0.624	0.033	0.753	0.066	0.447	0.095	0.328	0.069	0.813
43	SELCON3	0.051	0.953	0.048	0.747	0.086	0.659	0.034	0.746	0.073	0.382	0.110	0.181	0.072	0.802
	CDSSTR	0.064	0.929	0.042	0.792	0.081	0.704	0.028	0.843	0.067	0.462	0.089	0.444	0.065	0.833
	CONTIN/LL	0.053	0.942	0.048	0.756	0.084	0.674	0.031	0.781	0.076	0.373	0.096	0.262	0.069	0.817

[a] The 29-proteins form the reference protein set in the wavelength range 178–260 nm and a subset of the 37 and 43-protein reference sets, which have smaller wavelength ranges of 185–250 and 190–240 nm, respectively. The results presented illustrate the improvements in performance with increasing number of reference proteins. The secondary structures are: α_R, regular α helix; α_D, distorted α helix; β_R, regular β strand; β_D, distorted β strand; T, turns; and U, unordered.
[b] All six secondary structure fractions were considered collectively.

methods, but the difference is mainly due to improved performance for the unordered fraction; results for α, β, and turn fractions from all three methods are comparable. Also, the performance of a given method improves with the size of the reference set due to better representation of spectral variations. The perceived difficulty in the analysis of β-rich proteins with two types of characteristic CD spectra[51] is presumably overcome by the use of a larger reference set that provides adequate representation of both types of spectra, and variable selection which allows the inclusion of similar proteins.[52] The similarity of the performance of all three methods, despite the differences in the mathematical techniques or the means of implementation of variable selection, for α, β, and turn fractions gives a measure of the reliability of analysis. Improvements in both accuracy and reliability in the estimation of secondary structure from protein CD spectra are achieved by the use of multiple methods of analysis combined with a larger reference protein set.[53]

Other Structural Information

Methods have also been developed to estimate the number of segments of secondary structures and the tertiary structure class from protein CD spectrum. A matrix description of secondary structure segments,[79] with the number of α-helix, β-sheet, and coil segments forming the diagonal elements and the junctions between them forming the off-diagonal elements, has been used in conjunction with a neural network–based analysis of the protein CD spectra to estimate the number of α and β segments. The number of distorted residues in α helices and β sheets obtained from CD analysis has also been used to estimate the number of segments in a given protein,[71] considering four and two residues distorted in an α helix and a β sheet, respectively. The results from these two methods were comparable. The latter method is implemented in CDPro.

The tertiary structure class of a protein can be identified from its CD spectrum by representing proteins as vectors in a multidimensional space of CD data.[80] In such a hyperspace, proteins belonging to different tertiary structure classes form clusters separated by hyperplanes. The equations of the hyperplanes separating the different clusters are used to assign any new CD spectrum to a tertiary structure class (α-rich, β-rich, $\alpha\beta$, or denatured). A computer program, CLUSTER, which identifies the tertiary structure class and creates a reference set specific to the determined tertiary class, is available in CDPro. Such a tertiary class-specific reference set[77] provides

[79] P. Pancoska, V. Janota, and T. A. Keiderling, *Anal. Biochem.* **267**, 72 (1999).
[80] S. Y. Venyaminov and K. S. Vassilenko, *Anal. Biochem.* **222**, 176 (1994).

limited but specific information for the analysis of protein CD spectra, for example, an α-rich protein set for analyzing an α-rich protein CD, and improves the analysis.

Aromatic Contributions

Aromatic contributions can be considered weak in comparison with those from backbone amides.[81] However, changes in the environment of aromatic side chains in proteins caused by external perturbations, such as ligand binding, mutations, or experimental conditions, can lead to detectable changes in their CD signals. Such contributions[81–83] are generally monitored in the near-UV region, where backbone amides do not contribute. The analysis is, however, of a qualitative nature, although in some proteins aromatic CD bands can be assigned to specific residues to extract information about their environment. Protein-folding studies often utilize the changes in aromatic CD as an indicator of tertiary interactions.

In some proteins, even in the far UV, aromatic side chains make significant contributions. This is especially true for proteins with low α-helix content, such as lectins, immunoglobulins, and snake toxins. The positive CD bands in the 225- to 235-nm region reported for many proteins[84] are certainly due to Tyr or Trp side chains or to disulfides, since the amide contributions in this region are generally negative, and these side-chain contributions can interfere with protein CD analysis for secondary structure estimation. Although the protein CD analysis methods do not explicitly consider aromatic contributions, they are implicit in the analysis as a part of spectral variations represented in the reference set. Both aromatic contributions and secondary structure variations are accommodated in the analysis by the flexibility of the analysis, variably selecting proteins specific for a given CD spectrum. In the extreme case, where the anomalous aromatic contribution is not represented in the reference set, even flexible methods either give poor results or fail. Such situations can be identified with the use of multiple methods of analysis that give dissimilar results, and may be corrected by employing other spectroscopic techniques such as VCD and infrared (IR). Attempts to explicitly incorporate or extract aromatic CD contributions from a reference protein set[85] have been

[81] R. W. Woody and A. K. Dunker, in "Circular Dichroism and the Conformational Analysis of Biomolecules" (G. D. Fasman, ed.), p. 109. Plenum Press, New York, 1996.
[82] E. H. Strickland, *CRC Crit. Rev. Biochem.* **2**, 113 (1974).
[83] P. C. Kahn, *Methods Enzymol.* **61**, 339 (1979).
[84] R. W. Woody, *Eur. Biophys. J.* **23**, 253 (1994).
[85] I. A. Bolotina and V. Y. Lugauskas, *Mol. Biol.* [English translation of *Molekul. Biol.*] **19**, 1154 (1985).

of limited value, because of lack of structural patterns and characteristic CD bands, and cancellation of contributions from different aromatic side chains.

Aromatic side chains often form interacting pairs or clusters in proteins[86] and the coupled-oscillator interactions between them is a source of aromatic CD in proteins, which explains the absence of correlation between the number of aromatic side chains and aromatic CD bands. Coupling of two aromatic groups depends on the distance separating them and the relative orientation of the aromatic rings and, normally, shorter distances lead to stronger CD signals.[87] In spite of the apparent randomness of the aromatic CD bands due to the nature of the coupling, analysis of aromatic contributions can be performed by a combination of experimental and computational tools. Experimentally, contributions from an aromatic residue can be inferred by replacing it with a nonaromatic amino acid or by an aromatic amino acid with a weaker chromophore, and examining the difference CD spectra (wild-type CD − mutant CD). Using computational tools, one can perform a similar experiment in the computer to obtain aromatic contributions theoretically, assuming that the structure does not change with the mutation. Such analyses have been performed for bovine pancreatic trypsin inhibitor (BPTI),[88] bovine pancreatic ribonuclease,[28,82] dihydrofolate reductase (DHFR),[89] human interleukin 1β,[90] barnase,[91] carbonic anhydrase,[92] and gene 5 protein.[93] In most cases, the experimental difference CD spectra and theoretically calculated difference CD spectra are in good agreement, implying structural similarity of the wild-type and the mutant proteins. However, the difficulties involved in modeling structural changes due to the disruption of an aromatic cluster may lead to disagreement between theory and experiment.

Two examples of combined experimental and theoretical studies of mutants are given in Figs. 5 and 6. The theoretical CD spectra of Y → L and F → L mutant BPTI agree reasonably well with experiment[88] (Fig. 5).

[86] A. Thomas, R. Meurisse, B. Charloteaux, and R. Brasseur, *Proteins Struct. Funct. Genet.* **48**, 628 (2002).

[87] I. B. Grishina and R. W. Woody, *Faraday Discuss.* **99**, 245 (1994).

[88] N. Sreerama, M. C. Manning, M. E. Powers, J. Zhang, D. P. Goldenberg, and R. W. Woody, *Biochemistry* **38**, 10814 (1999).

[89] E. Ohmae, Y. Sasaki, and K. Gekko, *J. Biochem.* **130**, 439 (2001).

[90] S. Craig, R. H. Pain, U. Schmeissner, R. Virden, and P. T. Wingfield, *Int. J. Pept. Protein Res.* **33**, 256 (1989).

[91] S. Vuilleumier, J. Sancho, R. Loewenthal, and A. R. Fersht, *Biochemistry* **32**, 10303 (1993).

[92] P.-O. Freskgård, L.-G. Mårtensson, P. Jonasson, B.-H. Jonsson, and U. Carlsson, *Biochemistry* **33**, 14281 (1994).

[93] T. M. Thompson, B. L. Mark, C. L. Gray, T. C. Terwilliger, N. Sreerama, R. W. Woody, and D. M. Gray, *Biochemistry* **37**, 7463 (1998).

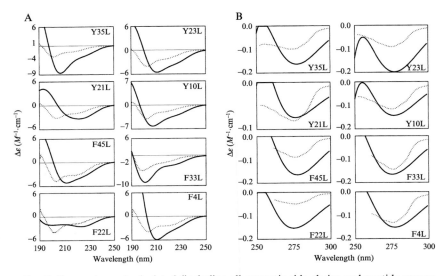

FIG. 5. Comparison of calculated (including all aromatic side chains and peptide groups, solid curve) and experimental (dotted curve) CD spectra for the Y → L and F → L mutants of bovine pancreatic trypsin inhibitor.[88] (A) Far-UV CD spectra. (B) Near-UV CD spectra.

In the near UV, theory predicts a negative CD band for all mutants, in agreement with the experiment. In the far UV, however, experimental and theoretical CD spectra show some disagreements that can be correlated to the aromatic side-chain interactions in different aromatic clusters in the BPTI structure.[88] The theoretical difference CD spectra[87] (wild-type − mutant) of specific mutants of DHFR (Fig. 6) are similar to those obtained experimentally.[89] In DHFR, two Trp side chains, W74 and W47, are in close proximity and give rise to strong coupling. Such a coupling is absent in the W74F and W47F mutants, so the interaction between W74 and W47 in the wild-type protein shows up as a positive couplet (Fig. 6) in both the theoretical and experimental difference CD spectra.[89] For the other three mutants, the theoretical difference CD spectra (Fig. 6B) show features observed in the experimental spectra (Fig. 6A). Molecular dynamics techniques can be used to model the local structural changes caused by a mutation, but large conformational changes may be difficult to reproduce.

Extrinsic Chromophores

Protein CD bands at wavelengths longer than 300 nm, in the near-UV and visible regions, are attributed to extrinsic chromophores. These are not part of the polypeptide chain, but bind to proteins either covalently or

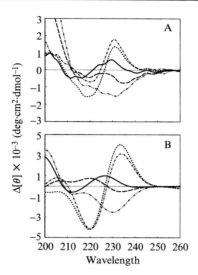

FIG. 6. Comparison of experimental and theoretical difference CD spectra (wild-type − mutant) for dihydrofolate reductase. (A) Experimental difference spectra[89] for W22L (—), W30L (– – –), W47L (· · ·), W74L (–--), and W133V (–· ·–). (B) Theoretical difference spectra for W22A (—), W30A (– – –), W47A (· · ·), W74A (–--), and W133A (–· ·–). The mutant structures used in theoretical calculations were obtained by replacing the side chain of Trp by that of Ala in the DHFR crystal structure (PDB code, 4dfr, molecule B). Theoretical CD spectra were obtained using all aromatic side chains and peptide groups.

noncovalently. Extrinsic chromophores by themselves, in the free state, may not show optical activity because they are either achiral or exist as enantiomeric mixtures, but their interaction with the chiral environment of the protein matrix on binding generates optical activity. The protein matrix may prefer a particular conformation of the chromophore, which may make it chiral and lead to intrinsic optical activity, or it may choose one enantiomer, enhancing its CD signal. The electronic transitions on the extrinsic chromophore in the bound state, in addition to having intrinsic CD, can also generate optical activity by mixing with transitions on protein chromophores by coupled-oscillator interactions. The $n\pi^*$ and $\pi\pi^*$ transitions on the extrinsic chromophore can also mix in the electrostatic field of the protein. An analysis of extrinsic contributions thus requires the knowledge of both the extrinsic chromophore conformation and the protein structure. Such analyses often involve a combination of computational and experimental studies in obtaining information about the conformation of extrinsic chromophore and the binding pocket.

The origin of the Soret CD band in heme proteins has been extensively studied.[94–98] The Soret band results from two degenerate, fully allowed $\pi\pi^*$

transitions of the porphyrin that appear in the 400-nm region. The coupled-oscillator interactions between the $\pi\pi^*$ transitions on the porphyrin and those in the protein aromatic side chains were identified as the source of positive Soret CD in sperm whale and horse metmyoglobins, and the calculated CD spectra agreed with experiment in sign and approximately in magnitude.[94] Coupled-oscillator calculations in hemoglobins from *Chironomus thummi thummi*[95] and lamprey predicted negative Soret bands, in agreement with experiment.[96] However, the coupled-oscillator mechanism alone cannot explain the observed CD differences between the two isomers of heme that differ by a 180° rotation about the α–γ axis. In the carbonmonoxy form of sperm whale myoglobin the two isomers exist in a 9:1 ratio. The major form has a strong positive band while the minor form has a weak negative band.[97] A simple rotation about the in-plane α–γ axis, giving rise to heme isomerism, does not change the net rotational strength for a planar heme in the coupled-oscillator picture. The solid-state and solution structures correspond to the major form, and attempts to model the minor form using computer simulations have not been successful. Molecular dynamics studies of the major form give a picture of nonplanar distortions of the heme and its interactions with the protein matrix, which are believed to be the source of Soret CD.[98] The intrinsic rotational strengths of the two Soret transitions, separated according to their polarization in the heme plane, show strong correlations with the ruffling deformation,[99] in which the opposite pyrrole rings are counterrotated, bringing the alternate *meso* carbons above and below the mean plane, and the average twist angle of the pyrrole rings, which measures the counterrotation of the opposite pyrroles. Rotational strengths generated by the mixing of heme transitions with both aromatic and peptide transitions were comparable to the intrinsic Soret rotational strengths.

The extrinsic CD in visual rhodopsins[100] has two positive bands, α and β, from the chromophore 11-*cis*-retinal linked to a lysine side chain via a protonated Schiff base (PSB). The extrinsic chromophore in bacterial rhodopsins[100] is all-*trans*-retinal–PSB and it gives rise to a positive α band

[94] M.-C. Hsu and R. W. Woody, *J. Am. Chem. Soc.* **93,** 3515 (1971).

[95] J. Fleischhauer and A. Wollmer, *Z. Naturforsch.* **27b,** 530 (1972).

[96] R. W. Woody, *in* "Protein–Ligand Interactions" (H. Sund and G. Blauer, eds.), p. 60. de Gruyter, Berlin, 1975.

[97] H. S. Aojula, M. T. Wilson, G. R. Moore, and D. J. Williamson, *Biochem. J.* **250,** 853 (1988).

[98] C. Kiefl, N. Sreerama, R. Haddad, L. Sun, W. Jentzen, Y. Lu, Y. Qiu, J. A. Shelnutt, and R. W. Woody, *J. Am. Chem. Soc.* **124,** 3385 (2002).

[99] J. A. Shelnutt, X. Z. Song, J. G. Ma, S. L. Jia, W. Jentzen, and C. J. Medforth, *Chem. Soc. Rev.* **27,** 31 (1998)

[100] R. R. Birge, *Biochim. Biophys. Acta* **1016,** 293 (1990).

and a negative β band. Theoretical calculations of retinal CD were performed[101] using the available crystal structures of both visual and bacterial rhodopsins, and the calculated rotational strengths for the two bands agree reasonably well with the experiment, reproducing the pattern of signs for the α and β bands in all five retinal-binding proteins studied. However, the calculated intrinsic contributions from retinal–PSB, in both visual and bacterial rhodopsins, were generally small owing to small nonplanar distortions along the polyene chain. The dominant contribution to the rotational strength was calculated from the coupling of retinal and aromatic transitions. The nonplanar distortions in retinal–PSB are the source of its intrinsic CD, which may have been suppressed by crystal interactions.

Acknowledgments

The work of our research group described in this review chapter was supported by NIH Grant EB002803. We thank Dr. K. Gekko (Hiroshima University, Hiroshima, Japan) for providing the experimental CD data of wild-type and mutant DHFR. Hirst *et al.*[102] have recently explored the effects of different factors (bandwidth of transitions, dielectric constant of the protein, etc.) on the calculated CD spectra of 47 proteins. Tetin *et al.*[103] demonstrated that the current CD analysis methods are resilient to distortions of far-UV CD caused by aromatic residues in immunoglobulins. The number of reference proteins used in protein CD analysis has been increased by the inclusion of membrane proteins.[104] The resulting soluble + membrane protein reference sets were found to improve membrane protein CD analysis.

[101] N. Sreerama, R. R. Birge, V. BuB, B. S. Hudson, D. Singh, and R. W. Woody, *Biophys. J.* **82,** 224A (2002).
[102] J. D. Hirst, K. Colella, and A. T. B. Gilbert, *J. Phys. Chem. B* **107,** 11813 (2003).
[103] S. Y. Tetin, F. G. Prendergast, and S. Yu. Venyaminov, *Anal. Biochem.* **321,** 183 (2003).
[104] N. Sreerama and R. W. Woody, *Protein Sci.* **13,** 100 (2004).

[14] Model Comparison Methods

By JAY I. MYUNG and MARK A. PITT

Introduction

The question of how one should choose among competing explanations (models) of observed data is at the core of science. Model comparison is ubiquitous and arises, for example, when a toxicologist must decide between two dose–response models or when a biochemist needs to determine which of a set of enzyme-kinetics models best accounts for observed data.

Copyright 2004, Elsevier Inc.
All rights reserved.
0076-6879/04 $35.00

Over the decades, a number of criteria that are thought to be important for model comparison have been proposed (e.g., Jacobs and Grainger[1]). They include (1) *falsifiability*[2]: whether there exist potential observations that are incompatible with the model; (2) *explanatory adequacy:* whether the theoretical account of the model helps to make sense of observed data but also established findings; (3) *interpretability:* whether the components of the model, especially its parameters, are understandable and are linked to known processes; (4) *faithfulness:* whether the ability of the model to capture the underlying regularities comes from the theoretical principles the model purports to implement, not from the incidental choices made in its computational instantiation; (5) *goodness of fit:* whether the model fits the observed data sufficiently well; (6) *complexity* or *simplicity:* whether the model's description of observed data is achieved in the simplest possible manner; and (7) *generalizability:* whether the model provides a good prediction of future observations.

Although each of these seven criteria is important in its own way, modern statistical approaches to model comparison consider only the last of these three (goodness of fit, complexity, and generalizability), largely because they lend themselves to quantification. The other four criteria have yet to be formalized and it is not clear how some even could be or should be (e.g., interpretability).

The purpose of this chapter is to provide a tutorial on state-of-the-art statistical model comparison methods. We walk the reader through the reasoning underlying their development so that how and why a method performs as it does can be understood. The chapter is written for researchers who are interested in computational modeling but are primarily involved in empirical work. We begin by discussing the statistical foundations of model comparison.

Statistical Foundations of Model Comparison

Notation and Definition

Statistically speaking, the data vector $\mathbf{y} = (y_1, \ldots, y_m)$ is a random sample from an unknown population, which represents the underlying regularity that we wish to model. The goal of modeling is to identify the model that generated the data. This is not in general possible because information in the data sample itself is frequently insufficient to narrow the choices down to a single model. To complicate matters even more, data

[1] A. M. Jacobs and J. Grainger, *J. Exp. Psychol. Hum. Percept. Perform.* **29,** 1311 (1994).
[2] K. R. Popper, "The Logic of Scientific Discovery." Basic Books. New York, 1959.

are inevitably confounded by random noise, whether it is sampling error, imprecision of the measurement instrument, or the inherent unreliability of the data collection procedure. It is usually the case that multiple models could have generated a single data sample.

In the field of engineering, this situation is referred to as an ill-posed problem because any solution (i.e., model) is not unique given what the data tell us about the underlying regularity. The statistician's way of dealing with ill-posedness is to use all of the information available (i.e., knowledge about the model and the data together) to make a best guess as to which model most likely generated the data. Stripped bare, model selection is an inference game, with selection methods differing in their rules of play.

Formally, a model is defined as a parametric family of probability distributions. Each distribution is indexed by the parameter vector $\mathbf{w} = (w_1, \ldots, w_k)$ of the model and corresponds to a population. The *probability (density) function,* denoted by $f(\mathbf{y}|\mathbf{w})$, specifies the probability of observing data $\mathbf{y}$ given the parameter $\mathbf{w}$. Given the fixed data vector $\mathbf{y}$, $f(\mathbf{y}|\mathbf{w})$ becomes a function of $\mathbf{w}$ and is called the *likelihood function* denoted by $L(\mathbf{w})$. For example, the likelihood function for binomial data is given by

$$L(\mathbf{w}) = \prod_{i=1}^{k} \binom{n_i}{y_i} w_i^{y_i} (1 - w_i)^{n_i - y_i} \tag{1}$$

where k is the number of conditions in an experiment, n_i is the number of Bernouilli trials (e.g., dichotomous observations, such as success or failure of a medical treatment, made n_i times), and w_i, as an unknown parameter, represents the probability of success on each trial, and $y_i (= 0, 1, \ldots, n_i)$ is the number of actually observed successes.

Given a data sample, the descriptive adequacy of a model is assessed by finding parameter values of the model that best fit the data in some defined sense. This procedure, called parameter estimation, is carried out by seeking the parameter vector $\mathbf{w}^*$ that maximizes the likelihood function $L(\mathbf{w})$ given the observed data vector $\mathbf{y}$—a procedure known as *maximum likelihood estimation*. The resulting maximized likelihood (ML) value, $L(\mathbf{w}^*)$, defines a measure of the model's *goodness of fit,* which represents a model's ability to fit a particular set of observed data.

Other examples of goodness-of-fit measures include the minimized sum of squared errors (SSE) between the predictions and observations of a model, the proportion variance accounted for or otherwise known as the coefficient of determination r^2, and the mean squared error (MSE) defined as the square root of SSE divided by the number of observations. Among these, ML is a standard measure of goodness of fit, most widely used in statistics, and all the model comparison methods discussed in the present chapter were developed using ML. In the rest of the chapter, goodness of fit refers to ML.

A Good Fit Can Be Insufficient and Misleading

Models are often compared on the basis of their goodness of fit. That is, among a set of models under comparison, the scientist chooses the model that provides the best fit (i.e., highest ML value) to the observed data. The justification for this choice may be that the model best fitting the data is the one that does a better job than its rivals of capturing the underlying regularity. Although intuitive, such reasoning can be unfounded because a model can produce a good fit for reasons that have nothing to do with its ability to approximate the regularity of interest, as is described below.

Selecting among models using a goodness-of-fit measure would make sense if data were free of noise. In reality, however, data are not "pure" reflections of the population of interest, as mentioned above. Noisy data make the task of inferring the underlying model difficult because what one is fitting is unclear: is it the regularity, which we care about, or the noise, which we do not? Put another way, goodness of fit can be decomposed conceptually into two separate terms as follows:

$$\text{Goodness of fit} = \text{fit to regularity} + \text{fit to noise} \qquad (2)$$

We are interested only in the first of these, fit to regularity, but any goodness-of-fit measure contains a contribution from the model's ability to fit random error as well as its ability to approximate the underlying regularity. The problem is that both quantities are unknown because when fitting a data set, we obtain the overall value of their sum, that is, a single goodness of fit. This is why a good fit can be misleading. In the worst case, a good fit can be achieved by a model that is extremely good at fitting noise yet a poor approximation of the regularity being modeled. In the next section we describe how this state of affairs can come about. The remainder of the chapter then focuses on how to correct it.

Model Complexity and Why It Matters

It turns out that a model's ability to fit random noise is closely correlated with the *complexity* of the model. Intuitively, complexity (or flexibility) refers to the property of a model that enables it to fit diverse patterns of data. For example, a model with many parameters is more complex than a model with few parameters. Also, two models with the same number of parameters but different forms of the model equation (e.g., $y = w_1 x + w_2$ and $y = w_1 x^{w_2}$) can differ in their complexity.[3,4] Generally

[3] I. J. Myung, *J. Math. Psychol.* **44**, 190 (2000).
[4] M. A. Pitt, I. J. Myung, and S. Zhang, *Psychol. Rev.* **109**, 472 (2002).

speaking, the more complex the model, the more easily it can absorb random noise, thus increasing its fit to the data without necessarily increasing its fit to the regularity. That is, too much complexity is what causes a model to fit noise. The relationship between goodness of fit and complexity is illustrated in the three small graphs in Fig. 1. The data are the dots and the lines the models. The model on the left is least complex; that on the right is most complex. Complexity is what enables the model in the lower right graph to fit the data better than the less complex models in the left and middle graphs. In fact, one can always improve goodness of fit by increasing model complexity, such as adding more parameters. This is portrayed by the top curve in the large graph. An implication of this spurious phenomenon is that an overly complex model can provide a better fit than a simpler model even if the latter generated the data. That is, the more complex model will overfit the data.

Overfitting is illustrated in Table I using four dose–response models in toxicology.[5] Simulated data varying in three sample sizes were generated from model M_2, which is considered the "true" model in the sense that it

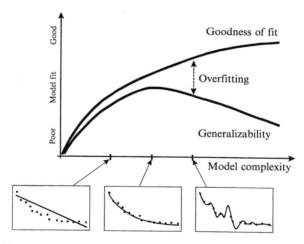

Fig. 1. An illustration of the relationship between goodness of fit and generalizability as a function of model complexity. The y axis represents any fit index, where a larger value indicates a better fit (e.g., maximized likelihood). The three smaller graphs provide a concrete example of how fit improves as complexity increases. In the left graph, the model (line) is not complex enough to match the complexity of the data (dots). The two are well matched in complexity in the middle graph, which is why this occurs at the peak of the generalizability function. In the right graph, the model is more complex than the data, capturing microvariation due to random error. Reprinted from Pitt and Myung.[6]

[5] Sand et al., Regulatory Toxicology and Pharmacology **36**, 184 (2002).
[6] M. A. Pitt and I. J. Myung, Trends Cogn. Sci. **6**, 421 (2002).

TABLE I
GOODNESS OF FIT AND GENERALIZABILITY OF MODELS DIFFERING IN COMPLEXITY[a]

	M_1 (1 parameter)	M_2 (true) (2 parameters)	M_3 (3 parameters)	M_4 (3 parameters)
Sample size: $n = 20$				
Goodness of fit	0	15	62	23
Generalizability	15	46	22	17
Sample size: $n = 100$				
Goodness of fit	0	31	56	13
Generalizability	0	64	27	9
Sample size: $n = 5000$				
Goodness of fit	0	96	4	0
Generalizability	0	98	2	0

[a] *Note:* The percentage of samples in which the particular model fitted the data best. The four models are as follows: M_1, $p = 1 - \exp(-0.1 - ax)$; M_2, $p = 1 /[1 + \exp(a - bx)]$; M_3, $p = c + (1 - c)[1 - \exp(-ax^b)]$; M_4, $y = c + (1 - c)/(1 + ax^{-b})$ where $a, b > 0$ and $0 < c < 1$. One thousand pairs of samples were generated from M_2 (true model) using $a = 2$ and $b = 0.2$ on the sample 10 points for $(x_1, \ldots, x_{10}) = (0.001, 1, 2, 4, 7, 10, 13, 16, 20, 30)$. For each probability, a series of $n(= 20, 100,$ or $5000)$ independent binary outcomes (0 or 1) were generated according to the binomial probability distribution. The number of ones in the series was summed to obtain an observed count being denoted by $y_i(i = 1, \ldots, 10)$ in Eq. (1). In this way each sample consisted of 10 observed counts. Goodness of fit was evaluated by the maximized likelihood of each model and generalizability was estimated through cross-validation an described in text.

generated the data. In each test, 1000 samples were generated and sampling error was added to each data set. The model M_2, as well as three other models differing in complexity, were fitted to these data. The first row under each sample-size condition in Table I shows the percentage of samples in which the particular model fitted the data best using the goodness-of-fit measure ML. First consider the results obtained with a small sample size of $n = 20$. M_1, with only one parameter, is clearly inadequate to capture the main trend in the data and thus never fitted the data best. It is analogous to the left-hand model in Fig. 1 and is an example of underfitting. Next consider the results for M_3 and M_4, which, with one extra parameter than M_2, are more complex than M_2. They provided a better fit much more often (62 and 23%) than the true model, M_2 (15%), even though M_2 generated the data. This is because the one extra parameter in the models enabled them to absorb sampling error above and beyond the underlying regularity. These models are analogous to the right-hand model in Fig. 1. Importantly also, note that M_3 provided a better fit more often than M_4, despite the fact that both have the same number of parameters (three). This difference in fit between the models is due to the effects of the functional form dimension of model complexity. As can be seen in the middle and lower portions of Table I, overfitting persisted even for a relatively large sample size of $n = 100$, before it all but disappeared when sample size was increased to an unrealistic level of $n = 5000$.

These simulation results should highlight the dangers of selecting models using only a goodness-of-fit measure. The potential exists for choosing the wrong model, which makes goodness of fit risky to use as a method of model selection.

Generalizability: A Solution That Embodies Occam's Razor

Returning to Eq. (2), the ability of a model to fit the regularity in the data, represented by the first term on the right-hand side of the equation, defines its *generalizability*. Generalizability, or predictive accuracy, refers to how well a model predicts the statistical properties of future, as yet unseen, samples from a replication of the experiment that generated the current data sample. That is, when a model is evaluated against its competitors, the goal should be not to assess how much better it fits a single data sample, but how well it captures the process that generated the data. This is achieved when generalizability, not goodness of fit, is the goal of model selection.

Statistically speaking, generalizability is defined in terms of a *discrepancy function* that measures the degree of approximation or dissimilarity between two probability distributions.[7] Specifically, a discrepancy between two distributions, f and g, is any well-behaved function, $D(f, g)$, that satisfies $D(f,g) \geq D(f,f) = 0$ for $f \neq g$. The larger the value of $D(f, g)$, the less one probability distribution approximates the other. The following equation gives a formal definition of generalizability of a given model f_M:

$$\text{Generalizability} = E_T[D(f_T, f_M)] = \int D[f_T, f_M(\mathbf{w}^*(\mathbf{y}))] f_T(\mathbf{y}) d\mathbf{y} \quad (3)$$

In Eq. (3), f_T represents the probability distribution from which all data are generated (i.e., true model), and $f_M(\mathbf{w}^*(\mathbf{y}))$, as a member of the model family under investigation, is the best-fitting probability distribution given a data vector $\mathbf{y}$, and thus represents a goodness-of-fit measure. According to Eq. (3), generalizability is a mean discrepancy between the true model and the model of interest, averaged across all data that could possibly be observed under the true model.

---

[7] H. Linhart and W. Zucchini, "Model Selection." John Wiley & Sons, New York, 1986.

Whereas goodness of fit is monotonically related to complexity, the relationship between generalizability and complexity is not so straightforward. An illustration of this is shown in Fig. 1. Generalizability increases positively with complexity only up to the point where the model is optimally complex to capture the regularities in the data. Any additional complexity beyond this point will cause generalizability to diminish as the model begins to overfit the data by absorbing random noise. Shown in the middle inset is the model with the highest generalizability. It captures the main trends in the data, but little if any of the random fluctuations of the points that in all likelihood are due to sampling error. In short, a model must be sufficiently complex to capture the underlying regularity yet not too complex to cause it to overfit the data and thus lose generalizability.

The improvements gained by using generalizability in model selection are shown in Table I. Generalizability was assessed using two independent data samples. Each model was fitted to the first sample to obtain the best-fitting parameter values. The resulting fit defines goodness of fit of a model, as shown in the first row under each sample-size condition. Next, with the best-fitting parameters of each model fixed, the models were fitted to the second data sample. The quality of this second fit is a measure of generalizability known as cross-validation.*

The second row under each sample-size condition in Table I shows the percentage of samples in which the particular model generalized the best. Comparison with the goodness-of-fit row reveals that an overly complex model (e.g., M_3) generalizes poorly (22% for $n = 20$ and $n = 27\%$ for $n = 100$), whereas the simpler, true model (e.g., M_2) generalizes better (46% for $n = 20$ and $n = 64\%$ for $n = 100$). This example demonstrates that the cost of the superior fit of a model to a particular data sample can result in a loss of generalizability when fitted to new data samples generated by that same process. This is because the fit of the model to the first sample was far too good, absorbing random error in the data, not just the regularity.

To reiterate, in model comparison, one would like to choose the model, among a set of competing models, that maximizes generalizability. Unfortunately, generalizability is not directly observable because noise always distorts the regularity of interest. Generalizability, therefore, must be estimated from a data sample. It is achieved by weighting the goodness of fit of a model relative to its complexity. An overly complex model is penalized to the extent that the extra complexity merely helps the model fit the random error, as illustrated in the right inset in Fig. 1. Generalizability is viewed as

---

* This method of estimating generalizability is known as cross-validation and is discussed in the next section. It is used in this example because of its simplicity. We have found that it performs poorly compared with more sophisticated measures such as MDL and BMS.

a formal implementation of Occam's razor: goodness of fit is traded off with complexity. A number of generalizability measures that implement this basic principle have been developed and are discussed in the next section.

Model Comparison Methods

Measures of Generalizability

The foremost goal of model comparison is achieving good generalizability by striking the right balance between two opposing pressures, goodness of fit and complexity. In this section we introduce five representative measures of generalizability that achieve this balance. They are the Akaike information criterion (AIC),[8,9] the Bayesian information criterion (BIC),[10] the minimum description length (MDL),[11–14] cross-validation (CV),[15,16] and finally, Bayesian model selection (BMS).[17,18] For a comprehensive treatment of these and other comparison methods, the reader is directed to a special issue of the *Journal of Mathematical Psychology*[19] and Burnham and Anderson.[20]

The five model comparison criteria are defined as follows:

$$
\begin{aligned}
\text{AIC} &= -2\ln L(\mathbf{w}^*) + 2k \\
\text{BIC} &= -2\ln L(\mathbf{w}^*) + k\ln(n) \\
\text{MDL} &= -\ln L(\mathbf{w}^*) + \frac{k}{2}\ln\left(\frac{n}{2\pi}\right) + \ln\int\sqrt{\det(\mathbf{I}(\mathbf{w}))}\,d\mathbf{w} \\
\text{CV} &= -\ln f(\mathbf{y}_{\text{val}}|\mathbf{w}^*(\mathbf{y}_{\text{cal}})) \\
\text{BMS} &= -\ln\int L(\mathbf{w})\pi(\mathbf{w})d\mathbf{w}
\end{aligned} \tag{4}
$$

[8] H. Akaike, *in* "Second International Symposium on Information Theory" (B. N. Petrox and F. Cask, eds.), pp. 267–281. Akademia Kiado, Budapest, Hungary, 1973.

[9] H. Bozdogan, *J. Math. Psychol.* **44,** 62 (2000).

[10] G. Schwarz, *Ann. Stat.* **6,** 461 (1978).

[11] J. Rissanen, *Ann. Stat.* **11,** 416 (1983).

[12] J. Rissanen, *IEEE Trans. Inform.* **42,** 40 (1996).

[13] P. Grunwald, *J. Math. Psychol.* **44,** 133 (2000).

[14] M. H. Hansen and B. Yu, *J. Am. Stat. Assoc.* **96,** 746 (2001).

[15] M. Stone, *J. R. Stat. Soc. B* **36,** 111 (1974).

[16] M. W. Browne, *J. Math. Psychol.* **44,** 108 (2000).

[17] R. E. Kass and A. E. Raftery, *J. Am. Stat. Assoc.* **90,** 773 (1995).

[18] L. Wasserman, *J. Math. Psychol.* **44,** 92 (2000).

[19] I. J. Myung, M. Forster, and M. W. Browne, eds., *J. Math. Psychol.* **44,** (2000).

[20] L. S. Burnham and D. R. Anderson, "Model Selection and Inference: A Practical Information–Theoretic Approach," 2nd Ed. Springer-Verlag, New York, 2002.

where ln $L(\mathbf{w}^*)$ is the natural logarithm of the model's maximized likelihood, k is the number of parameters of the model, and n is the sample size. Also in Eq. (4), $\pi(\mathbf{w})$ is the parameter prior density and $I(\mathbf{w})$ is the Fisher information matrix in mathematical statistics (e.g., Schervish[21]), det denotes the determinant of a matrix, and finally, $\mathbf{y}_{val}$ and $\mathbf{y}_{cal}$ are defined later when discussing the CV criterion. These comparison methods prescribe that the model minimizing the given criterion should be preferred.

AIC, BIC, and MDL

For AIC, BIC, and MDL, the first term represents a lack of fit measure and the remaining terms represent a model complexity measure. Each criterion defines complexity differently. AIC considers the number of parameters (k) as the only relevant dimension of complexity, whereas BIC considers sample size (n) as well. In MDL, the second and third terms together represent a complexity measure. The second term of MDL is essentially the same as that of BIC. It is the third term that is unique in MDL, and it accounts for the effects of complexity, due to functional form. Functional form is reflected through the Fisher information matrix $I(\mathbf{w})$. To grasp this, note that the matrix $I(\mathbf{w})$ is defined in terms of the second derivative of the log-likelihood function $\{\ln L(\mathbf{w})\}$, the value of which depends on the form of the model equation, for instance, whether $y = w_1 x + w_2$ or $y = w_1 x^{w_2}$.

Why are there such different measures of generalizability (AIC, BIC, and MDL)? How do they differ from one another and in what sense? AIC was derived as a large sample approximation of the discrepancy between the true model and the fitted model in which the discrepancy is measured by the Kullback–Leibler distance.[22] As such, AIC purports to select the model, among a set of candidate models, that is closest to the truth in the Kullback–Leibler sense. BIC has its origin in Bayesian statistics and seeks the model that is "most likely" to have generated the data in the Bayesian sense. BIC can be seen as a large sample approximation of a quantity related to BMS, which is discussed below. Finally, MDL originated from algorithmic coding theory in computer science. The goal of MDL is to select the model that provides the shortest description of the data in bits. The more the model is able to compress the data by extracting the regularities or patterns in it, the better the generalizability of the model because these uncovered regularities can then be used to predict accurately future data. As noted earlier, unlike AIC and BIC, MDL considers the

[21] M. J. Schervish, "The Theory of Statistics." Springer-Verlag, New York, 1995.
[22] S. Kullback and R. A. Leibler, *Ann. Math. Stat.* **22,** 79 (1951).

functional form of a model, and thus is designed for comparing models that differ along this dimension. Given this additional sensitivity, MDL is expected to perform more accurately than its two competitors. The price to pay for the superior performance of MDL is the computational challenge in its calculation; the evaluation of the integral term generally requires use of numerical integration techniques (e.g., Gilks *et al.*[23]).

Cross-Validation

CV is a sampling-based method in which generalizability is estimated directly from the data without an explicit consideration of model complexity. The method works as follows. First, the observed data are split into two subsamples of equal size. We then fit the model of interest to the first, calibration sample ($\mathbf{y}_{cal}$) and find the best-fitting parameter values, denoted by $\mathbf{w}^*(\mathbf{y}_{cal})$. With these values fixed, the model is fitted to the second, validation sample ($\mathbf{y}_{val}$). The resulting fit of the model to the validation data $\mathbf{y}_{val}$ defines the generalizability estimate of the model.

CV can easily be implemented using any computer programming language as its calculation does not require sophisticated computational techniques, in contrast to MDL. The ease of use of CV is offset by the unreliability of its generalizability estimate, especially for small sample sizes. On the other hand, unlike AIC and BIC, CV takes into account the functional form effects of complexity, although the implicit nature of CV makes it difficult to discern how this is achieved.

Bayesian Model Selection

BMS, a sharper version of BIC, is defined as the minus *marginal likelihood*. The marginal likelihood is the probability of the observed data given the model, averaged over the entire range of the parameter vector and weighted by the parameter prior density $\pi(\mathbf{w})$. As such, BMS aims to select the model with the highest mean likelihood of the data. The often cited *Bayes factor*[17] is a ratio of marginal likelihoods between a pair of models being compared. As in other Bayesian methods, the prior $\pi(\mathbf{w})$ in the marginal likelihood is to be determined by utilizing available information (i.e., informative priors) or otherwise as a noninformative prior such as a uniform density.[24]

BMS does not yield an explicit measure of complexity, although complexity is taken into account and is hidden in the integral. It is through

---

[23] W. R. Gilks, S. Richardson, and D. J. Spiegelhalter, "Markov Chain Monte Carlo in Practice." Chapman & Hall, New York, 1996.

[24] R. E. Kass and L. W. Wasserman, *J. Am. Stat. Assoc.* **91,** 1343 (1996).

the integral that the functional form dimension of complexity, as well as the number of parameters and the sample size, is reflected. It is also the integral that makes it nontrivial to implement BMS. As in MDL, the integral in BMS must be evaluated numerically.

Among the five comparison methods discussed above, BMS and MDL represent state-of-the-art techniques that will generally perform more accurately across a range of modeling situations than the other three criteria (AIC, BIC, and CV). On the other hand, the latter three are attractive given their ease of use, and are likely to perform adequately under certain circumstances. In particular, if the models being compared do differ in number of parameters and further, sample size is sufficiently large, then one may be able to use AIC or BIC with confidence instead of the more sophisticated BMS and MDL.

Relations to Generalized Likelihood Ratio Test

Although the generalized likelihood ratio test (GLRT) is often used to test the adequacy of a model in the context of another model, it is not an appropriate method for model comparison. In this section we briefly comment on GLRT and its relation to the model comparison methods discussed above. The generalized likelihood ratio test[25,26] is a null hypothesis significance test and is based on the G^2 statistic defined as

$$G^2 = -2 \ln \frac{\mathrm{ML_A}}{\mathrm{ML_B}} \tag{5}$$

In Eq. (5) $\mathrm{ML_A/ML_B}$ is a ratio of the maximized likelihoods of two nested models, A and B, with A being nested within B. A model is said to be nested within another if the latter yields the former as a special case. For instance, $y = w_1 x$ is nested within $y = w_1 x + w_2 x^2$.

The sampling distribution of G^2 under the null hypothesis that the reduced model A holds follows a χ^2 distribution with $\mathrm{df} = k_B - k_A$, where k_B and k_A are the numbers of free parameters of models B and A, respectively. When the null hypothesis is retained (i.e., the p value does not exceed the α level), the conclusion is that the reduced model A provides a sufficiently good fit to the observed data and therefore the extra parameters of the full model B unnecessary. If the null hypothesis is rejected, one concludes that model A is inadequate and the extra parameters are necessary to account for the observed data.

---

[25] S. S. Wilks, *Ann. Math. Stat.* **9,** 60 (1938).

[26] R. A. Johnson and D. W. Wichern, "Applied Multivariate Statistical Analysis," 4th Ed., pp. 234–235. Prentice Hall, Upper Saddle, NJ, 1998.

There are several crucial differences between GLRT and the five model comparison methods. First and importantly, GLRT does not assess generalizability, which is the goal of model comparison. Rather, it is a hypothesis-testing method that simply judges the descriptive adequacy of a given model. (For contemporary criticisms of null hypothesis significance testing, see, e.g., Berger and Berry[27] and Cohen[28]). Even if the null hypothesis is retained under GLRT, the result does not necessarily indicate that model A is more likely to be correct or generalize better than model B, or vice versa. Second, GLRT requires the nestedness assumption hold of the two models being tested. In contrast, no such assumptions are required for the five comparison methods, making them much more versatile. Third, GLRT was developed in the context of linear regression models with normally distributed error, further restricting its use. GLRT is also inappropriate for testing nonlinear models and models with nonnormal error. Again, no such restrictions are imposed on the preceding model comparison methods. In short, given its limited applicability and narrow interpretability, GLRT is a poor method of model comparison.

Model Comparison at Work

In this section, we present a model-recovery test to demonstrate the relative performance of two comparison methods, AIC and MDL. These two were chosen because they differ from each other in how model complexity is defined. AIC considers only the number of parameters, whereas MDL takes into account the functional form dimension as well. The ML, solely a goodness-of-fit measure, was included as a baseline from which the improvements in model recovery could be evaluated as the two dimensions of complexity are introduced into the selection method.

Three models, M_2–M_4 from Table I, were compared. One thousand data sets (with sampling noise added) were generated from each model, and all three models were then fitted to each group of 1000 data sets. The selection methods were compared on their ability to recover the model that generated the data. A good method should be able to identify the true model (i.e., the one that generated the data) 100% of the time. Any deviations from perfect recovery reveal a bias in the selection method.

The simulation results are reported in Table II. The top 3×3 matrix shows model recovery performance under ML. The result in the first column of the matrix, which is essentially the same as that in the first row

[27] J. O. Berger and D. A. Berry, *Am. Sci.* **76,** 159 (1998).
[28] J. Cohen, *Am. Psychol.* **49,** 997 (1994).

TABLE II

MODEL RECOVERY PERFORMANCE OF MODEL COMPARISON METHODS FOR THREE MODELS
DIFFERING IN COMPLEXITY[a]

		Data were generated from:		
Model comparison method	Model fitted	M_2	M_3	M_4
ML	M_2	30	0	2
	M_3	58	61	8
	M_4	12	39	90
AIC	M_2	76	0	7
	M_3	16	61	5
	M_4	8	39	88
MDL	M_2	90	0	15
	M_3	7	92	6
	M_4	3	8	79

[a] *Note:* The percentage of samples in which the particular model fitted the data best. The three models, M_2, M_3, and M_4, are defined in Table I. One thousand samples were generated from each model using the same 10 points of $(x_1, \ldots, x_{10}) = (0.001, 1, 2, 4, 7, 10, 13, 16, 20, 30)$. For each probability, a series of $n = 100$ (sample size) independent binary outcomes (0 or 1) were generated according to the binomial probability distribution. The parameter values used to generated the simulated data were as follows: $(a, b) = (2, 3)$ for M_2; $(a, b, c) = (0.3, 0.5, 0.1)$ for M_3, and $(a, b, c) = (500, 0.3, 0.1)$ for M_4. In parameter estimation and also in the calculation of MDL, the following parameter ranges were used: $0 < a < 100, 0 < b < 10$ for M_2; $0 < a < 100, 0 < b < 10, 0 < c < 1$ for M_3; $0 < a < 10,000, 0 < b < 10, 0 < c < 1$ for M_4.

of the middle panel in Table I, shows that an overly complex model, M_3, was chosen more often than the true data-generating model, M_2 (58 versus 30%). This bias is not surprising given that model M_3 has one more parameter than model M_2, and that a goodness-of-fit measure such as ML does not consider this or any other dimension of complexity. The first column in the simulation using AIC shows that when the difference in the number of parameters was taken into account, the bias was largely corrected. Now, the true model M_2 was chosen 76% of the time, much more often than the more complex model, M_3 (16%).

To see the effects of functional form in model selection, consider the second and third columns. The two models, M_3 and M_4, have the same number of parameters (three) but differ from each other in functional form. For these two models, an asymmetric pattern of recovery performance was observed for the models under ML. Model M_4 was correctly recovered 90% of the time, whereas model M_3 was recovered much less often (61%). In other words, M_4 fitted the data of its competitor as well as its own data, but the reverse was not true. Essentially the same pattern of model recovery was obtained under AIC. That is, AIC also overestimated the generalizability of M_4 relative to M_3. Because the two models have the same number of parameters, this bias must be due to a different dimension of complexity, namely functional form. Calculation of the complexity of these models [using the two right-hand terms of MDL in Eq. (4)] shows M_4 to indeed be more complex than M_3 (9.36 versus 8.57), with a complexity difference of 0.79. This means that to be selected under MDL, M_4 must provide a higher value of the log ML than M_3 by at least 0.79 for it to be selected. Compared with AIC, MDL imposes a stiffer tariff to counteract the added flexibility of M_4 in fitting random error. When the effects of complexity due to functional form were neutralized by using MDL, recovery generally improved, especially for the data generated from M_3 (bottom 3 × 3 matrix).

This example demonstrates the importance of accounting for all relevant dimensions of complexity in model comparison. However, the reader is cautioned not to overgeneralize the above-described simulation results. They should not be regarded as indicative of how the selection methods will perform across all settings. Model comparison is an inference problem. The quality of the inference depends strongly on the characteristics of the data (e.g., sample size, experimental design, type of random error) and the models themselves (e.g., model equation, parameters, nested versus nonnested). For this reason, it is unreasonable to expect a selection method to perform perfectly all the time. Rather, like any statistical test, a comparison method is developed using an unlimited amount of data (i.e., asymptotically), meaning that the method may not work as well with a small sample of data, but should improve as sample size increases.

Conclusion

Computational modeling is currently enjoying a heyday as more and more scientists harness the power of statistics and mathematics to develop quantitative descriptions of the phenomenon under study. Progress in this endeavor depends on there being equally sophisticated methods for comparing such models. The aim of this chapter was to introduce the reader to contemporary model selection methods. As stated early on, the problem is ill-posed because in the absence of sufficient data, solutions to the problem are nonunique. To date, the preferred strategy for solving this problem has been to maximize generalizability. It is achieved by evaluating the amount of information in the data relative to the information capacity (i.e., complexity) of the model.

When using any of these selection methods, we advise that the results be interpreted in relation to the other criteria important in model selection.

It is easy to forget that AIC and MDL are just fancy statistical tools that were invented to aid the scientific process. They are not the arbiters of truth. Like any such tool, they are blind to the quality of the data and the plausibility of the models under consideration. They will be most useful when considered in the context of the other selection criteria outlined at the beginning of this chapter (e.g., interpretability, falsifiability).

Acknowledgments

Both authors were supported by NIH Grant R01 MH57472.

[15] Practical Robust Fit of Enzyme Inhibition Data

By Petr Kuzmič, Craig Hill, and James W. Janc

Introduction

The analysis of enzyme inhibition data in the context of preclinical drug screening presents unique challenges to the data analyst. The good news lies in the advances in laboratory robotics, miniaturization, and computing technology. However, the good news is also the bad news. Now that we *can* perform thousands of enzyme assays at a time, how do we sensibly manage and interpret the vast amount of generated information? New methods of biochemical data analysis are needed to match the improvements in research hardware.

Another challenge is presented by increasing constraints on material resources, given the need to assay an ever larger number (thousands or hundreds of thousands) of individual compounds in any given project. Of course, only a small number of hits advance from high-throughput screening, using a single-point assay, into the dose–response screening to determine the inhibition constant. Even so, the sheer number of enzyme inhibitors that need to be screened often leads to suboptimal experimental design.

For example, a strategic decision may have been made that all inhibitor dose–response curves will contain only 8 data points (running down the columns of a 96-well plate), and that the concentration–velocity data points will not be duplicated, so that each 96-well plate can accommodate up to 12 inhibitors. With only eight nonduplicated data points, and with as many as four adjustable nonlinear parameters (e.g., in the four-parameter logistic

Copyright 2004, Elsevier Inc.
All rights reserved.
0076-6879/04 $35.00

equation), the experimental data better be extremely accurate and the concentrations optimally chosen. Alas, too often neither is the case.

This chapter is concerned with one particular nuisance arising in secondary preclinical screening of enzyme inhibitors, namely, the presence of gross outliers. For our purposes, outliers are data points that are affected by gross errors caused by malfunctioning volumetric equipment, by a human error in data entry, or by countless other possible mishaps. It is shown that Huber's Minimax approach to robust statistical estimation is particularly preferable over the conventional least-squares analysis.

Theory

Iteratively Reweighted Least Squares

Assume that the dependent variable y is related to the independent variable x through the functional relationship $y = f(x, \mathbf{p})$, where $\mathbf{p}$ is the m vector of adjustable model parameters to be estimated from the available data pairs $\{\{x_1, y_1\}, \{x_2, y_2\}, \ldots, \{x_n, y_n\}\}$. The usual ordinary least-squares[1-3] (OLS) estimation problem can be formulated as is shown in Eq. (1).

$$\min_{\mathbf{p}} S = \sum_{i=1}^{n} (y_i - f(x_i, \mathbf{p}))^2 \tag{1}$$

Many efficient computational methods exist to accomplish this minimization. Unfortunately, the OLS estimate of the model parameters is sensitive to the presence of outliers,[4] which has led to the design of various alternatives. For example, according to Eq. (2), instead of minimizing the sum of squared deviations, one might minimize the sum of absolute deviations.

$$\min_{\mathbf{p}} S = \sum_{i=1}^{n} |y_i - f(x_i, \mathbf{p})| \tag{2}$$

Computationally, the least absolute deviation (LAD) fit is more difficult than OLS. One approach[4] is to resort to derivative-free methods, such as the Nelder–Mead simplex algorithm.[5] A more feasible approach, leading

[1] M. L. Johnson and S. G. Frasier, *Methods Enzymol.* **117**, 301 (1985).
[2] M. L. Johnson and L. M. Faunt, *Methods Enzymol.* **210**, 1 (1992).
[3] M. L. Johnson, *Anal. Biochem.* **206**, 215 (1992).
[4] M. L. Johnson, *Methods Enzymol.* **321**, 417 (2000).
[5] J. A. Nelder and R. Mead, *Computer J.* **7**, 308 (1965).

to the same results, is to use *iteratively reweighted least squares*. This is based on the fact that the LAD fit can be accomplished as a sequence of weighted LS fits.

$$\text{repeat} \quad \min_{\mathbf{p}} S = \sum_{i=1}^{n} w_i(y_i - \hat{y}_i)^2 \tag{3}$$

In each step, the best-fit values $\hat{y}_i = f(x_i, \hat{\mathbf{p}})$ from the previous LS fit are used to recompute weights, such that $w_i = 1/|y_i - \hat{y}_i|$. Throughout this chapter the "hat" accent ($^\wedge$) represents "best-fit" quantities. After a sufficient number of reweighted LS minimizations, the model parameters $\mathbf{p}$ converge to the same values that would be obtained by LAD minimization using, e.g., the simplex method.[5]

A similar iteratively reweighted least-squares procedure forms the basis of the robust fit method discussed in this chapter.

Huber's Method

The LAD fit has been used occasionally for data analysis in biochemical kinetics.[6] It does solve the outlier problem, but it is probably not appropriate in most experimental situations arising in biochemistry. As had been pointed out,[4] the LAD fit does not provide *maximum likelihood* parameter estimates (ML, or M estimates) if the underlying statistical distribution of random errors is Gaussian, according to probability density function (4):

$$f(x) = \frac{1}{\sigma\sqrt{2\pi}} \exp\left[-\frac{1}{2}\left(\frac{x-\mu}{\sigma}\right)^2\right] \tag{4}$$

The LAD fit does produce ML parameter estimates if (and only if) the underlying error distribution function is a double-sided exponential, but such distribution is seen only infrequently.[6] On the other hand, the very presence of outliers in real-world experimental data proves the fact that strictly Gaussian error distribution is also unrealistic. What is the solution to this quandary?

Huber[7] proposed that random experimental errors arising in the physical sciences could be described by using the *contaminated Gaussian distribution* [Eq. (5)], where $\Phi(x)$ is the cumulative normal distribution:

[6] I. B. C. Matheson, *Comput. Chem.* **14**, 49 (1990).
[7] P. J. Huber, "Robust Statistics." John Wiley & Sons, New York, 1981.

$$F(x) = (1 - \varepsilon)\Phi\left(\frac{x - \mu}{\sigma}\right) + \varepsilon\Phi\left(\frac{x - \mu}{3\sigma}\right) \tag{5}$$

$$\Phi(x) = \frac{1}{\sqrt{2\pi}} \int_{-\infty}^{x} e^{-y^2/2} dy \tag{6}$$

According to Huber,[7] ε is typically in the range between 0.01 and 0.1, which does not imply that between 1 and 10% of all experiments necessarily are affected by gross errors, although this may be true in particular circumstances. The assumption that $0.01 \leq \varepsilon \leq 0.1$ merely implies the existence of two distinct categories of measurements, the majority are "good" points with the standard deviation σ, and a few are "bad" points drawn from another Gaussian distribution, characterized by the standard deviation several times larger (e.g., 3σ).

Starting from similar distributional assumptions, and from the central role of statistical *influence functions*,[8,9] a rigorous theory of robust estimation had been built.[7,10–12] Regardless of the particular form of the influence function, many computational algorithms for robust regression analysis rely on iteratively reweighted least squares,[13–16] as does Huber's method used here.

Huber's influence function[7] is constructed as follows. All "good" data points (to be defined below) are assigned the same weight in the iteratively reweighted series of LS estimations, exactly as they are in OLS. In contrast, deviant or "bad" points are progressively deemphasized, by being assigned progressively smaller weights according to Eq. (7). Here, a "good" data point is one for which the *standardized residual* R_i, defined in Eq. (8), is smaller in absolute value than a certain multiple of the estimated standard deviation of fit. The cutoff criterion c serves as an empirical tuning constant.

[8] D. A. Belsley, E. Kuh, and R. E. Welsch, "Regression Diagnostics: Identifying Influential Data and Sources of Collinearity." John Wiley & Sons, New York, 1980.

[9] R. D. Cook and S. Weisberg, "Residuals and Influence in Regression." Chapman & Hall, New York, 1982.

[10] W. J. J. Rey, "Introduction to Robust and Quasi-Robust Statistical Methods." Springer-Verlag, New York, 1983.

[11] F. R. Hampel, E. M. Roncheti, P. J. Rousseeuw, and W. A. Stahel, "Robust Statistics." John Wiley & Sons, New York, 1986.

[12] P. J. Rousseeuw and A. M. Leroy, "Robust Regression and Outlier Detection." Wiley Interscience, New York, 1987.

[13] P. W. Holland and R. E. Welsch, *Commun. Stat. Theory Methods* **A6**, 813 (1977).

[14] D. Coleman, P. Holland, N. Kaden, V. Klema, and S. C. Peters, *ACM Trans. Math. Software* **6**, 327 (1980).

[15] J. O. Street, R. J. Carroll, and D. Ruppert, *Am. Stat.* **42**, 152 (1988).

[16] R. Heiberger and R. A. Becker, *J. Comput. Graphics Stat.* **1**, 181 (1992).

$$w_i = \begin{cases} 1 & \text{if} \quad |R_i| \le c \\ c/|R_i| & \text{if} \quad |R_i| > c \end{cases} \tag{7}$$

$$R_i = \frac{y_i - \hat{y}_i}{\hat{\sigma}\sqrt{1 - h_i}} \tag{8}$$

It should be noted that different authors (including manual writers for advanced statistical software packages, such as S-PLUS, SAS, and MATLAB) variously refer to R_i either as standardized residuals or as *Studentized residuals*. This confusion is clearly explained by Rawlings.[17]

Importantly, Huber established that with $c = 1.345$, the M estimator so defined is 95% efficient. By "efficiency" we mean the ratio of variances from Huber's M estimate relative to normal regression models, assuming that the underlying error distribution is in fact normal.

The standard deviation of fit, $\hat{\sigma}$, is estimated from the median absolute deviation (MAD), computed as the median absolute deviation of the residuals from their median. In Eq. (9), MAD is divided by the factor relating the probable error (E_{50}) to the standard deviation (SD): $E_{50} = 0.6745 \times SD$. Note that MAD relates to the mean square error (MSE) as MAD $\approx$ (MSE)$^{1/2}$.

$$\hat{s} = \frac{\text{med}\{|(y_i - \hat{y}_i) - \text{med}\{y_i - \hat{y}_i\}|\}}{0.6745} \tag{9}$$

"Hat" Matrix and Nonlinear Leverages

The quantity h_i appearing in the denominator of Eq. (8) is the *leverage* of the ith data point in the nonlinear least-squares regression. It is the diagonal element of the $n \times n$ "hat" matrix $\mathbf{H}$ defined in Eq. (10), where $\mathbf{J}$ is the familiar $n \times m$ Jacobian matrix of first derivatives.[18] Recall that m is the number of adjustable parameters in the fitting model:

$$\mathbf{H} = \mathbf{J}(\mathbf{J}^\mathrm{T}\mathbf{J})^{-1}\mathbf{J}^\mathrm{T} \tag{10}$$

$$\{J\}_{ij} = \frac{\partial f(x_i, \hat{\mathbf{p}})}{\partial p_j} \tag{11}$$

[17] J. O. Rawlings, "Applied Regression Analysis—A Research Tool." Wadsworth, Belmont, CA, 1988.
[18] M. L. Johnson, *Methods Enzymol.* **321**, 425 (2000).

In linear regression, the experimental values $\mathbf{y}$ and the least-squares fit values $\hat{\mathbf{y}}$ are related through the simple matrix equation $\hat{\mathbf{y}} = \mathbf{H}\mathbf{y}$, thus the term "hat" matrix. The matrix $\mathbf{H}$ has interesting mathematical properties, many of which are practically useful for checking the computation of h_i. For example, $\mathbf{H}$ is a symmetric and idempotent projection matrix, that is, $\mathbf{HH} = \mathbf{H}$. It has m eigenvalues equal to 1 and $n - m$ eigenvalues equal to 0. The trace is $tr\{H\} = p$, and for all diagonal elements, we must have $0 \le h_i \le 1$.

Because we are interested only in the diagonal elements of the hat matrix, we can compute them directly by using Eq. (12), where $\mathbf{j}_i$ is the ith row vector in the Jacobian matrix $\mathbf{J}$.

$$h_i = \mathbf{j}_i (\mathbf{J}^\mathrm{T} \mathbf{J})^{-1} \mathbf{j}_i^\mathrm{T} \tag{12}$$

The $m \times m$ matrix inverse $(\mathbf{J}^\mathrm{T}\mathbf{J})^{-1}$ is hardly ever computed as written. Instead, in our work we utilize the QR decomposition[19] $\mathbf{J} = \mathbf{QR}$, where $\mathbf{R}$ is an $m \times m$ upper triangular invertible matrix with positive entries in its diagonal. Many good implementations of the QR decomposition algorithm are available as canned software.[20]

Kinetic Model

The kinetics of tight-binding enzyme inhibition[21,22] is described here by Eq. (13), where V_b is the baseline or background reaction rate, V_0 is the reaction rate observed in the absence of inhibitor ("negative control"), [E] and [I] are, respectively, the concentrations of the enzyme and the inhibitor, and K_i is the apparent inhibition constant.[23]

$$v = V_\mathrm{b} + V_0 \frac{[\mathrm{E}] - [\mathrm{I}] - K_\mathrm{i} + \sqrt{([\mathrm{E}] - [\mathrm{I}] - K_\mathrm{i})^2 + 4[\mathrm{E}]K_\mathrm{i}}}{2[\mathrm{E}]} \tag{13}$$

Numerical Example

The experimental data in the first three columns of Table I represent the micromolar concentration of an inhibitor (x_i) and the corresponding initial velocities of an enzyme reaction (y_i).

[19] D. C. Lay, "Linear Algebra and Its Applications." Addison-Wesley, Reading, MA, 1994.
[20] W. H. Press, S. A. Teukolsky, W. T. Vetterling, and Brian P. Flannery, "Numerical Recipes in C." Cambridge University Press, Cambridge, 1992.
[21] J. F. Morrison, *Biochim. Biophys. Acta* **185,** 269 (1969).
[22] J. W. Williams and J. F. Morrison, *Methods Enzymol.* **63,** 437 (1979).
[23] S. Cha, *Biochem. Pharmacol.* **24,** 2177 (1975).

TABLE I
RESULTS OF FIT FOR REPRESENTATIVE ENZYME INHIBITOR[a]

	Data		Least-squares fit				Robust fit			
i	x_i	y_i	$\hat{y}_i$	r_i	R_i	h_i	$\hat{y}_i$	r_i	R_i	w_i
1	0	133.0	143.4	10.4	−0.51	0.51	139.8	−6.8	−0.27	1
2	0.0031	143.6	135.5	8.1	0.35	0.36	136.8	6.8	0.24	1
3	0.0122	132.8	115.9	16.9	0.68	0.27	128.6	4.2	0.14	1
4	0.0488	34.0	71.3	−37.3	**−1.95**	0.57	103.3	**−69.3**	**−2.98**	0.12
5	0.195	65.8	27.0	**38.8**	1.55	0.26	57.2	8.6	0.28	1
6	0.781	13.5	7.6	5.9	0.2	0.03	20.3	−6.8	−0.19	1
7	3.125	3.4	2.0	1.4	0.05	0	5.6	−2.2	−0.06	1
8	12.5	1.1	0.5	0.6	0.02	0	1.5	−0.4	−0.01	1
9	50	0	0.1	−0.1	0	0	0.4	−0.4	−0.01	1

[a] Values in boldface represent the maximum absolute value in the given column.

Ordinary Least-Squares Fit

According to Huber's method, the robust regression analysis always begins with the ordinary least-squares fit, summarized in columns 4 through 7 in Table I. In the OLS fit to Eq. (13), the enzyme concentration ($[E] = 10 \ nM$) and the background rate ($V_b = 0$) were treated as fixed constants; the apparent inhibition constant K_i and the control rate V_0 were treated as adjustable model parameters. The best fit values are shown in the first row of Table II.

It is important to note the difference between ordinary residuals $r_i = y_i - \hat{y}_i$ and the standardized residuals R_i defined by Eq. (8). Examining the ordinary residuals r_i, one might be tempted to conclude that the fifth data point ($i = 5$) could be an outlier, because it has the largest absolute deviation. In contrast, the standardized residuals R_i suggest that the fourth data point could be an outlier, because it is associated with the largest (in absolute value) standardized residual.

This difference between r_i and R_i is caused by the nonlinear leverages h_i for each data point. Note that the leverage for the fourth data point (0.57) is more than twice the leverage for the fifth data point (0.26), which means that in the iteratively reweighted least squares the fourth data point will be initially given larger weight $[1/(1 - 0.57)^{1/2} = 1.52]$, compared with the fifth data point $[1/(1 - 0.26)^{1/2} = 1.16]$.

The leverages h_i for data points 7 through 9 are practically zero, which means that these data points contribute practically no useful information.

TABLE II
RESULTS OF FIT USING VARIOUS ANALYSIS METHODS[a]

Method	K_i (nM)	V_0	$n_{w<1}$	$\sum w_i$
Least squares	43.3 ± 25.1	143.4 ± 15.8	0	9
Huber ($c = 10$)	43.3 ± 25.1	143.4 ± 15.8	0	9
Huber ($c = 1.345$)	131.0 ± 47.0	139.8 ± 8.3	1	8.12
Huber ($c = 1$)	68.5 ± 17.2	151.4 ± 4.4	3	6.23
Huber ($c = 0.1$)	75.2 ± 3.2	149.8 ± 1.2	9	2.67
Huber ($c = 0.01$)	73.5 ± 3.1	150.7 ± 1.2	9	0.63
Huber ($c = 0.001$)	71.1 ± 0.2	151.9 ± 1.3	9	0.14
Absolute deviations (simplex)	77.0	148.7	—	—
Point deletion	146.1 ± 23.0	140.8 ± 3.7	0	8

[a] $n_{w<1}$ is the number of data points for which the final weight [Eq. (7)] was lower than unity; $\sum w_i$ is the sum of all final weights in the iteratively reweighted least squares.

This indicates a suboptimal experimental design. Huber[7] points out that large values of h_i should "serve as warning signals that the ith observation may have a decisive, yet hardly checkable, influence. Values $h_i \leq 0.2$ appear safe, values between 0.2 and 0.5 are risky, and if we can control the design at all, we had better avoid values above 0.5."

From this discussion it is clear the fourth data point is what Huber calls a *leverage point* (i.e., a data point associated with a dangerously high value of h_i), whereas data points 7 through 9 are useless. This is yet another unpleasant consequence of one-size-fits-all experimental designs traditionally seen in inhibitor screening. Most often, the same dilution ratio and the same maximum concentration are used for all inhibitors on the same 96-well plate, but if the inhibitors differ significantly in their inhibition constants, this uniform design generates a large number of data points with low information value.

Robust Fit

In the second stage of this analysis, the leverages h_i computed during the preliminary OLS fit were used in the iteratively reweighted OLS regression, employing the default value of Huber's tuning constant ($c = 1.345$). After several repeated OLS fits, the adjustable parameters converged to the values listed in the third row of Table II. The results of the fit are shown graphically in Fig. 1.

Note in Table I that the Huber reweighted regression ended with assigning unit weights (that is, giving the ordinary least-squares treatment)

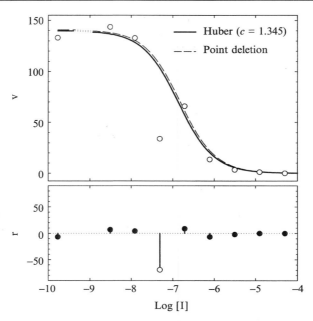

FIG. 1. *Top:* The open circles are data points (inhibitor concentrations versus initial velocities) for a particular enzyme inhibitor. The left most data point is the negative control, observed at zero inhibitor concentration. The thicker, solid curve represents the robust fit to rate Eq. (13) by using Huber's method with tuning constant $c = 1.345$. The thinner, dashed curve represents the results of the ordinary least-squares fit after the fourth data point ($\log[I] \approx -7.3$) was deleted. *Bottom:* The residuals for data points that were assigned the full unit weight ($w_i = 1$) in Huber's method are shown as solid circles. The residual shown as an open circle belongs to the (single) data point, which ended up with less than unit weight ($w_4 = 0.12$) in the iteratively reweighted least-squares fit.

to all data points except data point 4, which is assigned a small weight ($w_i = 0.12$). This result strongly suggests that the fourth data point is an outlier. Its standardized residual is almost equal to three ($R_i = 2.98$), which is yet another strong indication that the data point is affected by gross error.

Some authors[8] recommend that data points with $R_i > 2$ should simply be deleted. Others recommend a two-stage robust regression, starting with Huber's influence function (7) followed by Tukey's biweight scheme (14), where the 95% asymptotic efficiency of the standard normal distribution is achieved with $c = 4.6851$.

$$w_i = \begin{cases} 0 & \text{if} \quad |R_i| > c \\ [1 - (R_i/c^2)]^2 & \text{if} \quad |R_i| \leq c \end{cases} \qquad (14)$$

Note that Tukey's outliers are given zero weights, and thus are effectively excluded from analysis. We have experimented with Tukey's biweight and found that, too often, it deleted too many data points from our small data sets. Instead, in the context of inhibitor screening, we formulated a heuristic policy for data exclusion as follows. If and only if the Huber method produces a single data point with the final weight $w_i < 1$, this single data point is deleted (by being assigned $w_i = 0$), and the analysis is repeated one last time as ordinary least squares. For our example inhibitor, where this condition does apply, the results are shown graphically as the thin dashed curve in Fig. 1.

The solid and dashed curves in Fig. 1 do appear pleasingly similar, suggesting that Huber's fit with $c = 1.345$ and OLS fit with the fourth data point deleted are consistent with each other. The best-fit values of adjustable parameters, shown for the OLS fit with deletion in the last row in Table II, are also comparable for the two methods, although they are not identical. The difference is caused by the outlier point being assigned a nonzero weight, $w_4 = 0.12$, in Huber's method. However, the difference between $K_i = 131$ nM (Huber) and $K_i = 146$ nM (OLS with deletion) is only about 10%, whereas the OLS method with full data set produced an inhibition constant ($K_i = 43$ nM) that is off by more than a factor of three. Thus, the application of Huber's method alone produced two desirable effects. First, it reduced the systematic error in K_i due to a single outlier, from 330 to 10%. Second, it clearly diagnosed the presence of this outlier so it could be deleted.

Variations in Huber's Tuning Constant

Although the particular value $c = 1.345$ for Huber's tuning constant is rooted in statistical theory (it has been chosen because it leads to an M estimate that is 95% efficient), it is important to examine in practice how variations in c might affect the outcome of robust regression analyses in our particular experimental setting.

On the basis of theoretical considerations [see Eq. (7)], we can predict that as c becomes *very* large all data points will be assigned unit weights and the Huber regression turns into OLS. On the other hand, as c becomes small, we expect the Huber algorithm to resemble the LAD fit, because the weighting factors in the influence function (7) simply become reciprocal absolute residuals.

Figure 2 and the fourth row in Table II show the results of Huber regression with $c = 1.0$. Although this is only marginally lower than the recommended value $c = 1.345$, the results of fit are strikingly different. As is illustrated by the bottom panel in Fig. 2, the robust fit is dominated by six of the nine data points, which are assigned unit weights. Data points 1, 4, and 5 are deemphasized with weights $w_1 = 0.11$, $w_4 = 0.03$, and $w_5 = 0.10$. In other words, the Huber fit discovered too many outliers.

Further decreasing c to 0.1 created another problem. With $c = 0.1$ or lower, no points were assigned full weights in reweighted least squares. The best-fit values of model parameters remained approximately the same between $c = 0.1$ and $c = 0.001$, but the variances of the model parameters decreased drastically. This is expected from theory, because the Huber M estimate loses asymptotic efficiency as c moves away from its 95% efficient value of 1.345. Thus in any software system in which the Huber tuning

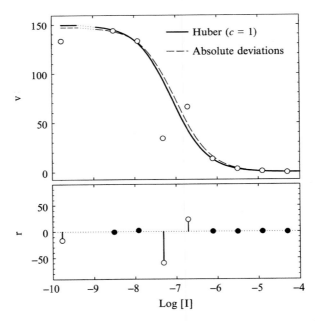

FIG. 2. *Top:* The thicker, solid curve represents the robust fit to rate Eq. (13) by using Huber's method with tuning constant $c = 1.0$. The thinner, dashed curve represents the results of the least-absolute deviation fit [Eq. (2)] using the Nelder–Mead simplex algorithm.[5,20] *Bottom:* For explanation of solid and open circles, see Fig. 1. Please note that the fit is completely dominated by six of the nine data points.

constant c is adjustable by the user (e.g., SAS, S-PLUS, MATLAB, or in the software built by us for inhibitor screening), one must proceed with caution. Lowering c might not only pick up too many "outliers" if the data set is small, it will also unrealistically shrink parameter variances.

As is expected from theoretical considerations, when we increased the tuning constant c above its 95% efficient value ($c = 1.345$), the algorithm simply turned into OLS. This is seen from Fig. 3 and the second row in Table II.

In some respects, these results are disconcerting. At least for this particular inhibitor, decreasing c only slightly (in fact, from 1.345 to 1.3; see Fig. 4) has led to the false identification of too many "outliers." In contrast, increasing c eventually missed the single outlier altogether. An important question then concerns how wide a range c can have for the Huber method to remain useful for analyzing data sets as small as ours are (nine data points, two to four adjustable parameters).

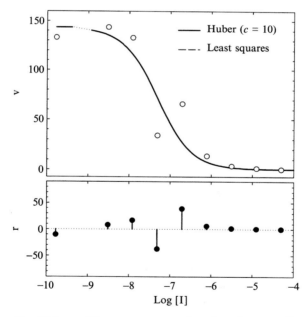

FIG. 3. *Top:* The thicker, solid curve represents the robust fit to rate Eq. (13) by using Huber's method with tuning constant $c = 10.0$. The thinner, dashed curve represents the results of the ordinary least-squares fit [Eq. (1)]. *Bottom:* For explanation of solid and open circles, see Fig. 1. Please note that two regression analyses yielded exactly identical results, as the two best-fit curves are indistinguishable.

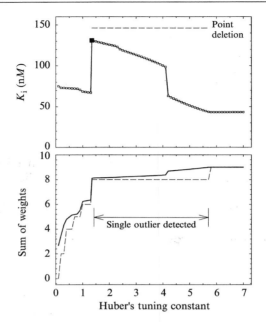

FIG. 4. *Top:* Variation in the best-fit value of the apparent inhibition constant K_i depending on the value of the Huber tuning constant c in Eq. (7). The large solid square is the value obtained at the default recommended value, $c = 1.345$. *Bottom:* The solid curve shows the sum of weights $\sum_{i=1}^{n} w_i$ obtained for the nine data points ($n = 9$) in the Huber regression, depending on the value of c. The dashed curve is the sum of all weights while counting only data points with full weights ($w_i = 1$).

To answer this question, we have varied c systematically between 0.1 and 10.0, stepping by $\Delta c = 0.1$ (100 different values), and performed the Huber regression at each point. The results are summarized in Fig. 4.

It is encouraging that the range of c values, within which the Huber algorithm successfully detected only a single data point with the final weight $w_i < 1$, is relatively wide ($c = 1.345$ through $c = 5.6$). Unfortunately, as the weight of this data point increases, the outlier progressively distorts the best-fit value of the inhibition constant.

Also note that the default value of the tuning constant ($c = 1.345$) precariously sits at the lower end of this interval. Thus, perhaps a minuscule change in the data could cause the algorithm to tip over that edge, and suggest falsely that the data set contains three "outliers" instead of one. These are serious challenges for the designer and administrator of a software system designed for automatic, robust, high-throughput analysis of enzyme inhibition data.

Implementation Notes

In the course of our ongoing work on new methods for automated data analysis in preclinical screening,[24–26] we have tested Huber's robust regression method on tens of thousands of enzyme inhibitors. The efficiency of the method in handling occasional outliers was good. This is in contrast with the OLS fit, where data points with large deviations have unduly large influence, and with the LAD fit,[4,6] where data points with small deviations dominate the fitted curve.

The latter statement is consistent with the fact that, in LAD regression implemented as iteratively reweighted least squares, the weights are $1/|y_i - \hat{y}_i|$, implying that a data point with zero deviation has infinite weight. In working with relatively small data sets and with nonlinear fitting models, this feature of LAD is particularly dangerous. A sufficiently flexible non-linear model with four parameters will always go through four data points, completely ignoring the remaining data, which become deemphasized in LAD fit.

The practical success of Huber's method applied even to relatively small data sets, such as those arising in preclinical screening, is due to the fact that the method behaves as OLS does if the data are "good," but at the same time it gives the LAD treatment to suspected outliers, while maintaining 95% asymptotic efficiency. The following are a few of many possible implementation issues, which were encountered in translating the theory of Huber's regression into a practically useful software system.

Replicated Measurements

Huber's robust regression method is suitable for the analysis of either replicated or nonreplicated data. However, when the number of replicates is small, it is best not to automatically average these replicates (e.g., duplicates) and then analyze the averaged data. Consider the hypothetical example where the replicated initial velocities are [100, 99], [81, 79], [62, **30**], [40, 38], [19, 21], and so on. These are 10 data points, only one of which

[24] P. Kuzmič, S. Sideris, L. M. Cregar, K. C. Elrod, K. D. Rice, and J. W. Janc, *Anal. Biochem.* **281,** 62 (2000).

[25] P. Kuzmič, K. C. Elrod, L. M. Cregar, S. Sideris, R. Rai, and J. W. Janc, *Anal. Biochem.* **286,** 45 (2000).

[26] P. Kuzmič, C. Hill, M. P. Kirtley, and J. W. Janc, *Anal. Biochem.* **319,** 372 (2003).

is clearly an outlier. By averaging we obtain five data points [99.5], [80], [**46**], [39], [20], among which the outlier would be more difficult to detect if the fitting model is nonlinear.

In this hypothetical example, a better alternative to robust regression might be *weighted least squares* (WLS) fit, where the weighting factors are reciprocal SDs from each replicate. Our experience shows that WLS with a small number of replicates can be treacherous for the following reason. If the inhibitor dose–response curve contains only a small number of (replicated) data points, it is possible that one particular duplicate might be fortuitously accompanied by a small standard error, much smaller than the standard errors from other averages. In such a case, the particular data point would be assigned a disproportionately large weight, and consequently it might unduly influence the regression. In a production software system, the user or the administrator should have a choice to decide whether to use robust regression or WLS, but preferably not both at the same time.

Outliers versus Deviations from Fitting Model

Practical experience with Huber's regression shows that in many cases the method will assign weights smaller than one to more than one data point, and sometimes even to all data points in the given dose–response curve. Obviously not all data points can be "outliers" if our distributional assumption [Eq. (5)] is correct. Rather, too many "outliers" (data points with $w_i < 1$) simply suggest that the fitted model is incorrect.

In such cases, a sensible software system would disregard the robust fit and revert to OLS with a suitable warning. Alternately, if only one of the weights is much smaller than the others, it might make sense to delete the corresponding data point (by setting its $w_i = 0$) and run one final OLS fit. A further refinement of this policy would be to take into account the total number of data points with $w_i < 1$, or the sum of weights for the entire data set (see Fig. 4).

One might be willing to accept the results of Huber's robust regression analysis only if the majority of data points end up with $w_i = 1$, or otherwise conclude that the model is inadequate, issue a warning, and end with OLS as the last resort. However the software system is constructed, it is important to avoid situations illustrated in Fig. 2, where few data points completely dominate the regression while the remaining data points are effectively excluded through weighting. Again, in a production software system, the user or the administrator should be able to control these policies.

Conclusions

We have discussed several mathematical quantities that should be of interest to the biochemical data analyst, but, to our knowledge, are hardly ever mentioned in the mainstream biochemical literature. First, standardized residuals defined by Eq. (8) are significantly more informative than ordinary residuals $(y - \hat{y}_i)$. Standardized residuals are more helpful than ordinary residuals not only for outlier detection, but also for model diagnostics.

Second, nonlinear leverages, which are diagonal elements of the "hat" matrix **H** [Eq. (10)], are useful for quickly assessing the presence of unduly influential data points (whether outliers or not) and the optimality of experiment design. If, after a least-squares fit, we find that a particular data point is associated with $h_i > 0.7$, it means that this is a leverage point. Small, random changes in this single data value might have a large effect on model parameters, which is undesirable.

On the other hand, if we find that too many data points are associated with zero leverages ($h_i = 0$), it means these data points were wasted, because they contribute no useful information at all about the model parameters. In such case, one should seriously consider improving the experimental design (in the case of inhibitor screening, the layout of concentrations) for the next round of experiments.

Both standardized residuals and leverages play a role in the Huber's method of robust regression analysis, implemented as iteratively reweighted least squares. We found that it is a good alternative to ordinary least squares when the goal is to exclude a single gross outlier from a relatively small data set. This approach can increase productivity in preclinical screening laboratories, faced with determining inhibition constants for thousands of enzyme inhibitors in a single project.

Kinetic data analysis does present unique challenges in an extensively automated and robotized enzymology laboratory, where success means a smooth flow of the massive data stream connecting microtiter plate readers with structure–activity databases. At the present time, no technology can completely replace a well-qualified enzymologist supervising the process. However, our practical experience shows that a software system judiciously implementing Huber's variant of robust regression does help in the specific task of objectively identifying grossly outlying data points.

[16] Measuring Period of Human Biological Clock: Infill Asymptotic Analysis of Harmonic Regression Parameter Estimates

By Emery N. Brown, Victor Solo, Yong Choe, and Zhenhua Zhang

Introduction

For many years it has been accepted that the intrinsic period of the human biological clock or circadian pacemaker is approximately 25 h, whereas the clocks of most other animals are known to have intrinsic periods closer to 24 h. Theoretical and experimental studies of how light affects the human circadian oscillator now suggest that the intrinsic period of the human pacemaker may also be closer to 24 h instead of 25 h.[1] The currently accepted overestimate is due to a failure of earlier protocols to consider the effects of environmental light on the clock's resetting mechanism.[2,2a] As a consequence, circadian physiologists are actively working to measure precisely the period of the human biological clock.

One approach to this question applies a new version of the forced desynchrony (FD) protocol. Under this protocol a subject is monitored for an extended time in an isolated environment whose light–dark cycle is maintained outside the interval between 23 and 27 h.[1] This 4-h interval, termed the range of entrainment, defines the set of light–dark cycle periods to which the human circadian pacemaker may be synchronized.[3] During two-thirds of the light–dark cycle the lights are maintained continuously on at a fixed intensity while during the remaining one-third of the cycle the subject is maintained in total darkness. For FD studies a light–dark cycle of either 20 or 28 h is typically chosen, the light intensity level is set at 10 lx, and the behavior of the clock is monitored by recording marker rhythms such as core temperature, plasma cortisol, and plasma melatonin levels. Because the light levels during FD are low and because the clock cannot synchronize to a light–dark cycle whose period is outside the range of entrainment, the circadian pacemaker oscillates at its intrinsic period.

[1] C. A. Czeisler, J. F. Duffy, T. L. Shanahan, E. N. Brown, J. F. Mitchell, D. W. Rimmer, J. M. Ronda, E. Silva, J. S. Allan, J. S. Emens, D. J. Dijk, and R. E. Kronauer, *Science* **284,** 2177 (1999).

[2] C. A. Czeisler, R. E. Kronauer, J. S. Allan, J. E. Duffy, M. E. Jewett, E. N. Brown, and J. M. Ronda, *Science* **244,** 1328 (1989).

[2a] R. E. Kronauer, C. A. Czeisler, and E. N. Brown, *Sleep Res.* **18,** 424 (1989).

[3] R. Wever, *in* "The Circadian System of Man." Springer-Verlag, Berlin, 1979.

Copyright 2004, Elsevier Inc.
All rights reserved.
0076-6879/04 $35.00

The FD study interval T is chosen to be long relative to the clock's expected period τ and the sampling interval Δ may be taken arbitrarily small since core temperature measurements are recorded electronically. For example, the circadian period is in the range from 24 to 25.5 h, T may be as long as 30 days, and Δ may range from 10^{-2} to 1 min. Hence, the number of core temperature measurements made on a single subject with just 1 min of sampling can exceed 40,000. Because there are many data points and the study interval is large relative to τ, a subseries is often sampled from the data at an interval larger than 1 minute in order to speed computations in parameter estimation using a harmonic regression model. The increased computational speed comes necessarily at the expense of a decrease in the precision with which τ and the other parameters can be estimated. On the other hand, the thermoregulatory mechanism of the body causes nearby core temperature measurements to have high serial dependence so that a diminishing amount of new information accrues as the sampling interval is made arbitrarily small. Quantifying the magnitude of the tradeoff between computational costs and measurement precision is of significant interest since the objective of current FD studies is to determine the intrinsic period of the clock with the highest possible precision.

Standard large-sample time series methods cannot be used directly to investigate the effect of the sampling interval on the variance of the harmonic regression parameter estimates because these techniques require that the study interval and number of data points tend to infinity with the sampling interval fixed.[4,5] Under FD, the study interval is large and fixed, and the number of data points is increased by decreasing the sampling interval. That is, under the FD protocol data sampling follows infill asymptotic behavior (Cressie,[6] p. 101). It is therefore important to define how precision in the estimation of τ and the other harmonic regression parameters change as a function of Δ when T is fixed and the data are serially dependent.

To address this question, we derive the approximate infill asymptotic covariance matrix of the maximum likelihood parameter estimates for a harmonic regression plus continuous first-order autoregressive model of core temperature data collected under FD. We find the limiting form of the matrix as the sampling interval tends to zero for a fixed study interval. From this approximation we develop an analytic description of how the sampling interval, the study interval, the time constant of the autoregressive process and the core temperature signal-to-noise ratio

[4] E. N. Brown, *Biometrika* **77**, 653 (1990).
[5] E. N. Brown, Y. Choe, H. Luithardt, and C. A. Czeisler, *Am. J. Physiol.* **279**, E669 (2000).
[6] N. A. C. Cressie, *in* "Statistics for Spatial Data." John Wiley & Sons, New York, 1992.

jointly affect the precision of the period estimate. Our infill analysis differs from current approaches to this type of asymptotics in that it applies the concept of scaling discussed in Solo.[7] We also discuss in this section the relation of our results to the classic formulas for the asymptotic covariance matrix of the harmonic regression parameter estimates. In Section III the implications of these results for determining the precision with which the period of the human biological clock may be estimated are studied in an analysis of core temperature data from five male subjects monitored under FD. In a later we briefly relate our findings to some previous work on parameter estimation under infill asymptotic conditions and mention some directions of future research.

Theory

We assume that core temperature data $y_{t_1}, \ldots, y_{t_N}$ are collected in the interval $[0, T]$, where $t_n = n\Delta, n = 1, \ldots, N$, and $N\Delta = T$ is the study interval. Forced desynchrony core temperature data are well described by the following harmonic regression plus correlated noise model,

$$y_{t_n} = s_{t_n} + x_{t_n} + v_{t_n} \tag{1}$$

where

$$s_{t_n} = \mu + \sum_{r=1}^{d} A_r \cos(\omega r t_n) + B_r \sin(\omega r t_n)$$

$$x_{t_n} = \sum_{r=1}^{4} C_r \cos\left(\frac{2\pi r t_n}{28}\right) + D_r \sin\left(\frac{2\pi r t_n}{28}\right)$$

where s_{t_n} is the circadian component; $\omega = 2\pi/\tau$, where τ is the intrinsic period of the circadian oscillator; and x_{t_n} is the effect of the FD protocol on the core temperature rhythm for a 28-h light–dark cycle.[5] The number of terms, d, in the harmonic regression representation of the circadian signal is set equal to either 2 or 3. The choice of $d = 2$ follows from the harmonic regression analysis of core temperature data from the constant routine circadian protocol,[8] whereas the choice of $d = 3$ makes the harmonic regression model equivalent to an asymptotic representation of the circadian pacemaker as a van der Pol oscillator.[5] Four harmonics have been shown to approximate well the square wave effect of the FD protocol on

[7] V. Solo, in "Infill Asymptotics: Time-Series, Splines, Kriging and Nonparametric Regression." Technical Report. Department of Statistics, Macquerie University, North Ryde, Australia, 1995.

[8] E. N. Brown and C. A. Czeisler, J. Biol. Rhythms **7**, 177 (1992).

the subject's core temperature data.[1] The random variable v_{t_n} is a discrete sample from a continuous first-order autoregressive process and represents serial dependence in the observed temperature data due primarily to the body's thermoregulatory mechanism. We define the continuous first-order autoregressive process as

$$v(t) = e^{-\alpha t}v(0) + \sigma \int_0^t e^{-\alpha(t-u)}dW(u)$$

where $W(t)$ is a Wiener process with $\mathrm{var}(v(0)) = \sigma^2/2\alpha = \sigma_v^2$. Therefore, $\mathrm{var}(v(t)) = \sigma_v^2$ for all t and $v(t)$ is covariance stationary. We express the sampled process as the discrete first-order autoregressive process

$$v_{t_n} = e^{-\alpha\Delta_n}v_{t_{n-1}} + \varepsilon_{\Delta_n}$$

where $\varepsilon_{\Delta n} = \int_{t_{n-1}}^{t_n} e^{-\alpha(t-u)}dW(u)$ is the Gaussian process with $E(\varepsilon_{\Delta n}) = 0$ and $\mathrm{var}(\varepsilon_{\Delta n}) = \sigma_v^2(1 - e^{-2\alpha\Delta_n})$. We will assume $\Delta_n = \Delta$ for all n and we set $\mathrm{var}(\varepsilon_{\Delta n}) = \sigma_\varepsilon^2$ and $\rho = e^{-\alpha\Delta}$.

The joint probability density of $y = (y_{t_1}, \ldots, y_{t_N})^T$ is

$$f(y|\theta_1,\theta_2,\alpha,\sigma^2) = \left(\frac{1}{2\pi\sigma_\varepsilon^2}\right)^{\frac{N-1}{2}}\left(\frac{1}{2\pi\sigma_v^2}\right)^{\frac{1}{2}}\exp\left[-\frac{1}{2}\left\{\frac{1}{\sigma_v^2}(y_{t_1} - s_{t_1} - x_{t_1})^2\right.\right.$$
$$\left.\left. +\frac{1}{\sigma_\varepsilon^2}\sum_{n=2}^N\{(y_{t_n} - s_{t_n} - x_{t_n}) - \rho(y_{t_{n-1}} - s_{t_{n-1}} - x_{t_{n-1}})\}^2\right\}\right] \quad (2)$$

where $\theta_1 = (\mu, A_1, B_1, \ldots, A_d, B_d, \omega)^T$ and $\theta_2 = (C_1, D_1, \ldots, C_4, D_4)^T$. Let $\theta = (\theta_1^T, \theta_2^T)$. The following four propositions establish our principal results. In the first two we derive the approximate Fisher information matrix and show that it is maximized at $\Delta = 0$ for T large and fixed. In the second two we derive the approximate covariance matrix of the maximum likelihood estimate of θ and show that the lower bound of this matrix is also obtained at $\Delta = 0$ for T large and fixed.

PROPOSITION 1. The approximate Fisher information matrix of the parameters θ_1, θ_2 and α is

$$I(\theta, \alpha, \Delta) = \begin{bmatrix} I(\theta_1, \Delta) & I(\theta_1, \theta_2, \Delta)^T & 0 \\ I(\theta_1, \theta_2, \Delta) & I(\theta_2, \Delta) & 0 \\ 0 & 0 & I(\alpha, \Delta) \end{bmatrix}$$

where $I(\theta_1, \Delta)$ is the $(2d + 2) \times (2d + 2)$ matrix whose entries are defined as

$$I(\theta_1, \Delta)_{1,1} = \frac{T}{2\pi f(0, \alpha, \Delta)}$$

for $r = 1, \ldots, d$,

$$I(\theta_1, \Delta)_{2r, 2r} = I(\theta_1, \Delta)_{2r+1, 2r+1} = \frac{T}{4\pi f(\omega r, \alpha, \Delta)}$$

$$I(\theta_1, \Delta)_{2d+2, 2r} = I(\theta_1, \Delta)_{2r, 2d+2} = \frac{T^2 r B_r}{8\pi f(\omega r, \alpha, \Delta)}$$

$$I(\theta_1, \Delta)_{2d+2, 2r+1} = I(\theta_1, \Delta)_{2r+1, 2d+2} = \frac{-T^2 r A_r}{8\pi f(\omega r, \alpha, \Delta)}$$
(3)

$$I(\theta_1, \Delta)_{2d+2, 2d+2} = T^3 \sum_{r=1}^{d} \frac{r^2 (A_r^2 + B_r^2)}{12\pi f(\omega r, \alpha, \Delta)},$$

and $I(\theta_1, \Delta)_{r,s} = O(1)$ otherwise. The submatrix $I(\theta_2, \Delta)$ is the 8×8 approximate diagonal matrix whose elements are defined by

$$I(\theta_2, \Delta)_{2r-1, 2r-1} = I(\theta_2, \Delta)_{2r, 2r} = \frac{T}{4\pi f\left(\frac{2\pi}{28} r, \alpha, \Delta\right)}$$
(4)

for $r = 1, \ldots, 4$ and $I(\theta_2, \Delta)_{r,s} = O(1)$ otherwise, where

$$f(\omega, \alpha, \Delta) = \frac{\sigma_v^2 (1 - e^{-2\alpha\Delta})\Delta}{2\pi(1 - 2e^{-\alpha\Delta}\cos(\omega\Delta) + e^{-2\alpha\Delta})}$$
(5)

is the spectrum of a discrete first-order autoregressive process. The submatrix $I(\theta_1, \theta_2, \Delta)$ is $8 \times (2d + 2)$ and its elements are

$$I(\theta_1, \theta_2, \Delta)_{2j-1, 1} = I(\theta_1, \theta_2, \Delta)_{2j, 1} = 0$$

$$I(\theta_1, \theta_2, \Delta)_{2j-1, 2d+2} = T\Delta \left[\sum_{r=1}^{d} \frac{r}{8\pi f\left(\frac{2\pi j}{28}, \alpha, \Delta\right)} \right.$$

$$\left[B_r \left\{ \frac{\sin\left((T+\Delta)\left(\frac{2\pi j}{28} + \omega r\right)\right)}{\sin\left(\left(\frac{2\pi j}{28} + \omega r\right)\Delta\right)} + \frac{\sin\left((T+\Delta)\left(\frac{2\pi j}{28} - \omega r\right)\right)}{\sin\left(\left(\frac{2\pi j}{28} - \omega r\right)\Delta\right)} \right\} \right.$$

$$\left. + A_r \left\{ \frac{\cos\left((T+\Delta)\left(\frac{2\pi j}{28} + \omega r\right)\right)}{\sin\left(\left(\frac{2\pi j}{28} + \omega r\right)\Delta\right)} - \frac{\cos\left((T+\Delta)\left(\frac{2\pi j}{28} - \omega r\right)\right)}{\sin\left(\left(\frac{2\pi j}{28} - \omega r\right)\Delta\right)} \right\} \right] \right]$$

$$+ \frac{s'(T)}{\sigma_\varepsilon^2} \rho \left\{ \cos\left(\frac{2\pi j}{28}(T+\Delta)\right) - \rho\cos\left(\frac{2\pi j}{28} T\right) \right\}$$

$$
I(\theta_1, \theta_2, \Delta)_{2j,\,2d+2} = T\Delta \left[\sum_{r=1}^{d} \frac{r}{8\pi f\left(\dfrac{2\pi j}{28}, \alpha, \Delta\right)} \right.
$$

$$
\left[-B_r \left\{ \frac{\cos\left((T+\Delta)\left(\dfrac{2\pi j}{28} + wr\right)\right)}{\sin\left(\left(\dfrac{2\pi j}{28} + wr\right)\Delta\right)} + \frac{\cos\left((T+\Delta)\left(\dfrac{2\pi j}{28} - wr\right)\right)}{\sin\left(\left(\dfrac{2\pi j}{28} - wr\right)\Delta\right)} \right\} \right.
$$

$$
\left. \left. + A_r \left\{ \frac{\sin\left((T+\Delta)\left(\dfrac{2\pi j}{28} + wr\right)\right)}{\sin\left(\left(\dfrac{2\pi j}{28} + wr\right)\Delta\right)} - \frac{\sin\left((T+\Delta)\left(\dfrac{2\pi j}{28} - wr\right)\right)}{\sin\left(\left(\dfrac{2\pi j}{28} - wr\right)\Delta\right)} \right\} \right] \right]
$$

$$
+ \frac{s'(T)}{\sigma_\varepsilon^2} \rho \left\{ \sin\left(\frac{2\pi j}{28}(T+\Delta)\right) - \rho\sin\left(\frac{2\pi j}{28}T\right) \right\} \tag{6}
$$

where $j = 1, \ldots, 4$ and $s'(T)$ is the derivative of s_t with respect to ω evaluated at T and $I(\theta_1, \theta_2, \Delta)_{r,s} = O(1)$ otherwise. The entry $I(\alpha, \Delta)$ is defined as

$$
I(\alpha, \Delta) = \frac{1}{2\alpha^2} + \frac{\Delta e^{-2\alpha\Delta}T}{1 - e^{-2\alpha\Delta}} + \frac{2T}{\Delta}\left(\frac{1}{2\alpha} - \frac{\Delta e^{-2\alpha\Delta}}{1 - e^{-2\alpha\Delta}}\right)^2 \tag{7}
$$

The result is established by taking the expectation of the negative Hessian of the log likelihood of Eq. (2). The proof is outlined in Appendix I.

Since T in the FD protocol is large relative to Δ, we treat all $O(1)$ terms in the information matrix as equal to zero in our analyses. To study the limiting behavior of the information matrix as Δ tends to 0 we establish two lemmas and state definitions for upper and lower bounds of a matrix sequence modified from Anderson[9] (p. 80).

LEMMA 1. As Δ tends to 0,

1. The spectral density function $f(\omega, \alpha, \Delta)$ converges to $f(\omega, \alpha, 0)$ where

$$
f(\omega, \alpha, 0) = \lim_{\Delta \to 0} f(\omega, \alpha, \Delta) = \frac{\sigma^2}{2\pi(\omega^2 + \alpha^2)} \tag{8}
$$

[9] T. W. Anderson, in "An Introduction to Multivariate Statistical Analysis," 2nd Ed. John Wiley & Sons, New York, 1984.

is the spectrum of the continuous first-order autoregressive process and $f(\omega, \alpha, \Delta) > f(\omega, \alpha, 0)$ for all $\Delta > 0$.

2. The entry $I(\alpha, \Delta)$ converges to $I(\alpha, 0)$ where

$$I(\alpha, 0) = \lim_{\Delta \to 0} I(\alpha, \Delta) = \frac{\alpha T + 1}{2\alpha^2} \tag{9}$$

and $I(\alpha, 0) > I(\alpha, \Delta)$ for all $\Delta > 0$.

3. The elements in the $2d + 2$ column of $I(\theta_1, \theta_2, \Delta)$ converge to

$$I(\theta_1, \theta_2, 0)_{2j-1,\, 2d+2} = \lim_{\Delta \to 0} I(\theta_1, \theta_2, \Delta)_{2j-1,\, 2d+2}$$

$$= T \sum_{r=1}^{d} \frac{r}{8\pi f\left(\frac{2\pi j}{28}, \alpha, 0\right)} \left[B_r \left\{ \frac{\sin\left(T\left(\frac{2\pi j}{28} + \omega r\right)\right)}{\left(\frac{2\pi j}{28} + \omega r\right)} + \frac{\sin\left(T\left(\frac{2\pi j}{28} - \omega r\right)\right)}{\left(\frac{2\pi j}{28} - \omega r\right)} \right\} \right.$$

$$\left. + A_r \left\{ \frac{\cos\left(T\left(\frac{2\pi j}{28} + \omega r\right)\right)}{\left(\frac{2\pi j}{28} + \omega r\right)} - \frac{\cos\left(T\left(\frac{2\pi j}{28} - \omega r\right)\right)}{\left(\frac{2\pi j}{28} - \omega r\right)} \right\} \right]$$

$$+ s'(T) \left\{ \frac{-\frac{2\pi j}{28} \sin\left(\frac{2\pi j}{28} T\right) + \alpha \cos\left(\frac{2\pi j}{28} T\right)}{\sigma^2} \right\}$$

$$I(\theta_1, \theta_2, 0)_{2j,\, 2d+2} = \lim_{\Delta \to 0} I(\theta_1, \theta_2, \Delta)_{2j,\, 2d+2}$$

$$= T \sum_{r=1}^{d} \frac{r}{8\pi f\left(\frac{2\pi j}{28}, \alpha, 0\right)} \left[-B_r \left\{ \frac{\cos\left(T\left(\frac{2\pi j}{28} + \omega r\right)\right)}{\left(\frac{2\pi j}{28} + \omega r\right)} + \frac{\cos\left(T\left(\frac{2\pi}{28}j - \omega r\right)\right)}{\left(\frac{2\pi j}{28} - \omega r\right)} \right\} \right.$$

$$\left. + A_r \left\{ \frac{\sin\left(T\left(\frac{2\pi j}{28} + \omega r\right)\right)}{\left(\frac{2\pi j}{28} + \omega r\right)} - \frac{\sin\left(T\left(\frac{2\pi j}{28} - \omega r\right)\right)}{\left(\frac{2\pi j}{28} - \omega r\right)} \right\} \right]$$

$$+ s'(T) \left\{ \frac{\frac{2\pi j}{28} \cos\left(\frac{2\pi j}{28} T\right) + \alpha \sin\left(\frac{2\pi j}{28} T\right)}{\sigma^2} \right\} \tag{10}$$

for $j = 1, \ldots, 4$.

The proof of this lemma is given in Appendix II.

DEFINITION 1. Let $M(\Delta)$ be a family of symmetric matrices defined for all $\Delta\varepsilon\Omega$, where Ω is a set in R^1. If there exists a symmetric matrix M such that $M - M(\Delta)$ is nonnegative definite for all $\Delta\varepsilon\Omega$, then M is an upper bound of $M(\Delta)$ on Ω. If $M(\Delta) - M$ is nonnegative definite for all $\Delta\varepsilon\Omega$, then M is a lower bound of $M(\Delta)$ on Ω.

LEMMA 2. If A is a $p \times p$ symmetric matrix with the following properties:

1. $A_{ii} > 0$, for $i = 1, \ldots, p$
2. $A_{ij} = 0$ for $i \neq p$ and $i \neq j$
3. A_{pp} can be decomposed as $A_{pp} = b_1 + \cdots + b_{p-1}$ such that $b_i > 0$ and $A_{ii}b_i > A_{pi}^2$ for $i = 1, \ldots, p - 1$

then A is a positive definite matrix.

The proof of this lemma is given in Appendix III.

PROPOSITION 2. As Δ tends to 0, each element of $I(\theta_1, \Delta)$ and $I(\theta_2, \Delta)$ converges to a limit defined by replacing $f(\omega, \alpha, \Delta)$ with $f(\omega, \alpha, 0)$; the $2d + 1$st column of $I(\theta_1, \theta_2, \Delta)$ converges to the limits given in Eq. (10); and $I(\alpha, \Delta)$ converges to the limit in Eq. (9). $I(\theta, \alpha, \Delta)$ is defined on $[0, \Delta_0]$ for any fixed $\Delta_0 < 14\tau(\tau j_{max} + 28d)^{-1}$ where j_{max} is the number of FD harmonic components. $I(\theta, \alpha, 0)$ is an upper bound of $I(\theta, \alpha, \Delta)$ on $[0, \Delta_0]$.

The restriction of Δ to the interval $[0, \Delta_0]$ is necessary because even though the elements of $I(\theta_1, \Delta)$, $I(\theta_2, \Delta)$, and $I(\alpha, \Delta)$ are defined for $\Delta \geq 0$, the elements of $I(\theta_1, \theta_2, \Delta)$ are undefined if any of the arguments of the sine terms in the denominator of Eq. (6) equals π or one of its multiples. The upper bound on the largest positive interval in the neighborhood of zero in which all the elements of $I(\theta_1, \theta_2, \Delta)$ are defined is given by Δ', where Δ' satisfies $(2\pi j_{max}28^{-1} + 2\pi d\tau^{-1})\Delta' = \pi$.

Proof. The limits follow directly from Lemma 1. Given Δ_0 we first show that $I(\theta, \alpha, \Delta)$ is positive definite for $\Delta \in [0, \Delta_0]$. Let $I(\theta, \Delta)$ be the upper $(2d + 10) \times (2d + 10)$ submatrix of $I(\theta, \alpha, \Delta)$. We define $\theta' = (\mu, A_1, B_1, \ldots, A_d, B_d, C_1, D_1, \ldots, C_4, D_4, \omega)^T$ and consider the information matrix $I(\theta', \Delta)$ which is equivalent to $I(\theta, \Delta)$ for all Δ in $[0, \Delta_0]$. The elements of $I(\theta', \Delta)$ are determined by rearranging the elements of $I(\theta, \Delta)$ according to the order of θ'. To show that $I(\theta, \alpha, \Delta)$ is positive definite it suffices to apply Lemma 2 and show that $I(\theta', \Delta)$ is positive definite since $I(\theta, \Delta)$ and $I(\theta', \Delta)$ are equivalent matrices and $I(\alpha, \Delta) > 0$ for all $\Delta \geq 0$. The lower $(2d + 9) \times (2d + 9)$ submatrix of $I(\theta', \Delta)$ satisfies the conditions of Lemma 2 since by Proposition 1 and Lemma 1 $I(\theta', \Delta)_{j,j} > 0$ for $j = 2, \ldots, 2d + 10$ and $I(\theta', \Delta)_{2d+10, 2d+10}$ can be expressed as the sum of $2d + 8$ terms defined as

$$b_{2r} = \frac{T^3 r^2 B_r^2}{15\pi f(\omega r, \alpha, \Delta)}, \qquad b_{2r+1} = \frac{T^3 r^2 A_r^2}{15\pi f(\omega r, \alpha, \Delta)}$$

for $r = 1, \ldots, d$ and

$$b_{2d+1+j} = T^3 \sum_{r=1}^{d} \frac{r^2 (A_r^2 + B_r^2)}{480\pi f(\omega r, \alpha, \Delta)}$$

for $j = 1, \ldots, 8$. It follows that $b_{2r} I(\theta', \Delta)_{2r, 2r} > I(\theta', \Delta)_{2d+10, 2r}^2$, $b_{2r+1} I$ $(\theta', \Delta)_{2r+1, 2r+1} > I(\theta', \Delta)_{2d+10, 2r+1}^2$ for $r = 1, \ldots, d$. For T large $b_{2d+1+j} I$ $(\theta', \Delta)_{2d+1+j, 2d+1+j} > I(\theta', \Delta)_{2d+10, 2d+1+j}^2$ for $j = 1, \ldots, 8$ because the former term is $O(T^4)$, whereas the later is $O(T^2)$. Thus, by Lemma 2 the lower $(2d + 9) \times (2d + 9)$ submatrix of $I(\theta', \Delta)$ is positive definite. Hence, $I(\theta', \Delta)$ is positive definite since $I(\theta', \Delta)_{1,1} > 0$ and is the only nonzero element in its row and column. Therefore, $I(\theta, \alpha, \Delta)$ is positive definite. To prove that $I(\theta, \alpha, 0)$ is an upper bound we define for Δ in $(0, \Delta_0) M(\Delta) = I(\theta, \alpha, 0) - I(\theta, \alpha, \Delta)$. Lemma 1 along with the analysis applied above to $I(\theta, \alpha, \Delta)$ establish that $M(\Delta)$ is positive definite for Δ in $(0, \Delta_0)$. Hence by Definition 1 $I(\theta, \alpha, 0)$ is an upper bound on the information in the likelihood on $(0, \Delta_0)$.   □

PROPOSITION 3. The approximate infill asymptotic covariance matrix of $\hat{\theta}_1, \hat{\theta}_2$, and $\hat{\alpha}$, the respective maximum likelihood estimates of θ_1, θ_2, and α, is

$$G(\Delta) = \begin{bmatrix} G_1(\Delta) & G_{12}(\Delta) & 0 \\ G_{21}(\Delta) & G_2(\Delta) & 0 \\ 0 & 0 & G_3(\Delta) \end{bmatrix} \qquad (11)$$

where $G_1(\Delta) = I(\theta, \Delta)^{-1}$, is the $(2d + 2) \times (2d + 2)$ symmetric matrix whose lower triangle is

$$4\pi \begin{bmatrix} K_0/T & & & & & & & \\ 0 & (K_1 + Ca_1^2)/T & & & & & & \\ 0 & Cb_1 a_1/T & (K_1 + Cb_1^2)/T & & & & & \\ & \vdots & \vdots & & & & & \\ 0 & Ca_r a_1/T & Ca_r b_1/T & \cdots & (K_r + Ca_r^2)/T & & & \\ 0 & Cb_r a_1/T & Cb_r b_1/T & \cdots & Cb_r a_r/T & (K_r + Cb_r^2)/T & & \\ & \vdots & \vdots & & \vdots & \vdots & & \\ 0 & Ca_d a_1/T & Ca_d b_1/T & \cdots & Ca_d a_r/T & Ca_d b_r/T & \cdots & (K_d + Ca_d^2)/T & \\ 0 & Cb_d a_1/T & Cb_d b_1/T & \cdots & Cb_d a_r/T & Cb_d b_r/T & \cdots & Cb_d a_d & (K_d + Cb_d^2)/T & \\ 0 & -Ca_1/T^2 & -Cb_1/T^2 & \cdots & -Ca_r/T^2 & -Cb_r/T^2 & \cdots & -Ca_d/T^2 & -Cb_d/T^2 & C/T^3 \end{bmatrix}$$

$$(12)$$

where

$$K_0 = \frac{1}{2} f(0, \alpha, \Delta), \qquad K_r = f(\omega r, \alpha, \Delta), \qquad a_r = \frac{1}{2} r B_r, \qquad b_r = -\frac{1}{2} r A_r$$

for $r = 1, \ldots, d$ and

$$C = C(\Delta) = 12 \Big/ \left\{ \sum_{r=1}^{d} r^2 \frac{(A_r^2 + B_r^2)}{f(\omega r, \alpha, \Delta)} \right\} \qquad (13)$$

and $G_2(\Delta)$, is the 8×8 approximate diagonal matrix whose elements are

$$G_2(\Delta)_{2r-1,\, 2r-1} = G_2(\Delta)_{2r,\, 2r} = 4\pi f\left(\frac{2\pi r}{28}, \alpha, \Delta\right) T^{-1} + O(T^{-3}) \qquad (14)$$

for $r = 1, \ldots, 4$, $G_{12}(\Delta)$ is the $(2d+2) \times 8$ matrix whose elements are

$$\begin{aligned} G_{12}(\Delta)_{r,j} &= -G_1(\Delta)_{r,\, 2d+2} I(\theta_1, \theta_2, \Delta)_{2d+2,\, j}\, I(\theta_2, \Delta)_{j,j}^{-1} \\ &= \begin{cases} 0 & r = 1 \\ O(T^{-2}) & r = 2, \ldots, 2d+1 \\ O(T^{-3}) & r = 2d+2 \end{cases} \end{aligned} \qquad (15)$$

for $j = 1, \ldots, 8$, $G_{21}(\Delta) = G_{12}^T(\Delta)$, and $G_3(\Delta) = I(\alpha, \Delta)^{-1}$.

Proof. Applying the standard formula for the inverse of a partitioned matrix (Anderson,[9] p. 594) and the definitions of the components of $I(\theta, \alpha, \Delta)$ given in Proposition 1 it is easy to verify that the components of $G(\Delta)$ are as stated above. To compute the explicit form of $G_1(\Delta)$ we re-express $I(\theta_1, \Delta) = \Lambda(T) I^*(\theta, \Delta) \Lambda(T)^T$, where $\Lambda(T)$ is a $(2d+2) \times (2d+2)$ diagonal matrix whose first $2d+1$ diagonal entries are $T^{1/2}$ and whose $2d+2$ entry is $T^{3/2}$. The matrix $I^*(\theta_1, \Delta)$ has the same entries as $I(\theta_1, \Delta)$ except that the T values have been omitted. The approximate covariance matrix of $\hat{\theta}_1$ is

$$G_1(\Delta) = \Lambda(T)^{-1} I^*(\theta_1, \Delta)^{-1} \Lambda(T)^{-1} \qquad (16)$$

The elements of $\Lambda(T)^{-1}$ are the reciprocals of the diagonal elements of $\Lambda(T)$ and the elements of $I^*(\theta_1, \Delta)^{-1}$ can be computed directly from Brown[4] (p. 655). □

The following lemma is proved as part of Theorem 12.2.14 in Graybill[10] (pp. 408–410). It will allow us to characterize the behavior of the infill asymptotic covariance matrix of the maximum likelihood parameter estimates as Δ tends to 0.

LEMMA 3. If A, B and $A - B$ are positive definite matrices, then $B^{-1} - A^{-1}$ is positive definite.

PROPOSITION 4. The covariance matrix

$$G(0) = \lim_{\Delta \to 0} G(\Delta)$$

[10] F. A. Graybill, *in* "Matrices with Applications to Statistics," 2nd Ed. Wadsworth, Belmont, CA, 1983.

is defined by substituting $f(\omega, \alpha, 0)$ for $f(\omega, \alpha, \Delta)$ in $G_1(\Delta)$, $G_2(\Delta)$, and $G_{12}(\Delta)$; by substituting $I(\theta_1, \theta_2, 0)_{2d+1, j}$ for $I(\theta_1, \theta_2, \Delta)_{2d+1, j}$ and $I(\theta_2, 0)_{j, j}$ for $I(\theta_2, \Delta)_{j, j}$ in Eq. (15); and by computing $G_3(0)$ as

$$G_3(0) = \lim_{\Delta \to 0} G_3(\Delta) = \frac{2\alpha^2}{\alpha T + 1}$$

Furthermore, $G(0)$ is a lower bound of $G(\Delta)$ for all Δ in $[0, \Delta_0]$ where Δ_0 is defined in Proposition 2.

Proof. Given Δ_0 and Δ satisfying $0 < \Delta \leq \Delta_0$, the matrices $I(\theta, \alpha, 0)$, $I(\theta, \alpha, \Delta)$, and $I(\theta, \alpha, 0) - I(\theta, \alpha, \Delta)$ are positive definite by the proof of Proposition 2. Since $G(\Delta) = I(\theta, \alpha, \Delta)^{-1}$ and $G(0) = I(\theta, \alpha, 0)^{-1}$, $G(\Delta) - G(0)$ is positive definite by Lemma 3 for $0 < \Delta \leq \Delta_0$. Hence, by Definition 1 $G(0)$ is the stated lower bound. □

Remark 1: The upper bound on Δ_0 in Proposition 2 for τ near 24 h, $j_{max} = 4$ and $d = 3$ is 1.87 h. Hence, as we show in the next section, $I(\theta, \alpha, \Delta)$ and $G(\Delta)$ are defined for a reasonable range of sampling intervals for the purpose of data analysis.

Remark 2: By Proposition 3, the estimates $\hat{\theta}$ and $\hat{\alpha}$ are uncorrelated, whereas the correlation between $\hat{\theta}_2$ and $\hat{\mu}$ is zero, and between $\hat{\theta}_2$ and the remaining components of $\hat{\theta}_1$ is $O(T^{-1})$. It follows from Proposition 4 that $\hat{\theta}$ and $\hat{\alpha}$ are not consistent as Δ tends to 0.

Remark 3: If $\Delta = 1$, the $2d + 2, 2d + 2$ entry of G agrees with the variance formula for $\hat{\omega}$ in the harmonic regression plus stationary correlated noise models reported by Hannan,[11,12] Walker,[13] and Brown.[4] If $d = 1$, $\Delta = 1$ and $\rho = 0$, that is, $\alpha \to \infty$, then this entry of G simplifies to the formula for the variance of $\hat{\omega}$ reported by Walker.[13] Equations (A1) to (A4) in the Appendix I provide a straightforward derivation of the well-known formula for the $O(N^{-3})$ dependence of the variance of $\hat{\omega}$ on the sample size for the case of a discrete first-order autoregressive Gaussian noise process.

Remark 4: The approximate variance of $\hat{\tau}$, the maximum likelihood estimate of τ, can be computed from the $2d + 2, 2d + 2$ entry of G_1 by the delta method and is given as

$$\text{var}(\hat{\tau}) = \frac{12\tau^4}{\pi T^3 \sum_{r=1}^{d} r^2 \dfrac{(A_r^2 + B_r^2)}{f(\omega r, \alpha, \Delta)}} \tag{17}$$

To study the factors that affect the precision with which $\hat{\tau}$ estimates τ we scale the standard error of $\hat{\tau}$ by τ.[7] That is, we consider the ratio

[11] E. J. Hannan, *J. Appl. Prob.* **8,** 767 (1971).
[12] E. J. Hannan, *J. Appl. Prob.* **10,** 510 (1973).
[13] A. M. Walker, *Biometrika* **58,** 21 (1971).

$$\frac{se_\Delta(\hat{\tau})}{\tau} = m^{-1}\{\pi T\alpha \text{SNR}(0)\lambda(\Delta)\}^{-\frac{1}{2}} \qquad (18)$$

where $m = T/\tau$, $\text{SNR}(0) = 1/\alpha C(0)$, and $\lambda(\Delta) = \text{SNR}(\Delta)/\text{SNR}(0)$. $\text{SNR}(0)$ is the intrinsic signal-to-noise ratio defined by the circadian and autoregressive processes and $\lambda(\Delta)$ measures the loss of precision due to sampling. Equation (18) shows that the precision of $\hat{\tau}$ is affected by four factors: (1) the intrinsic signal-to-noise ratio; (2) the information loss due to sampling; (3) m, the number of periods of τ that are observed; and (4) $T\alpha$, the number of noise time constants observed.

This result may be viewed as an infill asymptotic extension of the findings of Rice and Rosenblatt.[14] They considered the model in Eq. (1) with $\Delta = 1, d = 1, x_t = 0$, for all t and the ν_t being independent, identically distributed Gaussian random variables. Their simulation results showed that for $\hat{\omega}$ estimated by nonlinear least squares, the $O(N^{-3})$ asymptotic behavior for its variance was not realized unless the product of the amplitude of the harmonic component and N was large. Alternatively, if $\hat{\tau}$ were a consistent or an approximately unbiased estimate of τ, then Eq. (18) could be interpreted as an approximate coefficient of variation.

Remark 5: Equation (17) shows that the FD component does not enter explicitly into the formula for the variance of $\hat{\tau}$. The FD component estimate clearly affects this variance by reducing $\hat{\sigma}_\varepsilon^2$.

Remark 6: From Eq. (12) it follows that

$$\text{var}(\hat{\mu}) = \frac{\sigma_\nu^2(1 - e^{-2\alpha\Delta})\Delta}{T(1 - 2e^{-\alpha\Delta} + e^{-2\alpha\Delta})}$$

and that

$$\lim_{\Delta \to 0} \text{var}(\hat{\mu}) = \frac{\sigma^2}{T\alpha^2} \qquad (19)$$

Equation (19) is a continuous time analog of (5.3.9) in Priestley[15] and gives the correct expression for the var($\hat{\mu}$) that is incorrectly reported in Eq. (7) of Morris and Ebey.[16] Morris and Ebey studied the model in Eq. (1) with $s_t = \mu, x_t = 0$ for all t, and $T = 1$. They derived a formula for the variance of the sample mean, studied its behavior as Δ tends to 0 and concluded that there was an optimal finite sampling interval at which the variance of the sample mean is minimized. Our results show that their findings are incorrect. As shown in Proposition 4, the variance of each parameter estimate

[14] J. A. Rice and M. Rosenblatt, *Biometrika* **75**, 477 (1988).
[15] M. B. Priestley, *in* "Spectral Analysis and Time Series." Academic Press, London, 1981.
[16] M. D. Morris and S. F. Ebey, *Am. Stat.* **38**, 127 (1984).

is a minimum at $\Delta = 0$. The error in the Morris and Ebey calculations arises because in analyzing the sampled continuous autoregressive process, they did not appreciate that $\rho = e^{-\alpha\Delta}$ and that this parameter cannot remain fixed but must converge to 1 as Δ goes to 0.

Data Analysis

To investigate the implications of our results for the study of core temperature data collected under FD, we analyze data from five subjects studied on this protocol. The data were recorded at 1-min intervals on five healthy male subjects , ages 21 to 25 years, monitored on a 28-h FD protocol for 22 to 30 days. An analysis of these data, comparing use of the harmonic regression model with a van der Pol model for representing the circadian signal, has been given in Brown et al.[5] Here we report further analysis using only the harmonic regression model. For computational purposes the log likelihood of each subject's data is written as

$$\log f(y|B, \omega, \alpha, \sigma^2) = -\frac{N}{2}\log(2\pi) - \frac{1}{2}\log|\Gamma| - \frac{N}{2}\log(\sigma_\varepsilon^2)$$
$$-\frac{1}{2\sigma_\varepsilon^2}(y - ZB)^T\Gamma^{-1}(y - ZB) \qquad (20)$$

where $B = (\mu, A_1, B_1, \ldots, A_d, B_d, C_1, D_1, \ldots, C_4, D_4)^T$, Z is the $N \times (2d + 9)$ design matrix containing the cosine and sine terms of the circadian and FD components, and Γ is the covariance matrix of $\nu = (\nu_{t_1}, \ldots, \nu_{t_n})^T$. The log likelihood is efficiently maximized with a Newton's procedure in which the Kalman filter is used to evaluate $|\Gamma|$ and Γ^{-1},[17] and $\hat{B}$ is computed for given values of α and ω as the generalized least-squares estimate

$$\hat{B} = (Z^T\Gamma^{-1}Z)^{-1}Z^T\Gamma y \qquad (21)$$

Computing $\hat{B}$ as a generalized least-squares estimate means that the Newton procedure has only two nonlinear parameters. The model in Eq. (1) was fit to each subject's core temperature series for $d = 3$ and $\Delta = 1$, 10, 20, and 30 min. All four sampling intervals gave similar parameter estimates for each subject with the exception of α estimated with $\Delta = 1$ min. For this interval the estimates of α corresponded to values of ρ that are numerically equal to 1. Numerical performance of the model-fitting algorithm could possibly be improved for this small interval by using the continuous instead of the discrete representation of the first-order autoregressive process. We report here the results of the analyses with $\Delta = 20$ min.

[17] E. N. Brown and C. Schmid, Methods Enzymol. 240, 171 (1994).

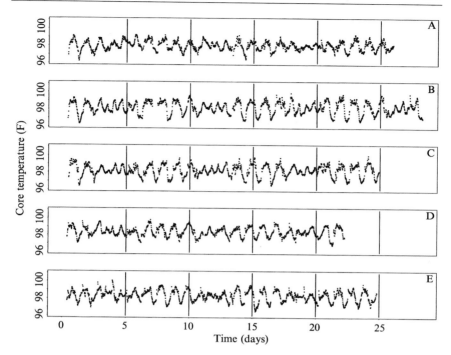

FIG. 1. The FD core temperature data of subjects 1 through 5 in order in (A) through (E). Reproduced from the *American Journal of Physiology, Endocrinology,* used by permission of the American Physiological Society.

It is clear from Fig. 1 that the core temperature series of each subject consists of a circadian, an FD, and a correlated noise component. The strong interaction between the two periodic processes is evidenced by the destructive interference on days 4, 11, and 18 in each subject's data (Fig. 1). This approximate 7-day modulation period is expected since the two periods are 28 h and approximately 24 h. Each subject's core temperature series was readily decomposed into its circadian, FD, and autoregressive components as the data from subject 1 illustrate (Fig. 2). Large values of the residuals occur at the points of maximum excursion of the temperature series (Fig. 2A). The log spectra, computed by periodogram smoothing with a span 100 modified Daniell filter using 20% tapering, were used along with their associated 95% confidence intervals computed by χ^2 approximation (Bloomfield,[18] pp. 195–197) to evaluate model goodness of

[18] P. Bloomfield, *in* "Fourier Analysis of Time Series: An Introduction." John Wiley & Sons, New York, 1976.

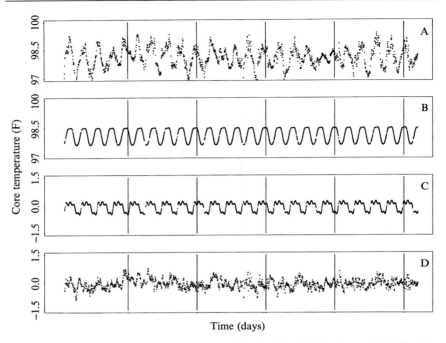

FIG. 2. Analysis of FD core temperature series of subject 1 with the model defined in Eq. (1). The original series (A) is decomposed into its estimated circadian component (B), estimated FD component (C), and estimated autoregressive component (D).

fit. No subject had any statistically significant frequencies in the spectrum of his residual series, suggesting that the model in Eq. (1) captures the salient structure in each core temperature series.[5,19]

For the estimated circadian harmonic coefficients four of the $\hat{A}_2$ values, four of the $\hat{A}_3$ values, and two of the $\hat{B}_3$ values are less than twice their estimated standard errors in absolute value (Table I). Among the FD parameter estimates, three $\hat{C}_2$ values, four $\hat{C}_3$ values, one $\hat{D}_3$, and one $\hat{D}_4$ are less than twice their estimated standard errors in absolute value (Table II). The approximation of the circadian and FD components as simple sine waves gives their respective amplitude estimates as $\hat{Amp}_{c_1} = (\hat{A}_1^2 + \hat{B}_1^2)^{1/2}$ and $\hat{Amp}_{FD_1} = (\hat{C}_1^2 + \hat{D}_1^2)^{1/2}$, and by the delta method, the approximate variances of these estimates as

---

[19] E. N. Brown and H. Luithardt, *J. Biol. Rhythms* **14**, 609 (1999).

TABLE I
CIRCADIAN HARMONIC REGRESSION PARAMETER ESTIMATES[a]

Parameter	Subject				
	1	2	3	4	5
$\hat{\mu}$	98.194	98.360	98.405	98.572	98.503
	(0.027)	(0.027)	(0.027)	(0.037)	(0.029)
$\hat{A}_1$	−0.372	−0.256	−0.350	−0.324	−0.521
	(0.049)	(0.044)	(0.044)	(0.053)	(0.041)
$\hat{B}_1$	0.326	0.410	0.400	0.416	0.308
	(0.047)	(0.038)	(0.041)	(0.047)	(0.051)
$\hat{A}_2$	0.009^b	0.020^b	0.028^b	0.067	0.029^b
	(0.025)	(0.024)	(0.032)	(0.025)	(0.031)
$\hat{B}_2$	0.061	0.062	0.137	0.056	0.131
	(0.022)	(0.021)	(0.024)	(0.026)	(0.025)
$\hat{A}_3$	-0.022^b	-0.012^b	-0.016^b	-0.030^b	0.044
	(0.019)	(0.018)	(0.024)	(0.019)	(0.022)
$\hat{B}_3$	−0.037	-0.031^b	−0.068	-0.035^b	−0.056
	(0.017)	(0.016)	(0.019)	(0.018)	(0.021)
$\hat{\tau}$	24.25	24.22	24.27	24.13	24.07
	(1.71)	(1.58)	(1.44)	(2.08)	(1.37)
$\hat{\mathrm{Amp}}_{C_1}$	0.495	0.483	0.531	0.527	0.605
	(0.031)	(0.030)	(0.032)	(0.036)	(0.034)
$\hat{\mathrm{Amp}}_C$	0.462	0.483	0.539	0.514	0.619
	(0.032)	(0.033)	(0.033)	(0.036)	(0.036)

[a] The numbers in parentheses are the standard errors of the estimates. The estimate $\hat{\tau}$ is reported in hours whereas its standard error is in minutes.
[b] A parameter estimate that does not exceed its standard error by more than 2 in absolute value.

$$\mathrm{var}(\hat{\mathrm{Amp}}_{C_1}) = 4\pi f\left(\frac{2\pi}{\tau}, \alpha, \Delta\right) T^{-1}$$

$$\mathrm{var}(\hat{\mathrm{Amp}}_{FD_1}) = 4\pi f\left(\frac{2\pi}{28}, \alpha, \Delta\right) T^{-1}$$

Since the circadian and FD components of the model are not simple sine waves, we also follow the convention in chronobiology of estimating the amplitudes of these rhythms as half the difference between the estimated maximum and minimum values of s_t and x_t within one of their respective cycles.[8] That is,

TABLE II
FD Harmonic Regression and Autoregression Parameter Estimates[a]

Parameter	Subject				
	1	2	3	4	5
$\hat{C}_1$	0.144 (0.033)	−0.370 (0.033)	−0.342 (0.033)	−0.237 (0.038)	−0.169 (0.035)
$\hat{D}_1$	0.267	0.647	0.516	0.391	0.436
$\hat{C}_2$	0.042^b (0.024)	0.060 (0.024)	0.027^b (0.025)	$−0.038^b$ (0.025)	0.060 (0.026)
$\hat{D}_2$	0.086	0.340	0.255	0.223	0.218
$\hat{C}_3$	$−0.010^b$ (0.018)	−0.053 (0.018)	$−0.003^b$ (0.020)	0.008^b (0.018)	$−0.010^b$ (0.020)
$\hat{D}_3$	0.035	−0.023	0.000^b	0.044	0.105
$\hat{C}_4$	−0.061 (0.014)	−0.136 (0.014)	−0.086 (0.016)	−0.066 (0.014)	−0.106 (0.016)
$\hat{D}_4$	0.089	0.024^b	0.087	0.052	0.062
$\hat{\text{Amp}}_{FD_1}$	0.303 (0.033)	0.662 (0.031)	0.619 (0.033)	0.602 (0.039)	0.468 (0.036)
$\hat{\text{Amp}}_{FD}$	0.353 (0.033)	0.860 (0.036)	0.648 (0.035)	0.552 (0.039)	0.614 (0.039)
$\hat{\alpha}$	0.357 (0.034)	0.344 (0.032)	0.433 (0.038)	0.251 (0.029)	0.390 (0.036)

[a] The numbers in parentheses are the standard errors of the estimates. By Eq. (14), for $i = 1, \ldots 4$, the standard error for $\hat{D}_i$ exactly equals the standard error of $\hat{C}_i$ and is therefore omitted.

[b] A parameter estimate that does not exceed its standard error by more than 2 in absolute value.

$$\hat{\text{Amp}}_C = (\hat{s}_{t_{\max}} - \hat{s}_{t_{\min}})2^{-1}$$
$$\hat{\text{Amp}}_{FD} = (\hat{x}_{t_{\max}} - \hat{x}_{t_{\min}})2^{-1}$$

where $t_{\min}$ and $t_{\max}$ are, respectively, the times at which the indicated model component assumes its minimum or maximum value. For comparison we compute the standard errors of these amplitude estimates by Monte Carlo methods under the assumption that the circadian and FD parameters are approximate Gaussian random variables with mean $\hat{\theta}_1, \hat{\theta}_2$, and $\hat{\alpha}$, and covariance matrix G. All the amplitude estimates are between 9 and 24 times their standard errors (Tables I and II). For subjects 2, 3, and 4, the FD amplitude estimates are larger than the circadian amplitude estimates for both definitions. This observation shows that the modulation of a subject's core-temperature rhythm induced by the FD protocol can be greater than the amplitude of the intrinsic circadian oscillation.

All subjects have physiologically reasonable estimates of α and τ (Table I). Each $\hat{\alpha}$ is at least eight times its standard error. These estimates show that there is strong serial dependence in the core-temperature data since they correspond to serial correlation coefficients ranging from 0.866 to 0.920. The reciprocals of the $\hat{\alpha}$ values are time constants for the autoregressive process and range from 2.31 to 3.98 h. All the period estimates are within 15 min of 24 h. The approximate 99% confidence intervals computed from Eq. (17) as $5.15 \times se(\hat{\tau})$ suggest that all the period estimates are significantly different from 25 h. These confidence intervals are 7.1 to 10.7 min in length. Together with the parameter estimates for $\hat{\tau}$ this finding offers strong evidence that the intrinsic period of the human biological clock is closer to 24 instead of 25 h.

To assess the effect of the sampling interval on the standard error of $\hat{\tau}$ we plot for each subject the standard error of $\hat{\tau}$ as a function of Δ (Fig. 3). These graphs show that for $\Delta \leq 60$ min the standard error of $\hat{\tau}$ is between 1.36 and 2.10 min for each subject. The factors contributing to these small standard errors are readily understood by evaluating the components of Eq. (18) for each subject (Table III). In this problem SNR(0) ranges from 0.103 to 0.205. However, for each subject $\lambda(\Delta)$ exceeds 0.996 for Δ less than 60 min, m ranges from 22 to 28 and $T\alpha$ from 132 to 254. Because m and $T\alpha$ are large and because $\lambda(\Delta)$ is close to $\lambda(0)$ for Δ less than 60 min, τ can be estimated with high precision despite the low intrinsic signal-to-noise ratios in the core temperature series.

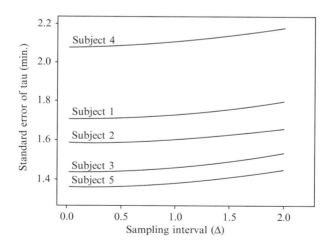

FIG. 3. The standard error of $\hat{\tau}$ as a function of Δ for each subject.

Subject	$\hat{m}$	$T\hat{\alpha}$	$\hat{SNR}(0)$	$\hat{\lambda}(\Delta)$
1	25.38	219.59	0.116	0.997
2	27.73	230.94	0.103	0.997
3	24.22	254.35	0.190	0.996
4	21.83	132.21	0.123	0.997
5	24.40	229.01	0.205	0.996

Evaluation of $|\Gamma|$ and Γ^{-1} in Eqs. (20) and (21) using the Kalman filter requires $O(N)$ operations. With $\Delta = 1$ min and T ranging from 22 to 30 days, N ranges from 31,680 to 43,200, whereas with $\Delta = 60$ min, N ranges from 528 to 720. Hence, important computational savings can be obtained by sampling each core temperature series at a 60-min interval with only a minimal loss in the precision with which τ is estimated. Alternatively stated, using only 1.67% of the observations τ can be estimated in a computationally efficient manner with near-maximal precision.

Discussion

We studied the infill asymptotic behavior of the maximum likelihood estimates of a harmonic regression plus continuous first-order autoregressive noise model in order to determine the precision with which the intrinsic period of the human biological clock can be estimated from core temperature measurements made under an FD protocol. We found that the intrinsic period of the human biological clock is closer to 24 h instead of 25 h and that it can be estimated with a standard error of 1.4 to 2.1 min. We also showed that for this FD protocol the intrinsic period of the clock can be estimated in a computationally efficient manner with near-maximal precision from a sampled series containing less than 2% of the original temperature data.

Our maximum likelihood parameter estimates are not consistent since their infill asymptotic covariance matrix is finite at $\Delta = 0$. Therefore, we characterized the factors, which affect the precision of the period estimate as Δ tends to 0 by scaling the standard error of $\hat{\tau}$ by τ. Our approach to the analysis of infill asymptotics differs from that of previous reports, which have studied conditions under which parameter estimates are consistent and asymptotically Gaussian.[20–24] Our core temperature analysis illustrates a case in which model parameters can be estimated with high precision

[20] S. J. Yakowitz and F. Szidarovsky, *J. Multivariate Anal.* **16,** 21 (1985).
[21] W. Hardle and P. D. Tuan, *J. Time Ser. Anal.* **7,** 191 (1986).

although not consistently because of the inherent properties of the biological process under study and the design of the experimental protocol required for the investigation. The scaling analysis applied to $\hat{\tau}$ may obviously be applied to any other model parameter estimate. A more detailed presentation on the use of scaling to study the infill asymptotic properties of other time series models as well as kriging, splines, and nonparametric regression models is given in Solo.[7] In future investigations we will extend our analysis of the model in Eq. (1) to cases in which the light–dark cycle interacts nonlinearly with the circadian signal and has pseudoperiodic behavior. We are also studying the problem of modeling the circadian signal directly as a weakly nonlinear differential equation.[5]

Appendix I: Outline of Proof of Proposition 1

We derive the $2d + 2, 2d + 2$ element of $I(\theta, \alpha, \Delta)$. The other entries of the matrix are computed using similar arguments. The second derivative of the log likelihood of Eq. (1) with respect to ω is

$$
\frac{\partial^2 \log f}{\partial \omega^2} = \left[\frac{1}{\sigma_v^2} \left\{ v_{t_1} \left(\frac{\partial^2 s_{t_1}}{\partial \omega^2} \right) - \left(\frac{\partial s_{t_1}}{\partial \omega} \right)^2 \right\} \right.
$$
$$
\left. + \frac{1}{\sigma_\varepsilon^2} \sum_{n=2}^{N} \left\{ - \left(\frac{\partial s_{t_n}}{\partial \omega} - \rho \frac{\partial s_{t_{n-1}}}{\partial \omega} \right)^2 + \left(v_{t_n} - \rho v_{t_{n-1}} \right) \left(\frac{\partial^2 s_{t_n}}{\partial \omega^2} - \rho \frac{\partial^2 s_{t_{n-1}}}{\partial \omega^2} \right) \right\} \right]
$$

$$(A1)$$

and the Fisher information for ω is

$$
E\left(-\frac{\partial^2 \log f}{\partial \omega^2} \right) = \left\{ \frac{1}{\sigma_v^2} \left(\frac{\partial s_{t_1}}{\partial \omega} \right)^2 + \frac{1}{\sigma_\varepsilon^2} \sum_{n=2}^{N} \left(\frac{\partial s_{t_n}}{\partial \omega} - \rho \frac{\partial s_{t_{n-1}}}{\partial \omega} \right)^2 \right\} \qquad (A2)
$$

Expanding the squares of the derivatives in Eq. (A2) and keeping terms of order n^2 we find that

$$
\left(\frac{\partial s_{t_n}}{\partial \omega} \right)^2 = (n\Delta)^2 \sum_{r=1}^{d} r^2 \frac{(A_r^2 + B_r^2)}{2} + O(n)
$$

$$
\frac{\partial s_{t_n}}{\partial \omega} \frac{\partial s_{t_{n-1}}}{\partial \omega} = (n\Delta)^2 \sum_{r=1}^{d} r^2 \frac{(A_r^2 + B_r^2)}{2} \cos(\omega r \Delta) + O(n)
$$

[22] M. L. Stein, *J. Am. Stat. Assoc.* **82,** 765 (1987).
[23] M. L. Stein, *Ann. Stat.* **16,** 55 (1988).
[24] M. L. Stein, *Ann. Stat.* **17,** 980 (1989).

Substituting these derivatives into Eq. (A2) and retaining terms up to order N^3 gives

$$E\left(-\frac{\partial^2 \log f}{\partial \omega^2}\right) = \frac{1}{\sigma_\varepsilon^2}\left[\sum_{n=2}^{N}(\Delta n)^2\left\{(1+\rho^2)\sum_{r=1}^{d}r^2\left(\frac{A_r^2 + B_r^2}{2}\right)\right.\right.$$

$$\left.\left. -2\rho\sum_{r=1}^{r}r^2\left(\frac{A_r^2 + B_r^2}{2}\right)\cos(\omega r\Delta)\right\}\right] + O(N^2)$$

$$= \frac{1}{\sigma_\varepsilon^2}\left[\Delta^2\frac{N^3}{6}\left\{\sum_{r=1}^{d}r^2(A_r^2 + B_r^2)(1+\rho^2-2\rho\cos(\omega r\Delta))\right\}\right] + O(N^2)$$

$$\text{(A3)}$$

Substituting $N = \frac{T}{\Delta}$, $\rho = e^{-\alpha\Delta}$, and $\sigma_\varepsilon^2 = \sigma_\nu^2(1-e^{-2\alpha\Delta})$ yields

$$E\left(-\frac{\partial^2 \log f}{\partial \omega^2}\right) = \frac{T^3}{6\sigma_\nu^2\Delta(1-e^{-2\alpha\Delta})}$$

$$\left\{\sum_{r=1}^{d}r^2(A_r^2 + B_r^2)(1-2e^{-\alpha\Delta}\cos(\omega r\Delta)+e^{-2\alpha\Delta})\right\} + O(T^2)$$

$$\text{(A4)}$$

The result is established be reexpressing Eq. (A4) using Eq. (5).

Appendix II: Proof of Lemma 1

The limits in Eqs. (8), (9), and (10) are established using l'Hôpital's rule.

1. To show that $f(\omega, \alpha, \Delta)$ assumes its minimum at $\Delta = 0$, we establish the more general result that

$$\Delta f_d(\eta\Delta) \geq f_c(\eta)$$

where $f_c(\eta)$ is the spectral density function of a second-order stationary process, and $f_d(\eta\Delta)$ is the spectrum of the associated discrete process sampled at an interval Δ. Our result will follow with $\Delta f_d(\eta\Delta)$ given in Eq. (5) and $f_c(\eta)$ given by Eq. (8). The autocovariance function of the continuous process is given by the inverse Fourier transform

$$\gamma(\tau) = \frac{1}{2\pi}\int_{-\infty}^{\infty}e^{i\eta\Delta k}f_c(\eta)d\eta$$

The autocovariance function of the sampled process is

$$\gamma(k\Delta) = \frac{1}{2\pi} \int_{-\infty}^{\infty} e^{j\eta\Delta k} f_c(\eta) d\eta$$

$$= \frac{1}{2\pi\Delta} \int_{-\infty}^{\infty} e^{j\zeta k} f_c(\zeta/\Delta) d\zeta$$

$$= \frac{1}{2\pi\Delta} \sum_{r=-\infty}^{\infty} \int_{2\pi r-\pi}^{2\pi r+\pi} e^{j\zeta k} f_c(\zeta/\Delta) d\zeta \qquad (A5)$$

$$= \int_{-\pi}^{\pi} e^{j\zeta' k} \sum_{r=-\infty}^{\infty} \frac{1}{2\pi\Delta} f_c\left(\frac{\zeta' + 2\pi r}{\Delta}\right) d\zeta'$$

where the interchange of summation and integration is justified by the bounded convergence theorem. By definition $\gamma(k\Delta)$ is also given as

$$\gamma(k\Delta) = \frac{1}{2\pi} \int_{-\pi}^{\pi} e^{j\zeta' k} f_d(\zeta') d\zeta' \qquad (A6)$$

Equating the corresponding terms of the integrands in Eqs. (A5) and (A6), and using the fact that $f_c(\eta) \geq 0$ for all η, gives

$$f_d(\zeta') = \frac{1}{\Delta} \sum_{r=-\infty}^{\infty} f_c\left(\frac{\zeta' + 2\pi r}{\Delta}\right) \geq \frac{1}{\Delta} f_c(\zeta'/\Delta)$$

or

$$\Delta f_d(\zeta'\Delta) \geq f_c(\zeta')$$

In particular, if $f_c(\eta)$ is the spectral density in Eq. (8), then $f_c(\eta) > 0$ for all η and we have the strict inequality $\Delta f_d(\eta\Delta) > f_c(\eta)$.

2. To show that $I(\alpha, \Delta)$ assumes its maximum at $\Delta = 0$, we establish that for $\alpha > 0, I(\alpha, \Delta)$ is a monotone decreasing function of Δ on $(0,\infty)$ by showing that its derivative is everywhere negative on this interval. The derivative of $I(\alpha, \Delta)$ with respect to Δ may be written as

$$\frac{\partial I(\alpha, \Delta)}{\partial \Delta} = -\frac{T\{g_1(\Delta) - g_2(\Delta)\}}{2\alpha^2 \Delta^2 (e^{2\alpha\Delta} - 1)^3} \qquad (A7)$$

where

$$g_1(\Delta) = e^{6\alpha\Delta} + 4(\alpha\Delta)^3 e^{4\alpha\Delta} + \left\{12(\alpha\Delta)^3 + 8(\alpha\Delta)^2 + 3\right\} e^{2\alpha\Delta} + 2(\alpha\Delta)^2$$

$$g_2(\Delta) = \left\{10(\alpha\Delta)^2 + 3\right\} e^{4\alpha\Delta} + 1$$

Since $e^{2\alpha\Delta} - 1 > 0$ for any $\Delta > 0$, the denominator in Eq. (A7) is strictly positive. Thus, to prove that $\partial I(\alpha, \Delta)/\partial\Delta < 0$, we need to show that for any $\Delta > 0$, $g_1(\Delta) - g_2(\Delta) > 0$. Using the Taylor series expansion we may rewrite

$$g_1(\Delta) = g(\Delta) + R_1(\Delta)$$

$$g_2(\Delta) = g(\Delta) + R_2(\Delta)$$

where

$$g(\Delta) = 4 + 12\alpha\Delta + 34(\alpha\Delta)^2 + 72(\alpha\Delta)^3 + 112(\alpha\Delta)^4 + \frac{1984}{15}(\alpha\Delta)^5$$

$$R_1(\Delta) = \sum_{n=6}^{\infty} \left(\frac{6^n}{n!} + \frac{4^{n-2}}{(n-3)!} + \frac{3 \times 2^{n-1}}{(n-3)!} + \frac{2^{n+1}}{(n-2)!} + \frac{3 \times 2^n}{n!} \right)(\alpha\Delta)^n$$

$$R_2(\Delta) = \sum_{n=6}^{\infty} \left(\frac{10 \times 4^{n-2}}{(n-2)!} + \frac{3 \times 4^n}{n!} \right)(\alpha\Delta)^n$$

Thus,

$$g_1(\Delta) - g_2(\Delta) = R_1(\Delta) - R_2(\Delta) = \sum_{n=6}^{\infty} h(n)(\alpha\Delta)^n \qquad \text{(A8)}$$

where

$$h(n) = \frac{1}{n!} \left\{ 6^n + \frac{1}{16}(n^3 - 13n^2 + 12n - 48)4^n + \left(\frac{3}{2}n^3 - \frac{5}{2}n^2 + n + 3 \right)2^n \right\}$$

for $n \geq 6$. Then it suffices to show that $h(n)$ is positive for all $n \geq 6$. For $n \geq 13$ we clearly have $h(n) > 0$ and for $n = 6, 7, \ldots, 12$ we have

$$h(6) = 5.333 \qquad h(7) = 13.333 \qquad h(8) = 17.956$$

$$h(9) = 17.126 \qquad h(10) = 12.902 \qquad h(11) = 8.139$$

$$h(12) = 4.459$$

By Eq. (A8) $g_1(\Delta) - g_2(\Delta) > 0$ and hence, by Eq. (A7) $\partial I(\alpha, \Delta)/\partial\Delta < 0$ for all $\Delta > 0$. Therefore, $I(\alpha, \Delta)$ is a strictly monotone decreasing function of Δ on $(0, \infty)$ and therefore, $I(\alpha, 0) > I(\alpha, \Delta)$ for all $\Delta > 0$.

Appendix III: Proof of Lemma 2

Let x by any nonzero vector in R^p. Then the quadratic form $x^T A x$ is

$$x^T A x = \sum_{i=1}^{p-1}(A_{ii}x_i^2 + 2A_{pi}x_ix_p) + A_{pp}x_p^2 = \sum_{i=1}^{p-1}(A_{ii}x_i^2 + 2A_{pi}x_ix_p + b_ix_p^2)$$

by properties *ii* and *iii*. To prove that A is positive definite it is sufficient to show that

$$g_i(x_i, x_p) = A_{ii}x_i^2 + 2A_{pi}x_ix_p + b_ix_p^2 \geq 0$$

for each $i = 1, \ldots, p-1$ and $g_k(x_k, x_p) > 0$ for some $1 \leq k \leq p-1$. Because $A_{ii} > 0$ by property *i* we may reexpress $g_i(x_i, x_p)$ as

$$g_i(x_i, x_p) = \left(A_{ii}^{1/2}x_i + \frac{A_{pi}}{A_{ii}^{1/2}}x_p\right)^2 + \frac{A_{ii}b_i - A_{pi}^2}{A_{ii}}x_p^2$$

which is nonnegative since $A_{ii}b_i > A_{pi}^2$ by *iii*. If $x_p \neq 0$, then $g_i(x_i, x_p) > 0$ for each $i = 1, \ldots, p-1$. If $x_p = 0$, we must have $x_k \neq 0$ for some $1 \leq k \leq p-1$, so $g_k(x_k, x_p) = A_{kk}x_k^2 > 0$. Therefore, A is positive definite.

Acknowledgments

    The authors are grateful to Charles A. Czeisler of the Division of Circadian and Sleep Disorders Medicine at the Brigham and Women's Hospital for helpful discussions and for providing the data analyzed in the example, to Richard E. Kronauer and Jude Mitchell for helpful discussions, and to Brenda Marshall for technical assistance in preparing the manuscript. This research was supported in part by Robert Wood Johnson Foundation Grant 23397; National Institutes of General Medical Sciences Grants GM53559, AG09975, AG06072, and MH45130; National Center for Research Resources General Clinical Research Grant RR02635; and the National Aeronautics and Space Administration through the NASA Cooperative Agreement NCC9-58 with the National Space Biomedical Research Institute.

[17] Bayesian Methods to Improve Sample Size Approximations

By CHRISTOPHER H. SCHMID, JOSEPH C. CAPPELLERI, and JOSEPH LAU

Introduction

Determining the sample size necessary to have a high probability of obtaining a statistically significant result is a key part of designing a study. Typically, a researcher uses scientific judgment to provide the statistician with estimates of model parameters, such as the mean and variance, and specifies small probabilities of false-positive (type I) and false-negative (type II) errors that can be tolerated. The statistician then calculates the necessary sample size assuming that the parameter estimates provided are correct. Because this approach will underestimate the number of observations required if the chosen parameter estimates are too optimistic, the calculations are often based on conservative estimates with a large sampling variance and the minimum difference considered scientifically important.

Although the standard calculations aim for sufficient power to account for sampling variation, they assume fixed values of the parameters and ignore the usually substantial prior uncertainty associated with them. The final sample size chosen then has sufficient power only if the true effect is not less than what is expected.

We can incorporate this uncertainty in a Bayesian analysis through a prior distribution describing our belief about the size of the study effect. The information for constructing this prior distribution may come from previous experiments, from a pilot study, or from scientific judgment. A formal meta-analysis may be available to quantify this prior information.[1] If so, we should like to be able to give sample size estimates that reflect these priors in order to plan more accurately.

In this chapter, we offer an approach to incorporating the prior uncertainty in our knowledge of the study effect into sample size calculations that may assume an informative prior distribution on the size of this effect. The classic sample size formulas are obtained as a special case by assuming a noninformative prior. Other authors have described variants of this approach for expressing the predictive power of a future sample.[2-4] These

[1] J. Lau, C. H. Schmid, and T. C. Chalmers, *J. Clin. Epidemiol.* **48,** 45 (1995).

[2] D. J. Spiegelhalter and L. S. Freedman, *Stat. Med.* **5,** 1 (1986).

[3] D. J. Spiegelhalter, L. S. Freedman, and P. R. Blackburn, *Control. Clin. Trials* **7,** 8 (1986).

[4] D. J. Spiegelhalter, L. S. Freedman, and M. K. B. Parmar, *J. R. Stat. Soc. Ser. A* **157,** 357 (1994).

Copyright 2004, Elsevier Inc.
All rights reserved.
0076-6879/04 $35.00

formulas will be of immediate use to a Bayesian analysis and will also be helpful for sensitivity analyses that use a range of priors to describe the initial skepticism or enthusiasm of interested parties.[4] Before tackling the sample size problem, we give a brief introduction to Bayesian inference for readers unfamiliar with this topic.

Bayesian Inference

Bayesian inference is an approach to statistical analysis based on Bayes' rule for probabilities in which model parameters θ are considered as random variables in the sense that we have uncertainty about them. Non-Bayesians consider them as fixed, but unknown, constants. We describe our belief about likely values of these parameters through a probability distribution that quantifies this uncertainty. For example, for a single parameter we might express this as a normal distribution with mean 10 and standard deviation 5, indicating that we believe the parameter to have a 95% chance of lying between 0 and 20 (2 standard deviations around the mean) and to be most likely equal to 10. This direct expression of the probability of a parameter value reflects the way most people understand statistical inference, in contrast to the non-Bayesian (classical or frequentist) approach that uses confidence intervals to express long-run frequency probabilities of statistical procedures.

Before collecting any data, we form a prior probability distribution $\pi(\theta)$ of our belief about the parameters. Combining this prior distribution with the observed study data in the form of a likelihood for these parameters leads to a posterior distribution $\pi(\theta|Y)$ of the belief about the location of the unknown parameters θ conditional on the data Y. This posterior distribution describes our belief about the location of the parameters after observing the data and represents a revision of our prior beliefs in light of the data. By combining the prior probability with the likelihood, we combine information about the parameters coming from the data with information from the prior distribution external to the data. The posterior distribution provides the best estimate of our knowledge about the treatment effect given the available information.

The likelihood is the probability of observing the data that were collected under the model proposed. Technically, the likelihood is defined at each possible combination of parameter values, so that it is a function of the parameters. For instance, the model may be that the data follow a normal distribution with mean μ and variance σ^2. In this case, μ and σ^2 are the model parameters and the likelihood is the joint probability of the observed data given specific values of μ and σ^2. When observations are collected independently of each other, this joint probability is simply the product of the individual probabilities of the occurrence of each sampled item. The values of μ and σ^2 that maximize the joint probability are called maximum likelihood estimates. Because the maximum likelihood estimates can be shown to be normally distributed in large samples, statistical inference about model parameters is often based on the assumption that they follow a normal distribution. This assumption can fail badly in small samples or in complex sampling schemes with nested designs.

The posterior probability that θ equals a particular value θ_0 is calculated from Bayes' rule by multiplying the prior probability of θ_0 by the likelihood of the data given the model with parameters equal to θ_0 and then dividing by the probability $f(Y)$ of the data Y. Symbolically,

$$\pi(\theta = \theta_0 | Y) = \frac{\pi(\theta = \theta_0) f(Y | \theta = \theta_0)}{f(Y)}$$

where the vertical line denotes the conditional probability of the quantity to the left of the line given that the quantity to the right occurs. The Bayes rule is simply a representation of a conditional probability. The probability that θ_0 occurs given that Y has occurred is the probability that both θ_0 and Y occur divided by the probability that Y has occurred. The product $\pi(\theta = \theta_0) f(Y | \theta = \theta_0)$ in the numerator of the expression above is the probability that θ_0 and Y occur jointly, while the denominator can be rewritten as $f(Y) = \int \pi(\theta) f(Y | \theta) d\theta$, where the integration is across all possible values that θ may take. The expression $f(Y)$ denotes the marginal distribution of the data and serves as a normalizing constant so that the posterior density integrates to one as required by the rules of probability densities.

The posterior distribution often has an appealing intuitive interpretation. For example, when the data are sampled from a normal distribution and the mean of this distribution has a normal prior, it turns out that the posterior distribution of the mean is normal with mean equal to the weighted average of the prior mean and the sample mean from the data. The weights are inversely proportional to the prior variance of the mean and the sampling variance of the data, respectively. Thus, if not much is known prior to collecting the data, most of the information about the mean will come from the sample. Conversely, if the sample is small and much is known prior to data collection, then the sample mean will have little weight in the posterior.

Bayesian methods have several advantages that follow from the direct quantification of the model parameters in the posterior distribution. The most important is that users can make statements about the probability of scientific hypotheses and parameter values conditional on the data observed. For example, we can state the chance that the study effect is less than zero, that it is at least a certain level, or that it is between two particular values. A 95% posterior probability interval gives the range within which the parameter is likely to lie with a probability of 0.95. In contrast, the classic confidence interval expresses only the probability that an interval will include the true value of the parameter. This is a subtle, but crucial distinction. The Bayesian can draw direct inferences about the parameter given the data actually observed, whereas the non-Bayesian can draw inferences only about the long-run frequency properties of the procedure. A 95% (non-Bayesian) confidence interval means only that constructing such an interval will include the true parameter 95% of the time if the model is correct (i.e., 95 of 100 possible confidence intervals will include the true parameter), but does not directly describe the probability that a hypothesized value of a parameter will lie within the particular interval constructed from the data.

Second, Bayesian inference provides a mechanism for formally incorporating information gained from previous studies into the current analysis. As the data collected accrue and are analyzed, conclusions can be revised to incorporate the new information. The information from new data is combined with knowledge based on previously processed information (possibly data) to reach a new state of knowledge.

Inference is also not just restricted to model parameters, but may be applied to functions and transformations of parameters through direct probability calculus. Such manipulations are simplified when the posterior distribution is computed numerically through Monte Carlo simulation because the new parameters are simulated by simply applying the appropriate function or transformation to the original simulated values.

Bayesian inferences need not rely on the large sample normal approximations required by maximum likelihood because the entire posterior distribution is available (whether analytically or numerically). This enables construction of scientifically plausible models for which parameters can be estimated exactly, rather than approximately. For example, a model for a binary trait sampled independently with probabilities that vary according to characteristics of the individuals may reflect both the Bernoulli distribution of each trait conditional on each individual's probability as well as the logistic regression model for the probabilities conditional on the individual characteristics. Markov chain Monte Carlo simulation makes this particularly simple because, provided the resampling algorithm is correctly designed, it will return draws from the correct joint posterior of the parameters for many potential priors and likelihoods.

This mode of inference is particularly powerful in multilevel structures in which some units are nested within others. Often, there are only a few units nested within each higher level, but inferences must be made to each level. Bayesian models then provide a way to borrow strength from like units to supplement the small sample sizes.

Although it might seem strange that a better estimate of a quantity may be obtained using data not collected on that quantity, this phenomenon, called shrinkage, is well known in statistical theory.[5,6] Intuitively, if we can assume that certain units are estimating the same thing, or are exchangeable in statistical terminology, we can gain more information about each one by using information from the others.[7] This phenomenon is familiar to doctors who supplement their patients' medical history with prior knowledge obtained through experience with similar patients. Another example involves predicting batting averages in baseball. A player's average after 1 month may not be a good indicator of his final average. Instead, we can better predict his level at the end of a 6-month season by combining the first month's statistics with his performance in previous seasons as well as with overall league performance. (See Efron and Morris[8] for a readable and enjoyable discussion of hierarchical Bayesian models in the context of baseball.)

Finally, Bayesian methods also permit sample size calculations to be adapted as data accrue. The stopping rule principle states that a study may be stopped and the posterior distribution calculated at any time as long as the reason for stopping does not depend on the parameter being estimated.[9] Although one of the strengths of Bayesian analysis is its flexibility in not requiring a prespecified sample size,[10,11] in practice, knowledge of study size is an essential ingredient in managing a clinical trial. We can, in principle, calculate the size of a future sample that combined with the current information about the treatment effect in the posterior will give a prespecified probability of reaching a firm conclusion. This sequential approach can be of great benefit to studies that have indicated a clinically important—but statistically inconclusive—treatment effect, leaving the researcher to decide whether the effect is not real and reflects chance variation alone or is real but suffers from an inadequate sample size. Increasing the sample size can reduce the posterior variability of the treatment

[5] B. Efron and C. N. Morris, *J. Am. Stat. Assoc.* **68,** 34 (1973).
[6] W. James and C. Stein, *in* "Proceedings of the Fourth Berkeley Symposium on Mathematical Statistics and Probability," Vol. 1, pp. 311–319. University of California Press, Berkeley, CA, 1961.
[7] D. Draper, J. S. Hodges, C. L. Mallows, and D. Pregibon, *J. R. Stat. Soc. Ser. A* **156,** 9 (1993).
[8] B. Efron and C. N. Morris, *Sci. Am.* **236,** 119 (1973).
[9] J. O. Berger and R. L. Wolpert, "The Likelihood Principle," 2nd Ed. Institute of Mathematical Statistics, Hayward, CA, 1988.
[10] D. A. Berry, *Stat. Med.* **4,** 521 (1985).
[11] D. A. Berry, *Am. Stat.* **41,** 117 (1987).

estimate, enabling the researcher to draw a more confident conclusion. While such adjustments are possible in the non-Bayesian framework using sequential design principles, they are not as natural and still must be planned before the data are collected and analyzed.[12,13]

Readers seeking more detail on Bayesian methods may wish to consult additional sources in addition to those cited above. Berry[14] offers a readable basic introduction to statistics from a Bayesian perspective. Good general texts at a higher mathematical level include Carlin and Louis,[15] Congden,[16] and Gelman et al.[17] Articles in the collection edited by Berry and Stangl[18] address Bayesian biomedical applications. In a previous volume of this series, Schmid and Brown[19] described hierarchical Bayesian models. Next, we apply these Bayesian techniques to the sample size estimation problem.

Deriving Sample Size Formulas

Typically, estimating a sample size for a study requires four types of information: (1) a model for the data, (2) specification of the confidence with which the conclusions may be drawn, (3) the sampling variability of the data, and (4) values for the true effects of the treatments.

In our derivation, we assume that a sample of size n will be drawn from a normally distributed random variable X having mean θ and known variance σ^2, that is, $X \sim N(\theta, \sigma^2)$. This model also includes the comparison of two means measured on samples of size n_i drawn from random variables $X_i \sim N(\theta_i, \sigma^2)$ if we set $\theta = \theta_1 - \theta_2$ and let $n = 1/(1/n_1 + 1/n_2)$. We also assume prior knowledge about the random variable of interest θ is summarized by the analytic prior $\theta \sim N(\mu_0, \sigma^2/n_0)$ with n_0 representing a "prior sample size."[4] The posterior mean and variance are then well known to be

$$\mu_{\text{post}} = E(\theta|X) = (n_0\mu_0 + n\bar{x})/(n_0 + n) \tag{1}$$

and

$$\sigma^2_{\text{post}} = V(\theta|X) = \sigma^2(1/n_0 + 1/n) \tag{2}$$

[12] P. C. O'Brien and T. R. Fleming, *Biometrics* **35,** 549 (1979).
[13] S. J. Pocock, *Biometrika* **64,** 191 (1977).
[14] D. A. Berry, *in* "Statistics: A Bayesian Perspective." Duxbury Press, Belmont, CA, 1996.
[15] B. P. Carlin and T. A. Louis, *in* "Bayes and Empirical Bayes Methods for Data Analysis," 2nd Ed. Chapman & Hall/CRC, New York, 2000.
[16] P. Congden, *in* "Bayesian Statistical Modelling." John Wiley & Sons, New York, 2002.
[17] A. Gelman, J. B. Carlin, H. S. Stern, and D. B. Rubin, *in* "Bayesian Data Analysis," 2nd Ed. Chapman & Hall/CRC, New York, 2004.
[18] D. A. Berry and D. K. Stangl, eds., *in* "Bayesian Biostatistics." Marcel Dekker, New York, 1996.
[19] C. H. Schmid and E. N. Brown, *Methods Enzymol.* **321,** 305 (2000).

where $\bar{x}$ is the sample mean. Therefore, the posterior distribution may be written $\theta|X \sim N(\mu_{\text{post}}, \sigma^2_{\text{post}})$. The marginal distribution of the sample mean is $\bar{x} \sim N[\mu_0, \sigma^2(1/n_0 + 1/n)]$.

Inferences about θ are often expressed by confidence statements that θ lies in an interval (θ_L, θ_U). Bayesians express these as posterior probabilities, calculating $\Pr\{\theta \in (\theta_L, \theta_U)\}$. By appropriate definition of the end points of the interval (θ_L, θ_U), inferences may be stated in terms of posterior probabilities by the Bayesian or as confidence intervals or hypothesis tests by the frequentist as

$$\Pr\{\theta_L < \theta < \theta_U\} \geq 1 - \alpha \tag{3}$$

where α is a small positive number (often taken as 0.05). For instance, if δ is the smallest acceptable positive treatment effect, the study should be designed to show that $\Pr\{\theta \mid X > \delta\} \geq 1 - \alpha$. This corresponds to setting $\theta_L = \delta$ and $\theta_U = \infty$ in Eq. (3). In an equivalence study to show that a new compound with fewer side effects is as effective as the standard, θ_L and θ_U would be the bounds that defined equivalency.

We can rewrite Eq. (3) in terms of the posterior mean, μ_{post}, and posterior standard deviation, σ_{post}, as

$$\Pr\left\{\frac{\theta_L - \mu_{\text{post}}}{\sigma_{\text{post}}} < \frac{\theta - \mu_{\text{post}}}{\sigma_{\text{post}}} < \frac{\theta_U - \mu_{\text{post}}}{\sigma_{\text{post}}}\right\} \geq 1 - \alpha$$

where we have dropped the conditioning on the data X to simplify notation. Under the assumption of normality, this inequality will be satisfied if

$$\frac{\theta_L - \mu_{\text{post}}}{\sigma_{\text{post}}} \leq z_{\alpha_1} \tag{4}$$

and

$$\frac{\theta_U - \mu_{\text{post}}}{\sigma_{\text{post}}} \geq z_{1-\alpha_2} \tag{5}$$

where z_α is the α quantile of the standard normal distribution and $\alpha_1 + \alpha_2 \leq \alpha$. In the two-sided case, we often take $\alpha_1 = \alpha_2 = \alpha/2$. In the one-sided case, in which either $\theta_L \to \infty$ or $\theta_U \to \infty$, only one of the inequalities is needed as the other is automatically satisfied.

Substituting μ_{post} and σ_{post} from Eqs. (1) and (2) into Eqs. (4) and (5) produces, after some algebra,

$$\frac{(n_0 + n)\theta_L - n_0\mu_0 + \sigma z_{1-\alpha_1}\sqrt{n_0 + n}}{n} \leq \bar{x} \leq \frac{(n_0 + n)\theta_U - n_0\mu_0 - \sigma z_{1-\alpha_2}\sqrt{n_0 + n}}{n}$$

or

$$A \leq \bar{x} \leq B \tag{6}$$

where

$$A = \frac{(n_0 + n)\theta_{\mathrm{L}} - n_0\mu_0 + \sigma z_{1-\alpha_1}\sqrt{n_0 + n}}{n}$$

and

$$B = \frac{(n_0 + n)\theta_{\mathrm{U}} - n_0\mu_0 - \sigma z_{1-\alpha_2}\sqrt{n_0 + n}}{n}$$

Since A must be no greater than B, it follows that

$$n_0 + n \geq \sigma^2 (z_{1-\alpha_1} + z_{1-\alpha_2})^2 / (\theta_{\mathrm{U}} - \theta_{\mathrm{L}})^2$$

which provides a lower bound on the total sample size. For one-sided intervals with $\theta_{\mathrm{L}} \to \infty$ or $\theta_{\mathrm{U}} \to \infty$, the bound is zero and provides no information.

Although Eq. (6) satisfies the requirement for an interval of size at least $1 - \alpha$, it does not take into account our uncertainty about the location of $\bar{x}$. To be confident that $\bar{x}$ will actually fall in this interval, we require a sample size so that Eq. (6) is true with a high probability, say $1 - \beta$. This probability depends on the predictive distribution of the sample mean $\bar{x}$ from a future trial designed with n observations. This is

$$\overline{X}|\mu_{\mathrm{p}}, n_{\mathrm{p}}, \sigma^2 \sim N[\mu_{\mathrm{p}}, \sigma^2(1/n_{\mathrm{p}} + 1/n)] \tag{7}$$

where μ_{p} and σ^2/n_{p}, respectively, the mean and variance of the design prior for θ, are distinct from the analytic prior mean μ_0 and variance σ^2/n_0.

Incorporating Eq. (7) into Eq. (6), we want

$$\mathrm{Pr}\left(\frac{A - \mu_{\mathrm{p}}}{\sigma\sqrt{1/n_{\mathrm{p}} + 1/n}} \leq \frac{\bar{x} - \mu_{\mathrm{p}}}{\sigma\sqrt{1/n_{\mathrm{p}} + 1/n}} \leq \frac{B - \mu_{\mathrm{p}}}{\sigma\sqrt{1/n_{\mathrm{p}} + 1/n}}\right) \geq 1 - \beta$$

This produces two inequalities to be solved for n and we need to apportion β into two parts such that $\beta_1 + \beta_2 \leq \beta$ as

$$\frac{A - \mu_{\mathrm{p}}}{\sigma\sqrt{1/n_{\mathrm{p}} + 1/n}} \leq z_{\beta_1} \tag{8}$$

and

$$\frac{B - \mu_{\mathrm{p}}}{\sigma\sqrt{1/n_{\mathrm{p}} + 1/n}} \geq z_{1-\beta_2} \tag{9}$$

We choose the smallest n that satisfies both inequalities. When the hypothesis is one sided, so that either $\theta_{\mathrm{L}} = -\infty$ or $\theta_{\mathrm{U}} = +\infty$, one of the inequalities is trivial so the allocation is automatic. If $\theta_L = -\infty$, then β_1 may be set arbitrarily close to zero, so that $\beta_2 = \beta$; if $\theta_U = +\infty$, then β_2 may be set arbitrarily close to zero, so that $\beta_1 = \beta$. Note that standard references for equivalence designs such as Makuch and Simon[20] use a one-sided hypothesis.

Choosing Prior Distributions

The analytic and design priors serve different purposes and thus need not necessarily be the same. The analytic prior reflects prior information from other studies believed pertinent to knowledge of θ. It describes how we want prior information to affect our analysis of the data. If there is little prior information or that available is contradictory, the analytic prior must be vague, having a large variance reflecting this uncertainty. The design prior reflects our belief about what the data will show through the sampling distribution of the mean. When a vague analytic prior gives substantial probability to a size of treatment effect that is not of interest, we will need to choose a more informative design prior that describes our belief about which values of θ are important. We might want to tailor the design prior to features of the population to be studied, perhaps by eliciting a prior from clinicians.[21] To convince a skeptic, we may want to use a conservative analytic prior so that the data must produce overwhelming favorable evidence to lead to a convincing posterior. But we would not want to conduct a study in which the true effect was as small (or nonexistent) as the skeptic believes.

Investigators would hardly design a new study with such uncertainty about the result. Their design prior would reflect their strong beliefs that enable them to undertake the considerable effort of a new study. In effect, the design prior reflects the investigator's belief about the knowledge the study will impart about the parameter, while the analytic prior can also reflect others' beliefs and other evidence that may be quite different and therefore lead to much greater uncertainty about the true effect. Competing hypotheses may even lead to using several different design or analytic priors.

Speigelhalter $et\ al.$[4] have catalogued several types of priors that may be usefully employed as either design or analytic priors. Skeptical priors reflect the belief of a skeptic (perhaps a regulatory agency) that the treatment under study is unlikely to be successful. These priors require the data to

---

[20] R. Makuch and R. Simon, *Cancer Treatment Rep.* **62,** 1037 (1978).
[21] K. Chaloner, T. Church, T. A. Louis, and J. P. Matts, *Statistician* **42,** 341 (1993).

give strong evidence of treatment efficacy in order for a posterior inference to be conclusive. Enthusiastic priors describe the opinion of those who are inclined to believe that the treatment will be effective. Such individuals require only a small amount of evidence from the data to be convinced that the treatment works. Priors that are neither skeptical nor enthusiastic may be called realistic. Noninformative priors can serve as a reference to calibrate the conclusions of different analyses.

A useful type of realistic prior that may lead to more widely acceptable conclusions is that derived objectively from data analysis. The more alike the prior data source is to the present study, the better such data-derived priors will represent the effect being measured. Prior distributions developed from pilot studies that reflect controlled conditions exactly like the ones planned are best. Next best are prior distributions derived from analysis of similar studies such as with meta-analysis. Because past subjects are not usually exchangeable with current subjects, it is wise to add additional uncertainty to capture the extra sources of variability.

The appropriate prior distribution will always depend on the problem, but it is always a good idea to try different prior distributions to quantify their influence on the posterior distribution. Because a sample size calculation must be conservative to ensure that the study will be large enough to answer the question posed, a certain degree of skepticism is probably wise and so we might choose to temper our data-derived prior by mixing it with a skeptical prior.

Clearly, the designer may have a different opinion of the outcome of the study than a consumer such as a regulatory agency. While the analysis must be able to withstand the initial skepticism expressed through a skeptical prior, in planning the study the scientist does not expect to need nearly as much information as the skeptic does. As a result, the design prior would be chosen much more optimistically. In doing so, the scientist is taking a risk that his or her opinion is more accurate than that of the skeptic, but on the other hand, the scientist may not be interested in a small effect at odds with the research.

Lest this seem contradictory, consider that the standard sample size formulas arise from assuming a noninformative analytic prior ($n_0 = 0$) and a design prior with zero variance ($n_p = \infty$) corresponding to a completely specified alternative hypothesis that $\theta = \mu_p$. This allows us to ask: "If the true mean were μ_p how big a sample would we need?" The answer depends on the accuracy of our estimate of the mean. If θ_0 is chosen too optimistically, the sample size estimate will be too small; if it is chosen pessimistically, too many patients may be randomized. Assuming an informative design prior averages over these conditional sample sizes.

Gain from Using Prior Information

To investigate the savings from using the prior information, we shall compare the sample sizes required given different amounts of prior evidence to examine a one-sided hypothesis that $\Pr(\theta > \theta_L) > 1 - \alpha$. Throughout, we take both α and β less than 0.5.

Define the degree of prior evidence $E_A = z_A/z_{1-\alpha}$ as the ratio of the z statistic from the analysis prior, $z_A = \sqrt{n_0}(\mu_0 - \theta_L)/\sigma$, to the standard normal quantile, $z_{1-\alpha}$. This quantity describes amount of information present a priori for examining the hypothesis. When $E_A \geq 1$, the prior evidence is sufficient to conclude that $\theta > \theta_L$ with probability at least $1 - \alpha$ (or in the classic framework to reject the null hypothesis that $\theta < \theta_L$ at level α). When E_A is near 0, there is little prior evidence that θ is different from θ_L. From the point of view of hypothesis testing, this last statement can be expressed as evidence of the truth of the null hypothesis. When this evidence is strong, no finite sample size will provide enough power for rejecting the null.

We also define a sample size multiplier $M_A = n/n_0$ as the ratio of the number of new observations n to the "prior sample size" n_0.

Complementary definitions apply to the design prior. Thus $E_D = z_D/z_{1-\alpha}$ and $M_D = n/n_p$, where $z_D = (n_p)^{1/2}(\mu_p - \theta_L)/\sigma$. Equation (8) for estimating the sample size may then be written as

$$\frac{\sqrt{1 + M_A} - E_A - E_D\sqrt{M_A M_D}}{\sqrt{M_A}(1 + M_D)} \leq z_\beta/z_{1-\alpha}$$

Simplification occurs if $\mu_0 = \mu_p$ and $n_0 = n_p$ and it is assumed that $M = M_A = M_D$ is large, for then $E = E_A = E_D$ and

$$M_A \geq \left(z_{1-\alpha}/(E_A z_{1-\alpha} + z_\beta)\right)^2$$

Assuming a fixed design prior such that $n_p \to \infty$, leads to

$$M_A \geq \left((z_{1-\alpha} - z_\beta)/E_A z_{1-\alpha}\right)^2$$

Figure 1 graphs M versus E as E varies between 0 and 1 for fixed values of α and β under both a random and a fixed design prior.

From the formulas above and from Fig. 1, we see that (1) given α and β, M is inversely proportional to E; (2) for all $\beta < 0.5$, M is smaller when the design prior mean is fixed; (3) for fixed E, the size of M increases with both the posterior probability $1 - \alpha$ and the power $1 - \beta$; (4) even when E is close to 1, M can be quite large; (5) for fixed β, the ratio of M_1 and M_2 corresponding to some E_1 and E_2 varies little with changing α; (6) as power increases for fixed α and E, sample sizes increase much slower when

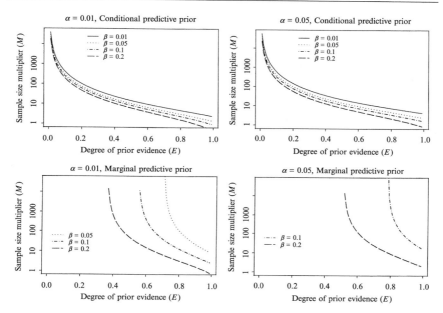

FIG. 1. Sample size multiplier, M, as a function of the degree of prior evidence, E. The four panels correspond to two levels of α (0.01 and 0.05), each with degenerate and nondegenerate design priors. Each panel shows a separate curve for four choices of $\beta = 0.01, 0.05, 0.10, 0.20$. Multipliers do not exist for some combinations of β and E under the design prior because of the boundary condition [Eq. (10)].

the design prior mean is fixed; and (7) the sample size can be infinite when the design prior is not fixed.

Point 1 follows because less prior evidence requires greater support in the data to support the hypothesis. Point 2 follows because of the added uncertainty when the design prior is unknown. Point 3 is well known; higher precision requires larger sample sizes. Point 4 arises because of the sampling variability in $\overline{X}$ and the considerable chance that it may shrink the prior mean toward θ_L with a small sample size. The invariance noted in point 5 arises because the term involving α cancels in the ratio E_1/E_2. Point 6 reflects the large effect of the uncertainty in the design prior on the sample size. The conditions under which no sample can achieve sufficient power as remarked in point 7 can be derived by noting that it is not possible to reach the conclusion that $\theta < \theta_L$ $|E| < 1$ and $\overline{X} < \theta_L$. As

$$\Pr(\overline{X} < \theta_L) = \Pr\left(\frac{\overline{X} - \mu_p}{\sigma\sqrt{1/n_p + 1/n}} < \frac{\theta_L - \mu_p}{\sigma\sqrt{1/n_p + 1/n}}\right) = \Phi\left(\frac{\theta_L - \mu_p}{\sigma\sqrt{1/n_p + 1/n}}\right)$$

a finite sample under a design prior is achieved only if

$$\beta > \Phi\left(\frac{\theta_L - \mu_p}{\sigma\sqrt{1/n_p + 1/n}}\right) \to \Phi\left(\frac{\theta_L - \mu_p}{\sigma/\sqrt{n_p}}\right) \quad \text{as} \quad n \to \infty \qquad (10)$$

This also implies that $E \geq z_{1-\beta} / z_{1-\alpha}$. If E is too small, then a lack of prior evidence implies substantial uncertainty about $\overline{X}$. In such cases, it may be wise to use a more optimistic design prior in order to obtain reasonable (at least finite!) sample sizes.

How do changes in E affect the total sample size N? Expressing N in terms of the prior sample size n_0 as

$$N = n_0 + n = (1 + M_A)n_0 = (1 + M_A)[E_A z_{1-\alpha}\sigma/(\mu_0 - \theta_L)]^2$$

shows that N is proportional to $E_A^2(1 + M)$ for fixed μ_0, θ_L, σ, α, and β. Because M is implicitly defined as a function of E, the total sample size N is a function of E alone. Figure 2 gives the percentage reduction in the total sample size as a function of β for $\alpha = 0.01$ and $\alpha = 0.05$ when E changes from 0.01 to 0.99 using a conditional predictive distribution with μ_0, θ_U, and σ fixed. For the most common choices of β between 0.01 and 0.20, the total reduction is about 10 to 20%.

Examples

We now present three examples drawn from actual randomized trials to demonstrate the effect of incorporating prior information in sample size calculations.

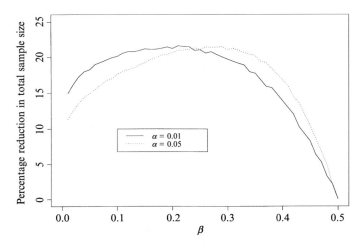

FIG. 2. Percentage reduction in the total sample size when $E = 0.99$ compared with $E = 0.01$ for $0 < \beta \leq 0.5$ at $\alpha = 0.01$ and $\alpha = 0.05$.

Example 1: Use of Calcium Supplements to Control Blood Pressure

Evidence of an inverse relationship between dietary calcium intake and blood pressure levels has led to randomized trials testing the effect of calcium supplements in reducing high blood pressure. One such trial[22] published in 1992 measured the difference in standing systolic blood pressure at baseline and at the end of the study, 24 weeks later, for 237 patients in the calcium group and 234 patients in the control group. Blood pressure dropped by 0.46 mmHg on average with a standard error of 0.67, a statistically nonsignificant change. Using these results as prior information, we might ask what size we should make the next study to have a high probability of obtaining a statistically significant reduction in blood pressure.

Letting θ be the change in blood pressure, we want to show that $\theta < 0$. The trial results give us $\mu_0 = -0.46$, $\sigma = 7.26$, and $n_0 = 1/(1/234 + 1/237) = 117.75$ for the analytic prior distribution. When $\alpha = 0.05$ and $\beta = 0.2$, using a design prior with θ_0 fixed at -0.46 leads to a required sample size of 3602 patients per group. If, instead, the analytic prior is noninformative, the number is 3912 per group. In this example, the prior information does not reduce the sample size much because the prior distribution is vague and supplies only a small amount of information. With either analytic prior, the numbers of patients suggested are far higher than those the trial actually enrolled.

Choosing θ_0 equal to the mean of the prior distribution of θ is a logical choice. Nevertheless, if we want to minimize the chance of a negative study, we may want to choose θ_0 far enough from the center of its prior distribution so that our sample will be sufficiently large if the true value of θ is different from what we expect.

Figure 3 displays the results of a sensitivity analysis showing the sample sizes required when $-1.9 \leq \theta_0 \leq -0.25$ with the same analytic prior. Underneath, we plot the $N(-0.46, 0.67^2)$ predictive prior distribution for θ to show the relative prior probability of θ for these values of θ_0. We observe the 3602 patients per group required when $\theta_0 = -0.46$ as calculated above, but note that the sample sizes increase sharply for smaller values of θ_0 that are not unlikely according to the predictive prior. If $\theta = -0.25$, for instance, which is only one-third of 1 standard deviation below the prior mean, the sample size increases to greater than 12000 per group. For any $\theta_0 > 0$, of course, the sample size is not finite.

When a design prior is used with $\beta = 0.2$, no sample size will be large enough because by Eq. (10), a finite sample size is achievable only if $\beta > \Phi \sqrt{117.75}(-0.46)/7.26] = 0.246$. In other words, if $\overline{X} \sim N(-0.46, 0.67^2 + 7.26^2/n)$ there is almost a 25% chance that the sample mean will be greater than zero and that the posterior probability of $\theta < 0$ will

[22] J. A. Cutler, P. K. Whelton, L. Appel *et al.*, *J. Am. Med. Assoc.* **267**, 1213 (1992).

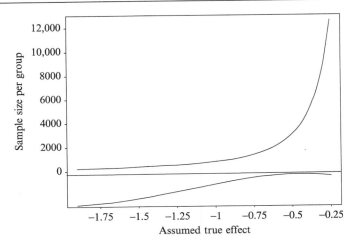

FIG. 3. Sample sizes required per group, using an N $(-0.46, 0.67^2)$ analytic prior with $\alpha = 0.025$ and $\beta = 0.20$ as the true effect θ_0 for calcium varies in the design prior. The prior distribution of θ (bottom) shows likely values of θ_0.

approach zero for large n. Thus there is no more than a 75% chance that any sample will give the desired result.

What are we to do in this situation? Using a design prior with θ fixed at the prior mean may leave us uncomfortable that we have not properly accounted for the uncertainty in θ, but this uncertainty is so large that a nondegenerate design prior leads to no finite estimate at all. To get a finite sample size for $\beta = 0.2$, we need to choose a design prior indicative of a larger difference in blood pressure. This choice may also be more clinically relevant if we believe that blood pressure reductions of 0.5 mmHg are unimportant. Let us assume that a 2-mmHg difference (slightly greater than the upper 95% confidence limit on θ in the published study) is clinically useful and therefore use a design prior for θ of $N(-2, 1)$ indicating that an increase in blood pressure with calcium supplements is unlikely. Then, 370 patients per group are needed under the informative analytic prior and 372 patients per group are needed under the noninformative analytic prior. In this example, the prior information does not gain us much of a reduction in sample size because the predictive prior conflicts with and is more optimistic than the analytic prior.

Example 2: Use of Streptokinase for Treating Acute Myocardial Infarction

For an example in which data are available to develop a prior, consider Table I showing short-term mortality rates in the first six randomized controlled trials (reported before 1972) of streptokinase versus placebo in the treatment of acute myocardial infarction.[23] This treatment is now standard, but only became universally accepted after two large trials, each with more than 10,000 patients, were completed in the late 1980s.[24,25] We use these six studies to develop a prior distribution for the treatment effect in order to help estimate the sample size that would have been required for a new study in 1972.

Nonasymptotic approaches using exact binomial or Poisson models could also be developed along similar lines as our methods. See Gould[26] for one approach that uses β priors and binomial likelihoods for equivalence studies with binary outcomes.

TABLE I

SIX EARLY RANDOMIZED CONTROLLED TRIALS OF INTRAVENOUS STREPTOKINASE THERAPY IN ACUTE MYOCARDIAL INFARCTION, USING SHORT-TERM [a] MORTALITY AS OUTCOME

Study (year)	No. of deaths/No. enrolled		Relative risk	95% CI	
	Treated	Control		Low	High
Fletcher (1959)[b]	1/12	4/11	0.23	0.03	1.75
Dewar (1963)[c]	4/21	7/21	0.57	0.20	1.66
Amery (1969)[d]	20/83	15/84	1.35	0.74	2.45
EWP (1971)[e]	69/373	94/357	0.70	0.53	0.92
Heikinheimo (1971)[f]	22/219	17/207	1.22	0.67	2.24
Dioguardi (1971)[g]	19/164	18/157	1.01	0.55	1.85

[a] Up to 3 months.

[b] A. P. Fletcher, S. Sherry, N. Alkjaersig, F. E. Smyrniotis, and S. Jick, *J. Clin. Invest.* **38,** 1111 (1959).

[c] H. A. Dewar, P. Stephenson, A. R. Horler, and A. J. Cassells-Smith, *Br. Med. J.* **1,** 915 (1963).

[d] A. Amery, G. Roeber, H. J. Vermeulen, and M. Verstraete, *Acta Med. Scan.* **505**(Suppl.), 1 (1969).

[e] N. Dioguardi, A. Lotto, G. F. Levi, M. Rota, C. Proto, P. M. Mannucci, P. Rossi, G. Lomanto, G. Mattei, G. Fiorelli, and A. Agostoni, *Lancet* **2,** 891 (1971).

[f] European Working Party, *Br. Med. J.* **3,** 325 (1971).

[g] R. Heikinheimo, P. Ahrenberg, H. Honkapohja, E. Iisalo, V. Kallio, Y. Konttinin, O. Leskinin, H. Mustaneimi, M. Reinikainin, and L. Siitonin, *Acta Med. Scan.* **189,** 7 (1971).

[23] J. Lau, E. M. Antman, J. Jimenez-Silva, B. Kupelnick, F. Mosteller, and T. C. Chalmers, *N. Engl. J. Med.* **327,** 248 (1992).

[24] Gruppo Italiano per lo Studio della Streptochinasi nell Infarto Miocardico, *Lancet* **i,** 397 (1986).

[25] ISIS-2 Collaborative Group, *Lancet* **ii,** 349 (1988).

[26] A. L. Gould, *Stat. Med.* **12,** 2009 (1993).

We construct two prior distributions for these data. The first prior is based on a fixed-effects model of all six studies using the log relative risks as the study outcomes and weighting each by the inverse of its asymptotic within-study variance.[27] The logarithmic transformation is used because it removes skewness in the relative risk and therefore is more appropriate for assuming normality.

The pooled mean (on the log scale) for this fixed-effects model gives $\gamma_0 = -0.18$ with a standard error of $\sigma_{\gamma_0} = 0.11$. Taking antilogarithms, this translates to a pooled relative risk of 0.83 with a 95% confidence interval of (0.68, 1.03). Although suggesting that streptokinase might be effective, this result is not statistically significant and indicates the need for further study. The heterogeneity of the relative risk estimates in these studies (which range from 0.23 to 1.35) suggests that a fixed-effects model that assumes a common treatment effect for each study might not be appropriate.

Therefore, we develop a second prior based on a Bayesian random effects model that postulates distinct study treatment effects and incorporates all uncertainty in our knowledge of the parameters of the distribution from which these distinct estimates are drawn. The pooled mean log relative risk is -0.12 with a standard error of 0.22. Taking antilogarithms, this translates into a pooled relative risk estimate of 0.88 with a 95% confidence interval of (0.57, 1.36). The heterogeneity in the results of these trials leads to a much wider confidence interval for the random-effects model that appropriately incorporates the between-study variation.

Figure 4 gives the sample sizes per group (assuming equal group sizes) required to achieve different levels of power for $\alpha = 0.05$ under degenerate and nondegenerate design priors with the two analytic priors described above and a noninformative analytic prior. On the basis of sample proportions of events in the control and treatment groups (0.185 and 0.155, respectively) we calculate $\sigma^2 = 9.85$.

The numbers in Fig. 4 bear out the observations noted in the preceding section. Sample sizes decrease as β increases and are always smaller when the design prior is fixed at a single value. The design prior fixed at the mean of the fixed-effects model generates smaller sample sizes than the one centered at the random-effects mean because the fixed-effects mean provides more prior evidence of treatment efficacy. For each predictive distribution, numbers are lowest for the most informative analytic prior, that based on the fixed-effects model, and highest for the noninformative prior. Use of the random-effects prior gives numbers close to the noninformative prior. Of course, the fixed-effects prior ignores the between-study heterogeneity of treatment effect and may therefore underestimate the

[27] A. Whitehead and J. Whitehead, *Stat. Med.* **10**, 1665 (1991).

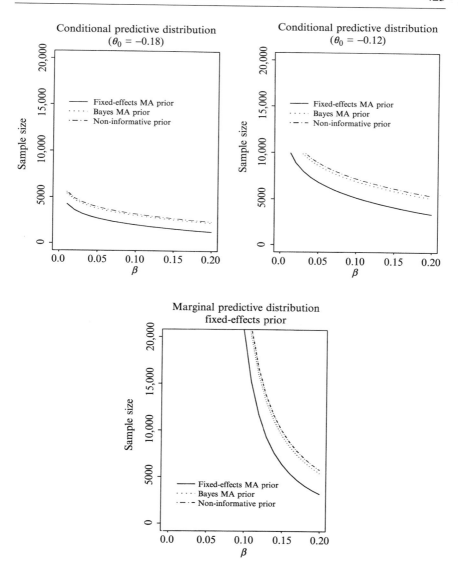

Fig. 4. Sample sizes required when $\alpha = 0.025$ as a function of β for a new study of streptokinase under different prior distributions. Each panel shows sample size curves for each of three analytic priors: (1) an $N (-0.18, 0.11)$ prior from a fixed-effects meta-analysis; (2) an $N(-0.12, 0.22)$ prior from a Bayesian random-effects meta-analysis; and (3) a noninformative prior. The left panel uses a design prior with $\theta_0 = -0.18$; the central panel uses $\theta_0 = -0.12$; and the right panel uses the $N(-0.18, 0.11)$ design prior from the fixed-effects meta-analysis.

total variability in the system. In the third panel of Fig. 4, the sample sizes for low values of β are not finite because the maximal power for the fixed-effects design prior is 94.9%. If we had used the predictive prior based on the random effects model, none of the sample sizes would have been finite as the maximal power is 70.7%.

To have an 80% probability ($\beta = 0.2$) that a 95% highest posterior density interval would show streptokinase to be an effective treatment ($\alpha = 0.05$) requires a sample size of 3175 per group with a nondegenerate design prior and an informative analytic prior based on the fixed-effects model. Changing to a noninformative analytic prior increases the numbers to 5755 per group. The analogous numbers for a degenerate design prior centered at the prior mean of $\theta_0 = 0.18$ are lower, 1268 and 2387, respectively. For this β, the sample sizes are close to their counterparts, with the nondegenerate design prior having a mean of -0.12. This value for θ_0 is close to the mean of the prior from the random-effects model, suggesting that, at least in this problem, the added uncertainty introduced by use of the design prior is equivalent to the uncertainty of the random-effects model relative to the fixed-effects model.

Example 3: Use of Lidocaine for Treating Acute Myocardial Infarction

Now consider a problem for which an equivalence trial is indicated. The effectiveness of lidocaine in controlling arrhythmia led to its prophylactic use for patients presenting with acute myocardial infarction. Table II shows the short-term mortality outcomes of 15 randomized controlled trials of intravenous or intramuscular lidocaine versus placebo conducted between 1970 and 1988.[23] It is apparent from the data that prophylactic use of lidocaine did not save lives. As the negative evidence accumulated, researchers should have realized that if lidocaine were an effective treatment, the effect must be small and therefore that larger trials would be needed. Yet only one study larger than 500 patients was ever performed.[28]

If the investigators had wanted to show that lidocaine was not a useful treatment, however, the use of the prior negative evidence would have been important for keeping the sample sizes lower. Combining all 15 studies by meta-analysis gives an average log relative risk of 0.1 with a standard error of 0.14. Using the average mortality in the treatment and control groups we estimate $\sigma^2 = 91.5$. Then, assuming that the true effect was a 10% increase in mortality from lidocaine (i.e., a design prior with $\theta_0 = 0.1$) and using an analytic prior based on the meta-analysis, a study would require 7490 patients per group to have an 80% chance ($\beta = 0.2$)

[28] R. W. Koster and A. J. Dunning, N. Engl. J. Med. **313**, 1105 (1985).

TABLE II
SHORT-TERM[a] MORTALITY FOR 15 RANDOMIZED CONTROLLED TRIALS OF PROPHYLACTIC
INTRAVENOUS/INTRAMUSCULAR LIDOCAINE FOR ACUTE MYOCARDIAL INFARCTION

Study (year)	No. of deaths/No. studied		Relative risk	95% CI	
	Treated	Control		Low	High
Mogensen (1970)[b]	9/44	7/44	1.29	0.53	3.15
Bennett (1970)[c]	10/110	8/106	1.20	0.49	2.93
Chopra (1971)[d]	7/39	4/43	1.93	0.61	6.09
Pitt (1971)[e]	9/107	16/110	0.58	0.27	1.25
Baker (1971)[f]	2/21	0/23	5.45	0.28	107.47
Darby (1972)[g]	12/103	11/100	1.06	0.49	2.29
O'Brien (1973)[h]	11/154	4/146	2.61	0.85	8.00
Bleifield (1973)[i]	2/41	4/48	0.59	0.11	3.03
Lie (1974)[j]	8/107	10/105	0.79	0.32	1.91
Singh (1976)[k]	0/27	0/27	1.00	0.02	48.66
Sandler (1976)[l]	0/91	0/90	0.99	0.02	49.32
Lie (1978)[m]	0/147	0/153	1.04	0.02	52.10
Koster (1985)[n]	22/2987	22/3037	1.02	0.56	1.83
Dunn (1985)[o]	3/207	1/195	2.83	0.30	26.94
Wyse (1988)[p]	6/168	2/165	2.95	0.60	14.39

[a] Less than 3 months.
[b] L. Mogensen, *Acta Med. Scan.* **513**(Suppl.), 1 (1970).
[c] M. A. Bennett, J. M. Wilner, and B. L. Penecoat, *Lancet* **2**, 817 (1970).
[d] M. P. Chopra, U. Thadani, R. W. Portal, and C. P. Aber, *Br. Med. J.* **3**, 668 (1971).
[e] A. Pitt, H. Lipp, and S. T. Anderson, *Lancet* **1**, 612 (1971).
[f] I. A. Baker, J. V. Collins, and T. R. Evans, *Guys. Hosp. Rep.* **120**, 1 (1971).
[g] S. Darby, J. C. Cruickshank, M. A. Bennett, and B. L. Pentecoat, *Lancet* **1**, 817 (1972).
[h] K. P. O'Brien, P. M. Taylor, and R. S. Croxson, *Med. J. Aust.* **2**(Suppl.), 36 (1973).
[i] W. Bleifield, K. W. Merx, K. W. Heinrich, and S. Effert, *Eur. J. Clin. Pharmacol.* **6**, 119 (1973).
[j] K. I. Lie, H. J. Wellens, F. J. van Capelle, and D. Durrer, *N. Engl. J. Med.* **291**, 1324 (1974).
[k] J. B. Singh and S. L. Kocot, *Am. Heart J.* **91**, 430 (1976).
[l] G. Sandler, N. Dey, and J. Amonkar, *Curr. Ther. Res.* **20**, 563 (1976).
[m] K. I. Lie, K. L. Liem, W. J. Louridtz, M. J. Janae, A. F. Willebrands, and D. Durrer, *Am. J. Cardiol.* **42**, 456 (1978).
[n] H. W. Koster and A. J. Dunning, *N. Engl. J. Med.* **313**, 1106 (1985).
[o] H. M. Dunn, J. M. McComb, and C. D. Kinney, *Am. Heart J.* **110**, 853 (1985).
[p] J. Wyse, *J. Am. Coll. Cardiol.* **12**, 507 (1988).

of concluding with 95% probability ($\alpha = 0.05$) that the true effect was no more than a 10% decrease in the mortality rate. If the prior information were ignored, the study size would double to 14,143 patients per group.

The corresponding numbers using a nondegenerate design prior, 24,596 and 42,940, are rather high. Instead, we might choose to reduce the variability of the predictive prior to a standard error of 0.05, say, to make the measured treatment effect of 0.1 significant. In that case, the sample sizes would be 8443 using the informative analytic prior and 16,125 using the noninformative analytic prior.

Conclusion

We have described several methods for determining the sample size of a prospective study within the Bayesian paradigm. These methods are determined by specifying a prior distribution for the resulting analysis and whether or not prior uncertainty about the true value of the effect will be considered when determining the predictive distribution of the sample statistic. Standard methods of calculating sample sizes ignore these types of uncertainty. For the non-Bayesian, our methods are still useful for quantifying the effect of prior uncertainty with a noninformative analytic prior distribution.

The use of the prior information has two effects on sample size estimates relative to the classical formulas that ignore it. To the extent that the analytic prior is informative about the true location of θ, its use will reduce the additional data that will need to be collected. Incorporating the prior uncertainty about θ into the predictive distribution of $\overline{X}$ will, conversely, increase the amount of data required because we must take account of prospective samples with extreme results. As the required power of the study grows, these sample sizes may become large. Bayesian sample size methods account for this uncertainty and prior information, and can lead to substantially different answers from the usual non-Bayesian approach.

Two limitations to the methods proposed here are their reliance on the assumption of normality and the assumption of known variance σ^2. In practice, many outcomes are not continuous and the variance is rarely known. However, posterior distributions are asymptotically normal and sufficient data (or prior studies) may provide a reasonably good estimate of σ^2. As sample sizes grow large, the uncertainty in our estimate of σ^2 will have an insignificant effect on our results. Thus, the procedure should work well in large samples for many types of outcomes.

We believe that a Bayesian approach can foster both a better appreciation of the uncertainty of our prior knowledge of treatment effects and a better use of the results from preliminary studies. Too often, studies are performed in isolation as if no other related study has ever or will ever be undertaken. Bayesian methods allow us to interpret our work in the light of other research. Application of these techniques can increase the chance that a study will be large enough to be able to reach the desired conclusion.

Acknowledgment

Supported by Grant R01-HS10064 from the Agency for Healthcare Quality and Research of the United States Public Health Service.

[18] Distribution Functions from Moments and the Maximum-Entropy Method

By Douglas Poland

Introduction

In this chapter we show how one can use experimental data such as that contained in titration curves and the temperature variation of the heat capacity to obtain distribution functions for proteins and nucleic acids. For example, we show how one can use a titration curve, which gives the average number of ligands bound as a function of ligand concentration, to calculate how many molecules have one ligand bound, two ligands bound, and so on, giving the distribution function for ligand binding. This distribution function gives a detailed picture of the probability of all possible different states of ligand binding. The process of going from the original titration curve, which gives only the average state of ligand binding, to the complete distribution function greatly increases one's knowledge of the different states of binding present in the system. In a similar manner we show how knowledge of the temperature dependence of the heat capacity of a protein or nucleic acid can be used to obtain a distribution function for the enthalpy content of the molecule. The enthalpy distribution function tells what fraction of the molecules have a given enthalpy value, and the temperature variation of this distribution gives a detailed picture of the process of denaturation in terms of the probabilities of different enthalpy states of the molecule.

An important aspect of the approach outlined here is that the distribution functions we obtain, ligand binding distributions or enthalpy distributions, are determined solely by experimental data. While models play an important role in understanding biological macromolecules, it is also important to have results that are independent of any assumed model and the distribution functions we obtain are examples of such knowledge.

The starting point for our calculation of distribution functions from experimental data is the realization that information on the average number of ligands bound as a function of ligand concentration or heat capacity as a function of temperature can be used to calculate a set of moments for the

Copyright 2004, Elsevier Inc.
All rights reserved.
0076-6879/04 $35.00

appropriate distribution function. The equations that relate titration and heat capacity data to moments of a distribution function are obtained using basic relations from statistical mechanics. In particular, the relations we use require only the basic partition functions of statistical mechanics and are not based on any specific model. In this manner we convert one set of experimental data (titration curves or heat capacity) into another set (moments of a distribution function).

Given a finite set of moments (say, two to six) the problem then is to calculate the corresponding distribution function. To accomplish the construction of an approximate distribution function from a finite set of moments, we use an algorithm based on the maximum-entropy method due to Mead and Papanicolaou.[1] In this process one trades experimental knowledge of moments of the distribution function for knowledge of the parameters describing the functional form of the distribution function. An example of the construction of a distribution function from moments is given by the familiar bell-shaped Gaussian distribution function. The parameters required to construct this function are the mean value of the distribution and the standard deviation. The mean value is simply the first moment of the distribution while the square of the standard deviation is the second moment minus the first moment squared. Similarly, application of the maximum-entropy method gives an approximate distribution function using moments as input, but in this approach we are not restricted to the use of just the first two moments; the more moments used the better the approximation obtained.

The applications we present in this chapter are based on methodology contained in two publications. The first publication shows how enthalpy distribution functions can be constructed from knowledge of the temperature dependence of the heat capacity[2] while the second shows how ligand-binding distributions can be constructed using the data contained in titration curves.[3] An outline of the general approach that we use is as follows. One starts with experimental data on the system of interest (a titration curve or heat capacity data) and makes use of basic relations from statistical mechanics to obtain moments for the distribution function as outlined below:

$$
\begin{array}{ccc}
\text{Experimental data} & \text{Statistical} & \text{Moments of} \\
\text{(titration curve,} & \dfrac{\text{mechanics}}{\longrightarrow} & \text{distribution function}
\end{array} \quad (1)
$$
\text{heat capacity)}

[1] L. R. Mead and N. Papanicolaou, *J. Math. Phys.* **25**, 2404 (1984).

[2] D. Poland, *J. Chem. Phys.* **112**, 6554 (2000).

[3] D. Poland, *J. Chem. Phys.* **113**, 4774 (2000).

One then uses the maximum-entropy method to convert the set of moments into parameters of the respective distribution function:

$$
\begin{array}{ccc}
\begin{array}{c}\text{Moments of}\\ \text{distribution}\\ \text{function}\end{array} & \xrightarrow[\text{method}]{\text{Maximum entropy}} & \begin{array}{c}\text{Parameters of}\\ \text{distribution function}\end{array}
\end{array} \quad (2)
$$

The first step, obtaining moments of the distribution function from experimental data, is the more difficult of the two and we devote most of this chapter to describing this process. In Section II we show how the moments of ligand-binding distribution functions can be obtained from titration curves. Then, in Section III we outline the maximum-entropy method for determining approximate distribution functions from experimental knowledge of moments. We return to ligand binding in Section IV where, using the binding of protons to the protein lyzozyme as an example, we take the moments obtained in Section II and construct distribution functions from them, utilizing the maximum-entropy method outlined in Section III. In Section V we use this approach to obtain enthalpy distributions for proteins from the temperature dependence of the heat capacity of these molecules. We illustrate how the enthalpy distributions obtained in this manner give a rich picture of the thermal denaturation of proteins. Finally, in Section VI, we use this general method to obtain distribution functions for self-association, using the clustering of ATP as an example.

Ligand Binding: Moments

In this section we show how one can obtain moments of the distribution function for ligand binding from the binding isotherm (or titration curve) that gives the experimental measurement of the average number of ligands bound as a function of the ligand concentration. Our presentation here follows that given in two previous publications.[3,4] We begin by considering the reaction for adding a ligand to a macromolecule. We let P represent any molecule, in particular a biopolymer such as a protein or nucleic acid. We let L represent a general ligand such as Mg^{2+}, H^+, or any small molecule. We begin with the reaction for adding one additional ligand to a molecule with $(n-1)$ ligands already bound:

$$
PL_{n-1} + L \longleftrightarrow PL_n \quad (3)
$$

Taking K_n [with units $(\text{mol/liter})^{-1}$] as the equilibrium constant for the reaction in Eq. (3), the standard equilibrium constant expression for this reaction is (using square brackets to indicate concentrations)

[4] D. Poland, *J. Protein Chem.* **20**, 91 (2001).

$$K_n = \frac{[PL_n]}{[PL_{n-1}][L]} \tag{4}$$

The development is easier if we relate all states of binding to the same species. This can be accomplished by adding together appropriate reactions of the form given in Eq. (3) as illustrated in the following example:

$$
\begin{array}{ll}
P + L \longleftrightarrow PL_1 & (K_1) \\
PL_1 + L \longleftrightarrow PL_2 & (K_2) \\
PL_2 + L \longleftrightarrow PL_3 & (K_3) \\
\hline
P + 3L \longleftrightarrow PL_3 & Q_3 = K_1 K_2 K_3
\end{array}
\tag{5}
$$

Here we have added together the first three successive steps of binding to give the reaction for binding three ligands directly to the species P. The equilibrium constant for the new reaction, Q_3, is the product of the equilibrium constants for the three reactions used (in general when one adds together two reactions, the equilibrium constant for the new reaction is the product of the constants for the two reactions). The generalization of the process illustrated in Eq. (5) is

$$P + nL \longleftrightarrow PL_n \tag{6}$$

with

$$Q_n = \frac{[PL_n]}{[P][L]^n} \tag{7}$$

where

$$Q_n = \prod_{i=1}^{n} K_i \tag{8}$$

If the total concentration of the macromolecule is P_T, then the mole fraction of a given state of binding is given by

$$f_n = \frac{[PL_n]}{P_T} \tag{9}$$

Taking

$$c = [L] \tag{10}$$

we can rewrite Eq. (7) in terms of the f values and c, giving

$$\frac{f_n}{f_0} = c^n Q_n \tag{11}$$

Solving Eq. (11) for f_n, we have our fundamental relation for the mole fraction of P molecules having n ligands bound:

$$f_n = f_o c^n Q_n \tag{12}$$

Now by definition the sum of the mole fractions must add up to 1 (conservation of mole fractions),

$$\sum_{n=1}^{N} f_n = 1 \tag{13}$$

Using Eq. (12) in Eq. (13) and solving for f_0 gives

$$f_o = \frac{1}{1 + cQ_1 + c^2 Q_2 + \cdots + c^N Q_N} \tag{14}$$

It is useful to define the quantity in the denominator of Eq. (14) as

$$\Gamma = 1 + cQ_1 + c^2 Q_2 + \cdots + c^N Q_N \tag{15}$$

The quantity Γ in Eq. (15) is known as the binding polynomial and Schellman[5] has described the use of this quantity to treat ligand binding in biopolymers. The binding polynomial is an example of what is known in statistical mechanics as a partition function. Here it represents a sum over all possible states of binding. Using the relation for f_0 given in Eq. (14) in Eq. (12), we have the following general relation for the mole fraction of P with n ligands bound:

$$f_n = \frac{c^n Q_n}{\Gamma} \tag{16}$$

From Eq. (16) one sees that the mole fraction of molecules with n ligands bound (which is equal to the probability of picking a molecule at random that has n ligands bound) is simply given by the term in Γ corresponding to the state representing n ligands bound divided by Γ, that is, the sum over all states of binding.

Equation (16) is a general relation for the mole fractions or probabilities of all different states of binding as a function of the concentration, c, of ligand in solution. As such this equation is the distribution function for ligand binding and it is this function that we want to construct using moments and the maximum-entropy method. We now show how moments of this function can be obtained from a binding isotherm (titration curve). Notice that in the derivation of Eq. (16) we made no assumptions about the Q_n or the independence of binding sites. We have assumed that the

[5] J. A. Schellman, *Biopolymers* **14**, 999 (1975).

ligand concentration in solution is dilute so that the use of the equilibrium constant expression in Eq. (4) is valid.

Experimentally one measures the average number of ligands bound as a function of c or $\ln c$ and this set of data gives the binding isotherm or titration curve. From Eq. (16) we have the following relation for average n (average extent of binding),

$$\langle n \rangle = \sum_{n=0}^{N} n f_n \qquad (17)$$

To obtain $\langle n \rangle$ given in Eq. (17) in terms of Γ we observe that

$$\frac{\partial c^n Q_n}{\partial c} = n c^{n-1} Q_n \qquad (18)$$

We notice that the operation in Eq. (18) "brings down" a factor of n. This type of operation is central to our entire method. Multiplying both sides of Eq. (18) by c gives

$$c \frac{\partial c^n Q_n}{\partial c} = \frac{\partial c^n Q_n}{\partial \ln c} = n c^n Q_n \qquad (19)$$

To obtain the term $n f_n$ appearing in Eq. (17) we need only divide all terms in Eq. (19) by Γ. In this manner we obtain the basic relation

$$\langle n \rangle = \frac{1}{\Gamma} \frac{\partial \Gamma}{\partial \ln c} = M_1 \qquad (20)$$

The quantity $\langle n \rangle$ is the first moment of the distribution that we designate as M_1. It is often convenient to introduce the fraction of binding, which we define as θ:

$$\theta = \langle n \rangle / N \qquad (21)$$

or

$$M_1 = N\theta \qquad (22)$$

In an experimental study of ligand binding what one measures is $M_1(c)$ or equivalently $\theta(c)$. We now show how higher moments of the binding distribution can also be obtained from these data. The higher moments (in general the mth) are defined in analogy to the definition of the first moment given in Eq. (17),

$$M_m = \sum_{n=0}^{N} n^m f_n \qquad (23)$$

Letting

$$x = \ln c \qquad (24)$$

then the analog of the operation given in Eq. (18) for bringing down a single factor of n also works when we want m factors of n,

$$M_m = \frac{1}{\Gamma} \frac{\partial^m \Gamma}{\partial x^m} \tag{25}$$

If we take the derivative of M_m with respect to x we obtain the following relation between different moments:

$$\frac{\partial M_m}{\partial x} = \frac{1}{\Gamma} \frac{\partial^{m+1} \Gamma}{\partial x^{m+1}} - \frac{1}{\Gamma^2} \frac{\partial \Gamma}{\partial x} \frac{\partial \Gamma^m}{\partial x^m} = M_{m+1} - M_1 M_m \tag{26}$$

We define the following symbol for derivatives of M_1 with respect to x as

$$M_1^{(j)} = \frac{\partial^j M_1}{\partial x^j} \tag{27}$$

Then successively using Eq. (26) we obtain

$$M_1^{(1)} = M_2 - M_1^2$$
$$M_1^{(2)} = M_3 - 3M_1 M_2 + 2M_1^3 \tag{28}$$
$$M_1^{(3)} = M_4 - 4M_1 M_3 - 3M_2^2 + 12M_1^2 M_2 - 6M_1^4$$

We can then solve these equations consecutively for M_2, M_3, and M_4 in terms of the derivatives of M_1 with respect to x,

$$M_2 = M_1^{(1)} + M_1^2$$
$$M_3 = M_1^{(2)} + 3M_1 M_2 - 2M_1^3 \tag{29}$$
$$M_4 = M_1^{(3)} + 4M_1 M_3 + 3M_2^2 - 12M_1^2 M_2 + 6M_1^4$$

Thus knowledge of M_1, or equivalently of $\langle n \rangle$, as a function of $x = \ln c$ (where $c = [L]$) can be used to give higher moments of the binding distribution. In Eq. (29) we give the relations required to obtain the higher moments through M_4. Recall that $M_1 = \langle n \rangle$ is the experimental binding isotherm. Thus the derivatives of this quantity with respect to x, defined in Eq. (27), can also be obtained from experiment and then used in Eq. (29) to obtain higher moments of the ligand-binding distribution.

In Fig. 1A we show a typical binding isotherm giving $\langle n \rangle$ as a function of $\ln c (= x)$. As an example we take the case of 20 independent binding sites with a binding constant of $K = 2500 \ (\text{mol/liter})^{-1}$. The binding isotherm for this system is shown by the solid curve in Fig. 1A, while we take the solid dots as model experimental data. The derivatives required in Eq. (29) are evaluated at a particular value of x that we will denote as x_0. We are free to pick any value of x to serve as the reference point. We can then expand the

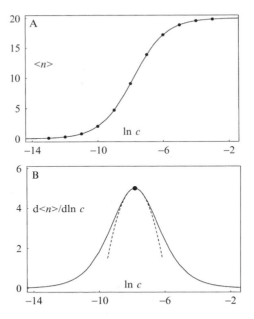

FIG. 1. (A) An example of a typical binding isotherm giving the average number of ligands bound as a function of the logarithm of the ligand concentration in solution. The curve was calculated assuming 20 independent sites with $K = 2500 \ (\text{mol/liter})^{-1}$. (B) A plot of the slope of the binding curve given in (A). The dashed curve represents the quadratic expansion given in Eq. (33) about the maximum in the curve, indicated by the solid dot.

function $M_1(x)$ in a Taylor series about $x = x_0$, using the variable

$$\Delta x = x - x_0 \tag{30}$$

giving (through the cubic term in Δx)

$$M_1(x) = a_0 + a_1 \Delta x + a_2 \Delta x^2 + a_3 \Delta x^3 + \cdots \tag{31}$$

which represents an empirical local fit of the experimental data. The a values in Eq. (31) are determined from experiment and have the following significance:

$$\begin{aligned} M_1^{(1)} &= a_1 \\ M_1^{(2)} &= 2a_2 \\ M_1^{(3)} &= 6a_3 \end{aligned} \tag{32}$$

But these are just the quantities required in Eq. (29) to give the first four moments of the binding distribution. Thus the local expansion of the binding isotherm, $M_1(x) = \langle n \rangle$, given in Eq. (31) allows us to calculate the first four moments of the ligand-binding distribution function.

Another way to obtain the information given in Eq. (32) is to use the derivative of $\langle n \rangle$ with respect to x, as a function of x. A plot of this function is shown in Fig. 1B, where we see that we now have a function that goes through a maximum at the value of x corresponding to the midpoint of the ligand-binding curve. A dominant feature of the curve shown in Fig. 1B is the width of the peak in the neighborhood of the maximum in the curve. The local series expansion of this function is given by [using Δx of Eq. (30)]

$$\frac{\partial \langle n \rangle}{\partial x} = M_1^{(1)}(x) = b_0 + b_1 \Delta x + b_2 \Delta x^2 + \cdots \tag{33}$$

where now

$$\begin{aligned}
M_1^{(1)} &= b_0 \\
M_1^{(2)} &= b_1 \\
M_1^{(3)} &= 2b_2
\end{aligned} \tag{34}$$

Thus the coefficients in Eq. (33) contain the same information (enough to calculate four moments of the ligand binding distribution) as does the expansion given in Eq. (31), but the coefficients now have simple physical interpretations:

b_0 is the gradient of the binding isotherm at x_0

b_1 is zero at x_0 $\hfill$ (35)

b_2 measures the width of the peak

The dashed curve in Fig. 1B shows the local quadratic expansion about the maximum in the curve as given by Eq. (33). From Eq. (35) we note that the construction of the quadratic curve shown in Fig. 1B requires only the value of the function at the maximum and the width of the peak. Since this is enough experimental information to calculate four moments of the ligand-binding distribution we see that there is no difficulty in obtaining a set of moments from experimental ligand-binding data. Given a finite set of moments, the next step is to turn information about the moments into parameters of the appropriate distribution function. To this end we use the maximum-entropy method that we describe in the next section.

Maximum-Entropy Distributions

It is an old problem in mathematics to construct an approximation to a distribution function given a finite set of moments of that function. In the preceding section we have seen how one can easily obtain a set of four to six moments for the distribution function for the binding of ligands to biological macromolecules from the binding isotherm (titration curve). In Section V we show how one can similarly obtain a finite set of moments for the enthalpy distribution function in proteins and other macromolecules from the temperature dependence of the heat capacity. The problem we address in this section is the use of these moments to construct an approximate distribution function.

The technique we use is the maximum-entropy method as applied to the moment problem by Mead and Papanicolaou[1] and by Tagliani.[6] To keep our discussion explicit we use a specific example of a distribution function, $f(X)$, which is illustrated in Fig. 2. Note that we are using upper-case X as the independent variable for this distribution; we are saving low-ercase x for another upcoming use. The function we have chosen as an example is bimodal with two unequal peaks. The explicit functional form of this distribution is the sum of two unequal Gaussian distributions, as shown:

$$f(X) = (1.6 \exp [-0.25(X - 5)^2] + \exp[-0.35(X + 1)^2])/A \quad (36)$$

where the value of A is chosen to give a normalized function

$$\int_{L_2}^{L_1} f(X) = 1 \quad (37)$$

The bounds L_1 and L_2 in Eq. (37) are taken as practical limits of the extent of the distribution function. From Fig. 2 we take these limits as the points

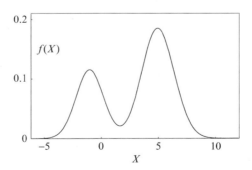

Fig. 2. An example of a bimodal distribution function with peaks of unequal height. The function shown is a sum of two Gaussian distributions as given by Eq. (36).

[6] A. Tagliani, *J. Math. Phys.* **34,** 326 (1995).

where the value of the distribution function have dropped essentially to zero. Thus we take

$$L_1 = 12 \quad \text{and} \quad L_2 = -6 \tag{38}$$

The choice of reasonable bounds to the range of the distribution function is an important part of the process. For our example given in Eq. (36) with the bounds of Eq. (38) the value of the normalizing constant is $A = 8.6678$.

The input for the calculation of a distribution function using the maximum-entropy method is a finite set of moments of the distribution function. For biological macromolecules these moments are obtained from experimental data such as titration curves for ligand-binding distributions or heat capacity data for enthalpy distributions. For the example at hand we know the exact distribution function as given in Eq. (36) and thus for this example we can calculate the moments precisely. The moments of $f(X)$ are given in general by the relation

$$M_m = \int_{L_2}^{L_1} X^m f(X) \, dX \tag{39}$$

For $f(X)$ of Eq. (36), the first four moments are

$$M_1 = 2.92618, \quad M_2 = 18.5069, \quad M_3 = 99.5994, \quad M_4 = 618.546 \tag{40}$$

These are the numbers that, in an actual application of the method, would be determined from experiment.

The next step is technical, but important. To apply the maximum-entropy method it is most convenient to scale the distribution function in question onto the unit interval so that rather than the limits L_2 and L_1 as given in Eq. (38) one has the limits zero and one. It is best to do this shift in two steps. First, we simply translate the function so that the lower limit is the origin. To do this we simply introduce a new variable,

$$y = X - L_2 \tag{41}$$

so that now when $X = -6$, the new variable has the value $y = 0$. The function now is $f(y)$ with y varying from 0 to $L_1 - L_2$. The moments for the shifted distribution function can easily be obtained form the former moments given in Eq. (40). Taking

$$L = L_1 - L_2 \tag{42}$$

one has

$$M'_m = \int_0^L y^m f(y)\, dy = \int_{L_2}^{L_1} (X - L_2)^m f(X) dX \qquad (43)$$

Defining

$$\alpha = -L_2 \qquad (44)$$

the first few moments of the scaled function are given:

$$M'_1 = M_1 + \alpha, \ M'_2 = M_2 + 2\alpha M_1 + \alpha^2$$
$$M'_3 = M_3 + 3\alpha M_2 + 3\alpha^2 M_1 + \alpha^3 \qquad (45)$$
$$M'_4 = M_4 + 4\alpha M_3 + 6\alpha^2 M_2 + 4\alpha^3 M_1 + \alpha^4$$

For our example, the above moments have the following numerical values (in our example $L = 18$ and $\alpha = 6$),

$$M'_1 = 8.92618, \qquad M'_2 = 89.621, \qquad M'_3 = 964.75, \qquad M'_4 = 10830.6 \quad (46)$$

We have now shifted the distribution so that the lower bound is at the origin, $y = 0$. The next step is to scale the function so that the upper bound is 1. To accomplish this we introduce yet another new variable,

$$x = y/L \qquad (47)$$

so that now when $y = L$ the new variable has the value $x = 1$. The moments for the distribution function expressed in terms of the new variable defined on the interval $x = 0$ to 1 are now simply given by

$$\mu_m = M'_m / L^m \qquad (48)$$

where L is given by Eq. (42). The final set of moments for the scaled distribution function in our example defined on the unit interval from 0 to 1 is now

$$\mu_1 = 0.495899, \qquad \mu_2 = 0.276608, \qquad \mu_3 = 0.165424, \qquad \mu_4 = 0.103172$$
$$(49)$$

Notice that the above moments form a moderate set of monotonically decreasing numbers and that in a sense the scaling process has "tamed" the values of the moments.

The above scaling process is really the most difficult part of the procedure. One problem in scaling experimentally determined moments is that one does not always have a clear sense of what the lower and upper bounds, the L_1 and L_2 of Eq. (38), are. For titration problems one usually knows the total number of proton-binding sites from the amino acid sequence of the protein, but for other distribution functions, such as the enthalpy distribution, one does not have such knowledge. If one knows the first and second moments of the distribution, then one can make a first estimate of the width of the distribution, using the standard Gaussian distribution:

$$f(x) = \frac{1}{\sigma\sqrt{2\pi}} \exp\left[-(M_1 - x)^2/2\sigma^2\right] \tag{50}$$

where σ is the standard deviation (giving the width of the distribution)

$$\sigma = \sqrt{M_2 - M_1^2} \tag{51}$$

Of course, for a distribution such as that illustrated in Fig. 2 the above function can give only a crude estimate of the overall range of the distribution function in question.

We now assume that one has a set of moments like those shown in Eq. (49) that have been scaled onto the unit interval from 0 to 1, starting with moments obtained from experiment. We now outline the approach of Mead and Papanicolaou[1] to obtain approximate distribution functions given a finite set of moments, using the maximum-entropy method.

The scaled distribution function on the unit interval will be referred to as $p(x)$, denoting a probability distribution function in the continuous variable x where x varies from 0 to 1. The probability distribution function $p(x)$ is normalized as follows:

$$\int_0^1 p(x)dx = 1 \tag{52}$$

We denote the moments of this function as

$$\mu_m = \int_0^1 x^m p(x)dx = \langle x^m \rangle \quad (\text{for} \quad m = 1 \text{ to } N) \tag{53}$$

where we indicate that a finite set of N moments is known. For the function given in Eq. (36) the first four of these moments are given in Eq. (49).

In the maximum-entropy method one defines an entropy-like quantity in terms of the function $p(x)$:

$$S = -\int_0^1 [p(x)\ln p(x) - p(x)]\, dx + \sum_{m=0}^{N} \lambda_m \left(\int_0^1 x^m p(x)\, dx - \mu_m \right) \tag{54}$$

where λ_m are Lagrange multipliers. Functional variation of S with respect to $p(x)$ plus the condition that the moments for $m = 0$ to N are given by Eq. (53) gives the result that $p(x)$ has the following functional form

$$p(x) = \exp\left[-g(x)\right] \tag{55}$$

where

$$g(x) = \sum_{n=0}^{N} \lambda_n x^n \tag{56}$$

Thus $g(x)$ is simply a finite polynomial in x. One sees that the more moments that are known (the larger the number N), the more terms in the polynomial of Eq. (56) one has and the closer the approximate distribution function given by Eqs. (55) and (56) will be to the actual distribution function. Note that the $g(x)$ polynomial given in Eq. (56) when used in Eq. (55) and then in Eq. (53) gives back exactly the first N known moments of the experimental distribution function.

To illustrate the maximum-entropy distribution functions obtained in this manner we first give the form for the distribution function when one knows only the first moment of the distribution (the average value of x). In this case the distribution is given by a simple exponential function,

$$\text{(one moment)} \quad p(x) = A \ \exp(-\lambda_1 x) \tag{57}$$

If the first two moments of the distribution function are known one obtains a quadratic (or Gaussian) distribution function:

$$\text{(two moments)} \quad p(x) = A \ \exp[-(\lambda_1 x + \lambda_2 x^2)] \tag{58}$$

Given the first four moments of the distribution function the function $g(x)$ of Eq. (56) will be a quartic polynomial in x giving the following approximate distribution function:

$$\text{(four moments)} \quad p(x) = A \ \exp[-(\lambda_1 x + \lambda_2 x^2 + \lambda_3 x^3 + \lambda_4 x4)] \tag{59}$$

The remaining technical problem is to determine the values of the parameters λ_n introduced in Eq. (56). Knowledge of the experimentally determined values of the first N moments as given in Eq. (53) gives one N known numbers. The task is to convert these N known quantities into known values of the N parameters λ_n for $n = 1$ to N (note that the parameter λ_0 simply acts as a normalization parameter for the distribution). This process is schematically shown as follows:

$$\begin{array}{ccc} \text{Experimental values} & & \text{Values of} \\ \text{of } \mu_m \text{ for} & \xrightarrow{\hspace{2cm}} & \text{distribution } \lambda_n \text{ for } n = 1 \text{ to } N \\ m = 1 \text{ to } N & & \end{array} \tag{60}$$

Mead and Papanicolaou[1] have given a general iterative algorithm using any number of moments to accomplish the process shown in Eq. (60) and that is the scheme that we will outline here. One initiates the iteration procedure by constructing three vectors having N elements each:

$$\boldsymbol{\mu} = (\mu_1, \mu_2, \ldots, \mu_N)$$
$$\boldsymbol{\lambda}_0 = (0, 0, \ldots, 0) \tag{61}$$
$$\mathbf{x} = (x, x^2, \ldots, x^N)$$

The $\boldsymbol{\mu}$ vector contains the known experimental values of the first N scaled moments while the $\boldsymbol{\lambda}_0$ vector contains the initial values (all zero) of the unknown $\boldsymbol{\lambda}_n$ parameters. The $\mathbf{x}$ vector contains the first N powers of the independent variable x. The iteration process is primed by setting a general vector $\boldsymbol{\lambda}$ equal to the initial vector $\boldsymbol{\lambda}_0$:

$$\boldsymbol{\lambda} = \boldsymbol{\lambda}_0 \tag{62}$$

One then begins the general iteration loop given.

General Iteration Loop

First, one forms the function $f(x)$, where the dot indicates the dot product of the two vectors:

$$f(x) = \exp[-\boldsymbol{\lambda} \cdot \mathbf{x}] \tag{63}$$

One then normalizes this function on the interval 0 to 1:

$$A^{-1} = \int_0^1 f(x)\, dx \tag{64}$$

giving the normalized approximation for the probability distribution

$$p(x) = Af(x) \tag{65}$$

Using this approximate distribution function one then calculates a set of approximate moments,

$$\mu_m* = \int_0^1 x^m p(x)\, dx \quad (m = 1, 2N) \tag{66}$$

Note that one calculates $2N$ moments. Using the first N of the preceding approximate moments one forms the vector

$$\boldsymbol{\mu}* = (\mu_1*, \mu_2*, \ldots, \mu_N*) \tag{67}$$

The next step is the formation of the following $N \times N$ matrix:

$$\mathbf{W} = (w_{ij}) \tag{68}$$

where the general matrix element is given by

$$w_{ij} = \mu_{i+j}^* - \mu_i^* \, \mu_j^* \tag{69}$$

This matrix is then inverted, giving

$$\mathbf{WI} = \mathbf{W}^{-1} \tag{70}$$

Finally, one forms a vector that gives the difference between the experimental values of the moments and the approximate moments calculated above:

$$\mathbf{v} = \boldsymbol{\mu} - \boldsymbol{\mu}* \tag{71}$$

One then forms the vector

$$\mathbf{a} = \mathbf{WI} \cdot \mathbf{v} \tag{72}$$

As a last step the improved estimate of the vector of λ values is given by

$$\boldsymbol{\lambda}_{new} = \boldsymbol{\lambda} - \mathbf{a} \tag{73}$$

One then iterates this procedure until there is no difference between $\boldsymbol{\lambda}_{new}$ and $\boldsymbol{\lambda}$. If there is a difference, then one sets

$$\boldsymbol{\lambda} = \boldsymbol{\lambda}_{new} \tag{74}$$

and goes back to the beginning of the general iteration loop above Eq. (63) and repeats the whole process.

The whole procedure is actually quite straightforward, and the computer program to carry out this iteration scheme can be written in about a dozen lines on the back of a postcard. The process usually converges to a limiting set of λ values in about 15 iterations. The only numerical problem that sometimes arises is that when one uses eight or more moments, there are sometimes precision problems in the matrix inversion step and one needs to increase the number of significant figures used in the program.

When one has run the iteration scheme outlined above and has found a set of λ values such that the new values are identical with the values obtained in the previous iteration round, one then wants to use the approximate distribution function $p(x)$ as given by Eqs. (55) and (56) to calculate the set of moments, μ_1 to μ_N, to be sure that the first N moments of the approximate distribution function indeed reproduce the experimental input. Mead and Papanicolaou[1] have proved that if the set of moments, μ_1 to μ_N, vary monotonically [such as the moments given in Eq. (49) do], then there is a unique set of λ values, λ_1 to λ_N, that results from the maximum-entropy method. This does not mean that the approximate maximum-entropy method function given in Eqs. (55) and (56) is the unique distribution function: it is the unique maximum-entropy approximation to the distribution function. As the number of moments used in the construction of the distribution function is increased, the distribution function, through Eq. (56), is described by a longer and longer polynomial until finally one has an infinite series (which, of course, is not attainable from experimental data, where the maximum number of moments that can reasonably obtained is about eight). In the unrealistic limit where one uses an infinite number of moments one would obtain a unique infinite series that would give the exact distribution function. Practically, the more moments used, the better the approximate distribution will be.

At the conclusion of the iteration process one has a distribution function, $p(x)$, defined on the interval $x = 0$ to 1 that reproduces the first N experimental moments that have been scaled using Eq. (43) and Eq. (47) to the unit interval. The only remaining task is to scale the distribution function back to the interval of interest for the given physical system being treated. Again, we use the function given in Eq. (36) and illustrated in Fig. 2 as an example. The upper and lower bounds for this function, L_1 and L_2, are given in Eq. (38) while the parameter L is given in Eq. (42). To scale back to the interval shown in Fig. 2 we simply invert the scaling process we have already used. First we scale the function from the unit interval $x = 0$ to 1 to the new interval $y = 0$ to L. This is achieved by the inverse of the relation given in Eq. (47), involving the substitution

$$y = xL \tag{75}$$

Then we shift the distribution by an amount L_2 going back to our original variable X (which describes the actual experimental distribution function of interest),

$$X = y + L_2 \tag{76}$$

where now the variable X varies from L_2 to L_1 (this is -6 to 12 for our example in Fig. 2). This double substitution takes the function $p(x)$ found by the maximum-entropy iteration and scales it back to the correct range of the experimental distribution, $f(X)$, which for our example is the function shown in Fig. 2. The moments calculated from the rescaled distribution give one the original set of moments, which in our example are the moments given in Eq. (39).

We now want to illustrate how this procedure works for our sample distribution function shown in Fig. 2 and given in Eq. (36). Notice that the function given in Eq. (36), a sum of two Gaussian distributions, does not have the same functional form as the maximum-entropy function given in Eqs. (55) and (56). Thus it would take an infinite number of moments for the maximum-entropy method to give the exact distribution function shown in Eq. (36). Nonetheless, the power of the maximum-entropy method is that for a finite set of moments, it can give an excellent approximation to many distribution functions such as that shown in Fig. 2.

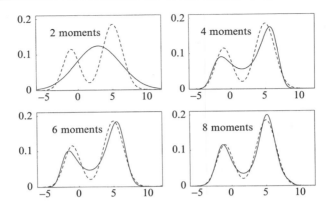

FIG. 3. Maximum-entropy approximations to the bimodal distribution shown in Fig. 2. The solid curves show, respectively, the maximum-entropy approximations obtained using two, four, six, and eight moments. The dashed curve in each box is the exact distribution function given by Eq. (36). In each graph the approximate distribution function is plotted as a function of X as in Fig. 2.

In Fig. 3 we show the maximum-entropy approximations to the function shown in Fig. 2 constructed using a variable number of moments. In the upper left-hand panel we show the result obtained using only two moments. In this case the approximate distribution is a single Gaussian and one obtains a rather poor approximation to the actual function (shown by the dashed curve in all the graphs). In the upper right-hand panel of Fig. 3 we give the result obtained when one uses four moments. Now the maximum-entropy method clearly resolves the bimodal character of the distribution (four is the minimum number of moments that will resolve bimodal behavior). In the other graphs we see that as the number of moments used is increased from six to eight the goodness of fit increases until the function obtained using eight moments gives an excellent fit to the original distribution function.

To test the accuracy of the maximum-entropy method outlined above[7] we have used exact distribution functions obtained from the two-dimensional Ising model. In that system we treat a model fluid when it is near the critical point, where it splits into two phases. In that case the density distribution for the fluid shows marked bimodal character that the maximum-entropy method, using a finite set of moments, reproduces with high accuracy.

[7] D. Poland, *Physica. A* **309,** 45 (2002).

Ligand Binding: Distribution Functions

In the previous section we outlined how an approximation to a molecular distribution function can be constructed from a finite set of the appropriate moments, the more moments used the better the approximation, as illustrated in Fig. 3. In Section II we showed how moments of the ligand-binding distribution function can be obtained from local expansions of the experimental binding isotherm. In particular we found that the local quadratic fit to the gradient of the binding isotherm, illustrated in Fig. 1B, was sufficient to give the first four moments of the ligand-binding distribution function. In this section we combine the two results and illustrate the ligand-binding distribution functions obtained in this manner.

As an example we use the titration curve of the protein lysozyme given by Tanford and Wagner.[8] In this case the ligand that binds to the macromolecule is a proton, H^+, and the distribution function gives the probability that an arbitrary number of protons are bound to the protein at a given pH. To discuss the binding of protons we need to briefly review a few facts about acid–base equilibria.

For a weak acid such as acetic acid the standard dissociation reaction in water (dissociation of a proton, the definition of an acid) is

$$CH_3COOH \leftrightarrow CH_3COO^- + H^+ \tag{77}$$

with the standard equilibrium constant expression

$$\frac{[CH_3COO^-][H^+]}{[CH_3COOH]} = K_a = 10^{-pK_a} \tag{78}$$

We note that Eq. (78) defines the quantity pK_a and that it is defined relative to base 10. For acetic acid at a temperature of $25°$ one has $pK_a = 4.54$, a typical value for a weak acid. We note that pH is also defined relative to base 10:

$$[H^+] = 10^{-pH} \tag{79}$$

For a weak base, such as methylamine, we have the following reaction when this molecule is placed in water:

$$CH_3NH_2 + H_2O \longleftrightarrow CH_3NH_3^+ + OH^- \tag{80}$$

which gives a slightly basic solution. The equilibrium constant for this reaction is expressed in terms of K_b and pK_b,

[8] C. A. Tanford and M. L. Wagner, *J. Am. Chem. Soc.* **76,** 3331 (1954).

$$\frac{[CH_3NH_3^+][OH^-]}{[CH_3NH_2]} = K_b = 10^{-pK_b} \tag{81}$$

where we follow the convention of leaving $[H_2O]$ out of the equilibrium constant expression in Eq. (81) since it is essentially constant. For methylamine at $25°$ one has $pK_b = 3.36$.

To treat the binding of protons to a protein it is more useful to express the reaction in Eq. (80) as the binding of a proton. We can achieve this by using the reaction for the self-ionization of water reaction,

$$H_2O \longleftrightarrow H^+ + OH^- \tag{82}$$

with

$$[OH^-][H^+] = K_W = 10^{-14} \tag{83}$$

If we invert Eq. (80) and add the reaction given in Eq. (82), we obtain the acid dissociation reaction,

$$CH_3NH_3^+ \longleftrightarrow CH_3NH_2 + H^+ \tag{84}$$

where now we view $CH_3NH_3^+$ as an acid with the acid dissociation constant,

$$K_a = K_w/K_b = 10^{-pK_a} \tag{85}$$

where $pK_a = 10.64$.

In our treatment of ligand binding we treated all the reactions involved as binding reactions. Thus we need to turn the reactions given in Eq. (77) and Eq. (84) around and write them as binding reactions. Thus we have in general for acids,

$$A^- + H^+ \longleftrightarrow AH, \ K = 10^{+pK_a} \tag{86}$$

where the binding constant now is 10 raised to the plus pK_a. For weak bases as written in Eq. (84) we have in general,

$$A + H^+ \longleftrightarrow AH^+, \ K = 10^{+pK_b} \tag{87}$$

If we consider each proton-binding site in isolation from the rest of the molecule, then in general we can treat each independent group that can bind a proton as having its own binding polynomial that is a sum over two states, proton not bound and proton bound. We write this function as follows:

$$\gamma = 1 + [H^+]10^{pK} \tag{88}$$

so that the mole fractions of the unbound and bound states are given by

$$f_A = \frac{1}{\gamma} \quad \text{and} \quad f_{AH} = \frac{[H^+]10^{pK}}{\gamma} \tag{89}$$

Nozaki and Tanford[9] have studied denatured ribonuclease (in 6 M guanidine-HCl) and they find that the proton-binding sites act as if they are independent. In that molecule they assign the following pK values to the appropriate groups:

$$
\begin{aligned}
&1.\ \alpha - \text{Carboxyl (1) p}K\ =\ \ \ 3.4 \\
&2.\ \beta - \text{Carboxyl (5) p}K\ =\ \ \ 3.8 \\
&3.\ \gamma - \text{Carboxyl (5) p}K\ =\ \ \ 4.3 \\
&4.\ \text{Imidazole (4) p}\quad\quad\ =\ \ \ 6.5 \\
&5.\ \alpha - \text{Amino (1) p}K\quad\ =\ \ \ 7.6 \\
&6.\ \text{Phenolic (3) p}K\quad\quad\ =\ \ \ 9.75 \\
&7.\ \text{Phenolic (3) p}K\quad\quad\ =\ 10.15 \\
&8.\ \varepsilon - \text{Amino (10) p}K\ \ =\ 10.35
\end{aligned}
\tag{90}
$$

There are also four guanidyl groups in the molecule with p$K > 12.5$ that are not included in this list. The listing given in Eq. (90) includes $N = 32$ protons that can bind to the protein. The α-carboxyl and α-amino groups are, respectively, the terminal carboxyl and amino groups of the molecule while the β- and γ-carboxyl groups are the carboxyl groups, respectively, on aspartic and glutamic acid (like acetic acid in our example). The imidazole and phenolic pK values refer to proton-binding sites, respectively, on histidine and tyrosine. Finally, the ε-amino groups are the amine groups on lysine (like methyl amine in our example).

For the special case where each binding site is independent (which is not the general case), the binding polynomial for the whole molecule is given by the following product:

$$\Gamma = \gamma_1 \gamma_2^5 \gamma_3^5 \gamma_4^4 \gamma_5 \gamma_6^3 \gamma_7^3 \gamma_8^{10} \tag{91}$$

If we expand this expression we obtain the polynomial

$$\Gamma = \sum_{n=0}^{N} Q_n [H^+]^n \tag{92}$$

where in this case $N = 32$. The probability of a given state of binding (number of protons bound) is simply

$$P_n = Q_n [H^+]^n / \Gamma \tag{93}$$

while the titration curve (average number of protons bound as a function of pH) is given by

[9] Y. Nozaki and C. Tanford, *J. Am. Chem. Soc.* **89**, 742 (1967).

$$\langle n \rangle = \sum_{n=0}^{N} n P_n = \frac{\partial \ln \Gamma}{\partial \ln [\mathrm{H}^+]} \tag{94}$$

where we note that

$$\ln[\mathrm{H}^+] = 2.303 \ \log[\mathrm{H}^+] = -2.303 \mathrm{pH} \tag{95}$$

Note that Q_n represents a sum over all ways that n protons can be bound.

In general the binding polynomial will not be a product of independent γ values. In the compact form of a native protein the charged groups will interact with one another and some will be buried, making them less accessible for binding. We now show how one can use moments of the experimental titration curve to obtain the complete binding polynomial for the case where the binding sites are not independent.

In Fig. 4 we show the experimental titration curve for the protein lysozyme based on the data of Tanford and Wagner.[8] This protein contains 22 sites for proton binding. The solid points shown in Fig. 4 were obtained by tracing the curve the authors drew through their experimental points and then taking evenly spaced points on this curve; we will explain the origin of the solid curve shortly. Using the set of experimental points shown in Fig. 4 one can then apply the techniques outlined in Section II to obtain a set of moments at different pH values.[4] First, we take seven different sets of contiguous points, each set containing seven points, and then we fit each set of data to a cubic polynomial centered at the middle point. The pH

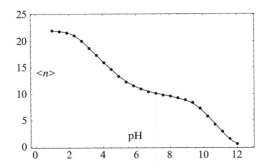

FIG. 4. The experimental titration curve of the protein lysozyme (solid dots) constructed using the original data of Tanford and Wagner.[8] The graph gives the average number of protons bound as a function of the pH of the solution. The maximum number of protons that can bind to this protein is 22. The solid curve represents the theoretical titration curve calculated from the binding polynomial as shown in Eq. (15) using the Q_n given in Fig. 5b. Reprinted from D. Poland, *J. Protein Chem.* **20,** 91 (2001), with kind permission of Kluwer Academic Press.

values at the midpoints of the seven sets of experimental points are as follows: pH 3.19, 4.07, 4.95, 6.28, 8.04, 9.80, and 10.68.

Once we have the set of polynomials, each representing a local expansion of the titration curve about a particular value of the pH, we then use the procedure outlined in Section II to obtain the first four moments of the proton-binding distribution from each of these polynomials (each polynomial representing a different pH value). Given these sets of moments we next use the maximum-entropy method outlined in the previous section to obtain seven different binding distribution functions, one for each of the seven different pH values indicated above. The seven distributions functions so obtained are shown in Fig. 5A. It turns out that for this system the distribution functions calculated using successively two, three, and four moments are virtually superimposable, so the distribution functions shown in Fig. 5A are essentially Gaussian distributions. Note that one needs to obtain four moments in order to confirm that the use of only

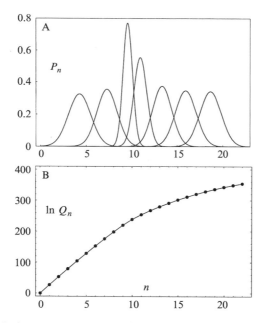

FIG. 5. (A) Maximum-entropy proton-binding distribution functions for the protein lysozyme constructed using moments obtained from local expansions of the titration curve given in Fig. 4. The distribution functions are for the pH values, from left to right: pH 3.19, 4.07, 4.95, 6.28, 8.04, 9.80, 10.68. (B) The values of $\ln Q_n$ for $n = 0$ to 22 (solid dots), giving the complete proton binding polynomial for the protein lysozyme. Reprinted from D. Poland, *J. Protein Chem.* **20**, 91 (2001), with kind permission of Kluwer Academic Press.

two moments gives an excellent representation of the proton-binding distribution function.

We now use the proton-binding distributions functions given in Fig. 5A to calculate all of the Q_n in the binding polynomial of Eq. (92). From Eq. (93) the ratio of the probabilities for successive species is given by

$$\frac{P_n}{P_{n-1}} = \frac{Q_n}{Q_{n-1}}[H^+] \tag{96}$$

Solving this equation for Q_n we have

$$Q_n = Q_{n-1}\left(\frac{P_n}{P_{n-1}}\right)[H^+] \tag{97}$$

Using the proton-binding distribution functions obtained from the maximum-entropy method, one can estimate the ratio P_n/P_{n-1} by evaluating the distribution function at integer values of n. The distribution function used refers to a specific value of the pH (one of the seven pH values used and listed previously) and so the value of $[H^+]$ is known. Thus Eq. (97) represents a recursion relation, giving Q_n in terms of Q_{n-1} and known quantities. To start the recursion process one must know the first term in the sequence. But from Eq. (15) the zeroth term in the binding polynomial is simply equal to 1 and hence we have

$$Q_0 = 1 \tag{98}$$

Thus we start the recursion process with the above value and then successively calculate all of the Q_n values from ratios P_n/P_{n-1} that are obtained from the proton-binding distribution functions, which in turn are obtained from moments of the titration curve.

In this manner we obtain the complete binding polynomial for the binding of protons to lysozyme. The quantities $\ln Q_n$ thus obtained are shown as a function of n in Fig. 5B. Given the proton-binding polynomial, Γ, one can then calculate the probability of any state of proton binding at any pH value. Thus all the possible proton-binding information for a given protein is contained in Γ through the coefficients Q_n. Given the complete binding polynomial Γ one can then use Eq. (94) to calculate the titration curve (the average extent of binding as a function of pH). Using the values of $\ln Q_n$ given in Fig. 5B, the calculated titration curve is shown by the solid curve in Fig. 4, which is seen to give an excellent fit to the solid points that are derived from the experimental data.

Thus the moment/maximum-entropy method can give one the complete proton-binding polynomial for a protein. This function in turn contains all the empirical information possible concerning the binding of protons to the protein. To dissect Q_n into terms that represent different microscopic sets binding constants requires a specific model. Knowledge of Q_n is the most information one can obtain empirically without any specific model.

One can also apply the method outlined in this section to other types of binding. In particular we have applied this approach to the binding of Mg^{2+} and small molecules to nucleic acids,[10] to the binding of denaturants to proteins,[11] and to the free energy of proton binding in a variety of proteins.[12]

Enthalpy Distributions

In this section we apply the use of moments to the calculation of enthalpy distribution functions in proteins. Just as there is a broad distribution of the state of ligand binding in a given biological macromolecule, there also is a broad distribution of enthalpies. In particular, the temperature dependence of the enthalpy distribution gives considerable insight into the process of the thermal unfolding of a protein.

We have already cited the reference[2] containing the basic outline of the use of the maximum-entropy method to obtain enthalpy distribution functions. In addition, this approach has been applied to the calculation of the density of states for a general substance,[13] to enthalpy distributions in proteins,[14] and to enthalpy distributions in the solid state of high polymers.[15] Most biological systems, including proteins, are usually studied at constant pressure (the system is simply open to the atmosphere). On the other hand, the calculation of the appropriate moments is simpler to describe for a system at constant volume, so we treat that case first and then indicate the minor changes in the formalism required to treat a constant pressure system. In any case, there is little difference between constant volume and constant pressure thermodynamics for condensed matter.

The starting point for the statistical thermodynamics of a constant volume system is the canonical partition function,[16,17] which is a sum over all energy levels of the system. Taking ε_i as a general energy level having degeneracy ω_i, the general canonical partition function is given by the following sum:

[10] D. Poland, *Biopolymers* **58,** 477 (2001).

[11] D. Poland, *J. Protein Chem.* **21,** 477 (2002).

[12] D. Poland, *Biopolymers*, in press (2004).

[13] D. Poland, *J. Chem. Phys.* **113,** 9930 (2000).

[14] D. Poland, *Biopolymers* **58,** 89 (2001).

[15] D. Poland, *J. Polym. Sci. B Polym. Phys.* **13,** 1513 (2001).

[16] T. L. Hill, *in* "An Introduction to Statistical Thermodynamics." Dover Publications, New York, 1986.

[17] D. Poland, *in* "Cooperative Equilibria in Physical Biochemistry." Oxford University Press, Oxford, 1978.

$$Z_v = \sum_i \omega_i \exp[-\beta \varepsilon_i] \tag{99}$$

where

$$\beta = 1/RT \tag{100}$$

In Eq. (100) T is the absolute temperature and R is the gas constant. We will measure energies in kilojoules per mole, in which case the gas constant has the value $R = 8.31451 \times 10^{-3} \text{ kJ mol}^{-1} \text{ K}^{-1}$. The connection between the partition function Z_v and thermodynamics is given by the relation

$$Z_v = \exp[-\beta A] \tag{101}$$

where A is the Helmholtz free energy.

Like the binding polynomial, this partition function is a sum over all states. The probability that the system is in a particular state is simply given by the term in the partition function for that state divided by the sum over all terms (the partition function). Thus our basic equation is

$$P_i = \frac{\omega_i \exp[-\beta \varepsilon_i]}{Z_v} \tag{102}$$

Given this expression for the probability of a general state we can then use it to obtain expressions for the moments of the energy distribution. The first moment is simply given by

$$\langle E \rangle = \sum_i \varepsilon_i P_i = E_1 \tag{103}$$

One can "bring down" the factor ε_i in Eq. (103) by taking the derivative with respect to $(-\beta)$,

$$\frac{\partial \omega_i \exp[-\beta \varepsilon_i]}{\partial(-\beta)} = (\varepsilon_i)\omega_i \exp[-\beta \varepsilon_i] \tag{104}$$

The above procedure is analogous to that used in Eq. (19) for the case of ligand binding. The first energy moment, as indicated in Eq. (103), is then given by the relation

$$E_1 = -\frac{1}{Z_v}\frac{\partial Z_v}{\partial \beta} \tag{105}$$

Higher moments are obtained in analogy with the procedure used in Eq. (25) for ligand binding,

$$E_m = \sum_i \varepsilon_i^m P_i = \frac{(-1)^m}{Z_v}\frac{\partial^m Z_v}{\partial \beta^m} \tag{106}$$

Finally, we note that the first moment is simply the internal energy of thermodynamics,

$$U = E_1 \tag{107}$$

The next step is to relate the moments given in Eq. (106) to derivatives of E_1 (or U) with respect to β (or, equivalently, with respect to T). One has

$$U = -\frac{Z_v^{(1)}}{Z_v}$$

$$\frac{\partial U}{\partial \beta} = \left(\frac{Z_v^{(1)}}{Z_v}\right)^2 - \left(\frac{Z_v^{(2)}}{Z_v}\right) \tag{108}$$

$$\frac{\partial^2 U}{\partial \beta^2} = -2\left(\frac{Z_v^{(1)}}{Z_v}\right) + 3\left(\frac{Z_v^{(1)}}{Z_v}\right)\left(\frac{Z_v^{(2)}}{Z_v}\right) - \left(\frac{Z_v^{(3)}}{Z_v}\right)$$

and so on, where

$$Z_v^{(j)} = \frac{\partial^j Z_v}{\partial \beta^j} \tag{109}$$

The first three moments of the energy distribution are then given by

$$E_1 = -\frac{Z_v^{(1)}}{Z_v}, \qquad E_2 = \frac{Z_v^{(2)}}{Z_v}, \qquad E_3 = -\frac{Z_v^{(3)}}{Z_v} \tag{110}$$

But since E_1 is the internal energy, U, one also has

$$E_1 = U, \qquad E_2 = -U_\beta^{(1)} + E_1, \qquad E_3 = U_\beta^{(2)} - 2E_1^3 + 3E_1 E_2 \tag{111}$$

where

$$U_\beta^{(j)} = \frac{\partial^j U}{\partial \beta^j} \tag{112}$$

We recall that $\beta = 1/RT$ and thus we see in Eqs. (111) and (112) that knowledge of the temperature dependence of U allows one to calculate moments of the energy distribution.

One can know the temperature dependence of the internal energy, U, without knowing the value of U itself. That is, one can know the derivatives of U with respect to temperature (such as the heat capacity) without knowing $U = E_1$. In that case one can construct central moments that are relative to the (unknown) value of E_1. In general one has

$$M_m = \langle (E - E_1)^m \rangle \tag{113}$$

We then have the simple results

$$M_1 = 0, \qquad M_2 = -U_\beta^{(1)}, \qquad M_3 = U_\beta^{(2)} \tag{114}$$

Given the value of E_1 one can then convert back to regular moments,

$$E_2 = M_2 + E_1^2, \qquad E_3 = M_3 + 3E_2E_1 - 2E_1^3 \tag{115}$$

The temperature dependence of U is given by the heat capacity (here at constant volume):

$$C_V = \left(\frac{\partial U}{\partial T}\right)_V \tag{116}$$

The experimental data concerning the energy moments are thus contained in the temperature dependence of the heat capacity, $C_V(T)$. One can express the temperature dependence of the heat capacity in the neighborhood of a reference temperature T_0 as an empirical Taylor series in ΔT where

$$\Delta T = T - T_0 \tag{117}$$

The general form for this expansion is given below

$$C_V(T) = c_0 + c_1 \Delta T + c_2 \Delta T^2 + \cdots \tag{118}$$

We can also formally write the internal energy as an expansion in ΔT,

$$U(T) = \sum_{m=0}^{\infty} \frac{1}{m!} \left(\frac{\partial^m U}{\partial T^m}\right)_{T_0} \Delta T^m \tag{119}$$

The coefficients in the $C_V(T)$ series given in Eq. (118) are then related to the derivatives used in Eq. (119) as follows:

$$U_T^{(0)} = U, \qquad U_T^{(1)} = c_0, \qquad U_T^{(2)} = c_1, \qquad U_T^{(3)} = 2c_2 \tag{120}$$

where

$$U_T^{(m)} = \frac{\partial^m U}{\partial T^m} \tag{121}$$

The equations for the energy moments require derivatives with respect to $\beta \, (= 1/RT)$. To convert from temperature derivatives to derivatives with respect to β we require the following relations:

$$T_\beta^{(m)} = \frac{\partial^m T}{\partial \beta^m} \tag{122}$$

the first few of which are

$$T_\beta^{(1)} = -\left(\frac{1}{R}\right)(RT)^2, \qquad T_\beta^{(2)} = \left(\frac{2}{R}\right)(RT)^3, \qquad T_\beta^{(3)} = -\left(\frac{6}{R}\right)(RT)^4 \quad (123)$$

Then finally we have

$$
\begin{aligned}
U_\beta^{(1)} &= T_\beta^{(1)} U_T^{(1)} \\
U_\beta^{(2)} &= T_\beta^{(2)} U_T^{(1)} + \left(T_\beta^{(1)}\right)^2 U_T^{(2)} \\
U_\beta^{(3)} &= T_\beta^{(3)} U_T^{(1)} + 3 T_\beta^{(1)} T_\beta^{(2)} U_T^{(2)} + \left(T_\beta^{(1)}\right)^3 U_T^{(3)}
\end{aligned}
\qquad (124)
$$

One sees that if one has the expansion of $C_V(T)$ in Eq. (118) through the c_2 term [quadratic fit of $C_V(T)$] this then gives $U_T^{(3)}$ from Eq. (120) and, finally, $U_\beta^{(3)}$ from Eq. (124). Thus knowledge of a quadratic fit of $C_V(T)$ contains enough information to calculate the first four energy moments. If one has the expansion of $C_V(T)$ through c_4 (quartic expansion) this is enough information to give the energy moments through E_6.

An example of a physical system where one knows the energy distribution function exactly (and hence all of the energy moments) is a fluid of hard particles. The only interaction between particles in such a fluid is that of repulsion between the hard cores of the particles. Since this fluid includes the effect of excluded volume it is not an ideal system, but it does not include any attractive interactions. Thus all of the internal energy U is kinetic energy. The distribution function for this system is most familiar as the Maxwell–Boltzmann velocity distribution. Here we express it as a distribution of the kinetic energy E,

$$P(E) = \frac{2\beta^{3/2}}{\sqrt{\pi}} \sqrt{E} \, \exp\left[-\beta E\right] \qquad (125)$$

with moments

$$E_1 = \frac{3}{2} RT, \qquad E_2 = \frac{15}{4}(RT)^2, \qquad E_3 = \frac{105}{8}(RT)^3, \qquad E_4 = \frac{945}{16}(RT)^4 \qquad (126)$$

where the first moment, $E_1 = 3RT/2$, is the familiar equipartition of energy result for a hard-core fluid. We[2] have used the moments/maximum-entropy method to approximate the exact distribution function given in Eq. (125). In this case the use of six moments gives a good fit to the exact distribution function. Note that the functional form of the distribution function given in Eq. (125) is not the same as the maximum-entropy distribution given in Eqs. (55) and (56), because of the preexponential square root of the E term. Thus it would take an infinite number of moments to reproduce the function given in Eq. (125) exactly. However, the use of a finite number of moments with the maximum-entropy method gives a good approximation to the exact distribution function.

We now indicate how moments and distribution functions are obtained for a system at fixed pressure. For such a system the appropriate partition function is the isobaric grand partition function[16,17]

$$Z_p = \int_V \sum_i w_i \exp[-\beta(\varepsilon_i + pV)]dV \tag{127}$$

which is related to the Gibbs free energy, G, as follows:

$$Z_p = \exp[-G/RT] \tag{128}$$

We can express Eq. (127) as a sum over enthalpy states, using the general definition of enthalpy,

$$H = U + pV \tag{129}$$

For a gas the pV term equals RT per mole that at room temperature gives a value of approximately 2.5 kJ/mol. Since the molar volume of condensed matter is approximately 10^{-3} that of a gas (e.g., 18 cm³/mol for liquid water versus 22.4 liters/mol for an ideal gas at STP), there is little difference between H and U for condensed matter.

We can define the enthalpy of a particular state as follows:

$$h_i = \varepsilon_i + pV \tag{130}$$

giving

$$Z_p = \int_V \sum_i w_i \exp[-\beta h_i]dV \tag{131}$$

The enthalpy moments are then obtained from Z_p in the same manner as the energy moments were obtained from Z_v of Eq. (99). Thus in analogy with the result given in Eq. (106) one has

$$H_m = \frac{(-1)^m}{Z_p} \frac{\partial^m Z_p}{\partial \beta^m} \tag{132}$$

We note that for the case $m = 1$, Eq. (132) reduces to the familiar equation from thermodynamics giving the average enthalpy (the first moment of the enthalpy distribution) as a temperature derivative of the Gibbs free energy,

$$H = \frac{(-1)}{Z_p} \frac{\partial Z_p}{\partial \beta} = \frac{\partial(G/T)}{\partial(1/T)} \tag{133}$$

Equation (132) is simply a generalization of this result to higher moments of the enthalpy distribution.

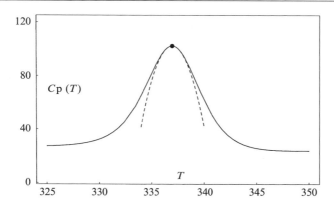

Fig. 6. The heat capacity of the protein barnase (kJ/mol) as constructed from the data of Makhatadze and Privalov.[18] The dashed curve shows a local quadratic expansion of the data about the maximum in the curve (solid dot).

For the constant pressure system the quantity determined experimentally is the heat capacity at constant pressure, $C_p(T)$. In analogy with Eq. (118) we can express this quantity as a series in ΔT,

$$C_p(T) = c_0 + c_1 \Delta T + c_2 \Delta T^2 + \cdots \qquad (134)$$

where the c values are empirical parameters determined by the fit of the heat capacity data with respect to temperature. The process of using the c values to calculate enthalpy moments is then exactly the same as the process for calculating energy moments from the expansion of $C_v(T)$ as given in Eq. (118).

We now turn to some actual heat capacity data for proteins. Figure 6 shows the heat capacity of the protein barnase in units of kJ mol^{-1} K^{-1} based on data of Makhatadze and Privalov.[18] Barnase is a small protein containing 110 amino acids with a molecular mass of 12,365 Da. The dashed curve gives the local quadratic expansion as shown in Eq. (134), where the expansion is taken about the maximum in the heat capacity curve (the point indicated by the solid dot in the graph; Fig. 6). We note that the heat capacity graph shown in Fig. 6 and the graph shown in Fig. 1B for the case of ligand binding are similar: in both cases a local quadratic expansion of the experimental curve gives four moments of the appropriate distribution function. Given the quality of the data shown in Fig. 6 one can easily obtain six enthalpy moments accurately. In this case the enthalpy distribution functions calculated using four or more moments are qualitatively

[18] G. I. Makhatadze and P. L. Privalov, *Adv. Protein Chem.* **47**, 307 (1995).

different from those calculated using only two moments. This is in marked contrast with the result we found for ligand binding, where the distribution functions based on two to four moments were virtually superimposable.

The enthalpy distribution functions constructed using six enthalpy moments obtained from the experimental heat capacity data for barnase shown in Fig. 6 are given in Fig. 7A for three different values of the temperature [corresponding to expansions as given by Eq. (134) centered around three different temperatures]. The temperatures used were the temperature of the maximum, T_m, and then $T_m - 1$ and $T_m + 1$, where $T_m = 337.1$ K. A natural interpretation of the presence of two peaks in the enthalpy distribution functions is that one peak (the one at lower enthalpy values) represents the native state of the molecule while the other peak (at higher enthalpy) represents the unfolded state of the molecule. It is clear that both species are in fact represented by broad distributions of enthalpy values. At $T_m - 1$ the low-enthalpy species is most probable

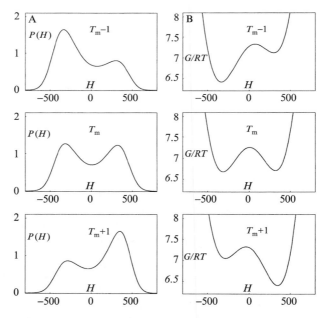

Fig. 7. (A) The enthalpy distribution function, $P(H)$, for the protein barnase near the melting temperature T_m. To simplify the scale the $P(H)$ functions shown have been multiplied by a factor of 1000. (B) The free energy distribution, G/RT, obtained from $P(H)$ using Eq. (135). A constant value, $\ln(1000)$, has been added to these curves reflecting the scaling used for $P(H)$. Reprinted from D. Poland, *Proteins Struct. Funct. Genet.* **45,** 325 (2001), with kind permission of Wiley-Interscience.

[higher $P(H)$ peak]. As the temperature is increased to T_m the two peaks become equal in height, indicating that both the native and unfolded species have approximately the same probability at this temperature. Finally, at $T_m + 1$, the high-enthalpy species is more probable, indicating that at this temperature most of the molecules are unfolded. Thus the temperature dependence of the enthalpy distribution gives a detailed view of the shift in populations between native and unfolded species. Note that in all cases the enthalpy distribution function of both the native and unfolded species is not given by a single enthalpy value (delta function in the distribution), but rather the enthalpy distribution function for each species is represented by a broad peak. The enthalpy distribution function gives one the probability that a molecule picked at random will have a given value of the enthalpy. From our construction of this function one sees that there is a great variability (even for the native form) in the enthalpy of a protein molecule, this variation being due to, among other causes, vibrations and breathing motions of the protein and variability in solvent structure around the protein.

Given the enthalpy distribution $P(H)$, one can define[19] a Gibbs free energy that is the potential for this distribution as follows:

$$P(H) = \exp[-G(H)/RT]/\exp[-G/RT] \tag{135}$$

where

$$\exp[-G/RT] = \int_H \exp[-G(H)/RT]dH \tag{136}$$

gives the total Gibbs free energy. From Eq. (135) the quantity $G(H)/RT$ is given by the relation

$$G(H)/RT = -\ln[P(H)] + C \tag{137}$$

where $C = G/RT$ is a constant that is independent of H but depends on T.

The functions $G(H)/RT$ obtained from the three different enthalpy distribution functions shown in Fig. 7A, using Eq. (137), are illustrated in Fig. 7B. For this function the most probable species is represented by the lowest valley in the $G(H)/RT$ curve. Thus at $(T_m - 1)$ the deepest minimum in the free energy curve corresponds to the native (low-enthalpy) species while at T_m there are two minima of equal depth, indicating that at this temperature the native and unfolded species are equal in probability. Then at $(T_m + 1)$ the deepest minimum in the free energy curve shifts to correspond to the unfolded (high-enthalpy) species. Recall that all the functions shown in Fig. 7 were determined from the single set of data

[19] D. Poland, *Proteins Struct. Funct. Genet.* **45**, 325 (2001).

shown in Fig. 6, namely, the temperature dependence of C_p. These results clearly illustrate the power of the moments/maximum-entropy method to construct, from standard experimental data, distribution functions that give detailed insight into the behavior of biological macromolecules.

The approach outlined in this section has been applied to the contribution of secondary structure in proteins to the enthalpy distribution,[20] free energy distributions for two different forms of myoglobin,[21] free energy distributions in tRNAs,[22] and the enthalpy distribution for the helix–coil transition in a model peptide.[23]

Self-Association Distributions

In this final section we consider distribution functions for the general clustering, or self-aggregation, of n monomers to give an n-mer or cluster containing n monomers. This process is similar to that of ligand binding treated in Section II except that in this case there is no parent molecule with a fixed set of binding sites. Rather, the monomers simply react with one another to form a cluster. We follow the treatment of self-association using the moments/maximum-entropy method that has been published.[24]

The general reaction for the addition of one monomer to a cluster containing $(n - 1)$ monomers with equilibrium constant K_m is

$$A_1 + A_{n-1} \longleftrightarrow A_n \qquad (K_n) \tag{138}$$

As was the case with ligand binding, it is useful to consider the formation of the cluster A_n directly from n monomer molecules,

$$nA_1 \longleftrightarrow A_n \qquad (Q_n) \tag{139}$$

The reaction in Eq. (139) is obtained by adding together the stepwise reactions of Eq. (138), thus giving the equilibrium constant Q_n for Eq. (139) as

$$Q_n = \prod_{m=2}^{n} K_m \tag{140}$$

For completeness, we have the null reaction

$$A_1 \longleftrightarrow A_1 \tag{141}$$

where

[20] D. Poland, *Biopolymers* **63,** 59 (2002).
[21] D. Poland, *J. Protein Chem.* **21,** 187 (2002).
[22] D. Poland, *Biophys. Chem.* **101,** 485 (2003).
[23] D. Poland, *Biopolymers* **60,** 317 (2001).
[24] D. Poland, *Biophys. Chem.* **94,** 185 (2002).

$$K_1 = Q_1 = 1 \tag{142}$$

From the general conservation of monomer units one has the relation

$$\sum_{n=1}^{\infty} n[A_n] = c \tag{143}$$

where c is the total original concentration of monomer units.

Using the equilibrium constant expression for the reaction in Eq. (139) one has

$$\frac{[A_n]}{[A_1]^n} = Q_n \tag{144}$$

or, solving for $[A_n]$,

$$[A_n] = [A_1]^n Q_n \tag{145}$$

We can now define the analog of the binding polynomial that was introduced for the case of ligand binding [see Eq. (15)]. We will call this function the association polynomial,

$$\Gamma = \sum_{n=1}^{\infty} [A_n] = \sum_{n=1}^{\infty} [A_1]^n Q_n \tag{146}$$

Physically, this is the sum over all cluster concentrations and gives the net concentration of clusters regardless of size. From this polynomial we obtain a general relation for the probability of a cluster containing n monomers,

$$P_n = [A_1]^n Q_n / \Gamma \tag{147}$$

This relation follows from the general rule for obtaining probabilities from partition functions: the probability of state n is the term in the partition function representing state n divided by the sum over all states (the partition function or, in this case, the association polynomial). Note that Eq. (147) gives the probability that a cluster picked at random contains n monomer units.

Using the probability of a given cluster size given in Eq. (147), the average cluster size is then given by

$$\langle n \rangle = \sum_{n=1}^{\infty} n P_n = \sum_{n=1}^{\infty} n[A_1]^n Q_n / \sum_{n=1}^{\infty} [A_1]^n Q_n \tag{148}$$

or

$$\langle n \rangle = \frac{[A_1]}{\Gamma} \frac{\partial \Gamma}{\partial [A_1]} \tag{149}$$

Higher moments are obtained in an analogous fashion,

$$\langle n^m \rangle = \sum_{n=1}^{\infty} n^m P_n \frac{1}{\Gamma} \frac{\partial^m \Gamma}{\partial y^m} \tag{150}$$

where

$$y = \ln[A_1] \tag{151}$$

One can then take derivatives of $\langle n \rangle$ as given in Eq. (149) with respect to y, giving

$$
\begin{aligned}
\frac{\partial \langle n \rangle}{\partial y} &= \langle n^2 \rangle - \langle n \rangle^2 \\
\frac{\partial^2 \langle n \rangle}{\partial y^2} &= \langle n^3 \rangle - 3\langle n \rangle \langle n^2 \rangle + 2\langle n \rangle^3
\end{aligned}
\tag{152}
$$

From these relations one obtains the higher moments, $\langle n^2 \rangle$ and $\langle n^3 \rangle$, in terms of the variation of $\langle n \rangle$ with respect to $[A_1]$ (or $y = \ln[A_1]$). But $[A_1]$ is the concentration of free monomer, not the total original concentration of monomer denoted by c and given in Eq. (143). The total original concentration of monomer, c, is the variable under experimental control and experiment gives $\langle n \rangle$ as a function of c. To obtain expressions for the moments of the distribution in terms of c one can expand $\langle n \rangle$ about a given value c_0,

$$\langle n(c) \rangle = n_0 + n'(c - c_0) + \frac{1}{2} n''(c - c_0)^2 + \cdots \tag{153}$$

where

$$n_0 = \langle n(c_0) \rangle, \qquad n' = (\partial \langle n \rangle / \partial c)_{c_0}, \qquad n'' = (\partial^2 \langle n \rangle / \partial c^2)_{c_0} \tag{154}$$

The preceding equations give a local quadratic fit to the experimental data: $\langle n \rangle$ as a function of c. As a result of this empirical fit one obtains the parameters n_0, n', and n'' evaluated at the point $c = c_0$.

To obtain derivatives with respect to $[A_1]$, as required in Eq. (152), we use the variable y defined in Eq. (151) and introduce the new variable w:

$$y = \ln[A_1] \qquad \text{and} \qquad w = \ln c \tag{155}$$

Note that $[A_1]$ is the concentration of free monomer and c is the total original concentration of monomer. We have the following relations between these variables:

$$\frac{\partial \langle n \rangle}{\partial y} = n^{(1)} \frac{\partial w}{\partial y}, \qquad \frac{\partial^2 \langle n \rangle}{\partial y^2} = n^{(2)} \left(\frac{\partial w}{\partial y} \right)^2 + n^{(1)} \frac{\partial^2 w}{\partial y^2} \tag{156}$$

where the $n^{(m)}$ are defined as follows:

$$n^{(1)} = \frac{\partial \langle n \rangle}{\partial w} = cn', \qquad n^{(2)} = \frac{\partial^2 \langle n \rangle}{\partial w^2} = cn' + c^2 n'' \qquad (157)$$

and are now given in terms of the experimentally determined quantities n' and n'' given in Eq. (154).

We can make the transformation from variable c to variable $[A_1]$, using the conservation relation:

$$c = \sum_{n=1}^{\infty} n[A_n] = \sum_{n=1}^{\infty} n[A_1]^n Q_n \qquad (158)$$

One then has

$$\frac{\partial w}{\partial y} = \frac{\langle n^2 \rangle}{\langle n \rangle}, \qquad \frac{\partial^2 w}{\partial y^2} = \frac{\langle n^3 \rangle}{\langle n \rangle} - \left(\frac{\langle n^2 \rangle}{\langle n \rangle} \right)^2 \qquad (159)$$

Defining the first three moments of the self-association distribution function as

$$M_1 = \langle n \rangle, \qquad M_2 = \langle n^2 \rangle, \qquad M_3 = \langle n^3 \rangle \qquad (160)$$

we finally have expressions for these moments in terms of experimentally measured quantities:

$$M_1 = n_0$$

$$M_2 = M_1^2 / (1 - n^{(1)}/M_1) \qquad (161)$$

$$M_3 = \left\{ (M_2/M_1)^2 (n^{(2)} - n^{(1)}) + 3M_1 M_2 - 2M_1^3 \right\} / (1 - n^{(1)}/M_1)$$

We will use as an example of this method the association of ATP to form linear clusters. This system has been studied using NMR techniques in the laboratory of H. Sigel.[25] The ends of the clusters have a characteristic signal in the NMR and hence one can measure the net concentration of clusters, $\langle n \rangle$, as a function of the total ATP concentration (our variable c). Obtaining the average cluster size as a function of total monomer concentration is the most difficult part of this approach. The curve giving this data for the self-association of ATP based on the work of Sigel et al.[25] is shown in Fig. 8.

Using the data on the average extent of clustering given in Fig. 8, one can then construct the first three moments of the cluster distribution

[25] K. H. Scheller, F. Hofstette, P. R. Mitchell, B. Prijs, and H. Sigel, J. Am. Chem. Soc. **103**, 247 (1981).

function using Eq. (161). The maximum-entropy distribution function obtained from these moments is shown in Fig. 9A. As was the case for ligand binding, given the cluster probability distribution function, one can

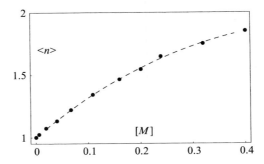

FIG. 8. The self-association of ATP giving the average number of ATP molecules in a cluster as a function of the total amount of ATP. The curve is based on the data of Scheller *et al.*[25] as constructed by Poland.[24] Reprinted from D. Poland, *Biophys. Chem.* **94,** 185 (2001), with kind permission of Elsevier Science.

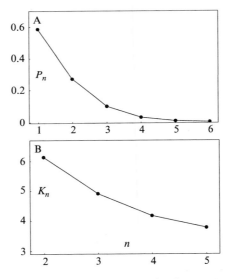

FIG. 9. (A) The maximum-entropy distribution function for aggregates of ATP constructed using moments obtained from local expansions of the self-association curve shown in Fig. 8. (B) The equilibrium constants, K_n, for adding an ATP molecule to a cluster containing $n - 1$ units. Reprinted from D. Poland, *Biophys. Chem.* **94,** 185 (2001), with kind permission of Elsevier Science.

then calculate the Q_n coefficients in the association polynomial of Eq. (146) and, in turn, the equilibrium constants for the successive binding of monomers as indicated in Eq. (138). The equilibrium constants obtained in this manner are plotted for $n = 2$ to 5 in Fig. 9B. One sees that for this system there is a subtle decrease in the magnitude of successive binding constants as n increases.

In summary, we have seen that the moments/maximum-entropy method outlined in this chapter is a straightforward way to obtain distribution functions for various molecular variables that characterize biological macromolecules. In this method one abstracts the appropriate set of moments from experimental data such as the titration curve for lyzozyme shown in Fig. 4, the heat capacity curve for barnase shown in Fig. 6, and the self-association curve for ATP shown in Fig. 8. One then uses the maximum-entropy method to convert knowledge of moments into parameters of the appropriate distribution function, giving, for the examples just cited, the distribution functions for the number of protons bound to lyzozyme, the enthalpy of barnase, and the extent of association of ATP. In each case one gains detailed knowledge about the distribution of the appropriate states in biological macromolecules.

Author Index

Numbers in parentheses are footnote reference numbers and indicate that an author's work is referred to although the name is not cited in the text.

A

Abramowitz, M., 249
Accerbi, M., 150(31), 151
Ackerman, M., 48
Ackers, G. K., 95
Adler, A. J., 282(3), 283
Adzhubei, A. A., 332, 335(41)
Aebersold, R., 151
Aguirre, J. L., 116
Akaike, H., 359
Akhtar, R. A., 150
Alameida, P. F. F., 227
Albinsson, B., 329
Albritton, D. L., 248, 249(21), 271(21)
Alcorn, S. W., 330, 331(32)
Alexov, E., 100, 104(27)
Allan, J. S., 382, 385(1)
Allen, M. P., 187
Allison, S. A., 116
Altar, W., 322
Altschul, S. F., 75
Amarakone, A., 150(28), 151
Anchin, J. M., 310
Anderson, D. R., 359
Anderson, T. W., 387, 391(9)
Andrade, M. A., 287, 340
Andreeva, A., 21
Anfinsen, C. B., 6, 48
Ansell, S. M., 153
Antman, E. M., 421
Antoch, M. P., 150(30), 151
Antosiewicz, J., 104(43), 105, 111, 175, 183(15)
Aojula, H. S., 350
Apostolakis, J., 120
Appel, L., 419
Applequist, J., 324, 333(16), 334(13; 16)

B

Arbeitman, M. N., 149
Arbuzova, A., 110
Arendall, W. B. I., 52
Armstrong, S. A., 150
Arrowsmith, C. H., 293
Asthagiri, D., 196
Auerbach, A., 230(3; 4), 231, 239(3)
Avondet, A. G., 277(34), 278
Awad, T. A., 150(28), 151
Axelsson, O., 101

Bagheri, B., 103(39), 104(43), 105, 112, 175, 183(15)
Bajaj, C. L., 97
Baker, B. S., 149
Baker, D., 19, 66, 66(4), 67, 69, 69(20; 21), 71(16; 17), 73(7; 12; 15; 19–21), 74(15), 75, 77(6), 79, 85(15), 86, 86(7; 8; 15), 88, 88(10), 92(12)
Baker, N. A., 94, 95(1; 4), 101, 101(1), 102(30; 36), 103, 103(36), 104(36; 38), 105(32), 106(36), 107(4; 36), 108(36), 109(1; 4), 110(4; 59), 111, 111(1), 175, 176
Bakker, P. I. W., 52
Balasubramanian, S., 10
Baldwin, R. L., 48, 305(80), 306, 307, 308, 331
Ball, F. G., 231, 233(7)
Ban, N., 108(49), 109
Bank, R. E., 105, 107(46)
Bannister, W. H., 342
Barker, V. A., 101
Barnett, J., 149
Barrett, C., 76
Barrick, D., 306
Bartfai, T., 149(8), 150

Subject Index

A

Akaike information criterion, generalizability measure in model comparison, 359–360, 363–365

AraC, CHARMM program analysis of mutant interactions with arabinose
  arabinose structure handling, 39–40
  energy determination script, 41–47
  input coordinate file preparation, 37–39
  overview, 36–37

Artificial neural network, secondary structure computation from circular dichroism data, 287

B

Bayesian inference, *see* Sample size approximation

Bayesian information criterion, generalizability measure in model comparison, 359–360

Bayesian model selection, generalizability measure in model comparison, 359, 361–362

BD, *see* Brownian dynamics

Bilayer modeling, *see* Membrane modeling

Binding energy, calculation using Poisson–Boltzmann methods, 11

Brownian dynamics
  algorithm, 171–174
  applications
    macromolecular crowding effects, 192–193
    protein diffusion in concentrated solutions, 191–192
    substrate channeling, 189–190
  assumptions, 171
  electrostatic interactions treatment, 174–177
  features of simulations, 168
  first solvation shell, 180–181

hydrophobic interactions treatment, 179–180
overview, 166–167
practical implementation
  electrostatic potential calculation, 183–185
  simulation running, 186–187
  structure finding, 181–183
  termination of simulations and analysis, 187–188
prospects, 194–198
purpose, 167–170
temporal resolution, 168
van der Waals interactions treatment, 177–179

C

CD, *see* Circular dichroism

CHARMM program
  advantages, 29
  AraC mutant interactions with arabinose
    arabinose structure handling, 39–40
    energy determination script, 41–47
    input coordinate file preparation, 37–39
    overview, 36–37
  capabilities
    model-building operations, 30–31
    molecular dynamics operations, 31–32
    simple structural questions, 30
  molecular dynamics theory, 32–35
  operation, 35–46

Circadian rhythm
  forced desynchrony measurements of human biological clock
    data sampling, 383
    infill asymptotic analysis of harmonic regression parameter estimates
      data analysis, 394–400
      performance, 400–401
      proofs of lemmas and propositions, 401–404

A

B

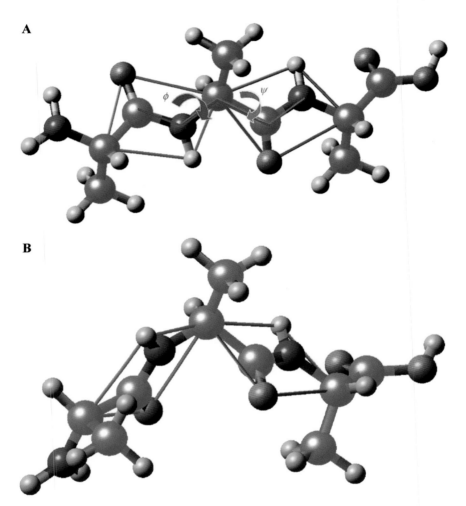

KRETSINGER *ET AL.*, CHAPTER 1, FIG. 1. Definition of ϕ and ψ with examples. (A) Trialanine. As drawn, $\phi_n = 180°$ and $\psi_n = 180°$. $C\alpha_n$, N_n, H_n, C_{n-1}, O_{n-1}, and $C\alpha_{n-1}$ are all contained in one plane; $C\alpha_n$, C_n, O_n, N_{n+1}, H_{n+1} and $C\alpha_{n+1}$ are contained in another. Positive rotation, increasing the value of the dihedral angle, is indicated by the arrows, proceeding along the peptide chain (N → C). Some find it easier to visualize ϕ by looking from Cα toward N; in that case, positive rotation would be indicated by the arrow in the opposite sense; see (B). $\psi_{n-1} = 180°$. At the N terminus ϕ_n is not defined without the amide group being involved in a peptide bond. $\phi_{n+1} = 180°$. At the C terminus, ψ_{n+1} is not defined without the carboxylate group being involved in a peptide bond. (B) $\phi_n = -109°$ and $\psi_n = 121°$ as seen in the R, or polyproline II, conformation for alanine (Fig. 2A).

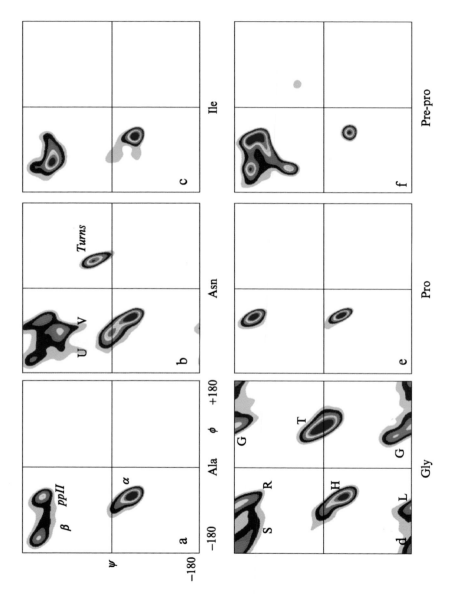

KRETSINGER ET AL., CHAPTER 1, FIG. 2. Ramachandran plot(s) for individual residues from structures determined at high resolution. (A) Alanine prefers to be in an α-helical conformation. Of the two regions in the β region, the left region (S, $\phi = -140°$ and $\psi = 140°$) is found in β sheets. (B) Asparagine has a complicated pattern of conformations with a large fraction in the turns, or left-handed α-helix, region (T at $\phi = 60°$ and $\psi = 35°$; see Fig. 1B) and in the two bridging areas U and V. It is rarely found in β strands. (C) Isoleucine prefers to be in β sheets (S, $\phi = -117°$ and $\psi = 128°$). (D) Glycine is the only amino acid without a Cβ and thus can have conformations that are sterically hindered in all other amino acids. Notice that the turn region, T, is the most occupied. The letters G, L, H, R, S, and T [together with U and V; see (B)] denote the eight discrete conformations assumed by amino acids in proteins. (E) Pro is the most restricted amino acid. Because its amine group is not available to form hydrogen bonds it is unusual in β sheets and is found in only the first three positions of α helices. (F) Although proline can have the conformation needed for an α helix, the amino acid just before proline nearly always has a β conformation, making proline a terminator of α helices (after Hovmöller et al.). About 5% of prolines occur with the peptide bond cis. This is not explicitly stated in the PDB file. In some lower-resolution structures a cis-proline has been built trans, thereby distorting the local geometry.

Kʀᴇᴛsɪɴɢᴇʀ *ET AL.,* Cʜᴀᴘᴛᴇʀ 1, Fɪɢ. 3. Illustration of alternative shapes of regions of the same protein. Residues His40–Met67 of G-actin form an α helix in one crystal structure and a β turn in another. Redrawn after Otterbein *et al.*

Kʀᴇᴛsɪɴɢᴇʀ *ET AL.,* Cʜᴀᴘᴛᴇʀ 1, Fɪɢ. 4. Summary of results of modeling an unknown structure. The sequence dahps-th, of target structure TH, is aligned with its closest homolog of known structure, 1 FWS (FWS), using BestFit, part of GCG. The two are 34% identical, 46% similar, over 238 residues of the target; hence, there is confidence in the general conformation of the predicted structure (see Fig. 5). The secondary structure of the target, dahps-th, was predicted by two different servers, PROF and PSIPRED; the two predictions agree in 257 of 339 positions, $Q_3 = 76\%$. More important, both predictions of secondary structure have similar patterns of helix (blue), extended (β strand; red), and coil as assigned by a PDB routine to the template FWS, using criteria both of general ϕ, ψ value and of hydrogen-bonding pattern. There is added faith in the general pattern of PROF and PSIPRED predictions because they map well onto the pattern actually observed in the template, a $(\beta/\alpha)_8$ or TIM barrel protein. There is an insertion of 11 residues, $_{186}$SVQLPGGLGDK$_{196}$, in the template, FWS, relative to the target, TH. This corresponds to a loop, residues 192–198, not visible in the crystal structure (see Fig. 5). The program that assigns secondary structure in the PDB assigns coil *(c),* often with justification, to such nonvisible regions. The top line, D8, assigns each residue to one of eight discrete conformations (see Fig. 2), as determined solely by ϕ, ψ. A BLAST search with the first 95 residues of the target indicates their optimal alignment with phospho-2-dehydro-3-deoxyheptonate synthase.

```
KDOPS (1fws, template) vs. DAHPS-thermatoga(target = TH)


  1 MIVVLKPGSTEEDIRKVVKLAESYNLKCHISKGQERTVIGIIGDDRYVVV 50 TH
    CEEEECCCCCHHHHHHHHHHHHHCCCCEEEEECCCCEEEEEEECCCCEEEE    PROF
    CEEEECCCCCHHHHHHHHHHHHHCCCCEEEEECCCCEEEEEEECCCCCCCH APSIPRED

 51 DKFESLDCVESVVRVLKPYKLVSREFHPEDTVIDLGDVKIGNGY 95  TH
    EECCHCCCCEEEEECCCCCCCECCCCCCCCCCCCEEEEECCEECCCCC    PROF
    HHHHHCCCCCHHHHCHHHHHHHHHHHHCCCCCCEEECCCEEECCCC   PSIPRED

    RRSSSSSRHSVHSHHHHHHHHHHHHHHHHHHHVHHSSRSRSSRHUVHURHSHH     D8
    CCEEEEEECSCCCHHHHHHHHHHHHHHHHHHCEEEEEEEEECCCCCCCCCCCC
  4 FLVIAGPCAIESEELLLKVGEEIKRLSEKFKEVEFVFKSSFDKANRSSIH  53 FWS
    | :||||||.:|  |:|:.    :  |   |    |:   |  |.| :
 96 FTIIAGPCSVEGREMLMETAHFLSELGVK......VLRGGAYKP.RTSPY 138 TH
    EEEEECCCCCCCHHHHHHHHHHHHHHHCH      HEEEEECCC CCCCC    PROF
    EEEEEECCCCCCHHHHHHHHHHHHHHHCCC      EEECCCCCC CCCCC PSIPRED

    RRHSUGHHHHHHHHHHHHHHHHHTSRSSSSSHSHHHHHHHHHHHHRHSSSSRHH    D8
    CCCCCHHHHHHHHHHHHHHHHHHHCEEEEEECCCCHHHHHHHCCEEECCCCC
 54 SFRGHGLEYGVKALRKVKEEFGLKITTDIHESWQAEPVAEVADIIQIPAF 103 FWS
    ||.|  |  |  |..  ||.  :.:|: : |:        ||| ||||||  |
139 SFQGLG.EKGLEYLREAADKYGMYVVTEALGEDDLPKVAEYADIIQIGAR 187 TH
    HHCHHE EEEECCCCCCCCCCCCCCEEEEECCCCCHHHHHHHHHHCCHH    PROF
    HCCCEE ECCCCCCCCCCCCCCCCEEECCCCHHHHHHHHHHCCEEECCCH PSIPRED

    VHHVHHHHHHHHHHHTRRSSSSRRHHRRHHHHHHHHHHHHHHHTRHUSSSSS     D8
    CCCHHHHHHHHHHCCCEEEEEECCCCCCCCHHHHHHHHHHHHCCCEEEEEE
104 LCRQTDLLLAAAKTGRAVNVKKGQFLAPWDTKNVVEKLKFGGAKEIYLTE 153 FWS
    .   || |   | .|:|       :      | :  | .| |
188 NAQNFRLLSKAGSYNKPVLLKRGFMNTIEEFLLSAEYIANSGNTKIILCE 237 TH
    HCCCHHHHHHHHHCCCEEEEEECCCCCCHHHHHHHHHHHHHHCCCCEEEEE    PROF
    HCCCHHHHHHHHHCCCEEEEEECCCCCCHHHHHHHHHHHHHHHCCCCCEEEEEC PSIPRED

    UG SRU GRTSSSSVHHHHHHHHHHUS  RSSSVHHHHHRS             D8
    CC CCC CCCEEEECCHHHHHHHCCC  EEEEEECCCCCCCcccccccccC
154 RG.TTF.GYNNLVVDFRSLPIMKQWA..KVIYDATHSVQLPGGLGDKSGG 199 FWS
    ||   ||        .|  ..||.:.   ::  | .|         |||
238 RGIRTFEKATRNTLDISAVPIIRKESHLPILVDPSH...........SGG 276 TH
    CCCCCCCCCCCCCCEEEEEHHHHHHCCCCEEEEEECC        CCC    PROF
    CCCCCCCCCCEECHHHHHHHHHCCCCEEEEECCCCCC        CCH PSIPRED

    HHHHHHHHHHHHHHHHGSHSSSSSSSSHUHHHRHSTHHHRRRHHHHHHHHHH     D8
    CCCHHHHHHHHHHHHHHCCEEEEEEEECCCCCCCCCCCCCCCEEECHHHHHHHH
200 MREFIFPLIRAAVAVGCDGVFMETHPEPEKALSDASTQLPLSQLEGIIEA 249 FWS
    |: :  ||  |||:||| |: .| |||||||||  |    . :::
277 RRDLVIPLSRAAIAVGAHGIIVEVHPEPEKALSDGKQSLDFELFKELVQE 326 TH
    CCCCCHHHHHHHHHHHCEEEEEEEECCCCCCCCCCCCCCCCCCHHHHHHHH    PROF
    CCHHHHHHHHHHHHHHCCCEEEEEECCCCCCCCCCCCCCCCCCHHHHHHHH PSIPRED


    HHHHHHHHHHHRRR        D8
    HHHHHHHHHHHCCC
250 ILEIREVASKYYETIPVK 267 FWS
    . .: :
327 MKKLADALGVKVN 339      TH
    HHHHHHHHHCCCC      PROF
    HHHHHHHHHCCCCC   PSIPRED
```

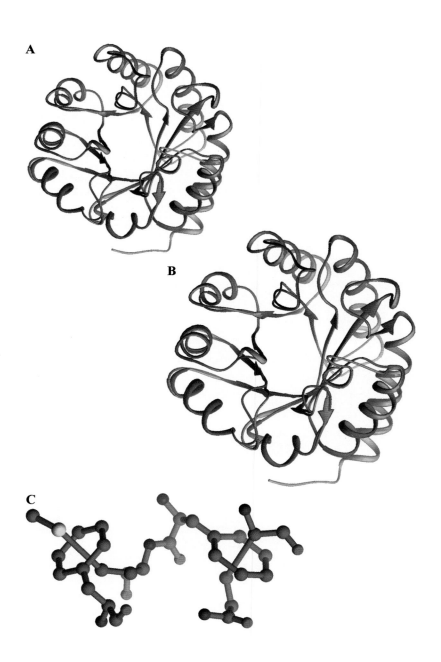

KRETSINGER *ET AL.,* CHAPTER 1, FIG. 5. Backbone drawing of the known template, kdops-1fw, with the automatically predicted model dahps-th, superimposed. The proteins are viewed approximately down the axis of the $(\beta/\alpha)_8$ barrel; the C termini of the β strands are near the viewer. (A) One element, residues ~40 to ~55 of the target, is obviously misplaced. Note that the loop not resolved in the electron density (residues 192–198) is represented by a green line connecting residue 191 to residue 199. The corresponding region of the target (see Fig. 4), is also guessed. (B) The misplaced element of the target has been manually corrected by shifting the alignment as described in text. (C) Many side chains of the model of (B) are placed in obviously unacceptable positions; for instance, the side chain of Gln190 is superimposed on the benzene ring of Phe211. Met212 is thrust through Phe218.

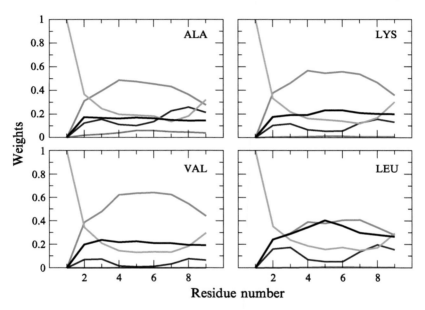

SRINIVASAN *ET AL.,* CHAPTER 3, FIG. 3. Plots of fractional distribution (weights) of residues in helix (red), strand (green), turn (blue), P_{II} (black), and coil (cyan) states for polyalanine, polylysine, polyvaline, and polyleucine in a hard sphere simulation.

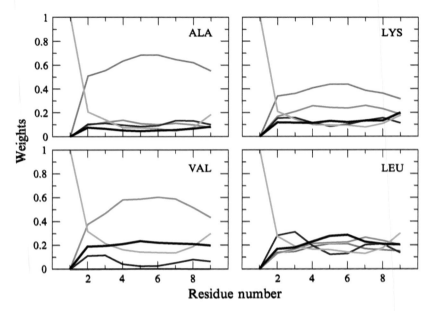

SRINIVASAN *ET AL.*, CHAPTER 3, FIG. 4. Plots of fractional distribution (weights) of residues in helix (red), strand (green), turn (blue), P_{II} (black), and coil (cyan) states for polyalanine, polylysine, polyvaline, and polyleucine with hydrogen bond energy calculation enabled.

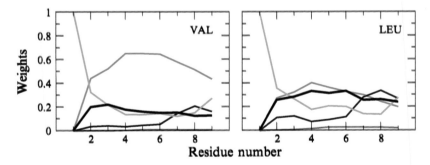

SRINIVASAN *ET AL.*, CHAPTER 3, FIG. 5. Plots of fractional distribution (weights) of residues in helix (red), strand (green), turn (blue), P_{II} (black), and coil (cyan) states for polyvaline and polyleucine with hydrophobic contact energy calculation enabled.

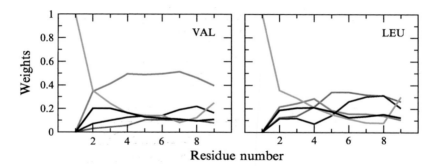

SRINIVASAN *ET AL.*, CHAPTER 3, FIG. 6. Plots of fractional distribution of residues in helix (red), strand (green), turn (blue), P$_{II}$ (black), and coil (cyan) states for polyvaline and polyleucine with hydrogen bond energy and hydrophobic contact energy calculations enabled.

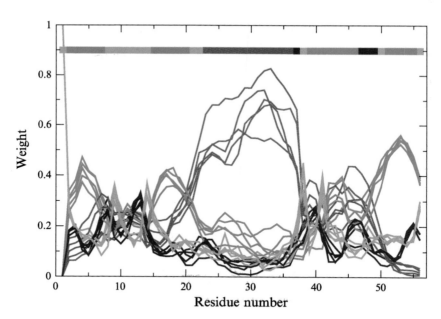

SRINIVASAN *ET AL.*, CHAPTER 3, FIG. 7. Plot of the fractional distribution (weights) of residues in helix (red), strand (green), turn (blue), and coil (cyan) states in protein G from five separate simulations with only local interactions enabled. The corresponding secondary structure segments for 1 pgb are indicated by the color-coded bar.

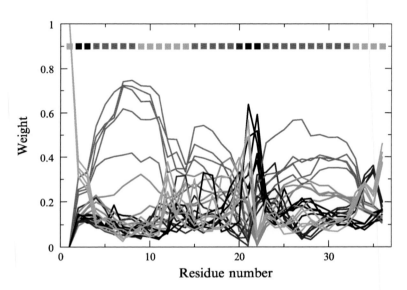

SRINIVASAN *ET AL.,* CHAPTER 3, FIG. 9. Plot of the fractional distribution (weights) of residues in helix (red), strand (green), turn (blue), P_{II} (black), and coil (cyan) states in villin headpiece subdomain from five separate simulations with only local interactions enabled. The corresponding secondary structure segments for 1 vii are indicated by the color-coded bar.

A

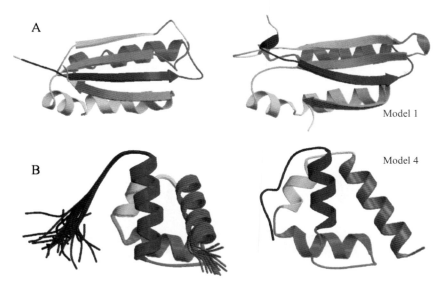

Model 1

Model 4

B

ROHL *ET AL.,* CHAPTER 4, FIG. 1. Rosetta-predicted protein structures for CASP 5 targets. *Right:* Models predicted using the *de novo* prediction protocol. *Left:* Experimental structure of each protein. Protein chains are colored in a blue-to-red gradient along the length of the chain to highlight correctly predicted secondary structure elements. (A) T0135. The predicted model has 54 residues (of 106 total) predicted at a Cα RMSD of 4 Å to the experimental structure. (B) T0171. The predicted model has 60 residues (of 69 total) predicted at a Cα RMSD of 4 Å to the experimental structure. The global Cα RMSD between the prediction and the experimental structure is 4.2 Å.

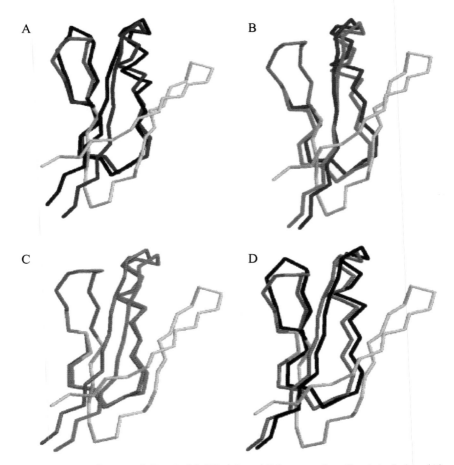

Rohl *et al.*, Chapter 4, Fig. 2. Modified "crank" fragment insertion into 1 dan. (A) Superposition of the protein conformations preceding (black) and following (blue) insertion of a nine-residue fragment. The fragment insertion window is shown in red. The portion of the chain unperturbed by insertion is shown in gray. (B) Superposition of the protein conformations preceding (blue) and following (green) optimization of angles at a wobble site (cyan) adjacent to the insertion window. (C) Superposition of the protein conformations preceding (green) and following (magenta) optimization of angles at a second wobble site (orange) nonadjacent to the insertion window. (D) Superposition of the original (black) and final (magenta) conformations.

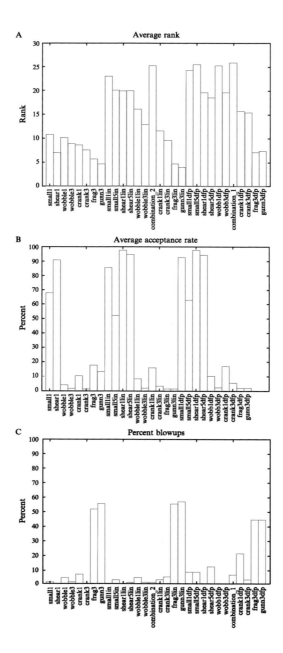

A Average rank

B Average acceptance rate

C Percent blowups

Rohl *ET AL.*, Chapter 4, Fig. 3. Comparison of move types in optimizing the all-atom energy function. Moves are named according to the type of perturbation made and the number of residues in the original perturbation (see text for details): small, random perturbation of one or more nonconsecutive (ϕ, ψ) pairs; shear, random compensating changes in a ϕ angle and the preceding ψ angle; wobble, insertion of a chuck fragment followed by a wobble of one residue; crank, insertion of a chuck fragment followed by a wobble of one residue adjacent to the insertion window and then by a wobble of two residues nonadjacent to the insertion window (illustrated in Fig. 2); frag, unmodified fragment insertion; gunn, insertion of a fragment selected using the gunn strategy. Addition of lin to the move indicates the move is followed by a single-line minimization along the gradient of the potential function before evaluation of the Metropolis criterion. Addition of dfp name indicates the move is followed by variable metric optimization of the potential function before evaluation of the Metropolis criterion. For combination 1, the attempted moves were cycled between small1dfp, small5dfp, shear5dfp, and wobble3dfp. For combination 2, the attempted moves were cycled between small1lin, shear5lin, wobble1lin, and wobble3lin. (A) Average rank of moves. For each starting decoy in the test set, the energies of the lowest energy decoy obtained from application of each move were sorted from highest energy (1) to lowest (30). The histogram reports the average overall decoys for each move type. (B) Percentage of moves accepted. Acceptance rates are reported for each move type, averaged over all decoys. The percentage was scaled on the basis of the percentage of independent simulations that resulted in an expanded structure, in order to account for the dramatic increase in acceptance rate into expanded models relative to compact models. (C) Frequency of simulations resulting in expanded structures.

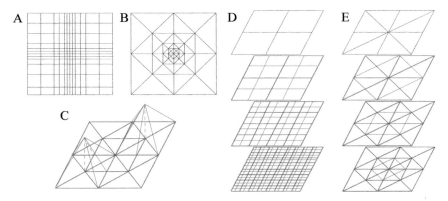

BAKER, CHAPTER 5, FIG. 2. Discretization schemes and hierarchies used in Poisson–Boltzmann (PB) solvers. (A) Cartesian mesh suitable for finite difference (FD) calculations. (B) Finite element (FE) mesh exhibiting adaptive refinement. (C) Examples of typical piecewise linear basis functions used in FE methods. (D) The multilevel hierarchy used to solve the PB equation for an FD discretization; red lines denote the additional unknowns added at each level of the hierarchy. (E) The multilevel hierarchy used to solve the PB equation for FE discretizations; red lines denote the simplex subdivisions used to introduce additional unknowns at each level of the hierarchy.

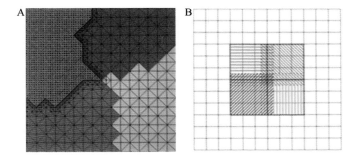

BAKER, CHAPTER 5, FIG. 4. Domain decomposition for parallel methods; colors denote mesh partitions belonging to individual processors of a parallel computer. (A) The Bank–Holst parallel FE method; the mesh shown has been refined from a coarser mesh by the green processor. (B) The parallel focusing method; each processor focuses from the larger coarse mesh to its particular smaller colored region.

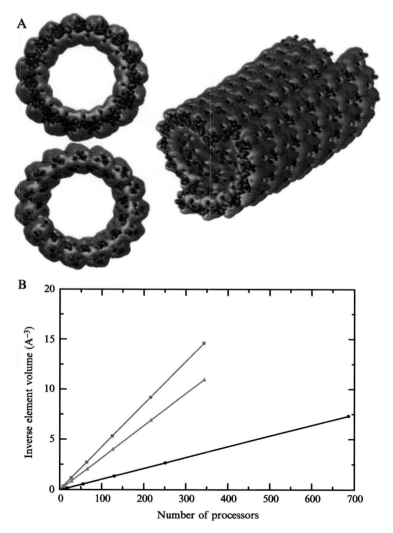

BAKER, CHAPTER 5, FIG. 5. Results from parallel focusing calculations. (A) Views of electrostatic isocontours for a 1.2-million atom microtubule structure (blue, $+1\ kT/e$; red, $-1\ kT/e$). (B) Graph illustrating the optimal efficiency of parallel focusing algorithm.

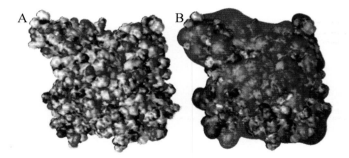

BAKER, CHAPTER 5, FIG. 6. Electrostatic potential of acetylcholinesterase (PDB ID 1MAH). Potential calculation by solution of the PB equation with APBS. (A) Electrostatic potential mapped onto molecular surface; regions with negative potential are shaded red, regions with positive potential are shaded blue. (B) Isocontours of electrostatic potential; blue contour at $+1kT/e$, red contour at $-1kT/e$.

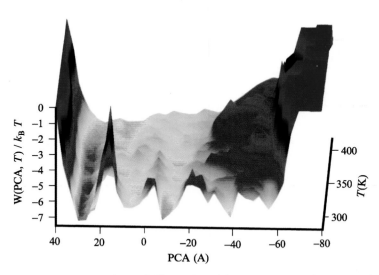

NYMEYER *ET AL.*, CHAPTER 6, FIG. 2. Illustration of the free energy surface sampled by the REMD method as a function of a structural reaction coordinate (PCA) and temperature (T). At a constant low temperature the energy landscape is rugged, with high-energy barriers separating local minima. However, replicas can move in temperature space and thereby avoid kinetic traps by moving around energetic barriers.